Volatiles in the Martian Crust

Volatiles in the Martian Crust

Edited by

Justin Filiberto
Lunar and Planetary Institute, USRA, Houston, TX, United States

Susanne P. Schwenzer
*The Open University, Milton Keynes, United Kingdom;
Lunar and Planetary Institute, Houston, TX, United States*

ELSEVIER

Elsevier
Radarweg 29, PO Box 211, 1000 AE Amsterdam, Netherlands
The Boulevard, Langford Lane, Kidlington, Oxford OX5 1GB, United Kingdom
50 Hampshire Street, 5th Floor, Cambridge, MA 02139, United States

British Library Cataloguing-in-Publication Data
A catalogue record for this book is available from the British Library

Library of Congress Cataloging-in-Publication Data
A catalog record for this book is available from the Library of Congress

ISBN: 978-0-12-804191-8

For Information on all Elsevier publications
visit our website at https://www.elsevier.com/books-and-journals

Working together
to grow libraries in
developing countries

www.elsevier.com • www.bookaid.org

Publisher: Candice Janco
Acquisition Editor: Marisa LaFleur
Editorial Project Manager: Hilary Carr
Cover Designer: Greg Harris

Crossed-polar image from the Lafayette meteorite (BM.1959,755). Image courtesy of the Trustees of the Natural
History Museum. Image: Virtual Microscope, The Open University.

For Scarlett, may this inspire you to reach for the stars.

Contents

List of Contributors

John C. Bridges
University of Leicester, Leicester, United Kingdom

Frances E.G. Butcher
The Open University, Milton Keynes, United Kingdom

Stephen M. Clifford
Lunar and Planetary Institute, Houston, TX, United States

Susan J. Conway
LPG, Université de Nantes, Nantes, France

William H. Farrand
Space Science Institute, Boulder, CO, United States

Justin Filiberto
Southern Illinois University, Carbondale, IL, United States; The Open University, Milton Keynes, United Kingdom

Heather B. Franz
NASA Goddard Space Flight Center, Greenbelt, MD, United States

Fabrice Gaillard
CNRS-Université d'Orléans, Orléans, France

Ralf Gellert
University of Guelph, Guelph, ON, Canada

Leon J. Hicks
University of Leicester, Leicester, United Kingdom

Bradley L. Jolliff
Washington University in St. Louis, St. Louis, MO, United States

Penelope L. King
Australian National University, Canberra, ACT, Australia

Andrew H. Knoll
Harvard University, Cambridge, MA, United States

Samuel P. Kounaves
Tufts University, Medford, MA, United States

Jérémie Lasue
IRAP-OMP, CNRS-UPS, Toulouse, France

Paul R. Mahaffy
NASA, Goddard Space Flight Center, Greenbelt, MD, United States

Nicolas Mangold
LPG, Université de Nantes, Nantes, France

Amy C. McAdam
NASA, Goddard Space Flight Center, Greenbelt, MD, United States

Francis M. McCubbin
NASA Johnson Space Center, Houston, TX, United States

Scott M. McLennan
State University of New York at Stony Brook, Stony Brook, NY, United States

Douglas W. Ming
NASA/Johnson Space Center, Houston, TX, United States

David W. Mittlefehldt
NASA/Johnson Space Center, Houston, TX, United States

John F. Mustard
Brown University, Providence, RI, United States

Elizabeth A. Oberlin
Tufts University, Medford, MA, United States

Karen Olsson-Francis
The Open University, Milton Keynes, United Kingdom

Ulrich Ott
MTA Atomki, Debrecen, Hungary; Max-Planck-Institut für Chemie, Mainz, Germany

Susanne P. Schwenzer
The Open University, Milton Keynes, United Kingdom; Lunar and Planetary Institute, Houston, TX, United States

Brad Sutter
Jacobs, NASA Johnson Space Center, Houston, TX, United States

Timothy D. Swindle
University of Arizona, Tucson, AZ, United States

G. Jeffrey Taylor
University of Hawai'i, Honolulu, HI, United States

Allan H. Treiman
Lunar and Planetary Institute, Houston, TX, United States

Tomohiro Usui
Tokyo Institute of Technology, Tokyo, Japan

Albert S. Yen
Jet Propulsion Laboratory, California Institute of Technology, Pasadena, CA, United States

Acknowledgments

We would like to thank Feargus Abernethy, David Catling, Natalie Cabrol, Shuo Ding, Jamie Gilmour, Timothy Glotch, Juliane Gross, Lydia Hallis, Suniti Karunatillake, Francis McCubbin, David Mittlefeldt, Bob Pepin, David Rothery, Mariek Schmidt, and G. Jeffrey Taylor, who were helpful in reviewing the chapters in this book.

LPI contribution # is 2093

INTRODUCTION TO VOLATILES IN THE MARTIAN CRUST

Justin Filiberto[1,2] and **Susanne P. Schwenzer**[2]
[1]Southern Illinois University, Carbondale, IL, United States [2]The Open University, Milton Keynes, United Kingdom

Society has long looked up at the sky to wonder if we are alone in the Solar System or the universe. Of the planetary bodies near the Earth, Mars has always inspired humankind's imagination, and owing to its close proximity to Earth attracted scientific investigations. Centuries of observation have seen enthusiasm for the planet wax and wane as evidence seemed to prove or disprove habitable conditions on the planet. As telescopes became better, observers captured their views of the planet on maps and drawings. One of the famous and far reaching maps was drawn during the Mars opposition of 1877 by Giovanni Virginio Schiaparelli (1835−1910) (Fig. 1.1). It sparked a wide interest and discussion among contemporary scholars, finally leading to Lowell's interpretation and book, "Mars, the Abode of Life." Many other observers drew similar features at similar places on Mars, making it likely for them to be actual Martian features (see, e.g., Sagan and Pollack, 1966; Jones, 2008 for reviews). However, their linear nature was not confirmed by many, a controversy that led Sagan and Pollack (1966, p. 117) to conclude that those interpretational discrepancies "have convinced many that the canals of Mars are a psychophysiological rather than an astronomical problem." However, like many scientific controversies, this also sparked new methods and thinking as it led to pioneering thoughts as to the ways in which understanding of geologic processes on Earth could be used to understand observations on Mars. In fact, Peal (1893) wrote, "The remarkable feature of the whole case seems to be that so far there has been little or no reference to terrestrial experience when discussing the problem of the distribution of land and water on Mars. The great recent geological discoveries bearing on the subject appear to have been overlooked, [. . .]." Looking at Earth analogs has, in fact, become an integral tool for the advancement of our understanding of Martian processes and environmental conditions.

In 1965 the first orbiter-based images taken by Mariner 4 (during a flyby of Mars) seemed to crush all hopes of Mars being a habitable world, since they showed a barren, cratered landscape (Fig. 1.2) with the linear features being interpreted as ridge systems (Sagan and Pollack, 1966). This picture of Mars has changed as our observational tools advanced. We will see in this book how our knowledge has evolved through Earth laboratory−based investigations of Martian meteorites, orbiter-based imagery and spectroscopy, and in situ investigations from landers and rovers on the surface.

In order to investigate whether a planetary body could have once been habitable, NASA has employed the "follow the water" strategy (e.g., Hubbard et al., 2002). Liquid water is required for life as we currently know it (e.g., Cockell et al., 2016), therefore finding evidence for water (either current or past) has been a high priority for planetary exploration. Looking at the planets in our

Volatiles in the Martian Crust. DOI: http://dx.doi.org/10.1016/B978-0-12-804191-8.00001-5

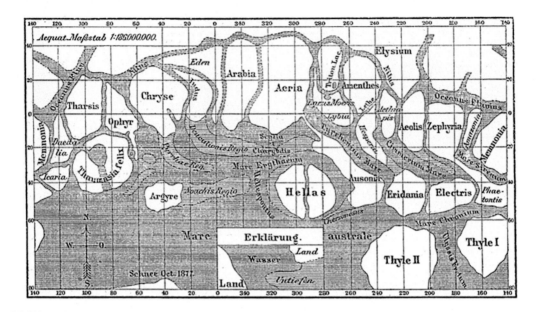

FIGURE 1.1

Giovanni Schiaparelli's 1877 map of the Martian surface.

Solar System, Mercury's close proximity to the sun means the surface temperature is too high for water to be present and original Mariner 10 mission images revealed a barren, cratered world that resembled the surface of the Moon (Murray et al., 1974; Hapke et al., 1975). Recent MESSENGER data have suggested that Mercury is much more complex and may even have relatively recent volcanism, a volatile-rich crust, and even ice in permanently shadowed polar craters, but the surface still lacks features of fluid flow (Prockter et al., 2010; McCubbin et al., 2012; Lawrence et al., 2013; Neumann et al., 2013; Murchie et al., 2015; Deutsch et al., 2016). Venus is covered in a thick dense atmosphere obscuring the surface from most measurements by orbital- or ground-based instruments (e.g., Esposito et al., 1997; Taylor et al., 1997). What we do know about the surface suggests a hot caustic environment outside the realm of most known organisms; on the other hand recent work has shown that there may be habitable environments within the clouds (e.g., Fegley and Treiman, 1992; Cockell, 1999; Schulze-Makuch et al., 2004). Interestingly on a side note, moons in the outer Solar System, such as Europa or Enceladus among others, may have oceans under their ice-crust (Kivelson et al., 2000; Keszthelyi et al., 2001; Postberg et al., 2009; Hsu et al., 2015). If there is internal heat being released into the ocean, there may be hydrothermal systems, which could be habitable environments similar to hydrothermal pipes on the Earth's seafloor (e.g., Kargel et al., 2000; Rothschild and Mancinelli, 2001; Zolotov and Shock, 2003; McKay et al., 2008). Two out of four planets in the inner Solar System lack evidence for once flowing water on the surface, a feature that is common on Earth and a prerequisite for most life as we know it. Mars is the second example with evidence for standing and flowing water at the surface in its ancient past.

(A)

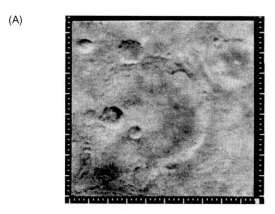

(B)

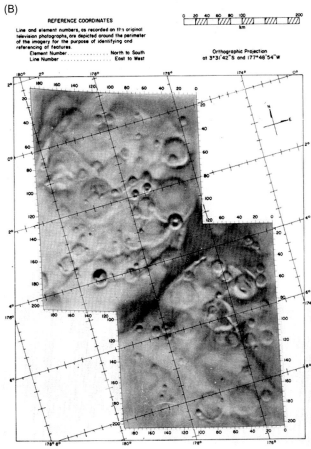

FIGURE 1.2

Mariner 4 images of (A) Mariner crater (named after the Mariner 4 mission) and (B) southern Amazonis Planitia region on Mars. *Image credit: NASA.*

Therefore in this book we explore the current state of knowledge for water and other volatiles in the Martian crust. We bring together constraints on volatiles in the Martian crust from vastly different data sets, with different footprints and constraints. Our understanding of the complexity and diversity of rocks in the Martian crust is constantly expanding with new meteorites being found every day, and new in situ and orbital investigations being conducted continually. Therefore the book is organized around these different data sets starting with the most detailed and building to a big picture understanding of volatiles in the Martian crust with implications for habitability of the surface. The book is divided into two parts with the first half focusing on detailed investigations of Martian meteorites and the second half focusing on spacecraft investigations (orbital, lander, and rover) of volatiles on Mars. All of the information from these diverse data sets is then synthesized to address the potential habitability of Mars through its geologic history.

1.1 SOURCES OF DATA

Here we will summarize some of the main sources of data used throughout this book, specifically focusing on recent missions and instrumentation. Some mission data (such as Viking and Pathfinder) have been discussed in great detail in previous publications (Kieffer et al., 1992) and are therefore not introduced in detail here.

1.1.1 MARTIAN METEORITES

Martian meteorites (Fig. 1.3) are pieces of Mars that have been blasted off of the surface, hurtled through space, and delivered to Earth (McSween and Treiman, 1998). We know they are from Mars because the noble gases, carbon, and nitrogen measured in shock melt pockets in certain meteorites have the same relative elemental abundances and similar isotopic ratios as the Martian atmosphere—a unique fingerprint (Bogard and Johnson, 1983). The rest of the Martian meteorites have similar chemical characteristics showing the roughly 100 unique stones that are from Mars (Clayton and Mayeda, 1996; Clayton, 2003). Martian meteorites present us with the ability to investigate these rocks similarly to how we would investigate a rock from Earth. However, Martian meteorites were not found in situ, presenting the problem that they are from unknown locations on the Martian surface. A few sites have been proposed but the exact location is not known (e.g., Hamilton et al., 2003; Tornabene et al., 2006; Lang et al., 2009; Ody et al., 2014; Werner et al., 2014). Most of the Martian meteorites are relatively young, with only a few representing rocks from the original billion years of Martian geologic time (Nyquist et al., 2001; Walton et al., 2008; Lapen et al., 2010; Humayun et al., 2013; Jones, 2015; Lapen et al., 2017). This is important because water is currently not stable on the surface of Mars but may have been stable 3−4 billion years ago (e.g., Clifford, 1993; Jakosky and Phillips, 2001). Further, they are all shock metamorphosed either from previous impacts into the Martian crust or from the excavation event that brought them to Earth (e.g., Duke, 1968; Beck et al., 2005; Fritz et al., 2005a,b; Sharp and DeCarli, 2006). Therefore, the textures have been altered, and in some cases parts of the rocks have been remelted (e.g., Walton and Spray, 2003; Beck et al., 2005; Sharp and DeCarli, 2006; Walton et al., 2014). That being said, they provide us with the ability to make the most detailed

FIGURE 1.3

Martian meteorite NWA 5789 showing a black glassy fusion crust. NWA 5789 is an olivine-phyric shergottite (Gross et al., 2011). Large green olivines are shown in the image. The brown color is terrestrial alteration—Sahara dust. *Image credit: Juliane Gross.*

measurements and provide us with the best clues about volatiles on Mars. The first few chapters of the book start with detailed investigations of these rocks structured by key volatile elements and what they can tell us about the different reservoirs (interior, crust, and atmosphere) of Mars. The book begins by focusing on volatiles in the Martian interior (Chapter 2: Volatiles in Martian Magmas and the Interior: Inputs of Volatiles Into the Crust and Atmosphere) because degassing of the interior, either through crystallization of a magma ocean or being brought to the surface dissolved in magmas, presumably provided the majority of the volatile elements we see in the crust (see review Filiberto et al., 2016). This then ties in with connecting the interior with the Martian crust and atmosphere and the different reservoirs of noble gases and water (Chapter 3: Noble Gases in Martian Meteorites: Budget, Sources, Sinks, and Processes; and Chapter 4: Hydrogen Reservoirs in Mars as Revealed by Martian Meteorites). These chapters lead to the two transition chapters on carbonates (Chapter 5: Carbonates on Mars) and sulfur (Chapter 6: Sulfur on Mars from the Atmosphere to the Core)—where both chapters combine Martian meteorite and data from in situ analyses (landers and rovers), as well as orbital data of the Martian crust.

1.1.2 MARS ORBITERS

Mars has a long history of missions in orbit examining the surface all the way back to Mariner 9 being the first mission to orbit Mars on November 14, 1971. Mariner 9 revolutionized our understanding of the Martian crust sending back images of a Martian surface that was shaped by impact, volcanic, tectonic, depositional, and erosional processes (e.g., McCauley et al., 1972). Orbital missions provide the capabilities to investigate large portions of the crust, but at significantly lower resolution than studying hand samples or in situ with landers and rovers. For the study of volatiles in the crust, Chapter 7, The Hydrology of Mars Including a Potential Cryosphere; and Chapter 8, Sequestration of Volatiles in the Martian Crust Through Hydrated Minerals: A Significant Planetary Reservoir of Water, rely on orbital data from NASA Mars Global Surveyor, NASA 2001 Mars Odyssey, ESA (European Space Agency) Mars Express, and NASA Mars Reconnaissance Orbiter missions.

Mars Global Surveyor flew five scientific instruments to investigate the crust: Mars Orbiter Camera, Mars Orbiter Laser Altimeter, Thermal Emission Spectrometer, Magnetometer, Ultrastable Oscillator, and Mars Relay (Albee et al., 1998; Christensen et al., 2001; Malin and Edgett, 2001). These measurements provided the topography (Fig. 1.4) and spectroscopy of the entire surface of Mars, which can be used to deduce geologic features and mineralogy, including hydrated minerals (Christensen et al., 2001; Smith et al., 2001).

Mars Odyssey is an active orbiter around Mars with three main instruments: Thermal Emission Imaging System (THEMIS), Gamma Ray Spectrometer (GRS), and Mars Radiation Environment Experiment (Saunders et al., 2004). The THEMIS and GRS measurements provide mineralogy and chemistry of the crust including measurements of hydrogen (which will be used throughout the book) (Boynton et al., 2002; Christensen et al., 2003; Boynton et al., 2004).

Mars Express is also an active orbiter around Mars (Chicarro et al., 2004). The three main instruments that are discussed in this book are the Observatoire pour la Minéralogie, l'Eau, les Glaces et l'Activité (OMEGA spectrometer), Mars Advanced Radar for Subsurface and Ionosphere Sounding (MARSIS), and High Resolution Stereo Camera (HRSC) instruments (Chicarro et al., 2004; Bibring et al., 2005; Jordan et al., 2009). Specifically MARSIS is discussed in Chapter 7,

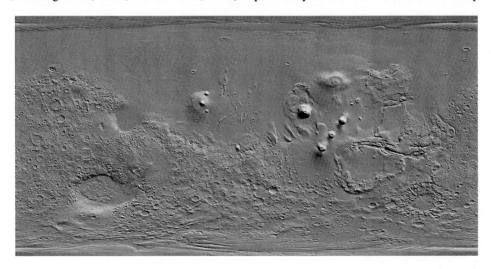

FIGURE 1.4

Mars Orbiter Laser Altimeter (MOLA) topography map of the crust of Mars (Smith et al., 1999). *Image Credit: MOLA Team, NASA, JPL.*

The Hydrology of Mars Including a Potential Cryosphere, to discuss a potential cryosphere on Mars, and OMEGA is discussed in Chapter 8, Sequestration of Volatiles in the Martian Crust Through Hydrated Minerals: A Significant Planetary Reservoir of Water, to investigate the hydrous mineralogy of the crust.

Finally, Mars Reconnaissance Orbiter includes three cameras (HiRISE, Context Camera, Mars Color Imager), two spectrometers (Compact Reconnaissance Imagine Spectrometer—CRISM and Mars Climate Sounder), and a radar (SHARAD—Shallow Subsurface Radar) to investigate the Martian crust (Graf et al., 2005).

Three recent missions [NASA MAVEN (Jakosky et al., 2015), ESA ExoMars Trace Gas Orbiter (Vago et al., 2015), and Indian Space Research Organization Mars Orbiter mission (Mangalyaan) (Sundararajan, 2013; Lele, 2014; Seetha and Satheesh, 2015)] that are not discussed in great detail here will investigate the volatile content of the Martian atmosphere and extend the discussion in this book.

1.1.3 MARS PHOENIX MISSION

The Mars Phoenix mission was a lander mission to near the north polar cap (Vastitas Borealis plains) to investigate the history of water and search for any potentially habitable environments (Smith, 2004; Smith et al., 2008). The primary instruments for measurement of volatiles were the Thermal and Evolved Gas Analyzer (TEGA), the Wet Chemistry Laboratory (WCL), and the Thermal and Electrical Conductivity Probe (TECP) (Smith, 2004; Smith et al., 2008). Kounaves and Oberlin (Chapter 9: Volatiles Measured by the Phoenix Lander at the Northern Plains of Mars) summarize the results from Phoenix and the analyses and characterization of near-surface water-ice surrounding the north polar cap.

1.1.4 MARS EXPLORATION ROVERS SPIRIT AND OPPORTUNITY

The Mars exploration rovers Spirit and Opportunity launched in 2003, landed in January 2004, and were originally scheduled as 90-day missions (Squyres et al., 2004a; Squyres et al., 2004b). Spirit lasted until 2010 before getting stuck, while Opportunity is still active and investigating the Martian surface. Spirit landed in Gusev Crater to investigate a potential dry lake bed (Cabrol et al., 1998; Golombek et al., 2003; Squyres et al., 2004a). Opportunity was sent to Meridiani Planum to investigate hematite, which can form in an aqueous environment (Arvidson et al., 2003; Golombek et al., 2003; Squyres et al., 2004b). The twin rovers were designed as field geologists with instruments mounted both on a mast above the rover, as well as on a moveable arm. The rovers had three instruments mounted on the mast: Panoramic Camera (Pancam), Navigation Camera (Navcam), and the Miniature Thermal Emission Spectrometer (Mini-TES). The rovers also had four main instruments on the robotic arm: Mössbauer spectrometer; Alpha particle X-ray spectrometer (APXS), Microscopic Imager; and the Rock Abrasion Tool (RAT). Chapter 10, Mars Exploration Rover Opportunity: Water and Other Volatiles on Ancient Mars; and Chapter 11, Alteration Processes in Gusev Crater, Mars: Volatile/Mobile Element Contents of Rocks and Soils Determined by the Spirit Rover, summarize the twin rovers' findings for the history of volatiles in the Martian crust at each landing site including confirming hematite and evidence of ancient fluid flow at Meridiani Planum and the lack of lake sediments and also the discovery of an ancient hot spring at Gusev Crater (Chapters 10 and 11).

1.1.5 MARS SCIENCE LABORATORY CURIOSITY

Building on the success of the MER mission Mars Science Laboratory—Curiosity rover (Fig. 1.5), is a roving geochemical bench top on the surface of Mars (Grotzinger et al., 2015). MSL Curiosity landed in Gale Crater in 2012 to investigate the geology and habitability of Gale Crater. It is carrying 10 instruments and in total 17 cameras, of which on the mast are the LIBS spectrometer and remote imager (ChemCam), rover environmental monitoring station (REMS), and the navigation and science cameras (NavCam, MastCam). In or on the rover itself are a combined gas chromatography, mass spectrometry, and tunable laser spectrometry instrument that can take solid and atmospheric sample (Sample Analyzer at Mars, SAM), X-ray diffraction instrument (CheMin), dynamic albedo of neutrons instrument (DAN), radiation assessment detector (RAD), descent camera (MARDI), and front and rear hazard cameras (HazCam). On the turret are APXS, Mars Hand Lens Imager (MAHLI) camera, the drill, brush, scoop, and sample sieving unit (Chimera). Using the full

FIGURE 1.5

The Mars Science Laboratory rover—Curiosity landed at Gale Crater, Mars, on August 6, 2012, 5:17 UTC. During its first 1950 sols on Mars it traveled about 18 km, investing the lake bed sediments in the crater. It found ample evidence for water in the crater's history. At the Noachian Hesperian boundary, rocks were deposited inside the crater that provide evidence for flowing water forming fan deposits and conglomerates, standing water forming a lake, and water within the rock formations causing alteration minerals to form. *Image credit: NASA/ JPL-Caltech/MSSS.*

suite of instruments, Curiosity has found evidence for sustained surface water flow, mineral alteration, and a wide variety of magmatic rocks. Data from Curiosity are used in almost all chapters of the book, but Chapter 12, Volatile Detections in Gale Crater Sediment and Sedimentary Rock: Results From the Mars Science Laboratory's Sample Analysis at Mars Instrument, focuses specifically on the SAM measurements of volatiles at Gale Crater.

REFERENCES

Albee, A.L., Palluconi, F., Arvidson, R., 1998. Mars global surveyor mission: overview and status. Science 279 (5357), 1671−1672.

Arvidson, R.E., et al., 2003. Mantled and exhumed terrains in Terra Meridiani, Mars. J. Geophys. Res. 108 (E12), 8073. Available from: doi:10.1029/2002je001982.

Beck, P., Gillet, P., El Goresy, A., Mostefaoui, S., 2005. Timescales of shock processes in chondritic and Martian meteorites. Nature 435 (7045), 1071−1074.

Bibring, J.-P., et al., 2005. Mars surface diversity as revealed by the OMEGA/Mars Express observations. Science 307 (5715), 1576−1581.

Bogard, D.D., Johnson, P., 1983. Martian gases in an Antarctic meteorite. Science 221 (4611), 651−654.

Boynton, W., et al., 2002. Distribution of hydrogen in the near surface of Mars: evidence for subsurface ice deposits. Science 297 (5578), 81−85.

Boynton, W., et al., 2004. The Mars Odyssey Gamma-Ray Spectrometer Instrument Suite, 2001 Mars Odyssey. Springer, Chichester, pp. 37−83.

Cabrol, N.A., Grin, E.A., Landheim, R., Kuzmin, R.O., Greeley, R., 1998. Duration of the Ma'adim Vallis/ Gusev Crater hydrogeologic system, Mars. Icarus 133 (1), 98−108.

Chicarro, A., Martin, P., Trautner, R., 2004. The Mars Express mission: an overview. In: Mars Express: The Scientific Payload. In: Wilson, A. (Ed.), Scientific coordination: Agustin Chicarro. ESA SP-1240, 2004. ESA Publications Division, Noordwijk, Netherlands, pp. 3−13. ISBN: 92-9092-556-6.

Christensen, P.R., et al., 2001. Mars Global Surveyor Thermal Emission Spectrometer experiment: investigation description and surface science results. J. Geophys. Res. Planets 106 (E10), 23823−23871.

Christensen, P.R., et al., 2003. Morphology and composition of the surface of Mars: Mars Odyssey THEMIS results. Science 300 (5628), 2056−2061.

Clayton, R.N., 2003. Oxygen isotopes in the solar system. Space Sci. Rev. 106 (1-4), 19−32.

Clayton, R.N., Mayeda, T.K., 1996. Oxygen isotope studies of achondrites. Geochim. Cosmochim. Acta 60 (11), 1999−2017.

Clifford, S.M., 1993. A model for the hydrologic and climatic behavior of water on Mars. J. Geophys. Res. Planets 98 (E6), 10973−11016.

Cockell, C.S., 1999. Life on Venus. Planet. Space Sci. 47 (12), 1487−1501.

Cockell, C.S., et al., 2016. Habitability: a review. Astrobiology 16 (1), 89−117.

Deutsch, A.N., et al., 2016. Comparison of areas in shadow from imaging and altimetry in the north polar region of Mercury and implications for polar ice deposits. Icarus 280, 158−171.

Duke, M.B., 1968. The Shergotty meteorite: magmatic and shock metamorphic features. In: French, B.V., Short, N.M., Goddard Space Flight Center, (Eds.), Shock Metamorphism of Natural Materials. Mono Book Corp., Baltimore, MD, pp. 612−621.

Esposito, L.W., Bertaux, J.-L., Krasnopolsky, V., Moroz, V., Zasova, L., 1997. In: Bougher, S.W., Hunten, D. M., Phillips, R.J. (Eds.), Chemistry of lower atmosphere and clouds, II. The University of Arizona Press, Tucson, Venus, pp. 415−458.

Fegley, B., Treiman, A.H., 1992. Chemistry of atmosphere-surface interaction on Venus and Mars. In: Luhmann, J.G., Tatrallyay, M., Pepin, R.O. (Eds.), Venus and Mars: Atmospheres, Ionospheres, and Solar Wind Interactions. American Geophysical Union, Washington, DC. Available from: doi:10.1029/GM066p0007.

Filiberto, J., et al., 2016. A review of volatiles in the Martian interior. Meteorit. Planet. Sci. 51 (11), 1935−1958.

Fritz, J., Artemieva, N., Greshake, A., 2005a. Ejection of Martian meteorites. Meteorit. Planet. Sci. 40 (9-10), 1393−1411.

Fritz, J., Greshake, A., Stöffler, D., 2005b. Micro-Raman spectroscopy of plagioclase and maskelynite in Martian meteorites: evidence of progressive shock metamorphism. Antarct. Meteorite Res. 18, 96−116.

Golombek, M.P., et al., 2003. Selection of the Mars Exploration Rover landing sites. J. Geophys. Res.: Planets 108 (E12), 8072.

Graf, J.E., et al., 2005. The Mars Reconnaissance Orbiter mission. Acta Astronaut. 57 (2), 566−578.

Gross, J., Treiman, A.H., Filiberto, J., Herd, C.D.K., 2011. Primitive olivine-phyric shergottite NWA 5789: petrography, mineral chemistry, and cooling history imply a magma similar to Yamato-980459. Meteorit. Planet. Sci. 46 (1), 116−133.

Grotzinger, J.P., et al., 2015. Curiosity's Mission of Exploration at Gale Crater, Mars. Elements 11 (1), 19−26.

Hamilton, V.E., Christensen, P.R., McSween, H.Y., Bandfield, J.L., 2003. Searching for the source regions of Martian meteorites using MGS TES: integrating Martian meteorites into the global distribution of igneous materials on Mars. Meteorit. Planet. Sci. 38 (6), 871−885.

Hapke, B., Danielson, G.E., Klaasen, K., Wilson, L., 1975. Photometric observations of Mercury from Mariner 10. J. Geophys. Res. 80 (17), 2431−2443.

Hsu, H.-W., et al., 2015. Ongoing hydrothermal activities within Enceladus. Nature 519 (7542), 207−210.

Hubbard, G.S., Naderi, F.M., Garvin, J.B., 2002. Following the water, the new program for Mars exploration. Acta Astronaut. 51 (1−9), 337−350.

Humayun, M., et al., 2013. Origin and age of the earliest Martian crust from meteorite NWA7533. Nature 503 (7477), 513−516.

Jakosky, B.M., Phillips, R.J., 2001. Mars' volatile and climate history. Nature 412 (6843), 237−244.

Jakosky, B.M., et al., 2015. The Mars atmosphere and volatile evolution (MAVEN) mission. Space Sci. Rev. 195 (1-4), 3−48.

Jones, B., 2008. Mars before the space age. Int. J. Astrobiol. 7 (2), 143−155. Available from: http://dx.doi.org/10.1017/S1473550408004138.

Jones, J.H., 2015. Various aspects of the petrogenesis of the Martian shergottite meteorites. Meteorit. Planet. Sci. 50 (4), 674−690.

Jordan, R., et al., 2009. The Mars express MARSIS sounder instrument. Planet. Space Sci. 57 (14), 1975−1986.

Kargel, J.S., et al., 2000. Europa's crust and ocean: origin, composition, and the prospects for life. Icarus 148 (1), 226−265.

Keszthelyi, L., et al., 2001. Imaging of volcanic activity on Jupiter's moon Io by Galileo during the Galileo Europa Mission and the Galileo Millennium Mission. J. Geophys. Res.: Planets 106 (E12), 33025−33052.

Kieffer, H.H., Jakosky, B.M., Snyder, C.W., Matthews, M.S., 1992. Mars. The University of Arizona Press, Tucson, AZ, p. 1455.

Kivelson, M.G., et al., 2000. Galileo magnetometer measurements: a stronger case for a subsurface ocean at Europa. Science 289 (5483), 1340−1343.

Lang, N.P., Tornabene, L.L., McSween Jr., H.Y., Christensen, P.R., 2009. Tharsis-sourced relatively dust-free lavas and their possible relationship to Martian meteorites. J. Volcanol. Geotherm. Res. 185 (1-2), 103−115.

Lapen, T.J., et al., 2010. A younger age for ALH84001 and its geochemical link to shergottite sources in Mars. Science 328 (5976), 347–351.

Lapen, T.J., et al., 2017. Two billion years of magmatism recorded from a single Mars meteorite ejection site. Sci. Adv. 3 (2). Available from: http://dx.doi.org/10.1126/sciadv.1600922.

Lawrence, D.J., et al., 2013. Evidence for water ice near Mercury's North Pole from MESSENGER neutron spectrometer measurements. Science 339 (6117), 292–296.

Lele, A., 2014. Mars Orbiter Mission, Mission Mars: India's Quest for the Red Planet. Springer, New Delhi, pp. 39–69.

Malin, M.C., Edgett, K.S., 2001. Mars global surveyor Mars orbiter camera: interplanetary cruise through primary mission. J. Geophys. Res.: Planets 106 (E10), 23429–23570.

McCauley, J.F., et al., 1972. Preliminary Mariner 9 report on the geology of Mars. Icarus 17 (2), 289–327.

McCubbin, F.M., Riner, M.A., Vander Kaaden, K.E., Burkemper, L.K., 2012. Is Mercury a volatile-rich planet? Geophys. Res. Lett. 39 (9), L09202.

McKay, C.P., Porco, C.C., Altheide, T., Davis, W.L., Kral, T.A., 2008. The possible origin and persistence of life on Enceladus and detection of biomarkers in the plume. Astrobiology 8 (5), 909–919.

McSween, H.Y., Treiman, A.H., 1998. Martian meteorites. In: Papike, J.J. (Ed.), Reviews in Mineralogy. Mineralogical Society of America, Chantilly, VA, pp. 6-01–6-40.

Murchie, S.L., et al., 2015. Orbital multispectral mapping of Mercury with the MESSENGER Mercury Dual Imaging System: evidence for the origins of plains units and low-reflectance material. Icarus 254, 287–305.

Murray, B.C., et al., 1974. Mercury's surface: preliminary description and interpretation from Mariner 10 Pictures. Science 185 (4146), 169–179.

Neumann, G.A., et al., 2013. Bright and dark polar deposits on mercury: evidence for surface volatiles. Science 339 (6117), 296–300.

Nyquist, L.E., et al., 2001. Ages and geologic histories of Martian meteorites. Space Sci. Rev. 96 (1), 105–164.

Ody, A. et al., 2014. Search for Analogue Sites of New Martian Shergottite Spectra Using NIR Data. Lunar and Planetary Institute Science Conference Abstracts, 45: Abstract #2207.

Peal, S., 1893. The canals of Mars. Science 535, 242–243.

Postberg, F., et al., 2009. Sodium salts in E-ring ice grains from an ocean below the surface of Enceladus. Nature 459 (7250), 1098–1101.

Prockter, L.M., et al., 2010. Evidence for young volcanism on Mercury from the third MESSENGER flyby. Science 329 (5992), 668–671.

Rothschild, L.J., Mancinelli, R.L., 2001. Life in extreme environments. Nature 409 (6823), 1092–1101.

Sagan, C., Pollack, J.B., 1966. On the nature of the canals of Mars. Nature 212 (5058), 117–121.

Saunders, R., et al., 2004. 2001 Mars Odyssey Mission Summary, 2001 Mars Odyssey. Springer, Rotterdam, pp. 1–36.

Schulze-Makuch, D., Grinspoon, D.H., Abbas, O., Irwin, L.N., Bullock, M.A., 2004. A sulfur-based survival strategy for putative phototrophic life in the Venusian atmosphere. Astrobiology 4 (1), 11–18.

Seetha, S., Satheesh, S., 2015. Mars Orbiter Mission. Curr. Sci. 109 (6), 1047.

Sharp, T.G., DeCarli, P.S., 2006. Shock effects in meteorites. In: Binzel, R.P., (Ed.), Meteorites and the Early Solar System II, vol. 943. The University of Arizona Press, Tucson, AZ, pp. 653–677.

Smith, D.E., et al., 1999. The global topography of Mars and implications for surface evolution. Science 284 (5419), 1495–1503.

Smith, D.E., et al., 2001. Mars Orbiter Laser Altimeter: experiment summary after the first year of global mapping of Mars. J. Geophys. Res.: Planets 106 (E10), 23689–23722.

Smith, P., et al., 2008. Introduction to special section on the phoenix mission: landing site characterization experiments, mission overviews, and expected science. J. Geophys. Res.: Planets 113 (E3), E00A18.

Smith, P.H., 2004. The Phoenix mission to Mars. 2004 IEEE Aerospace Conference Proceedings (IEEE Cat. No.04TH8720), vol. 1, pp. 1–342.

Squyres, S.W., et al., 2004a. The Spirit Rover's Athena science investigation at Gusev Crater, Mars. Science 305 (5685), 794–799.

Squyres, S.W., et al., 2004b. The Opportunity Rover's Athena Science Investigation at Meridiani Planum, Mars. Science 306 (5702), 1698–1703.

Sundararajan, V., 2013. Mangalyaan-overview and technical architecture of India's first interplanetary mission to Mars. AIAA SPACE 2013 Conference and Exposition, pp. 5503.

Taylor, F., Crisp, D., Bezard, B., 1997. In: Bougher, S.W., Hunten, D.M., Phillips, R.J. (Eds.), Near-infrared sounding of the lower atmosphere of Venus, II. The University of Arizona Press, Tucson, Venus, pp. 325–351.

Tornabene, L.L., et al., 2006. Identification of large (2–10 km) rayed craters on Mars in THEMIS thermal infrared images: implications for possible Martian meteorite source regions. J. Geophys. Res.: Planets 111 (E10). Available from: doi:10.1029/2005JE002600.

Vago, J., et al., 2015. ESA ExoMars program: the next step in exploring Mars. Solar Syst. Res. 49 (7), 518–528.

Walton, E.L., Spray, J.G., 2003. Mineralogy, microtexture, and composition of shock-induced melt pockets in the Los Angeles basaltic shergottite. Meteorit. Planet. Sci. 38 (12), 1865–1875.

Walton, E.L., Kelley, S.P., Herd, C.D.K., 2008. Isotopic and petrographic evidence for young Martian basalts. Geochim. et Cosmochim. Acta 72 (23), 5819–5837.

Walton, E.L., Sharp, T.G., Hu, J., Filiberto, J., 2014. Heterogeneous mineral assemblages in Martian meteorite Tissint as a result of a recent small impact event on Mars. Geochim. et Cosmochim. Acta 140 (0), 334–348.

Werner, S.C., Ody, A., Poulet, F., 2014. The Source Crater of Martian Shergottite meteorites. Science 343 (6177), 1343–1346.

Zolotov, M.Y., Shock, E.L., 2003. Energy for biologic sulfate reduction in a hydrothermally formed ocean on Europa. J. Geophys. Res.: Planets 108 (E4). Available from: doi:10.1029/2002KE001966.

VOLATILES IN MARTIAN MAGMAS AND THE INTERIOR: INPUTS OF VOLATILES INTO THE CRUST AND ATMOSPHERE

Justin Filiberto[1,2], Francis M. McCubbin[3], and G. Jeffrey Taylor[4]

[1]*Southern Illinois University, Carbondale, IL, United States* [2]*The Open University, Milton Keynes, United Kingdom*
[3]*NASA Johnson Space Center, Houston, TX, United States* [4]*University of Hawai'i, Honolulu, HI, United States*

2.1 INTRODUCTION

Since the 1970s evidence for water flowing on the Martian surface in the geologic past has captivated the public and scientists alike. From the discovery of landforms that resulted from flowing water to the discovery of secondary hydrous and carbonate minerals on its surface from orbital remote sensing studies, the history of volatiles on Mars is fundamental to understanding the geologic history of Mars. Furthermore, water is paramount to Martian habitability and the past presence of putative life on Mars.

Multiple observations from space missions (orbiters, landers, and rovers) to Mars have identified compelling evidence for abundant liquid water once flowing at the Martian surface in the ancient past (e.g., Bibring et al., 2006; McSween, 2006; Ehlmann et al., 2008; Carr and Head, 2010; Carter and Poulet, 2012; Williams et al., 2013) (these will be explored in more detail in the following chapters). However, liquid water is currently unstable at the Martian surface (e.g., Brass, 1980; Carr, 1996; Haberle et al., 2001).

One caveat is that there is evidence for recent brines in the Martian crust. Possible transient seeps involving brines have been suggested from satellite observations over the last decade (e.g., McEwen et al., 2011, 2014; Ojha et al., 2015) and geologically recent alteration seen in the nakhlite meteorites (Gooding et al., 1991; Swindle et al., 2000; Bridges et al., 2001; Swindle and Olson, 2004; Bridges and Schwenzer, 2012; Hicks et al., 2014) and regolith breccia NWA 7034 (Gattacceca et al., 2014; Muttik et al., 2014).

Changes in the abundances of volatile species on the surface may have been caused by diminishing basaltic eruptions through time and by diminishing concentrations of volatiles in the erupting materials (Owen et al., 1988; Poulet et al., 2005; Bibring et al., 2006; Carr and Head, 2010; Ehlmann and Mustard, 2012) as Martian basaltic magmas are presumably the ultimate mechanism of transfer of volatiles from the mantle to the surface and atmosphere (e.g., Greeley, 1987; Hirschmann and Withers, 2008; Johnson et al., 2008; Righter et al., 2009; Carr and Head, 2010; King and McLennan, 2010; Grott et al., 2011; Stanley et al., 2011; Balta and McSween, 2013). This has led to the

Volatiles in the Martian Crust. DOI: http://dx.doi.org/10.1016/B978-0-12-804191-8.00002-7

suggestion of a transition of surface conditions from water-dominated alteration within the Martian crust in the Noachian, to a sulfur-dominated hydrous alteration in the Hesperian, and finally anhydrous alteration through the Amazonian (e.g., Poulet et al., 2005; Bibring et al., 2006).

Therefore, in order to understand the concentrations of volatiles in the crust, it is vital to first constrain the volatile budget of the Martian interior, the preeruptive volatile content of Martian magmas, and evidence within the Martian meteorites for magmatic degassing. However, the concentrations of volatile species in ascending magmas and in their mantle source regions are highly uncertain and probably variable in space and time. To better understand the abundance and distribution of volatiles in the Martian interior, and to determine the volume and composition of fluids that could have once been on the Martian surface, many have looked to the Martian meteorites. Here, we discuss the constraints on the volatile content (water, halogens, sulfur, carbon, and nitrogen) of the Martian interior and basaltic magmas, and evidence for degassing of magmatic fluids into the Martian crust.

2.2 OVERVIEW FROM ORBIT

The Gamma-Ray Spectrometer (GRS) on the Mars Odyssey spacecraft measured the abundances of several elements from orbit (Boynton et al., 2008), including H, Cl, and S (Fig. 2.1). Data apply to the upper few tens of centimeters of the Martian surface. Concentrations were determined from gamma rays created by capture of thermal neutrons. Neutron fluxes are modulated mainly by H, which has a large cross section for absorbing thermal neutrons. Thus the effect of H abundance requires a correction involving modeling neutron transport and gamma-ray production. This procedure works well for equatorial regions, but begins to break down when H concentrations become higher. Thus the maps in Fig. 2.1 are limited to a region defined by H concentrations as described in detail by Boynton et al. (2007). Due to low counting rates, the data are smoothed over a radius of 10 degrees for H and Cl, and 25 degrees for S (which has a much weaker signal). As a practical matter, therefore, the maps are useful for assessing regional variations in elemental abundances.

The elemental maps in Fig. 2.1 show that H (expressed as wt% H_2O equivalent), Cl, and S are heterogeneously distributed on the Martian surface. In light of the ease of escape from lavas and different geochemical behavior during aqueous alteration, this is not surprising. H_2O ranges from 1.5 to 7 wt%, Cl from 0.2 to 0.8 wt%, and S varies from 0.7 to 3.2 wt%. H_2O is low (1.5−1.9 wt%) west and northwest of the Argyrebasin, and in Utopia Planitia. It is highest in and around Gusev crater (up to 7 wt%) and in Arabia Terra ($\sim$6 wt%). Like H_2O, Cl is highest around Gusev, but high Cl extends much farther to the east, including the Tharsis volcanic region. Cl is distinctively low where water is also low in Utopia Planitia, and somewhat low in the ancient western southern highlands where H_2O is lowest. High S areas do not correspond to the highest or lowest H_2O or Cl areas. In general, S is enriched in equatorial latitudes. The Argyre region in the western southern highlands is low in all three species. A weak correlation between H_2O and S might reflect the presence of water-bearing cation sulfates (McLennan et al., 2010).

The GRS data show that the Martian surface has distinctly higher concentrations of H_2O, Cl, and S than do Martian meteorites, perhaps indicating a significant enrichment in these mobile species during evolution of the Martian crust. Sediments and sedimentary rocks analyzed by the Spirit,

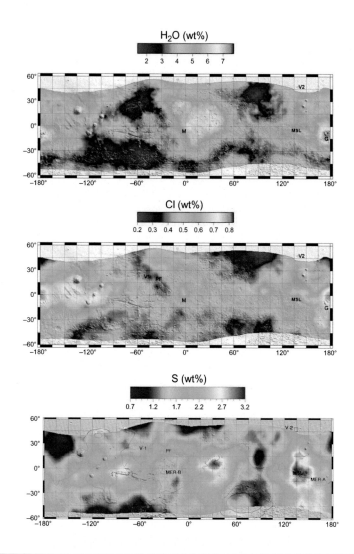

FIGURE 2.1

Maps of H_2O (stoichiometrically estimated from H content), Cl, and S concentrations (wt%) in the equatorial zone where H concentrations are low enough to allow for ready calculation of elemental abundances. The data are displayed over a shaded relief map of Mars. Landing sites are indicated: Viking 1 (V1), Viking 2 (V2), Pathfinder (PF), MER landing sites in Gusev crater (G or MER-A) and Meridiani (M or MER-B), and the MSL landing site in Gale crater (MSL). The S map shows broad topograhic features. All three elements are distributed heterogeneously on the Martian surface and do not correlate strongly with each other. *Image credit: 2001 Mars Odyssey Gamma-Ray Spectrometer (U. Arizona/NASA).*

Opportunity, and Curiosity rovers are consistent with the overall surface enrichment in volatiles compared to primary igneous rocks. However, the distribution of volatiles with depth in the crust is not known.

2.3 WATER

The Martian meteorites host a number of hydrous phases including apatite, amphibole, biotite, glass, scapolite, and a number of hydrated aqueous alteration minerals (Treiman, 1985; Gooding et al., 1991; Treiman et al., 1993; Bridges and Grady, 1999; Bridges et al., 2001; Patiño Douce and Roden, 2006; Filiberto and Treiman, 2009; Patiño Douce et al., 2011; Bridges and Schwenzer, 2012; Hallis et al., 2012a, 2012b; Gross et al., 2013; McCubbin et al., 2013, 2016a,b; Filiberto et al., 2014; Hicks et al., 2014; Giesting et al., 2015; McCubbin and Jones, 2015). Of these phases, glass, apatite, and amphibole have been the primary phases utilized to quantify the abundances of H_2O in Martian magmas based on known mineral-melt partitioning relationships. When applicable, bulk rock abundances of H_2O are also used as a record of magmatic water contents, but the bulk rock abundances are susceptible to several problems. One problem with bulk rock abundances of H_2O is terrestrial contamination by secondary alteration on Earth that could raise the H_2O content of the bulk rock. Another problem, for igneous rocks in particular, is that magmas could have experienced degassing during cooling and crystallization, which would effectively lower the H_2O content of the parental melt. The hydrous phases glass, apatite, and amphibole are not without their own caveats. Glasses, including glasses in melt inclusions, are also susceptible to degassing, either directly or via diffusion through a mineral host in the case of melt inclusions (e.g., Gaetani et al., 2012). Apatites in a single sample commonly exhibit variable compositions, and when that is the case, special considerations need to be made to understand which apatites will provide the most reliable estimate of the parental magmatic H_2O content (e.g., McCubbin and Nekvasil, 2008; Gross et al., 2013; McCubbin et al., 2013, 2016a,b; Howarth et al., 2015, 2016). Furthermore, bulk rock F data is required in order to use apatite to determine the H_2O content of the parental melt (McCubbin et al., 2015), which is not available for every sample (e.g., Filiberto et al., 2016a). The primary caveat with amphiboles is that we know little about mineral-melt partitioning relationships and how these relationships change as a function of amphibole crystal chemistry or composition of the melt (e.g., Giesting and Filiberto, 2014). Finally, the effect of shock from impact on the volatile content of hydrous minerals is not well constrained (e.g., Minitti et al., 2008a,b; Giesting et al., 2015). Nonetheless, a lot of careful work has been conducted on all three phases as well as on bulk rock data to estimate H_2O abundances for a number of Martian parental melts.

Once the H_2O abundance of the parental melt is known, there are a number of ways to determine the H_2O abundance in the source region. One reliable method is to use the fact that incompatible elements track each other during igneous processing. This allows us to determine the bulk planet abundances from the ratio of an incompatible volatile element to an incompatible refractory element. The closer their mineral-melt partition coefficients are for major minerals, the more similar their ratio remains during partial melting and fractional crystallization. It is reasonably well established that H_2O behaves as an incompatible element, with partition coefficients for major elements similar to those of Ce (Koga et al., 2003; Aubaud et al., 2004; Grant et al., 2007). Thus, if we can determine the water abundance in a magma (before outgassing and water loss occurs), we can use the H_2O/Ce ratio in the magma and the bulk Ce of Mars to determine the value of H_2O of Mars (or at least the mantle for a particular source region). The same exercise also applies to any relevant Martian crustal rocks. Although it is tempting to use the H_2O/Ce ratio of the basaltic breccia NWA 7034 to estimate the abundance of H_2O in the Martian crust, given that NWA 7034 is similar in composition and in spectral properties to the Martian surface (e.g., Beck et al., 2015), the primary source of H_2O in NWA 7034 is hydrated oxide phases that likely formed during secondary alteration of the breccia on Mars (Muttik et al., 2014; McCubbin et al., 2016b).

In Table 2.1, we provide estimated H_2O abundances for parental liquid compositions from the enriched shergottite Shergotty and depleted shergottite Yamato 980459 as well as their corresponding Ce abundances, which we used to estimate the amount of water in the Martian crust and mantle using the bulk Mars Ce value of 1.18 ppm estimated by Taylor (2013). The estimated H_2O concentrations in the mantle source regions for Y 980459 and Shergotty are similar and indicate a range of 15−70 ppm (assuming the parental melts formed by 10%−15% partial melting for the Shergotty source and 20%−40% melting for the Yamato 980459 source). However, because of their vastly different Ce abundances, the inferred bulk Mars abundances are very different: Y 980459 is more primitive (low Ce) and infers an H_2O bulk concentration of 210−320 ppm, whereas Shergotty is more evolved (high Ce) and infers an H_2O bulk of only 80−110 ppm. This large difference in estimated bulk water abundance suggests that processes operating during the petrogenesis of these two Martian basalts led to a decoupling of H_2O and Ce. In turn, this suggests that the petrogenesis of the two rocks (which are representatives of enriched and depleted shergottites) involved fluid exsolution or assimilation reactions between the Shergotty magma and water-bearing regions in the crust.

In addition to using Ce abundances to estimate the water content of the Martian mantle, other studies have taken a more petrological approach to such estimates. A number of studies have used the abundances of volatiles in volatile-bearing minerals, the bulk-rock abundances of fluorine, and

Table 2.1 H_2O Abundances for Source Regions within the Martian Mantle, Crust, and Bulk Mars Estimated from Parental Melt Abundances of H_2O and Ce in Shergottites

Parental Melt Values (Basis of Estimate)	H_2O	Ce	H_2O/Ce
Yamato 980459 (Ol-hosted melt inclusions)[a]	75−116	0.426	176−272
Shergotty (igneous apatite)[b]	363−484	5.18	70−93
Mantle, crust, and bulk Mars estimates			
H_2O in the Martian mantle from Y 980459[c]	15−47		
H_2O in Martian mantle from Shergotty[d]	36−72		
H_2O in bulk silicate Mars based on Y 980459 melt inclusions and bulk Mars Ce[e]	208−321	1.18	
H_2O in bulk silicate Mars based on Shergotty	83−110	1.18	

All values presented in ppm unless otherwise noted.
[a]Usui et al. (2012).
[b]McCubbin et al. (2016a,b).
[c]Calculated by assuming 20%−40% partial melting of mantle source[a].
[d]Calculated by assuming 10%−15% partial melting of mantle source[b].
[e]Calculated from bulk Ce in Mars (Taylor, 2013) and H_2O/Ce in melt inclusions.

mineral-melt partitioning relationships to determine parental melt volatile abundances. Subsequently, these parental melt values were combined with geochemical, petrological, and petrographic data to estimate the degree of fractionation, partial melting, and amount of crystal accumulation in each sample to estimate the abundance of H_2O in the source region (McCubbin et al., 2012, 2016a,b; Gross et al., 2013). In turn these source estimates can be used to estimate crustal abundances using the typical crust-mantle distributions of large ion lithophile elements of about 50% (Taylor, 2013). A compilation of estimates of H_2O, F, and Cl abundances in the Martian crust and mantle from hydrous phases is provided in Table 2.2. These methods generally result in comparatively lower abundances of H_2O in the Martian interior than estimates using the H_2O/Ce ratios. Moreover, these methods reveal that the Martian crust and mantle both have heterogeneous distributions of H_2O that seem to correlate with the geochemical characteristics of the source region (Filiberto et al., 2016a; McCubbin et al., 2016a,b). Importantly, we know the interior has a heterogeneous distribution of incompatible refractory lithophile elements (e.g., Laul, 1987; Longhi, 1991; Norman, 1999; Borg and Draper, 2003; Jones, 2003; Bridges and Warren, 2006; Shearer et al., 2008).

Other estimates of H_2O content of the Martian crust have been made based on fluvial features (Carr and Head, 2003). Carr and Head's (2003) preferred value is to use the volume of the northern plains, 196 m spread over the entire planet, corresponding to 31 ppm H_2O. Since this estimate applies to the crust only, bulk Mars would have contained double that amount (50% of the inventory of incompatible elements is in the crust; see Taylor et al., 2006 for a full discussion), so H_2O in bulk Mars is 62 ppm. A potentially important reservoir is deposits of phyllosilicates in the Martian crust. This was discussed by Carr and Head (2003), who used estimates of the crustal

Table 2.2 Computed Volatile Abundances for Source Regions within the Martian Interior				
Martian Reservoir	**Basis of Estimate**	**H_2O[a]**	**F[a]**	**Cl[a]**
Mantle	Yamato 980459 melt inclusions[e]	15–23	1–1.6	2.8–4.2
	Apatite within QUE 94201[f]	14–19	1	1.6–2.1
	Apatite within Shergotty[f]	36–72	3.6–5.4	12–23
	Kaersutitic amphibole in Chassigny[g]	140–250	–	–
Crust	GRS[c]	–	–	4200[c]
	GRS[c] + Martian meteorites[d]	350[d]	–	350
	NWA 7034[h]	6600	–	2200
	Enriched Shergottite mantle + LILE[b] crustal abundances[f]	1410	106	450
Bulk silicate Mars[f]		137	10	44

[a]Values presented in ppm unless otherwise noted.
[b]LILE, large ion lithophile elements, which have a crust-mantle distribution of ~50%[d].
[c]From Taylor et al. (2010a).
[d]From Taylor (2013).
[e]From Usui et al. (2012).
[f]From McCubbin et al. (2016a,b).
[g]From McCubbin et al. (2010).
[h]From Agee et al. (2013).

porosity to suggest that the amount of water contained in the Martian megaregolith could have ranged from a global equivalent layer between 540 and 1400 m thick. However, this porosity may have been replaced by phyllosilicates and other products of aqueous alteration (Chapter 8; Mustard, 2018). Making reasonable assumptions of the abundance of phyllosilicates and their H_2O contents, (Chapter 8; Mustard, 2018) suggests that phyllosilicates could have sequestered an amount of water equivalent to a global layer ranging from 15 to 930 m, which corresponds to 2.4−149 ppm H_2O. Taking the volume of the northern plains from Carr and Head (2003) as the lowest reasonable value and the highest value from (Chapter 8; Mustard, 2018) and doubling because half the water is in the crust and half in the mantle, we estimate that bulk Mars contained between 62 and 298 ppm H_2O. Our estimates from Martian meteorites (Table 2.2) are in this range.

Some of the discrepancies between methods may reflect heterogeneities within the Martian interior (as described earlier), or from different assumptions in the calculations. However, some may also reflect a sampling bias. All of the Martian meteorites are young rocks from a relatively small number of unknown locations on the Martian surface. Therefore the estimates from the shergottite meteorites represent estimates of the recent Martian mantle and not the water in the interior during the Noachian. Models suggest that Mars potentially had higher water contents in the interior early in Mars' history (e.g., Médard and Grove, 2006) and has subsequently lost volatiles from the interior during two later stages: early outgassing during differentiation (possibly during magma ocean solidification) and later degassing during volcanic eruptions (see Filiberto et al., 2016b for a review). This may help explain higher estimates of water in the interior based on Noachian-aged features and lower estimates for younger source regions; however, the GRS map (Fig. 2.1) shows low water concentrations in regions in the Noachian highlands and the northern Amazonian plains complicating any simple explanation.

Sample analysis indicates that the bulk silicate Mars H_2O concentration is in the range 100−200 ppm, consistent with the estimates from geologic and remote sensing studies, 6−300 ppm. This is less than the minimum value estimated for the bulk silicate Earth, 500 ppm (Mottl et al., 2007).

2.4 HALOGENS

Dreibus and Wänke in the 1980s pioneered the work to constrain the halogen content of the Martian interior. They showed that ratios of similarly incompatible elements, instead of the bulk composition alone, must be compared in order to investigate differences of the volatile content of different planetary bodies (Dreibus and Wänke, 1982, 1985, 1986, 1987). These comparisons of element ratios account for differences in magmatic processes such as partial melting and fractional crystallization, as well as degassing and secondary alteration, which a simple comparison of bulk compositions of planetary rocks does not account for (e.g., Dreibus and Wänke, 1987). Typically, halogens are compared with similarly incompatible elements that are fluid immobile, including La, Ti, or Th (Dreibus and Wänke, 1982, 1985, 1986, 1987; Treiman, 2003; Filiberto and Treiman, 2009; Taylor, 2013). Therefore, the elements should track each other during partial melting and fractional crystallization. Any deviations can, therefore, be explained by the volatile and fluid-mobile nature of most halogens through degassing (low values) and/or alteration (high values).

Chlorine: The Martian interior is thought to be 2−3 times enriched in halogens compared to the terrestrial mantle based on a comparison of the Cl/La ratio of Martian meteorites and terrestrial basalts (Dreibus and Wänke, 1985, 1987; Filiberto and Treiman, 2009; Filiberto et al., 2016a). Fig. 2.2 shows the Cl/La values for different classes of Martian meteorites. There are six meteorites that have significantly higher Cl/La ratios than the rest of the Martian meteorites: Nakhla, SaU 005, DaG 476, NWA 7034, Yamato 980459, and QUE 94201. Nakhla, SaU 005, DaG 476, and NWA 7034 have all experienced terrestrial or Martian alteration (Bridges and Grady, 1999; Bridges et al., 2001; Wadhwa et al., 2001; Gnos et al., 2002; Crozaz et al., 2003; Treiman 2005; Mohapatra et al., 2009; Filiberto et al., 2014; Hicks et al., 2014; Muttik et al., 2014). Therefore, we will ignore those for the rest of this discussion. Interestingly, depleted shergottites, Yamato 980459 and QUE 94201, have higher Cl/La ratios (342 ± 116) than the rest of the shergottites and values as high as some of the altered meteorites, which suggests that the depleted source region may have higher Cl than the average mantle. One sample, NWA 1068, has a significantly lower Cl/La ratio compared to the rest of the Martian meteorites. This can be explained by preferential loss of Cl either by volcanic degassing during eruption and emplacement or by a shock process (Fig. 2.2). By filtering the data to only those data that have not experienced Cl loss or gain, plus filtering out the depleted shergottites that have a higher Cl/La ratio, we calculate a Cl/La ratio for Martian basalts of 51 ± 17. Importantly,

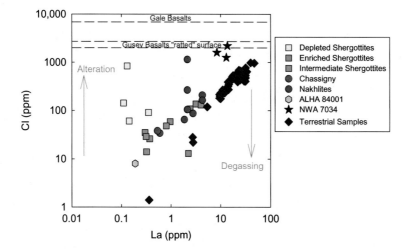

FIGURE 2.2

Cl/La ratio of Martian meteorites compared with terrestrial basalts updated from Filiberto and Treiman (2009) and Filiberto et al. (2016a). Shergottites are shown by *boxes*—depleted (yellow), intermediate (green), and enriched (blue), chassignites are shown as *pink circles*, nakhlites (altered and unaltered) are shown as *red circles*, ALHA 84001 as a *gray hexagon*, and breccia NWA 7034 as *black stars*. Bulk chlorine and lanthanum abundances in terrestrial basaltic rocks are shown as *blue squares*. Terrestrial samples chosen represent magmas and xenoliths that are relatively undegassed as in Filiberto et al. (2016a) and Filiberto and Treiman (2009). *Arrows* show how alteration (high Cl/La) and degassing (low Cl/La) affect the Cl/La ratio. *Modified from Filiberto and Treiman (2009) as in Filiberto et al. (2016a). Data sources for meteorite data as summarized in Filiberto et al. (2016a).*

Chassigny has a Cl/La ratio consistent with the Mars trend line and, based on noble gas data, may represent the Martian interior (Ott 1988), suggesting that the Cl/La ratio may represent the average Martian meteorite source region (Filiberto et al., 2016a). However, there may be source regions with different Cl/La ratios.

We can now use this comparison to calculate the Cl concentration of the source region of the Martian meteorites. Assuming a bulk mantle La value of 0.48 ppm (Dreibus et al., 1982) or 0.439 ± 0.048 ppm (Taylor, 2013), we calculate a bulk Cl abundance of 25 ± 8 ppm and 22 ± 9 ppm Cl, which is significantly lower than the bulk Mars Cl abundance calculated based on the crustal Cl/K ratio (390 ppm; Taylor et al., 2010a), and slightly lower than previous estimates based on the Martian meteorites 44 ppm (Dreibus and Wänke, 1985) and based on the Cl/Th bulk Mars abundance of 32 ± 9 ppm (Taylor, 2013). Interestingly, this estimate is similar to estimates for the terrestrial enriched mantle (30 ppm), but an order of magnitude higher than estimates for the terrestrial depleted mantle (1 ppm) (Michael and Cornell, 1998; Saal et al., 2002; Kendrick et al., 2012), and suggests a source region for the Martian meteorites with a chlorine content similar to the terrestrial enriched Mid-Ocean Ridge Basalt (MORB) source.

Chlorine in the crust (Fig. 2.1) was measured from orbit using the Gamma Ray Spectrometer instrument on Mars Odyssey (Boynton et al., 2007; Taylor et al., 2010a,b), and in situ from rover analyses (e.g., Gellert et al., 2004, 2006; Rieder et al., 2004; Schmidt et al., 2008; Litvak et al., 2014; McLennan et al., 2014; Mitrofanov et al., 2014; Farley et al., 2016). The average composition of the equatorial and mid-latitude crust is 0.49 wt% Cl, but varies across the surface by a factor of 4 (Keller et al., 2006), and at landing sites is up to ~ 2 wt% (e.g., Gellert et al., 2004, 2006; Rieder et al., 2004; Schmidt et al., 2008; Litvak et al., 2014; McLennan et al., 2014; Mitrofanov et al., 2014; Farley et al., 2016). Interestingly, Cl and H_2O concentrations from the Dynamic Albedo of Neutrons (DAN) instrument aboard MSL Curiosity weakly correlate and give an average H_2O:Cl ratio of $\sim 2.1 \pm 0.6$ (data from Litvak et al., 2014; Mitrofanov et al., 2014). Surprisingly, Martian magmas parental to Chassignites are thought to also have a 2:1 ratio of H_2O to Cl (Giesting et al., 2015), but this is magmatic concentrations and may be coincidental. NWA 7034 (and paired NWA 7533) has a H_2O:Cl ratio of 3−4.5 (data from Agee et al., 2013; Humayun et al., 2013; Santos et al., 2015). Mars Odyssey GRS data (using data obtained from July 8, 2002 and April 2, 2005) indicates a global H_2O/Cl ratio of 8.2 ± 0.6, substantially higher than meteorites and lander data. Interestingly, chlorine concentrations are elevated up to ~ 1 wt% in the crust around the volcanic provinces, specifically the Medusae Fossae Formation (Fig. 2.1), which potentially suggests volcanic degassing as a source for the Cl enrichment (Keller et al., 2006). The Martian meteorites contain evidence for degassing of such chlorine-rich fluids. Nakhla contains Cl-scapolite, which can only form from either a late stage Cl-rich, water-poor magma or a magmatic Cl-rich hydrothermal brine at a minimum temperature of 700 °C (Filiberto et al., 2014). Nakhlites MIL 03346, 090030, 090032, 090136, and NWA 5790 contain potassic-chloro-hastingsite, which could have only formed from a similarly Cl-rich, H_2O-poor hydrothermal brine; but, in order to form the amphibole, the fluid would have also had to be Fe and K-rich as well (McCubbin et al., 2013; Giesting and Filiberto, 2016). Finally, differences in apatite chemistry between those found in melt-inclusions and mesostasis in the Chassignites suggest exsolution of a Cl-bearing fluid phase and retention of this fluid phase within the melt inclusions into the hydrothermal regime; while interstitial apatite records fluid migration through the cumulus pile of a Cl-rich, H_2O-poor brine (McCubbin and Nekvasil, 2008).

Fluorine: Similar to chlorine, bulk fluorine contents of the Martian meteorites are also similar to terrestrial basalts, but there are very few analyses of fluorine in Martian meteorites (Dreibus and Wänke, 1985, 1987; Treiman, 2003), and fluorine cannot be measured in the bulk Martian crust. Filiberto et al. (2016a) calculated the F content of the Martian mantle by using the Cl/F ratio of Martian shergottites combined with terrestrial basalts to make up for the paucity of data. Shergottites have a Cl/F ratio of 0.80 ± 0.40 (Fig. 2.3). The scatter in the Cl/F ratio reflects degassing and alteration in some of the analyzed rocks, instead of igneous processing. F and Cl are both similarly incompatible during mantle melting and igneous crystallization and therefore igneous processes should not affect the Cl/F ratio (e.g., Fuge, 1977; Aiuppa et al., 2009; Dalou et al., 2012) until the crystallization of apatite (Boyce et al., 2014; McCubbin et al., 2015). However, Cl readily degasses from a basaltic magma, while F is compatible with a magma (e.g., Carroll and Webster, 1994; Webster and Rebbert, 1998; Webster, 2004; Aiuppa, 2009; Pyle and Mather, 2009), therefore degassing of the magma before crystallization causes the Cl/F ratio to be lower than the initial value. Further, alteration causes the Cl/F ratio to be higher than the initial value because Cl is more soluble in aqueous fluids (e.g., Pyle and Mather, 2009; Anazawa et al., 2011). Therefore, Filiberto et al. (2016a) assumed a ~1:1 correlation to calculate the Martian mantle F concentration, and calculated a bulk F abundance of the source of 25 ppm; however, the error on this calculation is significantly higher than for the Cl calculation and they assumed an error of 13 ppm based on the uncertainty in the data (Fig. 2.3). A Martian mantle with 25 ppm F is, within error, the same as estimates for the primitive terrestrial mantle of 25 ppm F (McDonough and Sun, 1995), and slightly higher than the terrestrial depleted mantle of 16 ppm F (Saal et al., 2002).

Interestingly, magma compositions calculated from Martian apatite (Fig. 2.4) suggest a parental magma with a Cl/F ratio of 2.1−40, which is significantly higher than the bulk rock analyses of

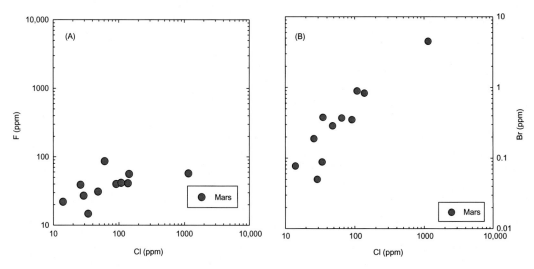

FIGURE 2.3

(A) F/Cl and (B) Br/Cl abundances of Martian meteorites (*red circles*). *Data as summarized in Filiberto et al. (2016a) and as in Fig. 2.2.*

Martian meteorites and terrestrial basalts, and suggests that Martian magmas had significantly less fluorine. Depleted shergottites have the lowest calculated Cl/F compositions, while enriched shergottites have the highest calculated Cl/F ratios (Fig. 2.4). This can be explained by two different processes to elevate the Cl/F ratio: (1) a source region with elevated Cl/F, depleted in F and/or enriched in Cl above terrestrial values; (2) the apatites analyzed in the Martian meteorites formed by fractional crystallization and the early-formed F-rich apatite is not sampled. Filiberto et al. (2016a) assumed that the calculated parental magma Cl/F ratio from apatite represents the igneous compositions and calculated a mantle source composition of 1 ± 1 ppm F, which suggests a Martian interior that is an order of magnitude depleted in F compared with the terrestrial mantle (McDonough and Sun, 1995; Saal et al., 2002).

Fluorine cannot be detected from orbit or by the APXS instruments on MER Spirit and Opportunity, and MSL Curiosity; however, ChemCam on MSL Curiosity has recently analyzed fluorine-rich spots in rocks and soils at Gale crater (Forni et al., 2015). Fluorine is thought to reside in fluorine-aluminosilicates, fluorapatite, or fluorite (Forni et al., 2015). Such fluorine mobility in the crust has previously been suggested for the Home Plate and surrounding regions at Gusev crater (Filiberto and Schwenzer, 2013). This suggests that, while fluorine cannot be easily analyzed and has not been analyzed across crustal materials, it is likely enriched in the crust and important for Martian low-T alteration processes (Filiberto and Schwenzer, 2013; Forni et al., 2015).

Bromine: There is a paucity of data for bromine in Martian meteorites. Martian rocks typically contain $0.05-1.0$ ppm Br (Dreibus and Wänke, 1985, 1986, 1987; Cartwright et al., 2013); while terrestrial rocks typically contain $0.1-2$ ppm Br (e.g., Yoshida et al., 1971). What little data exist shows that Br is approximately three orders of magnitude lower than chlorine in Martian meteorites (Fig. 2.3; Dreibus and Wänke, 1985, 1986, 1987; Cartwright et al., 2013), which is consistent with a three order of magnitude difference in terrestrial rocks as well (e.g., Yoshida et al., 1971). Following the approach to constrain fluorine in the Martian interior, we use the Cl/Br ratio from Martian meteorites. Cl/Br in Martian meteorites averages 224 ± 140; however, this is based on less than 20 analyses of bromine in bulk Martian meteorites (Fig. 2.3). Calculating a mantle composition based on this ratio, and the chlorine content of the interior, gives 0.11 ± 0.24 ppm Br in the mantle and more accurately states that the Martian interior bromine content is <0.35 ppm Br and is approximately three orders of magnitude lower than chlorine or fluorine. Similarly, Taylor (2013) calculated 0.19 ppm Br in the Martian interior based on the correlation of Br with Cl and assuming Cl is 32 ppm in bulk Mars.

2.5 ABUNDANCES OF S, C, AND N IN BULK SILICATE MARS

We estimate the abundances of sulfur, carbon, and nitrogen in bulk silicate Mars. It is almost certain that their concentrations are high in the Martian core, as all have siderophile tendencies when equilibrated with metallic iron. Nevertheless, the mantle inventories of these elements are pertinent to understanding core formation and are tracers of magmatic activity, including the extent to which volatile species have been transferred from the mantle to the crust.

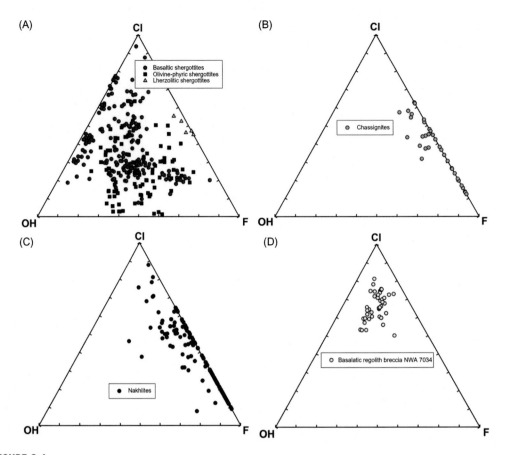

FIGURE 2.4

Ternary plots of apatite X-site occupancy (mol%) from many Martian meteorite types, including shergottites, nakhlites, chassignites, and regolith breccia NWA 7034. The data from the ternary plots were compiled from a number of literature sources (Gross et al., 2013; McCubbin and Nekvasil 2008; McCubbin et al., 2012, 2013, 2016a). Most of the apatites in the ternary plots were determined by electron probe microanalysis, so OH was not directly measured; however, OH was calculated assuming $1 - F - Cl = OH$. Electron microprobe data yielding $(F + Cl) > 1$ atom were plotted along the OH-free join assuming $1 - Cl = F$. (A) Apatites from shergottites, with distinction made between basaltic, olivine-phyric, and lherzolitic petrologic types. (B) Apatites from chassignites. (C) Apatites from nakhlites. (D) Apatites from basaltic regolith breccia NWA 7034. Apatites plotted by their respective geochemical groupings are available in McCubbin et al. (2016a).

The sulfur concentrations in Martian magmas or in the mantle are not well established. This is partly due to the wide range in S contents in Martian meteorites, which range from 34 to 2865 ppm in extrusive rocks (basaltic and olivine-phyric shergottites and nakhlites, Ding et al., 2015). Excluding the nakhlites, which have experienced significant mineral accumulation, the range is 408−2865 ppm. The large range probably reflects variations in the amount of S lost from magmas before or during eruption, variations in the amount of S in mantle source regions, and differences in the amounts of partial melting and fractional crystallization each magma experienced. In addition, in a given meteorite most of the S is present in small sulfide grains, so sampling problems can lead to scatter in the data.

Ding et al. (2015) used all available data on S in shergottites to model the S concentration in the mantle. They took an elaborate approach to calculate permissible ranges in S, taking into account crystallization of magma in the crust and the amount of isentropic partial melting of the mantle source. The results suggest that the melts produced are somewhat undersaturated in S and were produced in mantle source regions containing $<700-1000$ ppm S, assuming the amount of melting varied from 2% to 17%. This assumes, of course, that the shergottites are representative probes of the mantle. This estimate can be checked roughly by using the mean S content of shergottites (olivine-phyric and basaltic) and dividing by 4, the factor S is enriched in mafic magma compared to their mantle source regions on Earth (McDonough and Sun, 1995). Shergottites have an average S content of 1380 ± 680 ppm (using the data compilation reported in Ding et al., 2015), implying a mantle S content of $\sim 350 \pm 170$ ppm. Considering loss during eruption, this might be taken as a lower limit for the concentration of S in the mantle. The Ding et al. (2015) results are a much more reliable estimate for the S content of the Martian mantle.

Sulfur, carbon, and nitrogen are moderately to highly volatile and tend to be lost shortly before or during eruptions. One way of determining their abundances is to tie their concentrations to another element whose abundance we know (or think we know) and to use samples that have not outgassed significantly. The best candidates for this analysis are melt inclusions in the olivine phyric shergottite Yamato 980459. This rock represents a primitive magma (Usui et al., 2008) and H, S, and C have been measured in its melt inclusions (Usui et al., 2012). The concentrations in the three inclusions analyzed by Usui et al. (2012) are given in Table 2.3. Assuming that the S/H_2O ratio in the inclusion represents the S/H_2O ratio in bulk Mars and that H_2O in Mars ranges from 300 to 700 ppm, we estimate that S in bulk silicate Mars is in the range ~ 1100 to ~ 3500 ppm. These values are substantially greater than that Ding et al. (2015) estimated from careful modeling and the concentration of S in shergottites. However, if we assume that the H_2O in Mars is $100-300$ ppm (consistent with estimates from the shergottites reported earlier in Tables 2.1 and 2.2) we calculate S in bulk silicate Mars to be ~ 370 to ~ 1100 ppm, which is consistent with the estimates from Ding et al. (2015).

Table 2.3 Concentrations in Bulk Silicate Mars of S and C Estimated from H Concentration in Melt Inclusions in Y 980458; all Concentrations in ppm

Y 980459		$H_2O = 300$[a]	$H_2O = 700$[a]	Y 980459	$H_2O = 300$[a]	$H_2O = 700$[a]
H_2O	S	S in BSM[b]	S in BSM[b]	CO_2	CO_2 in BSM[b]	CO_2 in BSM[b]
146	634	1303	3040	265	545	1271
251	930	1112	2594	1601	1914	4465
183	902	1479	3450	582	954	2226

All concentration in ppm.
[a]Estimated range for bulk silicate Mars.
[b]Bulk silicate Mars.

Table 2.4 Concentrations in Bulk Silicate Mars of C and N, Assuming Chondritic Proportions[a] of H, C, and N

	C/H	C (BSM)	CO2 (BSM)[b]	N/H	N (BSM)
CI and CM[a]	2.07			0.106	
BSM (300)[c]		69.1	253		3.5
BSM (700)[d]		161	591		8.3

All concentration in ppm; ratios by weight.
[a]*Average of CI and CM chondrites (Alexander et al., 2012), weighted equally.*
[b]*CO_2 in bulk Mars calculated from C concentration assuming all C is in CO_2.*
[c]*Assumes bulk silicate Mars (BSM) contains 300 ppm H_2O.*
[d]*Assumes BSM contains 700 ppm H_2O.*

Using the same approach for carbon (expressed as CO_2) and assuming the same lower and upper limits to H_2O concentration in bulk silicate Mars, we estimate that CO_2 content in Mars is in the range $\sim$550 to $\sim$4500 ppm. This might overestimate CO_2 as the C/H in the inclusions ranges from 4 to 16, compared to 2.1 in the average of CI and CM chondrites (Alexander et al., 2012). This suggests that H has been lost preferentially from the inclusions (Gaetani et al., 2012), leaving behind a composition with a higher C/H than was originally present.

Another estimate for C and N can be done by using our estimates of H_2O in bulk silicate Mars and assuming that C/H and N/H are in chondritic proportions (Table 2.4). This approach assumes that C and H were scavenged quantitatively from the mantle during core formation and then replenished in the mantle by late addition of volatile-bearing planetesimals with CI−CM ratios of H/C and H/N. The calculations indicate that CO_2 in bulk silicate Mars is in the range $\sim$250 to $\sim$590 ppm (Table 2.4). This estimated range is lower than we find using the observed values in Yamato 980459 (Table 2.3).

The same approach suggests an N concentration between 3.5 and 8.3 ppm in bulk silicate Mars. Insufficient data are available to assess the veracity of this range, although for the purposes of this chapter, the precise N abundance is unlikely to have a significant effect on the phase equilibria or physical properties of magmas.

2.6 SUMMARY AND CONCLUSIONS

Volatile elements are distributed heterogeneously in the crust and mantle of Mars. This may be due to the combination of their distribution resulting from primary differentiation, presumably in a magma ocean, fractionation during magma genesis and evolution, loss from magmas at low pressure, and redistribution in the crust by aqueous processes.

Meteorite analyses indicate that water is at least a factor of four lower in abundance than in the bulk silicate Earth, about 140 ppm (Table 2.2 vs >500 ppm in bulk Earth; Mottl et al., 2007). In contrast, chlorine is 2.5 ±1 times more abundant in estimated bulk Mars compared to Earth. It is similar in abundance to enriched MORB mantle sources, but over an order of magnitude greater than in depleted MORB mantle sources. Average H_2O/Cl at the Curiosity site and in magmas parental to Chassignites is about 2. This ratio is somewhat greater in ancient crustal meteorites NWA 7034 and 7533, about 4. Average H_2O/Cl for the bulk crust (from the mid-equitorial region) as determined by GRS is much greater, about 8. This variation in the relative amounts of H_2O and Cl may indicate variations in volcanic degassing, but aqueous redistribution may have produced variations with depth in the crust. More analyses (and samples) are needed to understand the variation in H_2O/Cl. Fluorine data are scarce, though intriguing. Cl/F in parental magmas calculated from apatite compositions indicate that Cl/F ranges from 2.1 to 40. This range might indicate a range in Cl/F in mantle source regions or formation of apatite by fractional crystallization.

Sulfur concentrations vary on Mars, again indicating some combination of mantle heterogeneity and crustal processes. Sulfur concentrations in basaltic Martian meteorites range from ~ 400 to ~ 2900 ppm. These values are probably minima because sulfur gases are lost during eruptions. Taking this into account, the most reasonable estimates of S in bulk silicate Mars range from 350 to 1000 ppm.

REFERENCES

Agee, C.B., Wilson, N.V., McCubbin, F.M., Ziegler, K., Polyak, V.J., Sharp, Z.D., et al., 2013. Unique meteorite from early Amazonian Mars: water-rich basaltic breccia Northwest Africa 7034. Science 339 (6121), 780−785. Available from: http://dx.doi.org/10.1126/science.1228858.

Aiuppa, A., 2009. Degassing of halogens from basaltic volcanism: insights from volcanic gas observations. Chem. Geol. 263, 99−109.

Aiuppa, A., Baker, D.R., Webster, J.D., 2009. Halogens in volcanic systems. Chem. Geol. 263, 1−18.

Alexander, C.M.O.'D., Bowden, R., Fogel, M.L., Howard, K.T., Herd, C.D.K., Nittler, L.R., 2012. The provenances of asteroids, and their contributions to the volatile inventories of the terrestrial planets. Science 337 (6095), 721−723. Available from: http://dx.doi.org/10.1126/science.1223474.

Anazawa, K., Peter Wood, C., Browne, P.R.L., 2011. Fluorine and chlorine behavior in chlorine-rich volcanic rocks from White Island, New Zealand. J. Fluor. Chem. 132, 1182−1187.

Aubaud, C., Hauri, E.H., Hirschmann, M.M., 2004. Water partitioning coefficients between nominally anhydrous minerals and basaltic melts. Geophys. Res. Lett. 31, L20611. doi 10.1029/2004GRL021341.

Balta, J.B., McSween, H.Y., 2013. Water and the composition of Martian magmas. Geology 41, 1115−1118.

Beck, P., Pommerol, A., Zanda, B., Remusat, L., Lorand, J.P., Göpel, C., et al., 2015. A Noachian source region for the "Black Beauty" meteorite, and a source lithology for Mars surface hydrated dust? Earth Planet. Sci. Lett. 427, 104−111.

Bibring, J.-P., Langevin, Y., Mustard, J.F., Poulet, F., Arvidson, R., Gendrin, A., et al., 2006. Global mineralogical and aqueous Mars history derived from OMEGA/Mars express data. Science 312, 400−404.

Borg, L.E., Draper, D.S., 2003. A petrogenetic model for the origin and compositional variation of the Martian basaltic meteorites. Meteorit. Planet. Sci. 38, 1713−1731.

Boyce, J.W., Tomlinson, S.M., McCubbin, F.M., Greenwood, J.P., Treiman, A.H., 2014. The Lunar Apatite Paradox. Science 344, 400−402.

Boynton, W.V., Taylor, G.J., Evans, L.G., Reedy, R.C., Starr, R., Janes, D.M., et al., 2007. Concentration of H, Si, Cl, K, Fe, and Th in the low-and mid-latitude regions of Mars. J. Geophys. Res. 112. Available from: http://dx.doi.org/10.1029/2007JE002887.

Boynton, W.V., Taylor, G.J., Karunatillake, S., Reedy, R.C., Keller, J.M., 2008. Elemental abundances determined by Mars Odyssey GRS. In: Bell, J.F. (Ed.), The Martian Surface: Composition, Mineralogy, and Physical Properties. Cambridge University Press, Cambridge, pp. 105−124.

Brass, G.W., 1980. Stability of brines on Mars. Icarus 42, 20−28.

Bridges, J.C., Grady, M.M., 1999. A halite-siderite-anhydrite-chlorapatite assemblage in Nakhla: mineralogical evidence for evaporites on Mars. Meteorit. Planet. Sci. 34, 407−415.

Bridges, J.C., Schwenzer, S.P., 2012. The nakhlite hydrothermal brine on Mars. Earth Planet. Sci. Lett. 359−360, 117−123.

Bridges, J.C., Warren, P.H., 2006. The SNC meteorites: basaltic igneous processes on Mars. J. Geol. Soc. 163, 229−251.

Bridges, J.C., Catling, D.C., Saxton, J.M., Swindle, T.D., Lyon, I.C., Grady, M.M., 2001. Alteration assemblages in Martian meteorites: implications for near-surface processes. Space Sci. Rev. pp. 365−392.

Carr, M.H., 1996. Accretion and Evolution of Water, Water on Mars. Oxford University Press, New York, pp. 146−169.

Carr, M.H., Head, J.W., 2010. Geologic history of Mars. Earth Planet. Sci. Lett. 294, 185−203.

Carr, M.H., Head, J.W., 2003. Oceans on Mars: an assessment of the observational evidence and possible fate. J. Geophys. Res. 108 (E5), 5042. Available from: https://doi.org/10.1029/2002JE001963.

Carroll, M.R., Webster, J.D., 1994. Solubilities of sulfur, noble gases, nitrogen, chlorine, and fluorine in magmas. In: Carroll, M.R., Webster, J.D. (Eds.), Volatiles in Magmas. Mineralogical Society of America, Washington, DC, pp. 231−279.

Carter, J., Poulet, F., 2012. Orbital identification of clays and carbonates in Gusev crater. Icarus 219, 250−253.

Cartwright, J.A., Gilmour, J.D., Burgess, R., 2013. Martian fluid and Martian weathering signatures identified in Nakhla, NWA 998 and MIL 03346 by halogen and noble gas analysis. Geochim. Cosmochim. Acta 105, 255−293.

Crozaz, G., Floss, C., Wadhwa, M., 2003. Chemical alteration and REE mobilization in meteorites from hot and cold deserts. Geochim. Cosmochim. Acta 67, 4727−4741.

Dalou, C., Koga, K., Shimizu, N., Boulon, J., Devidal, J.-L., 2012. Experimental determination of F and Cl partitioning between lherzolite and basaltic melt. Contrib. Mineral. Petrol. 163, 591−609.

Ding, S., Dasgupta, R., Lee, C.-T.A., Wadhwa, M., 2015. New bulk sulfur measurements of Martian meteorites and modeling the fate of sulfur during melting and crystallization − implications for sulfur transfer from Martian mantle to crust-atmosphere system. Earth Planet. Sci. Lett. 409, 157−167. Available from: https://doi.org/10.1016/j.epsl.2014.10.046.

Dreibus, G., Wänke, H., 1982. Parent body of the SNC-meteorites: chemistry, size and formation. Meteoritics 17, 207−208.

Dreibus, G., Wänke, H., 1985. Mars, a volatile-rich planet. Meteoritics 20, 367−381.

Dreibus, G., Wänke, H., 1986. Comparison of Cl/Br and Br/I ratios in terrestrial samples and SNC meteorites. In: Proceedings of the MECA Workshop on the Evolution of the Martian Atmosphere. Lunar and Planetary Institute, Houston, TX, pp. 13−14.

Dreibus, G., Wänke, H., 1987. Volatiles on Earth and Mars—a comparison. Icarus 71, 225−240.

Dreibus, G., Palme, H., Rammensee, W., Spettel, B., Weckwerth, G., Wänke, H., 1982. Composition of Shergotty parent body: further evidence for a two component model of planet formation. Lunar Planet. Sci. 13, 186−187.

Ehlmann, B.L., Mustard, J.F., 2012. An in-situ record of major environmental transitions on early Mars at Northeast Syrtis Major. Geophys. Res. Lett. 39, L11202.

Ehlmann, B.L., Mustard, J.F., Murchie, S.L., Poulet, F., Bishop, J.L., Brown, A.J., et al., 2008. Orbital identification of carbonate-bearing rocks on Mars. Science 322, 1828−1832.

Farley, K.A., Martin, P., Archer Jr, P.D., Atreya, S.K., Conrad, P.G., et al., 2016. Light and variable $^{37}Cl/^{35}Cl$ ratios in rocks from Gale Crater, Mars: possible signature of perchlorate. Earth Planet. Sci. Lett. 438, 14−24.

Filiberto, J., Schwenzer, S.P., 2013. Alteration mineralogy of Home Plate and Columbia Hills—formation conditions in context to impact, volcanism, and fluvial activity. Meteorit. Planet. Sci. 48, 1937−1957.

Filiberto, J., Treiman, A.H., 2009. Martian magmas contained abundant chlorine, but little water. Geology 37, 1087−1090.

Filiberto, J., Treiman, A.H., Giesting, P.A., Goodrich, C.A., Gross, J., 2014. High-temperature chlorine-rich fluid in the Martian crust: a precursor to habitability. Earth Planet. Sci. Lett. 401, 110−115.

Filiberto, J., Gross, J., McCubbin, F.M., 2016a. Constraints on the water, chlorine, and fluorine content of the Martian mantle. Meteorit. Planet. Sci. 51 (11), 2023−2035. Available from: https://doi.org/10.1111/maps.12624.

Filiberto, J., Baratoux, D., Beaty, D., Breuer, D., Farcy, B.J., Grott, M., et al., 2016b. A review of volatiles in the Martian interior. Meteorit. Planet. Sci. 51 (11), 1935−1958. Available from: https://doi.org/10.1111/maps.12680.

Forni, O., Gaft, M., Toplis, M.J., Clegg, S.M., Maurice, S., Wiens, R.C., et al., 2015. First detection of fluorine on Mars: implications for Gale Crater's geochemistry. Geophys. Res. Lett. 42, 1020−1028.

Fuge, R., 1977. On the behavoir of fluorine and chlorine during magmatic differentiation. Contrib. Mineral. Petrol. 61, 245−249.

Gaetani, G.A., O'Leary, J.A., Shimizu, N., Bucholz, C.E., Newville, M., 2012. Rapid reequilibration of H_2O and oxygen fugacity in olivine-hosted melt inclusions. Geology 40, 915−918. Available from: http://dx.doi.org/10.1130/G32992.1.

Gattacceca, J., Rochette, P., Scorzelli, R., Munayco, P., Agee, C., Quesnel, Y., et al., 2014. Martian meteorites and Martian magnetic anomalies: a new perspective from NWA 7034. Geophys. Res. Lett. 41, 4859−4864.

Gellert, R., Rieder, R., Anderson, R.C., Brückner, J., Clark, B.C., Dreibus, G., et al., 2004. Chemistry of rocks and soils in Gusev crater from the alpha particle X-ray spectrometer. Science 305, 829−832.

Gellert, R., Rieder, R., Brückner, J., Clark, B.C., Dreibus, G., Klingelhöfer, G., et al., 2006. Alpha particle X-ray spectrometer (APXS): results from Gusev crater and calibration report. J. Geophys. Res. 111. Available from: http://dx.doi.org/10.1029/2005JE002555.

Giesting, P.A., Filiberto, J., 2014. Quantitative models linking igneous amphibole composition with magma Cl and OH content. Am. Mineral. 99, 852−865.

Giesting, P.A., Filiberto, J., 2016. Amphibole chemistry and the formation environment of potassic-chloro-hastingsite in the Nakhlites MIL 03346 and pairs and NWA 5790. Meteorit. Planet. Sci. 51, 2127−2153. Available from: https://doi.org/10.1111/maps.12675.

Giesting, P.A., Schwenzer, S.P., Filiberto, J., Starkey, N.A., Franchi, I.A., Treiman, A.H., et al., 2015. Igneous and shock processes affecting chassignite amphibole evaluated using chlorine/water partitioning and hydrogen isotopes. Meteorit. Planet. Sci. 50, 433−460.

Gnos, E., Hofmann, B., Franchi, I.A., Al-Kathiri, A., Huser, M., Moser, L., 2002. Sayh al Uhaymir 094: a new Martian meteorite from the Oman desert. Meteorit. Planet. Sci. 37, 835−854.

Gooding, J.L., Wentworth, S.J., Zolensky, M.E., 1991. Aqueous alteration of the Nakhla meteorite. Meteoritics 26, 135−143.

Grant, K.J., Kohn, S.C., Brooker, R.A., 2007. The partitioning of water between olivine, orthopyroxene and melt synthesized in the system albite-forsterite-H_2O. Earth Planet. Sci. Lett. 3260, 227−241.

Greeley, R., 1987. Release of juvenile water on Mars—estimated amounts and timing associated with volcanism. Science 236, 1653−1654.

Gross, J., Filiberto, J., Bell, A.S., 2013. Water in the Martian interior: evidence for terrestrial MORB mantle-like volatile contents from hydroxyl-rich apatite in olivine−phyric shergottite NWA 6234. Earth Planet. Sci. Lett. 369−370, 120−128.

Grott, M., Morschhauser, A., Breuer, D., Hauber, E., 2011. Volcanic outgassing of CO_2 and H_2O on Mars. Earth Planet. Sci. Lett. 308, 391−400.

Haberle, R.M., McKay, C.P., Schaeffer, J., Cabrol, N.A., Grin, E.A., Zent, A.P., et al., 2001. On the possibility of liquid water on present-day Mars. J. Geophys. Rese. Planets 106, 23317−23326.

Hallis, L.J., Taylor, G.J., Nagashima, K., Huss, G.R., 2012a. Magmatic water in the Martian meteorite Nakhla. Earth Planet. Sci. Lett. 359−360, 84−92.

Hallis, L.J., Taylor, G.J., Nagashima, K., Huss, G.R., Needham, A.W., Grady, M.M., et al., 2012b. Hydrogen isotope analyses of alteration phases in the nakhlite Martian meteorites. Geochim. Cosmochim. Acta 97, 105−119.

Hicks, L.J., Bridges, J.C., Gurman, S.J., 2014. Ferric saponite and serpentine in the nakhlite Martian meteorites. Geochim. Cosmochim. Acta 136, 194−210.

Hirschmann, M.M., Withers, A.C., 2008. Ventilation of CO_2 from a reduced mantle and consequences for the early Martian greenhouse. Earth Planet. Sci. Lett. 270, 147−155.

Howarth, G., Pernet-Fisher, J., Bodnar, R., Taylor, L., 2015. Evidence for the exsolution of Cl-rich fluids in Martian magmas: apatite petrogenesis in the enriched lherzolitic shergottite Northwest Africa 7755. Geochim. Cosmochim. Acta 166, 234−248.

Howarth, G.H., Liu, Y., Chen, Y., Pernet-Fisher, J.F., Taylor, L.A., 2016. Postcrystallization metasomatism in shergottites: evidence from the paired meteorites LAR 06319 and LAR 12011. Meteorit. Planet. Sci. 51, 2061−2072. Available from: http://dx.doi.org/10.1111/maps.12576.

Humayun, M., Nemchin, A., Zanda, B., Hewins, R.H., Grange, M., Kennedy, A., et al., 2013. Origin and age of the earliest Martian crust from meteorite NWA 7533. Nature 503, 513−516. Available from: http://dx.doi.org/10.1038/nature12764.

Johnson, S.S., Mischna, M.A., Grove, T.L., Zuber, M.T., 2008. Sulfur-induced greenhouse warming on early Mars. J. Geophys. Res. 113, E08005.

Jones, J., 2003. Constraints on the structure of the Martian interior determined from the chemical and isotopic systematics of SNC meteorites. Meteorit. Planet. Sci. 38, 1807−1814.

Keller, J.M., Boynton, W.V., Karunatillake, S., Baker, V.R., Dohm, J.M., Evans, L.G., et al., 2006. Equatorial and midlatitude distribution of chlorine measured by Mars Odyssey GRS. J. Geophys. Res. 111. Available from: http://dx.doi.org/10.1029/2006JE002679.

Kendrick, M.A., Kamenetsky, V.S., Phillips, D., Honda, M., 2012. Halogen systematics (Cl, Br, I) in Mid-Ocean Ridge Basalts: a Macquarie Island case study. Geochim. Cosmochim. Acta 81, 82−93.

King, P.L., McLennan, S.M., 2010. Sulfur on Mars. Elements 6, 107−112.

Koga, K., Hauri, E., Hirschmann, M., Bell, D., 2003. Hydrogen concentration analyses using SIMS and FTIR: comparison and calibration for nominally anhydrous minerals. Geochem. Geophys. Geosys. 4, 1019. Available from: http://dx.doi.org/10.1029/2002GC000378.

Laul, J.C., 1987. Rare earth patterns in shergottite phosphates and residues. J. Geophys. Res. 92, E633−E640.

Litvak, M.L., Mitrofanov, I.G., Sanin, A.B., Lisov, D., Behar, A., Boynton, W.V., et al., 2014. Local variations of bulk hydrogen and chlorine-equivalent neutron absorption content measured at the contact between the Sheepbed and Gillespie Lake units in Yellowknife Bay, Gale Crater, using the DAN instrument onboard Curiosity. J. Geophys. Res. Planets 119, 1259−1275.

Longhi, J., 1991. Complex magmatic processes on Mars-inferences from the SNC meteorites. Proceedings of the Lunar and Planetary Science Conference 21. Lunar and Planetary Institute, Houston, TX, pp. 695−709.

McCubbin, F.M., Jones, R.H., 2015. Extraterrestrial apatite: planetary geochemistry to astrobiology. Elements 11, 183−188.

McCubbin, F.M., Nekvasil, H., 2008. Maskelynite-hosted apatite in the Chassigny meteorite: insights into late-stage magmatic volatile evolution in Martian magmas. Am. Mineral. 93, 676–684.

McCubbin, F.M., Hauri, E.H., Elardo, S.M., Vander Kaaden, K.E., Wang, J., Shearer, C.K., 2012. Hydrous melting of the Martian mantle produced both depleted and enriched shergottites. Geology 40, 683–686.

McCubbin, F.M., Elardo, S.M., Shearer, C.K., Smirnov, A., Hauri, E.H., Draper, D.S., 2013. A petrogenetic model for the comagmatic origin of chassignites and nakhlites: inferences from chlorine-rich minerals, petrology, and geochemistry. Meteorit. Planet. Sci. 48 (5), 819–853.

McCubbin, F.M., Vander Kaaden, K.E., Tartese, R., Boyce, J.W., Mikhail, S., Whitson, E.S., et al., 2015. Experimental investigation of F, Cl, and OH partitioning between apatite and Fe-rich basaltic melt at 1.0–1.2 GPa and 950–1000 °C. Am. Mineral. 100, 1790–1802.

McCubbin, F.M., Boyce, J.W., Srinivasan, P., Santos, A.R., Elardo, S.M., Filiberto, J., et al., 2016a. Heterogeneous distribution of H_2O in the Martian interior: implications for the abundance of H_2O in the depleted and enriched mantle sources. Meteorit. Planet. Sci. 51, 2036–2060. Available from: https://doi.org/10.1111/maps.12639.

McCubbin, F.M., Boyce, J.W., Szabó, T., Santos, A.R., Tartèse, R., Domokos, G., et al., 2016b. Geologic history of Martian regolith breccia Northwest Africa 7034: evidence for hydrothermal activity and lithologic diversity in the Martian crust. J. Geophys. Res. Planets 121, 2120–2149.

McCubbin, F.M., Smirnov, A., Nekvasil, H., Wang, J., Hauri, E., Lindsley, D.H., 2010. Hydrous magmatism on Mars: a source of water for the surface and subsurface during the Amazonian. Earth Planet. Sci. Lett. 292, 132–138.

McDonough, W.F., Sun, S., 1995. The composition of the Earth. Chem. Geol. 120, 223–253.

McEwen, A.S., Ojha, L., Dundas, C.M., Mattson, S.S., Byrne, S., Wray, J.J., et al., 2011. Seasonal flows on warm Martian slopes. Science 333, 740–743.

McEwen, A.S., Dundas, C.M., Mattson, S.S., Toigo, A.D., Ojha, L., Wray, J.J., et al., 2014. Recurring slope lineae in equatorial regions of Mars. Nat. Geosci. 7, 53–58.

McLennan, S.M., Boynton, W.V., Karunatillake, S., Hahn, B.C., and Taylor, G.J., 2010. Distribution of sulfur on the surface of Mars determined by the 2001 Mars Odyssey Gamma Ray Spectrometer. In: 41st Lunar and Planetary Science Conference, abstract #2174.

McLennan, S.M., Anderson, R.B., Bell, J.F., Bridges, J.C., Calef, F., Campbell, J.L., et al., 2014. Elemental geochemistry of sedimentary rocks at Yellowknife Bay, Gale Crater, Mars. Science 343. Available from: http://dx.doi.org/10.1126/science.1244734.

McSween, H.Y., 2006. Water on Mars. Elements 2, 135–137.

Médard, E., Grove, T.L., 2006. Early hydrous melting and degassing of the Martian interior. J. Geophys. Res. 111, E11003. Available from: https://doi.org/10.1029/2006JE002742.

Michael, P.J., Cornell, W.C., 1998. Influence of spreading rate and magma supply on crystallization and assimilation beneath mid-ocean ridges: evidence from chlorine and major element chemistry of mid-ocean ridge basalts. J. Geophys. Res. 103, 18325–18356.

Minitti, M.E., Leshin, L.A., Dyar, M.D., Ahrens, T.J., Guan, Y., Luo, S.-N., 2008a. Assessment of shock effects on amphibole water contents and hydrogen isotope compositions: 2. Kaersutitic amphibole experiments. Earth Planet. Sci. Lett. 266, 288–302.

Minitti, M.E., Rutherford, M.J., Taylor, B.E., Dyar, M.D., Schultz, P.H., 2008b. Assessment of shock effects on amphibole water contents and hydrogen isotope compositions: 1. Amphibolite experiments. Earth Planet. Sci. Lett. 266, 46–60.

Mitrofanov, I.G., Litvak, M.L., Sanin, A.B., Starr, R.D., Lisov, D.I., Kuzmin, R.O., et al., 2014. Water and chlorine content in the Martian soil along the first 1900 m of the Curiosity rover traverse as estimated by the DAN instrument. J. Geophys. Res. Planets 119, 1579–1596.

Mohapatra, R.K., Schwenzer, S.P., Herrmann, S., Murty, S.V.S., Ott, U., Gilmour, J.D., 2009. Noble gases and nitrogen in Martian meteorites Dar al Gani 476, Sayh al Uhaymir 005 and Lewis Cliff 88516: EFA and extra neon. Geochim. Cosmochim. Acta 73, 1505−1522.

Mottl, M.J., Glazer, B.T., Kaiser, R.I., Meech, K.J., 2007. Water and astrobiology. Chem. Erde 67, 253−282.

Mustard, J.F., 2018. Sequestration of volatiles in the Martian crust through hydrated minerals: a significant planetary reservoir of water. This volume.

Muttik, N., McCubbin, F.M., Keller, L.P., Santos, A.R., McCutcheon, W.A., Provencio, P.P., et al., 2014. Inventory of H_2O in the ancient Martian regolith from Northwest Africa 7034: the important role of Fe oxides. Geophys. Res. Lett. 41, 8235−8244.

Norman, M., 1999. The composition and thickness of the crust of Mars estimated from REE and Nd isotopic compositions of Martian meteorites. Meteorit. Planet. Sci. 34, 439−449.

Ojha, L., Wilhelm, M.B., Murchie, S.L., McEwen, A.S., Wray, J.J., Hanley, J., et al., 2015. Spectral evidence for hydrated salts in recurring slope lineae on Mars. Nat. Geosci. 8, 829−832.

Ott, U., 1988. Noble gases in SNC meteorites: Shergotty, Nakhla, Chassigny. Geochim. Cosmochim. Acta 52, 1937−1948.

Owen, T., Maillard, J.P., De Bergh, C., Lutz, B.L., 1988. Deuterium on Mars: the abundance of HDO and the value of D/H. Science 240, 1767.

Patiño Douce, A.E., Roden, M., 2006. Apatite as a probe of halogen and water fugacities in the terrestrial planets. Geochim. Cosmochim. Acta 70, 3173−3196.

Patiño Douce, A.E., Roden, M.F., Chaumba, J., Fleisher, C., Yogodzinski, G., 2011. Compositional variability of terrestrial mantle apatites, thermodynamic modeling of apatite volatile contents, and the halogen and water budgets of planetary mantles. Chem. Geol. 288, 14−31.

Poulet, F., Bibring, J.P., Mustard, J.F., Gendrin, A., Mangold, N., Langevin, Y., et al., 2005. Phyllosilicates on Mars and implications for early Martian climate. Nature 438, 623−627.

Pyle, D.M., Mather, T.A., 2009. Halogens in igneous processes and their fluxes to the atmosphere and oceans from volcanic activity: a review. Chem. Geol. 263, 110−121.

Rieder, R., Gellert, R., Anderson, R.C., Brückner, J., Clark, B.C., Dreibus, G., et al., 2004. Chemistry of rocks and soils at Meridiani Planum from the alpha particle X-ray spectrometer. Science 306, 1746−1749.

Righter, K., Pando, K., Danielson, L.R., 2009. Experimental evidence for sulfur-rich Martian magmas: implications for volcanism and surficial sulfur sources. Earth Planet. Sci. Lett. 288, 235−243.

Saal, A.E., Hauri, E.H., Langmuir, C.H., Perfit, M.R., 2002. Vapour undersaturation in primitive mid-ocean-ridge basalt and the volatile content of Earth's upper mantle. Nature 419, 451−455.

Santos, A.R., Agee, C.B., McCubbin, F.M., Shearer, C.K., Burger, P.V., Tartèse, R., Anand, M., 2015. Petrology of igneous clasts in Northwest Africa 7034: Implications for the petrologic diversity of the martian crust. Geochimica et Cosmochimica Acta 157, 56−85.

Schmidt, M.E., Ruff, S.W., McCoy, T.J., Farrand, W.H., Johnson, J.R., Gellert, R., et al., 2008. Hydrothermal origin of halogens at Home Plate, Gusev Crater. J. Geophys. Res. 113. Available from: http://dx.doi.org/10.1029/2007JE003027.

Shearer, C.K., Burger, P.V., Papike, J.J., Borg, L.E., Irving, A.J., Herd, C.D.K., 2008. Petrogenetic linkages among Martian basalts: implications based on trace element chemistry of olivine. Meteorit. Planet. Sci. 43, 1241−1258.

Stanley, B.D., Hirschmann, M.M., Withers, A.C., 2011. CO_2 solubility in Martian basalts and Martian atmospheric evolution. Geochim. Cosmochim. Acta 75, 5987−6003.

Swindle, T.D., Olson, E.K., 2004. 40Ar-39Ar studies of whole rock nakhlites: evidence for the timing of formation and aqueous alteration on Mars. Meteorit. Planet. Sci. 39, 755−766.

Swindle, T.D., Treiman, A.H., Lindstrom, D.J., Burkland, M.K., Cohen, B.A., Grier, J.A., et al., 2000. Noble gases in iddingsite from the Lafayette meteorite: evidence for liquid water on Mars in the last few hundred million years. Meteorit. Planet. Sci. 35, 107−115.

Taylor, G.J., 2013. The bulk composition of Mars. Chem. Erde 73, 401−420. Available from: http://dx.doi.org/10.1016/j.chemer.2013.09.006.

Taylor, G.J., Boynton, W.V., Brückner, J., Wänke, H., Dreibus, G., Kerry, K., et al., 2006. Bulk composition and early differentiation of Mars. J. Geophys. Res. 111. Available from: http://dx.doi.org/10.1029/2005JE002645.

Taylor, G.J., Boynton, W.V., McLennan, S.M., Martel, L.M.V., 2010a. K and Cl concentrations on the Martian surface determined by the Mars Odyssey Gamma Ray Spectrometer: implications for bulk halogen abundances in Mars. Geophys. Res. Lett. 37, L12204. Available from: http://dx.doi.org/10.1029/2010gl043528.

Taylor, G.J., Martel, L.M.V., Karunatillake, S., Gasnault, O., Boynton, W.V., 2010b. Mapping Mars geochemically. Geology 38, 183−186.

Treiman, A.H., 1985. Amphibole and hercynite spinel in Shergotty and Zagami Magmatic water, depth of crystallization, and metasomatism. Meteoritics 20, 229−243.

Treiman, A.H., 2003. Chemical compositions of Martian basalts (shergottites): some inferences on basalt formation, mantle metasomatism, and differentiation in Mars. Meteorit. Planet. Sci. 38, 1849−1864.

Treiman, A.H., 2005. The nakhlite meteorites: augite-rich igneous rocks from Mars. Chem. Erde Geochem. 65, 203−270.

Treiman, A.H., Barrett, R., Gooding, J., 1993. Preterrestrial aqueous alteration of the Lafayette (SNC) meteorite. Meteoritics 28, 86−97.

Usui, T., McSween Jr, H.Y., Floss, C., 2008. Petrogenesis of olivine-phyric shergottite Yamato 980459, revisited. Geochim. Cosmochim. Acta 72, 1711−1730.

Usui, T., Alexander, C.M.O.D., Wang, J., Simon, J.I., Jones, J.H., 2012. Origin of water and mantle−crust interactions on Mars inferred from hydrogen isotopes and volatile element abundances of olivine-hosted melt inclusions of primitive shergottites. Earth Planet. Sci. Lett. 357−358, 119−129.

Wadhwa, M., Lentz, R.C.F., McSween, H.Y., Crozaz, G., 2001. A petrologic and trace element study of Dar al Gani 476 and Dar al Gani 489: twin meteorites with affinities to basaltic and Iherzolitic shergottites. Meteorit. Planet. Sci. 36, 195−208.

Webster, J.D., 2004. The exsolution of magmatic hydrosaline chloride liquids. Chem. Geol. 210, 33−48.

Webster, J.D., Rebbert, C.R., 1998. Experimental investigation of H_2O and Cl-solubilities in F-enriched silicate liquids; implications for volatile saturation of topaz rhyolite magmas. Contrib. Mineral. Petrol. 132, 198−207.

Williams, R.M.E., Grotzinger, J.P., Dietrich, W.E., Gupta, S., Sumner, D.Y., Wiens, R.C., et al., 2013. Martian fluvial conglomerates at Gale Crater. Science 340, 1068−1072.

Yoshida, M., Takahashi, K., Yonehara, N., Ozawa, T., Iwasaki, I., 1971. Fluorine, chlorine, bromine, and iodine contents of volcanic rocks in Japan. Bull. Chem. Soc. Jpn 44, 1844−1850.

NOBLE GASES IN MARTIAN METEORITES: BUDGET, SOURCES, SINKS, AND PROCESSES

3

Ulrich Ott[1,2], Timothy D. Swindle[3], and Susanne P. Schwenzer[4,5]

[1]MTA Atomki, Debrecen, Hungary [2]Max-Planck-Institut für Chemie, Mainz, Germany
[3]University of Arizona, Tucson, AZ, United States [4]The Open University, Milton Keynes, United Kingdom
[5]Lunar and Planetary Institute, Houston, TX, United States

3.1 INTRODUCTION: THE MARS—MARTIAN METEORITE CONNECTION

Noble gases have been instrumental in establishing a link between planet Mars and the group of meteorites otherwise referred to as SNC (shergottite, nakhlite, chassignite) meteorites, a group later complemented by the orthopyroxenite (O), ALH 84001. While an origin from a large body had been suspected before because of the relatively young formation ages of the SNC meteorites as igneous rocks (Wasson and Wetherill, 1979; McSween and Stolper, 1980; Wood and Ashwal, 1981), it was the detection of trapped noble gases in shock melt glass of the Antarctic shergottite EETA 79001 by Bogard and Johnson (1983) and their similarity to the abundance pattern measured by the Viking mission for the Martian atmosphere (MA) (Owen et al., 1977) that provided the major impetus for accepting Mars as the most likely shergottite parent body (SPB). This was further enforced by similar results obtained by Becker and Pepin (1984), which included nitrogen besides the noble gases, as well as by the results of Marti et al. (1995) on shock glass of the shergottite Zagami. Oxygen isotopes, characteristic for a given planetary body, then leave little doubt that, if the shergottites come from Mars, the same is true for the nakhlites and Chassigny as well (Clayton and Mayeda, 1983).

If the results obtained by Bogard and Johnson (1983) and Becker and Pepin (1984) are compared with the nominal Viking data, a remarkable fact appears: the volume concentrations (cm^3 STP of gas per cm^3 of impact glass or cm^3 of MA) are identical (Ott and Begemann, 1985; Pepin, 1985), within the admittedly large uncertainties of the Viking measurements (Fig. 3.1). Some skepticism about a Martian origin of (all) the SNC meteorites has remained nevertheless (e.g., Ott and Begemann, 1985; Pellas, 1987), partly due to the limited precision of the reference composition measured by Viking. Nevertheless the agreement, if not the exact 1:1 relationship between impact glass and MA has been beautifully confirmed by the new, high-precision measurements of MA by the SAM analyzer aboard the Mars Science Laboratory (MSL) (Fig. 3.1). This is even more convincingly demonstrated in Fig. 3.2, which compares several diagnostic element and isotope ratios, and where for the meteorites the most up-to-date refined results (mostly from Bogard et al., 2001) are used.

The number of recovered Martian meteorite specimens—mostly finds—has exploded since these early discoveries and, not taking pairing into account, has crossed the 100 mark (Llorca et al.,

Volatiles in the Martian Crust. DOI: http://dx.doi.org/10.1016/B978-0-12-804191-8.00003-9

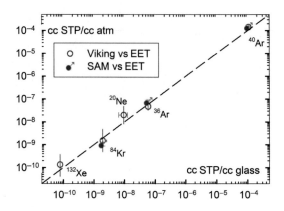

FIGURE 3.1

Volume concentrations (cm³ STP of gas per cm³ volume) of noble gases in EETA 79001 impact glass (horizontal axis) compared with those in the Martian atmosphere as measured by Viking and by SAM. See also similar figures in Pepin (1985) and Wiens and Pepin (1988), which compare the Viking and meteorite measurements only, but also include nitrogen and CO_2 besides the noble gases. *From Data sources for EETA 79001: Becker and Pepin (1984), combined with an $^{40}Ar/^{36}Ar$ ratio of 1800 (Bogard et al., 2001) with an assumed error of 200. Data source for Viking: Number densities from Hunten et al. (1987). Data sources for SAM: Volume fractions from Franz et al. (2015), with $^{40}Ar/^{36}Ar = (1.9 \pm 0.3) \times 10^3$ (Mahaffy et al., 2013), and from Conrad et al. (2014) were converted to ccSTP/cm³ for a total pressure of 7.2 mbar (Nier et al., 1976).*

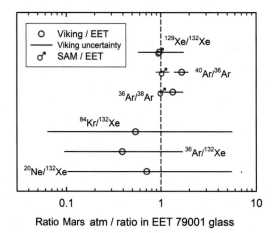

FIGURE 3.2

Comparison of diagnostic elemental and isotopic ratios measured for gas trapped in EETA 79001 impact glass and measurements of the Martian atmosphere (MA) proper by Viking and SAM. *From Data sources for EETA 79001: elemental ratios from Bogard et al. (2001) and, for $^{20}Ne/^{132}Xe$, Hunten et al. (1987). Isotopic ratios from Pepin (1991) for $^{36}Ar/^{38}Ar$, Bogard et al. (2001) for $^{40}Ar/^{36}Ar$ and Garrison and Bogard (1998) for $^{129}Xe/^{132}Xe$. Data sources for Viking: elemental ratios from Hunten et al. (1987), Ar and Xe isotopes from Owen et al. (1977). Data sources for SAM: Mahaffy et al. (2013), Atreya et al. (2013), error-weighted average of static measurements from Conrad et al. (2016); for the latter see also footnote to Table 3.4.*

2013; Meteoritical Bulletin, 2015). However, noble gas data obtained on these new specimens more or less conform to the basic observations and presence of noble gas components identified earlier (e.g., Busemann et al. 2015). Hence to a large extent the reviews by Bogard et al. (2001) and Swindle (2002) are still relevant, and thus they form the basis of the current review on noble gases in Martian meteorites. Only a few additional specimens like the unique possible soil sample NWA 7034 (Agee et al., 2013) are addressed here, while we also include and refer to some additional earlier observations not covered in the previous reviews.

In Section 3.2, which focuses on the budget, we will first simply (and at times critically) present and describe the various "trapped" noble gas components found in Martian meteorites. A discussion of ideas how these came about is the topic of Section 3.3.

3.2 COMPONENT OVERVIEW

As a rule, meteorites contain several noble gas components of different origin and different distribution among their mineral constituents, and Martian meteorites are no exception to this rule. Cosmogenic noble gases—produced in situ—allow the determination of the time of ejection from Mars and, by inference, the minimum number of ejection events/local Martian sources. Eugster (2003), based on 2 σ error limits for the individual exposure ages and assuming different ejection events for Chassigny and the nakhlites because of their different chemical makeup, concluded there were at least eight such events. To this we may have to add as another event, the one that put NWA 7034 and its siblings (paired specimens) on their path to Earth (Cartwright et al., 2014; Busemann et al., 2015). Also produced in situ are radiogenic noble gas nuclides such as ^{40}Ar from the decay of ^{40}K, which allow determination of formation ages and identification of metamorphic events. The "trapped" noble gases in the Martian meteorites, those that were acquired from gaseous reservoirs on Mars, are the topic of interest here. As discussed further, Martian meteorites contain a multitude of trapped noble gas components that were acquired from distinct reservoirs by a variety of trapping processes. During a typical analysis the various components are released in variable mixtures, and generally it is a nontrivial task to identify and characterize the individual components in such a mixture. A case in point is trapped Ne, which is mostly swamped by cosmogenic Ne. Similarly, in the case of the nakhlites, the cosmogenic gases dominate the ^{36}Ar and ^{38}Ar isotopes. Conversely, in case of the shergottites, significant ^{40}Ar trapped from the MA is a nuisance (to say the least) when the aim is to obtain an age based on the K-Ar system, even in the ^{40}Ar-^{39}Ar variant (e.g., Swindle, 2002; Walton et al., 2008). The task is additionally complicated by the ever-present contamination by terrestrial noble gases, in particular for the finds from hot and cold deserts (e.g., Mohapatra et al., 2009).

Among the major trapped components in the Martian meteorites, one is an atmospheric component that—as discussed in the Introduction—was instrumental in establishing the Martian origin of this group of meteorites and is clearly present in most shergottites. A useful representation for showing and identifying these and additional components is the three-nuclide plot ^{129}Xe/^{132}Xe versus ^{84}Kr/^{132}Xe, and altogether three major components have been identified in this way, first by the analysis of the type specimens Shergotty, Nakhla, and Chassigny (Fig. 3.3; Ott and Begemann, 1985; Ott, 1988). Apart from the atmospheric component that had

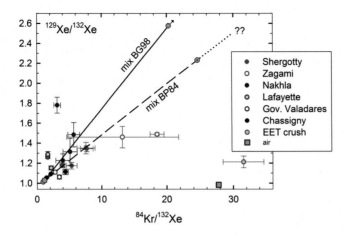

FIGURE 3.3

"Classic" plot for bulk Martian meteorites of $^{129}Xe/^{132}Xe$ versus $^{84}Kr/^{132}Xe$. To minimize the effect of terrestrial contamination, only data for falls—except for the nakhlites, where Lafayette and Governador Valadares are included—and for high-temperature releases (generally > 800 °C) are plotted. In addition, we show the composition of gas released by crushing EETA 79001 glass and, in order to visualize the effect of terrestrial contamination, the composition of (unfractionated) air. *From Data sources for bulk samples: Ott (1988), Ott et al. (1988), Schwenzer (2004), Schwenzer et al. (2007a), Mathew and Marti (2001). Besides bulk data also some mineral separates are included: Maskelynite from Shergotty and Zagami (Schwenzer et al., 2007a) and olivine from Chassigny (Mathew and Marti, 2001). Crush data from Wiens (1988), air from Ozima and Podosek (2002). Mixing lines are shown between endmembers (green dots) Chassigny PB4 (Ott 1988) and: (A) (BG98) the Martian atmosphere as inferred by Bogard and Garrison (1998) and listed in Table 3.1; (B) (BP84) the atmospheric composition calculated by Becker and Pepin (1984).*

been clearly seen in the impact glass of EETA 79001 (Bogard and Johnson, 1983; Becker and Pepin, 1984), these are (1) an "interior" component (often equated with mantle noble gas), which appears to be present in almost pure form in Chassigny, and in the shergottites appears to be mixed with the atmospheric gas; and (2) an apparently elementally fractionated atmospheric component, which is mixed with "interior" gas, in the case of nakhlites. However, as is obvious from Fig. 3.3 in particular from the Zagami data point of Ott et al. (1988), this may not be the whole story. A similar rightward shift as for Zagami in these element ratio plots is also seen even for several data points for impact glass (see corresponding figure in Bogard and Garrison, 1998; and Bogard et al., 2001). A preferred explanation for this shift is contamination by (terrestrial) air, the composition of which is also shown in Fig. 3.3. To minimize disturbances of this kind, we have plotted in Fig. 3.3 only data for falls (no data for desert finds) and only high-temperature (generally > 800 °C) extractions. It is also notable that the observed rightward shift may require contamination by elementally almost unfractionated air, a very unusual situation. Naturally, the interpretation of the Shergotty data points in Fig. 3.3 as a simple mixture will depend to some extent on knowledge of the true element ratio in the Martian atmospheric component—whether that obtained by Becker and Pepin early on in 1984 or the composition derived

Table 3.1 Relative Elemental Abundances of Ne, Ar, Kr, and Xe (Atom/Atom) in Mars Atmosphere as well as "Martian Interior" Derived from Analyses of Martian Meteorites. For Comparison the Corresponding Values for Earth Atmosphere, the Solar Wind, and the Sun Are Also Shown

	^{20}Ne/^{132}Xe	^{36}Ar/^{132}Xe	^{84}Kr/^{132}Xe
Mars atm. (met.)	*212 ± 32*[a]	900 ± 100[b]	20.5 ± 1.5[b]
Mars interior[b]	–	≤ 5	< 1.1
Earth atmosphere[c]	703	1343	27.8
Solar wind	$(9.97 \pm 0.59) \times 10^5$ [d]	$(2.37 \pm 0.14) \times 10^4$ [d]	9.9 ± 0.3[d]
			9.7 ± 1.6[e]
Solar[f]	2.13×10^6	5.45×10^4	22.1

Data sources:
[a]*Hunten et al. (1987).*
[b]*Bogard et al. (2001).*
[c]*Ozima and Podosek (2002).*
[d]*Vogel et al. (2011) for ^{36}Ar/^{132}Xe and ^{84}Kr/^{132}Xe (first entry), while ^{20}Ne/^{132}Xe has been calculated from ^{20}Ne/^{36}Ar in Heber et al. (2012) coupled with ^{36}Ar/^{84}Kr/^{132}Xe from Vogel et al. (2011).*
[e]*Meshik et al. (2014).*
[f]*Lodders et al. (2009).*

by Bogard and Garrison (1998) from a review of new and literature data at the time, which is listed in the reviews by Bogard et al. (2001) and Swindle (2002) as well as in Table 3.1. To show the impact, two mixing lines are included in Fig. 3.3, using these two endmembers for MA. Hopefully, results from the direct measurement of MA will at some time be able to settle the question. (Note that Conrad et al., 2014 and Conrad et al., 2016 report the isotopic compositions of Kr and Xe and the abundance in the MA of Kr, but not of Xe.) An even more extreme possibility exists, namely that the elemental composition of the MA could vary if clathrates form and dissociate on or near the surface of Mars (Musselwhite and Swindle, 2001; Swindle et al., 2009; Mousis et al., 2013), a possibility discussed in more detail later.

Currently suggested compositions for the atmospheric and the interior components based on the meteorite analyses are given in Table 3.1 for the element abundance ratios. The corresponding isotopic compositions of Ne and Ar are displayed in Table 3.2, while those for Kr and Xe are given in Tables 3.3 and 3.4, respectively. Results from the recent in situ measurements of MA are included for comparison, where available. In the following subsections (3.2.1–3.2.3) we will briefly comment on the compositions of the previously mentioned major components. This will be followed by a section on gases released by crushing the shergottites (Section 3.2.4) and a section about trapped noble gases in the "old" meteorites ALH 84001 and NWA 7034 (Section 3.2.5).

3.2.1 THE ATMOSPHERIC COMPONENT (MA)

Detailed information about the composition of the current MA has come recently also from in situ analyses by the SAM instrument aboard the MSL rover (Mahaffy et al., 2013; Atreya et al., 2013; Conrad et al., 2014; Franz et al., 2015). This includes isotopic analyses of Kr and Xe, which have

Table 3.2 Ne and Ar Isotopic Compositions Derived from Martian Meteorites for Mars Atmosphere as well as for Martian Interior Gas and Crush Component. Mars Atmosphere Measured In Situ, Earth Atmosphere, Solar Wind, and a Derived Solar Composition (Heber et al., 2012) are Listed for Comparison as well as the Composition of Q-Gases

	$^{20}Ne/^{22}Ne$	$^{36}Ar/^{38}Ar$	$^{40}Ar/^{36}Ar$
Mars atmosphere	$7-10$[a]	4.1 ± 0.2[b]	≈ 1900[a]
Mars atm. in situ		4.2 ± 0.1[c]	$(1.9 \pm 0.3) \times 10^3$ [d]
Mars interior[e]	–	≥ 5.26	≤ 212
Crush component[f]	–	4.9 ± 0.2	$430 - 680$
Earth atmosphere	9.80[g]	5.305[h]	298.56[h]
Solar wind	13.78 ± 0.03[i]		<1
	13.972 ± 0.025[j]	5.501 ± 0.005[j]	
		5.50 ± 0.01[k]	
	14.001 ± 0.042[l]	5.501 ± 0.014[l]	
		5.5005 ± 0.0040[m]	
Solar[n]	13.36	5.37	<1
Q[o]	$10.67 \pm 0.02; 10.11 \pm 0.02$	5.34 ± 0.02	<0.12

Data sources:
[a]*Swindle (2002); and Garrison and Bogard (1998).*
[b]*Wiens et al. (1986).*
[c]*Atreya et al. (2013).*
[d]*Mahaffy et al. (2013).*
[e]*Bogard et al. (2001); and Mathew and Marti (2001).*
[f]*Wiens (1988).*
[g]*Eberhardt et al. (1965).*
[h]*Lee et al. (2006).*
[i]*Heber et al. (2009).*
[j]*Meshik et al. (2007).*
[k]*Vogel et al. (2011).*
[l]*Pepin et al. (2012).*
[m]*Meshik et al. (2014).*
[n]*Heber et al. (2012).*
[o]*Ott (2014).*
As for the solar wind, note the differences for $^{20}Ne/^{22}Ne$ determined in different laboratories, which contrasts with the close agreement for $^{36}Ar/^{38}Ar$. Exact data for $^{40}Ar/^{36}Ar$ in the solar wind and the Sun are not given, because no blank-corrected final data from Genesis have been reported yet (cf. Meshik et al., 2014).

just become available at the time of writing (Conrad et al., 2016). The SAM results beautifully confirm the conclusion about the Martian origin of the SNC meteorites based on diagnostic isotope ratios such as $^{40}Ar/^{36}Ar$, $^{38}Ar/^{36}Ar$, and $^{129}Xe/^{132}Xe$. Differences occur, though, at ^{80}Kr and ^{82}Kr, and small, although apparently significant, differences are apparent also in the light Xe isotopes (in particular $^{124,126}Xe$), which are suggestive of the addition of neutron capture (Kr) and cosmogenic (Xe) components. Surprisingly, these contributions appear to be much higher in situ than in the meteoritic analysis; we will return to this topic at the end of Section 3.3. A consequence is that the Martian atmospheric composition determined from the Martian meteorites would be the more primitive one and we will focus later, as well as in the discussion of sources in Section 3.3.1, on this more primitive

Table 3.3 Krypton Isotopic Composition Derived from Martian Meteorites for Mars Atmosphere (Two Versions) as well as for Martian Interior Gas. Mars Atmosphere Measured In Situ, Earth Atmosphere, Solar Wind, and the Composition of Kr in Q-Gases are Included for Comparison. Errors in the Last Digits are Shown in Parentheses

	$^{78}Kr/^{84}Kr$	$^{80}Kr/^{84}Kr$	$^{82}Kr/^{84}Kr$	$^{83}Kr/^{84}Kr$	$^{86}Kr/^{84}Kr$
Mars atmosphere[a]	0.00662 (29)	0.0433 (6)	0.2056 (15)	0.2031 (14)	0.3005 (18)
Mars atmosphere[b]	*0.093 (2)*	0.0432 (2)	0.2099 (6)	0.2058 (8)	0.2975 (7)
Mars atm. in situ[c]	–	0.0723 (56;7)	0.2163 (47; 15)	0.2023 (36; 14)	0.3059 (70; 30)
Mars interior[d]	0.0071 (13)	0.0490 (28)	0.2076 (46)	$\equiv 0.2034$	0.3002 (52)
Earth atmosphere[e]	0.006087	0.03960	0.20217	0.20136	0.30524
Solar wind[f]	0.00642 (5)	0.0412 (2)	0.2054 (2)	0.2034 (2)	0.3012 (4)
Q (P1)[g]	0.00603 (3)	0.03937 (7)	0.2018 (2)	0.2018 (2)	0.3095 (5)

Data sources:
[a]*Weighted mean of Becker and Pepin (1984) and Swindle et al. (1986); similar to Pepin (1991) and Swindle (2002).*
[b]*Garrison and Bogard (1998), 1550 °C extraction of EET 79001, 8A; as for $^{78}Kr/^{84}Kr$, we list the given value but assume a typographical error in the report.*
[c]*Conrad et al. (2016), semistatic unweighted averages. The errors given in parentheses give separately first the statistical uncertainty, followed by the systematic uncertainty. Errors for mass discrimination had been mistakenly left out in Table 3.2 of Conrad et al. (2016), which shows only the statistical errors. The systematic errors shown here are based on a reassessment by R.O. Pepin. For $^{83}Kr/^{84}Kr$ the unrealistically small statistical error from averaging has been replaced by the average uncertainty of the individual measurements (Conrad et al., 2016).*
[d]*1800 °C extraction of Chassigny by Schwenzer (2004) after correction for spallation contributions assuming the solar wind ratio for trapped $^{83}Kr/^{84}Kr$ and the cosmogenic composition given by Bogard et al. (1971).*
[e]*Ozima and Podosek (2002).*
[f]*Meshik et al. (2014).*
[g]*Busemann et al. (2000).*

composition. Whenever in the following we will talk about Martian atmospheric composition without further qualification, it will be the one derived from the meteorite data.

Noble gases in the MA obtained from the analysis of Martian meteorites are centered on results from the analysis of shock glass in shergottites, in particular EETA 79001. They are remarkably similar in composition to those in the terrestrial atmosphere. This extends in particular to two properties in which both are distinctly different from other noble gas reservoirs like the Sun/giant planets or chondritic meteorites: (1) While—as found in the direct analysis of MA (Owen et al., 1977; Mahaffy et al., 2013; Conrad et al., 2014; Franz et al., 2015)—noble gases (not counting radiogenic ^{40}Ar) are lower in abundance by about two orders of magnitude compared to the terrestrial atmosphere, the elemental abundance *pattern* is quite similar (Table 3.2, Fig. 3.4). (2) The isotopic composition of xenon (excepting, again, radiogenic ^{129}Xe) in Earth's atmosphere, which again is quite distinct from solar or chondritic Xe, is remarkably similar to that in the MA as determined from the meteorite data. To first order, both are characterized by a strong mass-dependent fractionation favoring the heavy isotopes (Table 3.4, Fig. 3.5). There is a difference, however, between the meteorite data and the directly measured atmosphere in the lightest Xe isotopes (see discussion in Section 3.3.1.2).

Table 3.4 Xenon Isotopic Composition Derived from Martian Meteorites for Mars Atmosphere (Three Versions) as well as for Martian Interior Gas (Two Versions). Mars Atmosphere Measured In Situ, Earth Atmosphere, Solar Wind, and the Composition of Xe in Q-Gases Are Included for Comparison. Errors in the Last Digits Are Shown in Parentheses

	$^{124}Xe/^{132}Xe$	$^{126}Xe/^{132}Xe$	$^{128}Xe/^{132}Xe$	$^{129}Xe/^{132}Xe$	$^{130}Xe/^{132}Xe$	$^{131}Xe/^{132}Xe$	$^{134}Xe/^{132}Xe$	$^{136}Xe/^{132}Xe$
Mars atm.[a]	0.0038 (2)	0.0033 (2)	0.0735 (9)	2.40 (2)	0.1543 (9)	0.7929 (27)	0.4007 (15)	0.3514 (15)
Mars atm.[b]	0.00352 (17)	0.00322 (17)	0.0724 (11)	2.384 (10)	0.1532 (5)	0.7934 (36)	0.3959 (21)	0.3475 (27)
Mars atm.[c]	–	–	0.0751 (33)	2.590 (31)	0.1561 (22)	0.8100 (117)	0.4043 (59)	0.3533 (47)
Mars atm. in situ[d]	0.00468 (25; 4)	0.00403 (40; 3)	0.07466 (77; 47)	2.5221 (63; 149)	0.15374 (94; 91)	0.8123 (29; 44)	0.4021 (26; 25)	0.3452 (22; 25)
Mars int.[e]	0.00488 (18)	0.00422 (18)	0.0846 (9)	1.080 (7)	0.1656 (7)	0.8243 (61)	0.3651 (34)	0.2983 (20)
Mars int.[f]	0.00403 (95)	≡ 0.00420	0.0840 (37)	1.028 (19)	0.1656 (39)	0.8242 (144)	0.3697 (224)	0.3028 (54)
Earth atm.[g]	0.003537	0.003300	0.07136	0.9832	0.15136	0.7890	0.3879	0.3294
Solar wind[h]	0.00489 (6)	0.00420 (7)	0.0842 (2)	1.0405 (10)	0.1648 (3)	0.8256 (12)	0.3698 (6)	0.3003 (5)
Q-Xe[i]	0.00455 (2)	0.00406 (2)	0.0822 (2)	1.042 (2)	0.1619 (3)	0.8185 (9)	0.3780 (11)	0.3164 (8)

Data sources:
[a]Swindle et al. (1986).
[b]Marti and Mathew (2015).
[c]Garrison and Bogard (1998), 1550 °C extraction of EETA 79001.
[d]Conrad et al. (2016), static, weighted average. The errors given in parentheses give separately first the statistical uncertainty, followed by the systematic uncertainty. Errors for mass discrimination and hydride correction had been mistakenly left out in Table 3.2 of Conrad et al. (2016), which shows only the statistical errors. The systematic errors shown here are based on a reassessment by R.O. Pepin.
[e]Marti and Mathew (2015), Chass-S.
[f]Ott (1988), Chassigny PB4 after correction for spallation contributions, using the equations of Hohenberg et al. (1981), a (La + Ce + Nd)/Ba ratio of 0.30 from data in Lodders (1998) and a solar $^{126}Xe/^{130}Xe$ ratio (Meshik et al., 2015) for the trapped component.
[g]Ozima and Podosek (2002).
[h]Meshik et al. (2015).
[i]Busemann et al. (2000).

Compared to xenon, differences in krypton isotopic compositions between Solar System reservoirs are typically modest, and Kr on Mars follows this pattern (Table 3.3 and Fig. 3.6). Martian atmospheric Kr as seen in Martian meteorites to first order is either almost indistinguishable from solar wind Kr (Composition #1 in Table 3.3; Pepin, 1991), or slightly mass fractionated favoring the light isotopes (Composition #2 in Table 3.3; Garrison and Bogard, 1998). There appears to be an enhancement at ^{80}Kr. From the meteorite data, it is not clear whether this is a true atmospheric feature (see Section 3.3.1.2), but the conclusion is supported by the direct measurement of Martian atmospheric Kr by the SAM instrument (Conrad et al., 2016), where an even larger enhancement is seen, accompanied by somewhat enhanced ^{82}Kr. Unfortunately, no useful data for ^{78}Kr were obtained by SAM due to a dominating interference at mass 78 (Conrad et al., 2016).

Argon is completely different from the solar or terrestrial composition and characterized by a uniquely low ^{36}Ar/^{38}Ar ratio of ~ 4 suggestive of a fractionating loss process, and by highly radiogenic ^{40}Ar/^{36}Ar (Table 3.2).

Neon, finally, is hard to characterize, since in most specimens it is dominated by cosmogenic Ne, making it difficult to deduce a reliable composition for the trapped component. Values between ~ 7 and ~ 10 have been obtained from various analyses (see Fig. 3.1 in Garrison and Bogard, 1998; and more recently Park and Nagao, 2006). Note that these values have all been obtained by linear fits to temperature release data—assuming a two-component mixture of trapped and cosmogenic gas. In fact, the situation may be more complex: (1) as in the heavier noble gases, more than one trapped component (atmosphere and interior) may be present in Ne as well and (2) since it is dependent on target element composition, the composition of cosmogenic Ne may not be constant but may change with release temperature when different minerals release their cosmogenic gas (cf. Mohapatra et al., 2009).

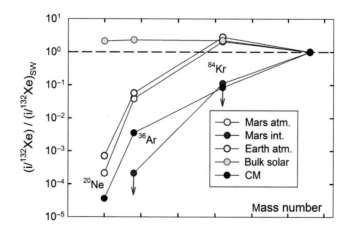

FIGURE 3.4

Comparison of elemental abundance patterns. Mars atmosphere, Mars interior, and Earth atmosphere, bulk Sun and as representative for a primitive meteorite, CM2 Maribo, are shown, normalized to ^{132}Xe and the abundances in the solar wind. *Data for Maribo from Haack et al. (2012), otherwise compositions as in Table 3.1.*

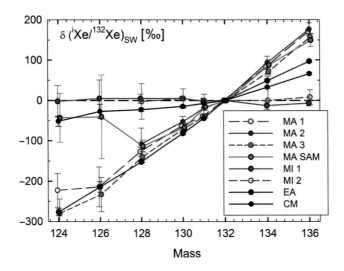

FIGURE 3.5

Xe isotopic compositions for Martian atmosphere (MA) and interior, shown as per mill deviations from the solar wind composition, are compared with Earth atmosphere and CM2 Maribo. MA1, MA2, and MA3 as well as MI1 and MI2 denote the various compositions derived from meteorite analyses listed in Table 3.4. *Data sources as given there, Maribo from Haack et al. (2012). MA SAM is the Martian atmospheric composition (error-weighted averages of static measurements) from in situ analysis of MA reported by Conrad et al. (2016). Errors plotted for these data are the sum of the statistical and reevaluated systematic uncertainties (Table 3.4).*

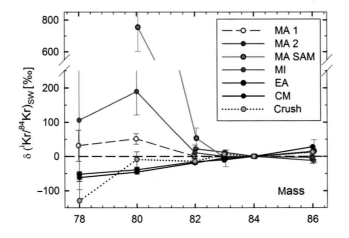

FIGURE 3.6

Kr isotopic compositions for Martian atmosphere (MA) and interior and the crush-released EETV component, shown as per mill deviations from the composition of solar wind, are compared with Earth atmosphere and CM2 Maribo. MA1 and MA2 denote the two compositions derived from meteorite analyses listed in Table 3.3. *Data sources as given there, Maribo from Haack et al. (2012), crush from Wiens (1988). MA SAM is the Martian atmospheric composition (unweighted average of semistatic measurements) from in situ analysis of MA reported by Conrad et al. (2016). Errors plotted for these data are the sum of the statistical and reevaluated systematic uncertainties (Table 3.3).*

3.2.2 THE "INTERIOR" COMPONENT (MI)

Gases found trapped in the Chassigny meteorite have generally been assumed to be representative of an "interior" Martian component, which is characterized by exceedingly low Ar/Xe and Kr/Xe ratios (Table 3.1; Fig. 3.3). In line with this, isotopic compositions (Tables 3.2–3.4; Figs. 3.5 and 3.6) have been assessed using Chassigny data with the lowest Ar/Xe and Kr/Xe ratios, since these are likely to be the least diluted with other gas components such as atmosphere. Xenon isotopic compositions listed in Table 3.3 and plotted in Fig. 3.5 include the composition measured by Mathew and Marti (2001) in the 500 °C pyrolysis step of an olivine separate (see also Marti and Mathew, 2015), and the 1600 °C release of Chassigny PB4 reported by Ott (1988), after correction for spallation contributions. Within the somewhat larger error bars for PB4, the two data sets agree except for the lower $^{129}Xe/^{132}Xe$ in the latter, which may indicate the interior component in the PB4 sample is purer. Besides this primitive interior component, also dubbed Chass-S by Mathew and Marti (2001), these authors identified an additional "evolved" interior component, Chass-E, that is characterized by enhanced abundances of the heavy isotopes ^{134}Xe and ^{136}Xe. Intriguingly, with the exception of $^{129}Xe/^{132}Xe$ in Chass-S, Chass-S and PB4 agree with solar wind Xe, as already noted in Ott and Begemann (1985).

The situation is less clear for the other noble gases. Isotopic data for Kr in Martian meteorites are sparse, and concerning Chassigny, the only reported data we are aware of are those of Ott (1988) and those obtained from the PhD work of Schwenzer (2004). Error bars in these analyses are somewhat large, in particular when considering the fact that variations in Kr isotopes between different Solar System reservoirs are modest compared to those in Xe. The composition given in Table 3.3 is from the 1800 °C extraction of Schwenzer (2004), where a solar wind $^{83}Kr/^{84}Kr$ ratio has been assumed in the correction for cosmogenic Kr. While $^{80}Kr/^{84}Kr$ is enhanced, probably as a result of neutron capture on Br, the remaining ratios (including $^{82}Kr/^{84}Kr$, also when corrected for a proportional neutron capture component) are compatible both with solar wind and current MA (Fig. 3.6).

The situation for interior Ar and Ne is more uncertain. Mathew and Marti (2001) give a lower limit of 5.26 for $^{36}Ar/^{38}Ar$ and an upper limit of 212 for $^{40}Ar/^{36}Ar$, which is what we have listed in Table 3.2. However, these data were obtained on a very small amount ($\sim 1\%$) of ^{36}Ar from the last release step of their bulk sample, not the olivine separate defining Chass-S, and furthermore they were not accompanied by the low Ar/Xe ratio characteristic of the unevolved "Chass-S" component. If real, these numbers are possibly more characteristic of the evolved interior component. As for neon, Nyquist et al. (2015) have proposed a $^{20}Ne/^{22}Ne$ ratio of 10.60 ± 0.16 for Martian mantle, based on data for DaG 476 (Mohapatra et al., 2009). To us it seems rather bold to make the connection and we tend to follow the arguments of Mohapatra et al. (2009) and consider the high abundance of neon found in DaG 476 as a rather specific kind of (terrestrial) air contamination (cf. the excess He and Ne found in terrestrial calcite by Scheidegger et al., 2010). Neon in Chassigny proper, unfortunately, is simply too strongly dominated by cosmogenic Ne (Ott, 1988) to draw any conclusions regarding trapped Ne in this meteorite. At this time, the exact compositions of Ne and Ar in the interior component still remain highly uncertain (Swindle, 2002).

3.2.3 THE "FRACTIONATED ATMOSPHERIC" COMPONENT

The nakhlites, like the shergottites, show enhanced $^{129}Xe/^{132}Xe$ ratios, but at a lower $^{84}Kr/^{132}Kr$; that is, in the $^{129}Xe/^{132}Xe$ versus $^{84}Kr/^{132}Kr$ plot of Fig. 3.3 they plot above the mixing line between Chassigny and Mars atmosphere. This has led to the suggestion that they contain mixtures of Mars interior and elementally fractionated atmospheric gases (Ott, 1988), which has remained a working hypothesis since then (e.g., Gilmour et al., 1999; Bogard et al., 2001; Swindle, 2002). If so, this should also be reflected in the isotopic compositions, which should lie between those of the two end-members. In fact, Gilmour et al. (2001) have argued that xenon in Nakhla can best be described as a mixture of atmospheric Xe on one hand and "an intimate mixture" of interior and spallation Xe on the other. Due to their chemical makeup and the comparably long cosmic ray exposure age of ~ 11 Ma (Eugster et al., 2002; Eugster, 2003), Nakhla and the other nakhlites have much higher abundances of cosmogenic Ar, Kr, and Xe than the shergottites and Chassigny, which makes it difficult (if not almost impossible in the case of Ar) to determine the detailed composition of the trapped gas.

Fig. 3.7 shows the composition of trapped Xe after correction for spallation. Data for two samples are shown: Nakhla H-1, which has the highest $^{129}Xe/^{132}Xe$ ratio, and thus presumably the highest proportion of MA among the Nakhla samples studied by Ott (1988), after spallation correction as listed by Mathew et al. (1998); and the Nak-D1 untreated/unirradiated Nakhla aliquot of Gilmour et al. (2001), which we have corrected for spallation as described in the Fig. 3.7 caption. Only isotopes $128-136$ (with the exception of the special case ^{129}Xe) are shown, because spallation contributions to ^{124}Xe and ^{126}Xe are too large to leave reliable trapped abundances. An exception to the general trend is ^{128}Xe in the NakD aliquot, which is clearly enhanced due to neutron capture on ^{127}I. Clear neutron capture effects have been seen in Kr in Nakhla (Ott, 1988), and this specific aliquot may have been unusually rich in iodine, since nothing comparable is seen in Nakhla H1 (nor in the second aliquot of NakD, which has larger uncertainties, however). Otherwise, though, both spectra for Nakhla follow closely the Martian atmospheric composition. This has also been noted by Gilmour et al. (2001), who, from fits to their complete data set of irradiated Nakhla samples, conclude that—for isotopes with no neutron capture contributions—trapped Xe in Nakhla is indistinguishable from that in MA, except for a slightly lower $^{129}Xe/^{132}Xe$ ratio of 2.35. Looking at the bulk level, this may appear somewhat surprising: with endmember $^{129}Xe/^{132}Xe$ ratios of ~ 1 (MI) and ~ 2.6 (MA) or a "true" nakhlite ratio of 2.35 (Gilmour et al., 2001) for that matter, one would expect from the measured ratios (Nakhla H1: ~ 1.5; NakD: ~ 2.05) that the observed compositions would plot only roughly 1/3 and 2/3 to 3/4, but not fully, displaced from the interior toward the atmospheric composition. Air contamination may provide a (partial) solution, since nonradiogenic air Xe is similar in isotopic composition to nonradiogenic Xe in the MA. Note that presence of air has been inferred for Nak-D by Gilmour et al. (2001) and that the Nakhla data for H1 shown in Fig. 3.7 are based on the totals, not the high-temperature release (Ott, 1988; Mathew et al., 1998). In any case, open questions remain and it remains to be confirmed whether trapped Xe in the nakhlites really conforms to the simple picture of being a mixture of Mars interior and fractionated atmosphere.

3.2.4 GASES RELEASED BY CRUSHING (EETV COMPONENT)

Results obtained by crushing samples of Martian meteorites, although meager, have been surprising. Early speculations (Wiens et al., 1986), inspired by the observation of microvesicles in the shock

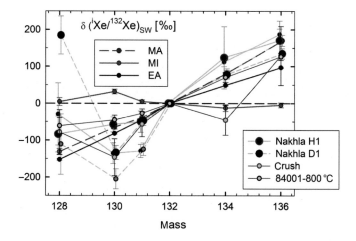

FIGURE 3.7

Xenon isotopic composition in Nakhla after correction for spallation, compared with Mars atmosphere (MA), Mars interior (MI), and Earth atmosphere (EA). In order not to clutter the diagram, weighted averages of the meteoritic data in Table 3.4 for only MA and MI are shown. Also shown are the composition of Xe released by crushing of EETA 79001 glass (Wiens, 1988) and the 800 °C release (maximum ^{129}Xe/^{132}Xe) in the analysis of ALH 84001 by Mathew and Marti (2001). Compositions are shown as per mill deviations from the solar wind composition. For Nakhla two compositions are shown: Nakhla H1 (Ott, 1988) as corrected for spallation by Mathew et al. (1998), and Nakhla D1 = the 7.12 mg aliquot #1 of unirradiated dry sample NakD from Gilmour et al. (2001). Corrections to the reported data for NakD have been performed using Mars atmosphere for the trapped component (uncritical) and a cosmogenic component according to Hohenberg et al. (1981) using element abundances from Lodders (1998). The atmospheric and interior compositions for Mars shown here are weighted averages of those shown individually in Fig. 3.5, for references see there and Table 3.4.

glass (Gooding and Muenow, 1986), were that the Martian atmospheric gas (presumably shock introduced) might be, to a large extent, present just in these vesicles. To test the idea, Wiens (1988) crushed a sample of EETA 79001 shock glass and analyzed the noble gases that were released. What was found to be released from vesicles of (probably) 10–100 μm diameter, however, was quite unlike MA, with ^{84}Kr/^{132}Kr as high as in the atmospheric component (or probably even higher), but ^{129}Xe/^{132}Xe low (Fig. 3.3). Assuming that the measured composition may also include some small fraction of MA, upper limits for ^{129}Xe/^{132}Xe and ^{40}Ar/^{36}Ar in "crush gas" are 1.21 and 680, respectively. A lower limit for ^{40}Ar/^{36}Ar derived by Wiens (1988) from Ar-Xe systematics is 430, while the measured ^{36}Ar/^{38}Ar ratio is 4.9 (Table 3.2). Obviously, if generally present in shergottites as a third component in addition to Mars atmosphere and interior, a crush component with such composition may be responsible for the rightward shift relative to the Chassigny-Mars atmosphere mixing line of shergottite data points in Fig. 3.3. Note also that, as argued by Wiens (1988), a major fraction of this crush (also named EETV) component in his experiment may have persisted after crushing, only being released in the later heating steps together with the MA.

Krypton in the EETV component (Fig. 3.6) is roughly similar to terrestrial krypton and does not show an enhanced ^{80}Kr abundance as MA (thought to be related to neutron capture on Br). Xe data are not very precise (Fig. 3.7) and—with the exception of ^{129}Xe, and a remarkably low

^{134}Xe/^{132}Xe—they are indistinguishable, within error, from either terrestrial or MA. While one might thus think of contaminating (terrestrial) air, from Ar−Xe elemental/isotopic considerations and elemental abundance ratios (see Fig. 3.3 in Wiens, 1988) this can be ruled out or appears highly unlikely at least (Wiens, 1988). In particular, contaminating air commonly is characterized by Ar/Xe and Kr/Xe ratios lower than air (e.g., Mohapatra et al., 2009), contrary to observation.

Unfortunately, the "crush component" has received little attention in subsequent work. The only other crush analysis of a Martian meteorite we are aware of is that of LEW 88516 (Mohapatra et al., 2009). Their sample was bulk meteorite, not pure glass, and interestingly, as shown in Fig. 3.8, data for their crush step fall on the mixing line between Chassigny (interior) and the EETV composition of Wiens (1988) in the common ^{129}Xe/^{132}Xe versus ^{84}Kr/^{132}Xe plot. The same is true in ^{129}Xe/^{132}Xe versus ^{36}Ar/^{132}Xe space (Mohapatra et al., 2009). Remarkably, gas released from LEW 88516 after crushing plots toward MA (or gas released after crushing EETA 79001) at 1200 °C, while that released at 1800 °C plots close to Chassigny, which is the reverse of the situation for the uncrushed sample of LEW 88516 analyzed by Mohapatra et al. (2009). Clearly, a more thorough investigation of crush-released gas and the effect of crushing on the siting of remaining gas components are warranted.

3.2.5 OLD MARTIANS

The first indications that the SNC meteorites come from Mars came from the realization that these are magmatic rocks with a young crystallization age, ~ 1.3 Ga for the nakhlites and Chassigny, and even younger for the shergottites (Wasson and Wetherill, 1979; McSween and Stolper, 1980;

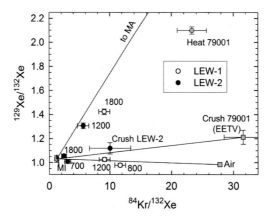

FIGURE 3.8

Plot of ^{129}Xe/^{132}Xe versus ^{84}Kr/^{132}Xe as in Fig. 3.3. Here we show results obtained for two samples of LEW 88516 (LEW-1 and LEW-2) obtained by crushing and heating (Mohapatra et al., 2009). For comparison the composition of gas released by crushing EETA 79001 glass (Wiens, 1988) is shown, as well the composition of the gas released from this sample by subsequent heating. The three mixing lines connect Martian interior MI (equated here with Chassigny PB4; Ott, 1988) with Martian atmosphere (MA; Bogard et al., 2001), the EETV composition (Wiens, 1988) and air, respectively.

Wood and Ashwal, 1981). Two more recently found specimens do not conform to this observation. One is the orthopyroxenite ALH 84001 with a crystallization age of ~4.5 Ga (Nyquist et al., 2001), and ^{40}Ar-^{39}Ar age of about 4 Ga (Turner et al., 1997; Bogard and Garrison, 1999) that are a record of shock processes and possibly also relate to carbonate formation (Nyquist et al., 2001), and thus aqueous processes on early Mars. The other is Northwest Africa 7034 (plus paired specimens), a basaltic breccia acquired in Morocco in 2011 (Agee et al., 2013; Stephen and Ross, 2014) that appears to be the most hydrated Martian meteorite found so far (MacArthur et al., 2015). Originally reported to have a formation age by Rb-Sr of 2.1 Ga (Agee et al., 2013), the age has more recently been revised to being even older at ~4.4 Ga using Sm-Nd dating (Nyquist et al., 2013) and dating of zircons (Humayun et al., 2013).

In line with the different provenance of the two ancient meteorites (magmatic rock vs fragmental breccia), the noble gas signatures of the two "ancients" are quite different from each other. In the classical ^{129}Xe/^{132}Xe versus ^{84}Kr/^{132}Xe plot, ALH 84001 plots above the MI–MA mixing line along with the nakhlites, other magmatic rocks, and thus apparently also contains fractionated (relative to the current composition) MA (Fig. 3.9). Ratios ^{84}Kr/^{132}Xe and ^{129}Xe/^{132}Xe of ~6 and ~2.16, respectively, for the fractionated component have been derived from these data (Mathew and Marti, 2001) and this has been inferred to be the signature of ancient MA trapped ~4 Ga ago (Gilmour et al., 1998; Mathew and Marti, 2001). Otherwise the isotopic composition of the Xe in the elementally fractionated component, as suggested by the most ^{129}Xe-rich release step (800 °C)

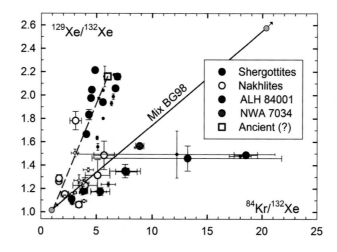

FIGURE 3.9

Plot of ^{129}Xe/^{132}Xe versus ^{84}Kr/^{132}Xe for the "old Martians" ALH 84001 and NWA 7034. *Data sources for ALH 84001: Swindle et al. (1995), Miura et al. (1995), Murty and Mohapatra (1997), and Mathew and Marti (2001). Small symbols are totals, while large symbols are without low-temperature steps, i.e., only gas released at >800 °C (Swindle et al., 1995), at >700 °C (Miura et al., 1995), at >1000 °C (Murty and Mohapatra, 1997) and at >700 °C (Mathew and Marti, 2001). For NWA 7034 the composition of gas released at >600 °C (Cartwright et al., 2014) is plotted. The nakhlites and shergottites shown in Fig. 3.3 are included for comparison, again both with (small symbols) and without (large symbols) low-temperature gas releases.*

in the analysis of Mathew and Marti (2001) appears to be close to that of the nakhlites, as shown in Fig. 3.7, but somewhat shifted toward solar wind.

If Xe in ALH 84001 is ancient, the fact that $^{129}Xe/^{132}Xe$ is lower than in modern atmosphere implies that it is "less evolved," but this conclusion hinges on the assumption that the highest *measured* ratio (actually slightly higher at 2.23 ± 0.01; Murty and Mohapatra, 1997) represents the true value and is not lowered by addition of another component like Martian interior or terrestrial contamination. Terrestrial contamination is widespread in Antarctic or hot desert finds (e.g., Schwenzer and Ott, 2006; Mohapatra et al., 2009; Schwenzer et al., 2013), and in order to explore the effect on the observed signature, we have plotted in Fig. 3.9 ratios also for the sum of high temperature releases (large symbols) in addition to the totals (small symbols), but this does not change the general picture. It is worthy of note in this context that also in the analysis of nakhlites, which presumably trapped elementally fractionated more recent atmosphere, the $^{129}Xe/^{132}Xe$ ratio as inferred for the MA from impact glass analyses of 2.60 (Table 3.1; Bogard and Garrison, 1998) or the 2.52 ratio measured in situ (Conrad et al., 2016) has not been reached. Conventionally extracted gases rarely exceed 2 and, from the higher values reached in laser heating steps, Gilmour et al. (2001), based on the mixing relation between trapped and cosmogenic gases conclude that the ratio in the "nakhlitic" component is only ~ 2.35—higher than observed in ALH 84001, but still lower than the atmospheric ratio (Bogard and Garrison, 1998; Conrad et al., 2016).

For NWA 7034, the only detailed noble gas analysis so far is that of Cartwright et al. (2014), who found trapped components similar to those in the much younger shergottites, i.e., consistent with a mixture of Martian interior and current MA (Fig. 3.9). NWA 7034 shows strong neutron capture effects in the Kr isotopes, and the conclusion by these authors concerning trapped components is reinforced by two additional observations: (1) after reasonable correction for neutron capture at ^{36}Ar, NWA 7034 may plot on the MI–MA mixing line also in $^{129}Xe/^{132}Xe$ versus $^{36}Ar/^{132}Xe$ space (their Fig. 3b) and (2) interpreting the heavy Xe isotopes as a mixture of MA, MI, and fission Xe, the fission Xe age is in the range $\sim 3.8-4.2$ Ga (the detailed value depending on whether only spontaneous fission of ^{238}U is taken into account or ^{244}Pu fission is also included), in agreement with the Sm–Nd age of Nyquist et al. (2013). If the MA component was trapped at this time, the data suggest little evolution since then, in apparent contrast to the conclusions based on ALH 84001 (see discussion in Section 3.3.3). It appears possible, though, that MA in NWA 7034 was introduced later (Cartwright et al., 2014). Note in this context that the arguments of Cartwright et al. (2014) rely on the data for total trapped gas. Looking at individual gas release steps suggests that the situation may be more complex and that, in fact, NWA 7034 may be a so far unique specimen, or an assemblage of unique specimens, given that it consists of a variety of clasts (Stephen and Ross, 2014).

3.3 SOURCES, SINKS, AND PROCESSES

3.3.1 ORIGIN AND MODIFICATION OF MARTIAN ATMOSPHERE

As noted previously and in many other reviews of the Martian noble gas story, the most intriguing property of the current Martian atmospheric composition as derived from meteorite analyses is its similarity to that of the terrestrial atmosphere—in particular the elemental abundance pattern and the distinctive fractionated isotopic composition of xenon. This clearly suggests that similar processes were acting on a similar source and to a similar extent.

3.3.1.1 Processes and modification

In the most comprehensive model developed along these lines (Pepin, 1991, 1994; Pepin and Porcelli, 2002), the stage is set by early (>4.3 Ga ago) noble gas loss from a primordial atmosphere by hydrodynamic escape (Hunten et al., 1987). The energy driving this process is thought to be in extreme ultraviolet (EUV) radiation from the young Sun, some $2-3$ orders of magnitude more intense than today. While the lighter noble gases (including Kr) are almost completely lost from the atmosphere at this stage, a remnant of xenon remains and shows the observed strongly fractionated isotopic pattern favoring the heavy isotopes. The lighter noble gases found in the atmosphere must have been replenished then by outgassing from the interior and infalling material. Larger projectiles may have led to impact erosion of the atmosphere, reducing overall abundances, but without fractionating the noble gases.

Another important loss process, of lesser (relative) significance in the early stage, when hydrodynamic escape ruled, but becoming important later on is loss by solar wind–induced sputtering (e.g., Jakosky et al., 1994). The major mechanism is sputtering of neutral atoms above the homopause by energetic (~ 1 keV) oxygen ions created by EUV irradiation and accelerated in the magnetic field accompanying the solar wind. The process is efficient on Mars (in contrast to Earth), because Mars has no intrinsic magnetic field, at least at present (Hutchins et al., 1997). Also, since the process acts above the homopause, where the atmosphere is no longer well-mixed, it results in preferential removal of the lighter isotopes (Jakosky et al., 1994; Hutchins and Jakosky, 1996; Pepin and Porcelli, 2002). The low $^{36}Ar/^{38}Ar$ ratio of the MA may be largely the result of this solar wind–induced sputtering (Atreya et al., 2013), and the same has recently been argued for the $^{20}Ne/^{22}Ne$ ratio (Nyquist et al., 2015). In the Ne case, however, it has to be kept in mind that the actual ratio in the MA is poorly constrained (see Section 3.2.1) and that the result is also dependent on the assumed initial composition of Martian neon.

What is somewhat unsatisfying in the hydrodynamic escape scenario is the similarity between Martian xenon, as determined from Martian meteorites (note, however, the differences in the light isotopes in the composition determined by SAM; Table 3.4, Fig. 3.5), and terrestrial xenon. Since hydrodynamic loss can be expected to be planet-specific (among others, EUV fluxes must have been different on the two planets, as included in the model of Pepin, 1991), this may appear to be accidental (Swindle, 2002). Considering this, creating the pattern in a preplanetary process, for example, diffusive separation in the gravitational field of a porous planetesimal (Zahnle et al., 1990), may be worth exploring in more detail, as well as fractionation during incorporation in ices that later (via impacting comets) contributed to the MA (Owen et al., 1992; Notesco et al., 2003). The former addresses, however, mainly the Xe isotopic composition (those planetesimals would contribute more or less just xenon), while the latter focuses on elemental abundances only (see also update of their Ar/Kr/Xe element plot in Ott, 2014). In a separate study, no significant isotopic fractionation of Xe was found during trapping in water-ice (Notesco et al., 1999), probably ruling out cometary water as a major source of Xe (but see also Section 3.3.1.2).

In the models discussed previously, isotopic fractionation of Xe would have been an early process, preplanetary or during hydrodynamic escape occurring within the first $100-200$ Ma of planetary history. As an interesting alternative in the case of Earth's atmosphere, a variant of the sputtering process has recently been suggested as an explanation for—remarkably—both the low abundance of Xe relative to Kr and the fractionated Xe isotopic pattern (Hébrard and Marty, 2014). This process would have been an ongoing one, more efficient at early times (the Archean), in line with the higher UV flux, but still operating also at later times and possibly identifiable in the composition of trapped ancient atmospheric Xe (Fig. 3.1 in Hébrard and Marty, 2014).

By inference, the process may also be relevant to Mars. The model is based on the fact that Xe is much more easily photoionized than the other noble gases, and that the process would happen at an altitude where in a CH_4-rich atmosphere an organic haze may be an efficient trap for Xe ions. The heavy isotopes would have been preferentially incorporated, and the isotopically light Xe remaining would have been lost from the atmosphere, thus explaining the apparent Xe deficit. Heavy xenon trapped in the haze would have been retained, resulting in the isotopically heavy Xe we find in the atmosphere of Earth (and possibly also of Mars, for that matter).

3.3.1.2 Sources

For a detailed and quantitative application of models, thorough knowledge not only of the current atmosphere (for which we here take that derived from the meteorite analyses), but also of the original (primordial) composition, is essential. In turn, if a given model is accepted, it can serve to constrain the original composition on which the corresponding processes acted. One natural choice for the starting composition is solar or solar wind. These may not be quite identical as discussed in the context of Martian interior later, but for now we use "solar" as shorthand for both of them. Another is the primordial (historically named "planetary") component found in primitive meteorites (Ott, 2014).

If the light noble gases (including Kr) currently in the atmosphere are primarily derived from outgassing of the interior after loss of the primary inventory during hydrodynamic escape, this may suggest a solar wind isotopic composition for these. The Martian interior component (Section 3.2.2) is characterized by extremely low Ar/Xe and Kr/Xe ratios, so it appears well possible in such a scenario that Xe has been largely retained in the interior, while most of the lighter gases have found their way into the MA (Pepin, 1994). Retention of Xe in the interior is essential since otherwise the fractionated isotopic composition established in the atmosphere during hydrodynamic escape or by porous planetesimals would have been diluted. The case of Earth, by comparison, seems more problematic. We have no evidence for a similar Xe-rich interior component on Earth (so far). Some promising ideas exist about xenon in the Earth's interior (e.g., existence at depth of xenon silicates) (Sanloup et al., 2002) or intermetallic compounds (Zhu et al., 2014), but all searches for "missing Xe" in (surficial) reservoirs have been unsuccessful (Ozima and Podosek, 2002; and references therein). If, alternatively, the "secondary" starting composition for Ne, Ar, and Kr on Mars was supplied by infalling primitive material (interplanetary dust, micrometeorites, etc.), neon would probably be dominated by implanted solar wind, while Kr would be of the "chondritic" type dominated by gas hosted mostly by "phase Q" (Ott, 2014), and Ar possibly intermediate. It appears difficult in this case to avoid bringing in the required amount of Kr without accompanying Xe (at $^{84}Kr/^{132}Kr \approx 1$); that is, it would be difficult (or impossible) to simultaneously keep the fractionated Xe isotopic signature from the loss process and arrive at a high Kr/Xe ratio. A way out may be if the volatiles were brought in by infalling comets instead, since trapping in ice at low temperatures appears to favor Kr over Xe (Owen et al., 1992; Notesco et al., 2003; Ott, 2014), and so the fractionated composition of the Xe that had remained from the primary atmosphere would have been only slightly diluted (see also Dauphas, 2003 for a similar scenario involving comets for the origin of the terrestrial atmosphere).

The $^{36}Ar/^{38}Ar$ ratio is only moderately different between the solar and chondritic components (Ott, 2014), so currently both appear viable candidates for the starting composition of Ar that was modified by sputtering loss (Jakosky et al., 1994) or equivalent processes. The diffusive separation factor that enters into fractionation by sputtering is the same for $^{36}Ar/^{38}Ar$ and $^{20}Ne/^{22}Ne$ (Table III in Jakosky et al., 1994). Accordingly with a factor ~ 1.3 ($= 0.77^{-1}$) implied from $^{38}Ar/^{36}Ar$, one would arrive at a low atmospheric $^{20}Ne/^{22}Ne$ of ~ 7.3 (Park and Nagao, 2006) when starting with a ratio of ~ 10.7. From this it has been argued that the original neon was from phase Q (Nyquist et al., 2015), where "Q"-gases are those that dominate the inventory of trapped Ar, Kr, and Xe in primitive meteorites (Ott, 2014). This appears not very likely. Ne as typically found in primitive meteorites (if they do not contain solar wind) is dominated by HL−Ne with a low $^{20}Ne/^{22}Ne$ ratio of ~ 8.5 (rather than by Ne−Q; Ott, 2014) and an appropriate mixture of Ne−HL with solar Ne is a more likely source having a ratio of ~ 10.7, if this is what it takes. If the true atmospheric ratio is higher (e.g., ~ 10; Table 3.2), solar-type Ne is probably a good starting point. This might be either solar wind or bulk solar Ne (Heber et al., 2012). Starting from fractionated solar wind previously known as SEP−Ne (Wieler, 2002) the result for the atmosphere would be intermediate.

In krypton, isotopic variations among Solar System reservoirs are typically small, but analyses of impact glass containing MA seem to have been affected by neutron capture of glass-precursor regolith materials (Rao et al., 2011), as also seen in the analysis of ancient NWA 7034 (Cartwright et al., 2014). In fact, an even much larger neutron capture contribution is seen in the measurement of current Martian atmospheric Kr by the SAM instrument (Conrad et al., 2016). After taking this into account, Martian atmospheric krypton is fully compatible with solar wind krypton (Table 3.3, Fig. 3.6; Conrad et al., 2016), which is to be expected if solar-type Kr is outgassed from the interior, while not being affected by the late processes of sputtering and photochemical loss (Pepin, 1994). Overall, thus, the evidence seems to favor a scenario where the current inventory of Ne, Ar, and Kr in the MA is ultimately derived from a reservoir having solar-like isotopic composition.

The case of xenon is more involved. Xenon in the MA (as in the terrestrial case) contains additions from radiogenic ^{129}Xe (easy to discern), but how much or even whether fission xenon arising from the decay of extinct ^{244}Pu is present at all is less clear. Lack of ^{244}Pu fission Xe might seem strange because ^{244}Pu has a longer half-life than ^{129}I. The latter would decay away first and thus absence of radiogenic ^{129}Xe would appear more likely than missing Pu fission products (Swindle, 2002). As for the Earth, the detailed models of Pepin (e.g., Pepin, 1991, 2000) call for a baseline primordial component U−Xe (identical to solar wind except for the two heaviest isotopes ^{134}Xe and ^{136}Xe), which became mass fractionated during the hydrodynamical loss episode and to which then the fission component (some %) was added. A similar conclusion was recently reached by Marti and Mathew (2015). U−Xe has been derived by Pepin and Phinney (1978) from multidimensional correlation analysis to stepwise heating data of carbonaceous chondrites. Until recently, its existence (in pure form at least) has not been reliably demonstrated in terrestrial or extraterrestrial materials in spite of various attempts (e.g., Busemann and Eugster, 2002; see also Ott, 2014). However, it appears to have now been found (with apparently even more extreme composition) in the coma of comet 67P/Churyumov-Gerasimenko (Marty et al., 2017).

Mars is different from Earth; that is, the model based on U−Xe works for Earth but not for Mars. Instead, different starting compositions for the two planets are required (e.g., Pepin, 2000), which then undergo essentially the same degree of mass fractionation (see also Marti and Mathew, 2015). This was already recognized by Swindle et al. (1986), who found that the composition of Martian atmospheric xenon as derived from the shergottites, in contrast to the terrestrial one, could be modeled assuming for the starting composition chondritic Xe (formerly also called AVCC). This is not a pure component, but rather a mixture, naturally occurring in primitive meteorites and shown as "CM" in Fig. 3.5, of Q−Xe (Table 3.4) and a small amount of presolar Xe−HL (see Ott, 2014 and references therein). A drawback of this choice is that it does not allow for the addition of ^{244}Pu fission xenon. If instead solar wind is assumed for the starting composition, addition of some fissiogenic Xe appears to be allowed (Swindle and Jones, 1997; Mathew et al., 1998), so this may be a more attractive solution. Nevertheless, also in this case the amount of fission Xe is remarkably small, and also may be even consistent with zero as shown by Conrad et al. (2016) in their assessment of the SAM and meteoritic Xe data. Obviously this has implications for the degassing history of the planet.

While starting with a solar Xe isotopic composition may also seem attractive because it is in line with our preferred starting composition for Ne, Ar, and Kr as discussed previously, it has to be kept in mind that Martian atmospheric xenon appears to date from an earlier epoch/reservoir than the lighter noble gases, which according to current models were essentially completely lost from the early atmosphere and replenished from a different reservoir. Naturally, the two different reservoirs need not have to have the same origin and composition. Finally, there is the high ^{129}Xe/^{132}Xe ratio in the current MA. Since this is due to the radiogenic contribution from ^{129}I with a half-life of 16 Ma, the high ratio requires either very early loss from the atmosphere or storage in some reservoir until after atmospheric loss (Swindle, 2002). A possibility suggested by Musselwhite et al. (1991) is that iodine was dissolved in water and sequestered in minerals that held back the radiogenic ^{129}Xe during the loss episode. All in all, apart from their fractionated nature, the Martian and terrestrial atmospheric xenon must have evolved quite differently.

More precise measurements of the isotopic composition of the noble gases, particularly Xe, should, in principle, make it possible to rule out some of the models discussed earlier. This may be not so straightforward, especially when it comes to the rare light Kr and Xe isotopes. It is obvious that in the meteorites their abundances may have been affected by cosmogenic and neutron capture contributions. Surprisingly, however, these contributions appear to be even more significant in the case of the SAM measurements of the current MA. A thorough understanding of the origin of the enhanced light Xe isotopes in particular, in comparison with the "meteoritic Martian Xe," will be required before far-reaching conclusions on the overall origin of Martian Xe can be drawn from the SAM data (see end of Section 3.3.4 for additional discussion).

3.3.1.3 Trapping

It is certainly fair to say that general consensus exists at this time that the atmospheric noble gases found (primarily) in the shock glass of Martian meteorites have been introduced by shock. Although not all details of shock implantation are well understood, the fact that it works is convincingly demonstrated by laboratory experiments (Bogard et al., 1986; Wiens and Pepin, 1988) in which samples of terrestrial basalt were shocked in controlled gas mixtures. Ar, Kr, and Xe were readily implanted in these experiments without any evidence for elemental or isotopic fractionation. Neon, while also implanted, showed a "leaky" behavior and in some samples was lost by diffusion after shock implantation. There appears to be also a correlation between Ne implantation efficiency and Ne release temperature. As noted by Wiens and Pepin (1988), samples with a high efficiency for Ne implantation release most of it at low temperature (500 °C in their analyses), while samples with less trapping efficiency showed their major release at higher temperature. These authors also found Ne nominal isotopic ratios indicative of diffusive loss, but unfortunately the results were not precise enough for a firm conclusion. Their helium isotopic data, in any case, appear suggestive of diffusive fractionation. Loss of (radiogenic) helium by shock is also indicated by the study of radiogenic He abundances in SNC meteorites by Schwenzer et al. (2008).

The leaky behavior of shock-implanted neon may offer an explanation for the variation in implied trapped $^{20}Ne/^{22}Ne$ ratios in the shergottites (Table 3.2). High values appear generally associated with analyses where the extrapolation to the trapped composition is largely determined by low-temperature releases (e.g., Swindle et al., 1986; Wiens et al., 1986, LEW 88516 in Mohapatra et al., 2009), while the Dho 378 analysis of Park and Nagao (2006) indicative of a low ratio of ~ 7.3 is based on a fit not including the lowest temperature, which plots above the regression line. While it cannot be excluded that the early releases are compromised by some terrestrial contamination, this nevertheless appears to be a subject deserving further scrutiny. Ultimately this may be resolved only if at some time definitive Ne isotopic data from the direct analysis of MA become available.

3.3.2 MARTIAN INTERIOR COMPONENT(S)

3.3.2.1 Chassigny—primitive and evolved

Our ideas about Martian interior gases are primarily based on analyses of the Chassigny meteorite. Only one other chassignite is known at this time, NWA 2737, but this meteorite is much more severely shocked than Chassigny, has very low abundances of Xe, lacks the solar-like (Table 3.4; Fig. 3.5) Xe composition seen in Chassigny, and appears to have shergottite-like trapped gases instead (Marty et al., 2006). The solar signature in Chassigny was first reported by Ott and Begemann (1985) and Ott (1988) and since it contains much less radiogenic ^{129}Xe than the atmosphere, contrary to the case of the Earth, it must be a sample of an undegassed interior reservoir (Swindle, 2002). A more detailed study of xenon in Chassigny has been performed by Mathew and Marti (2001), and these authors have identified additional "interior" Martian reservoirs, which are more evolved. In the terminology of Mathew and Marti (2001), the "pure solar" component seen by Ott and Begemann (1985) and Ott (1988) is called Chass-S, while the evolved component is Chass-E. Chass-E is prominent in the high-temperature (>1000 °C) releases in their analysis of Chassigny bulk, as well as an olivine separate, and can be derived from Chass-S by the addition of ^{244}Pu fission xenon, so this relict from early times must have been stored well-mixed with solar Xe in (some) interior reservoir(s) from which it was trapped.

While this appears straightforward, a closer look suggests that Chassigny is a more complicated rock. For one, the Chass-S Xe composition of Marti and Mathew (2015)—MI1 in Table 3.4 and Fig. 3.5—defined by the 500 °C pyrolysis from their olivine separate, is not fully identical to the solar wind composition, contrary to what Mathew and Marti (2001) state. The difference in $^{129}Xe/^{132}Xe$ is significant. On the other hand, while the data are generally less precise, what has been observed in the Chassigny bulk sample PB4 by Ott (1988) is fully compatible with solar wind, but is distinct from Chass-S in $^{129}Xe/^{132}Xe$ (only), so in essence sample PB4 might be a "purer Chass-S." Furthermore, there is no trace of Chass-E in the PB4 sample (a bulk sample, not an olivine separate) and about half of the "purer" Chass-S is retained after pyrolysis at 1200 °C, where in the Mathew and Marti (2001) analysis it is, if present at all, swamped by Chass-E. The "early release" of Chass-S has been attributed by Mathew and Marti (2001) to siting in melt inclusions, but obviously this is not the only siting, and moreover there are parts of the meteorite devoid of Chass-E.

Finally, an interesting question is whether the solar-type xenon apparently present as Chass-S is indeed "bulk solar" or just "solar wind," a difference little considered in past noble gas work. It has, however, become evident for the light noble gases (He, Ne, Ar) that their isotopes are fractionated during solar wind acceleration (Heber et al., 2012). If the formalism for inefficient Coulomb drag (ICD), which seems to describe well the empirically derived differences (Heber et al., 2012), is used to evaluate possible differences in Kr and Xe, they may be surprisingly large, but a better understanding is still required (Ott, 2014). As for Martian interior Xe, this would make a difference when trying to establish how the solar-type xenon was acquired. The lighter noble gases are not of help in this: interior Kr is not very precise (Table 3.3, Fig. 3.6), while the lower limit of 5.26 for ^{36}Ar/^{38}Ar given by Mathew and Marti (2001), even if problematic (see Section 3.2.2), does not give real constraints on this matter. Another problem with establishing the presence of the MI component is that its composition resembles that of extremely fractionated air (EFA) often present in desert meteorites (Mohapatra et al., 2009). This point will be shortly and separately addressed later (Section 3.3.4).

3.3.2.2 Argon in the shergottites

Given the low ^{36}Ar/^{132}Xe and ^{84}Kr/^{132}Xe in the "interior" (Chassigny like) noble gas component, ^{36}Ar in the shergottites is almost exclusively from the atmospheric component, even though both components are present (Fig. 3.3). In the case of Ar, there must be additional radiogenic ^{40}Ar from in situ decay of ^{40}K since the meteorite rock formed. If there are no other sources of ^{40}Ar, the radiogenic abundance then must be equal to the difference between the measured abundance and the atmospheric abundance (^{36}Ar abundance $\times$ ^{40}Ar/^{36}Ar ratio of the atmosphere). Nature is not so simple, however, and in ^{39}Ar-^{40}Ar dating of shergottites, Bogard and Garrison (1999) found apparent ages significantly higher than those obtained by other techniques. Similarly, Schwenzer et al. (2007a) found more "radiogenic" ^{40}Ar than could be accounted for by the K abundance of their pyroxene and maskelynite separates and the accepted age of the respective shergottites. The unspectacular interpretation is that ^{40}Ar was incompletely degassed when the rock formed. Another possibility is that the shergottites have been tapping into mantle reservoirs with variable ^{40}Ar/^{36}Ar (Bogard and Garrison, 1999; Schwenzer et al., 2007a) or possibly a more near-surface reservoir that was available during rock formation or ejection (Swindle, 2002). A more recent study of Zagami plagioclase and pyroxene combined with detailed calculations arrives at the conclusion that in this case, and by inference in the case of the other shergottites, (1) the excess ^{40}Ar is distributed through the crystal lattice and not contained in impact glass and (2) that it was most likely inherited from the source magma or from crustal rock that was assimilated by the magma (Bogard and Park, 2008). This magma seems to have possessed and then lost significant amounts of ^{40}Ar.

3.3.3 FRACTIONATED AND ANCIENT ATMOSPHERE

In this section we are going to treat the elementally fractionated (relative to current atmosphere) compositions observed in the nakhlites and the old orthopyroxenite ALH 84001 together. This is because characteristic elemental and isotopic signatures appear to be close, if not identical (Fig. 3.9). The similarly fractionated $^{84}Kr/^{132}Xe$ ratio naturally suggests a similar fractionation and thus the same, or at least a similar, trapping process. There is one process that would only work for ALH 84001, however. If, as in some models of atmospheric evolution (e.g., Pepin, 1994), Kr was replenished after early hydrodynamic escape on an extended time scale, e.g., by outgassing from the interior, at ~ 4 Ga ago, the Kr/Xe ratio simply may not have reached its current value yet (cf. Fig. 3.4 in Pepin, 1994). The isotopic composition of Kr may give a clue in this respect, but currently available data (Table 3.3) are not precise enough. In such a scenario, in any case, the very similar signatures seen in the nakhlites and ALH 84001 would be fortuitous, a not very attractive solution.

The search for host phases of fractionated MA and possible fractionation mechanisms has been an active research field (e.g., Gilmour et al., 1999, 2000; Swindle et al., 2000; Schwenzer, 2004). Ott et al. (1988), who performed a leaching experiment using HCl on Nakhla in which the high $^{129}Xe/^{132}Xe$ (i.e., atmospheric) component was removed, suggested olivine to be the main host phase, since the observed weight loss of $\sim 15\%$ closely matched the modal abundance of olivine. Later on the search centered on processes related to the action of water and Martian weathering products, which were initially called iddingsite (a weathering product of olivine; for simplicity we will use that name in the following), but have more recently found to be complex alteration veins consisting of a carbonate−clay−gel succession (Hicks et al., 2014). "Iddingsite" became an early suspect (Drake et al., 1994) as a carrier of the fractionated gas. This relies on the fact that Xe is more soluble in water than Kr by about a factor of two (Ozima and Podosek, 2002), roughly consistent with the data point array in Fig. 3.9 and an $^{84}Kr/^{132}Xe$ ratio of 8 ± 1 as estimated by Schwenzer and Ott (2006). Note, however, from the data array in Fig. 3.9, that there may well be variations in the Kr/Xe ratio of the fractionated component. This may not be surprising, since for a given process the extent of fractionation will vary in relation to exact conditions (as already pointed out by Drake et al., 1994).

While the first analysis of iddingsite (Drake et al., 1994) seemed compatible with the iddingsite idea, a later study by Swindle et al. (2000) of iddingsite in the nakhlite Lafayette showed the mass balance to be inconclusive. In fact, results obtained by Gilmour et al. (1999) indicate that the bulk of the fractionated component in Nakhla is hosted by the major minerals olivine and pyroxene as well as in mesostasis, with concentrations decreasing in the order mesostasis > olivine > pyroxene. Similar results were obtained by Bart et al. (2001), Schwenzer (2004), and Schwenzer et al. (2006). Nevertheless, there appears to be also a role for fluids as (original?) carriers of ^{40}Ar-rich (implied atmospheric) Martian gases in Nakhla. This is indicated by the combined Ar-Ar and halogen abundance study of Cartwright et al. (2013), although the fluid-related Ar, from this work at least, is a minor component.

The siting within the major minerals seems to require a different mechanism for fractionating and trapping Mars atmosphere from that envisioned by Drake et al. (1994) and Garrison and Bogard (1998). In particular, simple fractionation according to solubility in water will not do the trick (Bart et al., 2001). Other possibilities for trapping being considered (Gilmour et al., 1999,

2000; Bart et al., 2001), therefore are (1) magmatic incorporation and (2) implantation by shock. In both cases, the fractionation must have happened in a separate step before trapping. In (1), elementally fractionated gas may have been released from soil grain surfaces into the magma from which the nakhlites formed (Gilmour et al., 1999), while in (2) the gas may have become fractionated by adsorption at the low temperatures of Martian winters (Gilmour et al., 2000; Bart et al., 2001; Swindle, 2002). That the "atmospheric" component in nakhlites seems to be associated with leachable iodine (Gilmour et al., 2001) may provide a clue. As for the shock hypothesis for the nakhlites, a problem is that they appear to be mildly shocked only (14−20 GPa for Nakhla and maybe as low as 5 GPa for Lafayette; e.g., Fritz et al., 2005), and that, in the experiments of Bogard et al. (1986), implantation efficiency at such low shock pressure is significantly less than at the higher shock pressures typical for the shergottites. On the other hand, unlike at the higher shock pressures, emplacement efficiency for Ar and Kr is distinctly less than for Xe. Shock implantation at very low shock, if sufficient in quantity, might thus also account for the apparent elemental fractionation with no separate fractionation step necessary. In fact, low shock pressure for implantation relative to ALH 84001 has been suggested as an explanation for the lower release temperature of the nakhlites (Gilmour et al., 2001).

There is, of course, also the possibility that the "fractionated" gases in the nakhlites were, in fact, not elementally fractionated from the atmospheric composition, but simply reflect the composition of the MA at the time of trapping (Musselwhite and Swindle, 2001). This could be the case if polar clathrates, which preferentially incorporate Xe compared to Ar or Kr, form either seasonally or in eras with a different obliquity (Musselwhite and Swindle, 2001; Swindle et al., 2009; Mousis et al., 2013). If the clathrates form seasonally, it would be remarkable that "old" ALH 84001 apparently shows just the same Ar/Xe and Kr/Xe ratios as do the nakhlites (Fig. 3.9), although obliquity changes occur over timescales comparable to the differences in cosmic ray exposure ages of the nakhlites and ALH 84001 on one hand and the shergottites on the other.

ALH 84001, with its >4 Ga age, apparently suffered severe shock in its early history (Treiman, 1998) and has been little altered in the meantime, so this again makes implantation of the "nakhlite-like" noble gas component by shock more likely than the weathering hypothesis (if we are to assume the same process). Particularly interesting in this case is, of course, that it may contain a sample of Martian paleoatmosphere. If we, like Mathew and Marti (2001), accept that the maximum $^{129}Xe/^{132}Xe$ ratio of 2.16 observed by these authors is that of the atmosphere at the time of trapping (but see Section 3.2.3), it follows that the MA had not yet acquired its full complement of ^{129}I-derived ^{129}Xe at ∼4 Ga ago. Not only this, but based on Xe isotopic systematics Mathew and Marti (2001) also conclude that the composition of trapped Xe (excluding ^{129}Xe) in ALH 84001 is Chass-S-like, only being augmented by fission Xe from ^{244}Pu (see Fig. 3.7). Hence also the isotopic fractionation of Xe characteristic of modern MA would not yet have occurred at this time.

Because of its importance, this conclusion of Mathew and Marti (2001) deserves further scrutiny. Other analyses of xenon in ALH 84001 (Swindle et al., 1995; Miura et al., 1995; Murty and Mohapatra, 1997) are not of sufficient detail to rule on the question, so we take a closer look here using the spallation-corrected data of Mathew and Marti (2001). In Fig. 3.10 we show as an example a three-isotope plot of $^{130}Xe/^{132}Xe$ versus $^{136}Xe/^{132}Xe$, excluding the low-temperature (<600 °C) extractions. Unfortunately, in plots of $^{130}Xe/^{132}Xe$ as well as $^{131}Xe/^{132}Xe$ or $^{134}Xe/^{132}Xe$ versus $^{136}Xe/^{132}Xe$ a problem is that mixing between SW (Chass-S) and MA is

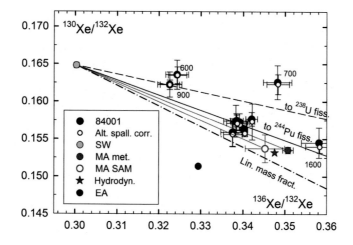

FIGURE 3.10

Three-isotope plot of spallation-corrected data for ALH 84001 (Mathew and Marti, 2001; *filled black symbols*). $^{130}Xe/^{132}Xe$ is plotted on the ordinate, versus $^{136}Xe/^{132}Xe$ on the abscissa. The *open symbols* plotted have been spallation-corrected using a different assumption for the trapped $^{126}Xe/^{130}Xe$ ratio (midway between Martian atmosphere (MA) and solar wind versus pure solar wind used by Mathew and Marti, 2001). Differences are minor. Solar wind (SW), Earth (EA), and Martian (MA) atmosphere are plotted for comparison as well as mixing lines for solar wind mixtures with Martian atmosphere, as well as ^{244}Pu and ^{238}U fission xenon. Two values are shown for MA: the error-weighted average of the compositions obtained from meteorite measurements as well as the in situ measured composition with updated uncertainties (Table 3.4). The *dash-dotted line* indicates the mass fractionation trend for a linear mass fractionation law. Mass fractionation through hydrodynamical escape as modeled in Conrad et al. (2016) leads to a composition similar to the SAM Mars atmosphere value.

virtually indistinguishable from the mixing of SW and ^{244}Pu fission xenon. However, the 1600 °C data point, with $^{136}Xe/^{132}Xe$ larger than the Mars atmosphere, clearly shows the need for extra ^{244}Pu fission xenon. The excess relative to SW composition of $\sim 3.5 \times 10^{-14}$ cm^3 STP/g can be accounted for by in situ decay over ~ 4.2 Ga for a U content of 11 ppb (Lodders, 1998) and assuming chondritic Pu/U (Hudson et al., 1989). Since Xe can be assumed to be more efficiently retained than Ar, and with a ~ 4 Ga Ar-Ar age, this may not be unreasonable (see also discussion of fission Xe in Mathew et al., 1998). The main release, however, is situated in a cluster around $^{136}Xe/^{132}Xe$ ~ 0.34 and $^{130}Xe/^{132}Xe$ ~ 0.157, in an area compatible with both mixtures of SW + MA and SW + ^{244}Pu fission. In essence, from Fig. 3.10 (as well as similar ones involving ^{131}Xe or ^{134}Xe) it appears that the results are compatible with a mixture of MI ($=$ Chass-S $=$ solar wind SW) and MA, augmented by some fission xenon.

Complexity is indicated if ^{128}Xe is considered. For one, a spread in $^{128}Xe/^{132}Xe$ is seen within the cluster for the main releases. This is not due to uncertainties introduced by the spallation corrections. We have checked the sensitivity to the assumptions used by Mathew and Marti (2001) in doing the correction and found only minor differences when using other reasonable input values. However, it may, in principle, be due to contributions from neutron capture on ^{127}I (Mathew et al.,

1998; note also that neutron capture in Kr has been seen by Swindle et al., 1995). More tellingly, though, is the small variation in the $^{128}Xe/^{130}Xe$ ratio, the ratio of the two isotopes that after spallation correction (and ignoring neutron capture) are from the trapped component(s) only (Fig. 3.11). Observed ratios are all very distinct from current MA and those for the major release (1100−1550 °C; with small errors) are actually close to solar wind; that is, Chass-S (Fig. 3.11; Mathew and Marti, 2001). This then is the strongest argument suggesting that the isotopic composition of trapped Xe in ALH 84001 is unlike that of the current MA.

If this observation can be confirmed, there are major consequences. Above all, this definitely poses a problem for the popular model of hydrodynamic escape at very early times (Pepin, 1994;

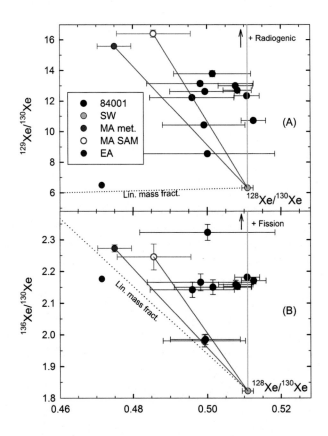

FIGURE 3.11

Three-isotope plots $^{129}Xe/^{130}Xe$ (A) and $^{136}Xe/^{130}Xe$ (B) versus $^{128}Xe/^{130}Xe$ for ALH 84001. The data have been corrected for spallation using the solar wind ratio for the trapped component and the cosmogenic composition used by Mathew and Marti (2001) and given in the footnote of their Table 3.1. The various lines indicate the trend for mass fractionation, mixing between solar wind (SW) and Martian atmosphere (MA) as well as the shifts due to addition of radiogenic/fissiogenic Xe. Two values are shown for MA: the error-weighted average of the compositions obtained from meteorite measurements as well as the in situ composition with updated uncertainties (Table 3.4).

Pepin and Porcelli, 2002) as the process mainly responsible for the observed isotopic fractionation of Martian (and terrestrial) atmospheric xenon. Possible alternatives that come to mind have problems of their own, however. A way out, for example, would be to relax the assumption that the "fractionated atmosphere" in ALH 84001 is actually trapped atmosphere. A possible alternative would be that it constitutes another "evolved" interior component, similar to Chass-E but much more evolved, with higher radiogenic and fissiogenic contributions. In turn, naturally, this would raise the question about the fractionated component in the nakhlites. Either the similarities in Kr/Xe and $^{129}Xe/^{132}Xe$ ratios (Figs. 3.3 and 3.9) are coincidental or the fractionated component in the nakhlites is also of interior rather than atmospheric origin, which in turn is contradicted by their lower (compared to ALH 84001) $^{128}Xe/^{130}Xe$ ratio (Mathew et al., 1998). In fact, as pointed out earlier (Section 3.2.3), the trapped composition of the nonradiogenic Xe isotopes in Nakhla appears to be closer to the atmospheric one than expected based on $^{129}Xe/^{132}Xe$ (Fig. 3.7), which is the opposite from ALH 84001. Clearly, establishing a precise composition of Xe in ALH 84001 and the nakhlites should be a major goal in future work.

3.3.4 COMPLICATIONS—FOR BETTER OR WORSE

There are two items we have not yet addressed in Section 3.3. One is the origin of the EETV component (Section 3.2.4), the other is the fact that most Martian meteorites in our collections are hot or cold desert finds, which at times makes it difficult to discern the proper Martian signatures (e.g., Schwenzer and Ott, 2006).

Only scant attention has been paid in the course of noble gas studies on Martian meteorites to the role of inclusions or vesicles. This is in spite of the fact that their possibly important role has been recognized early on. Wiens et al. (1986) suggested that microvesicles found in the impact glass of EETA 79001 (Gooding and Muenow, 1986) carried shock-implanted MA, while fluid inclusions were suggested as a host for (a minor portion of) MA in the nakhlites (Cartwright et al., 2013) and melt inclusions in Chassigny as host of the solar wind–like Chass-S Xe (Mathew and Marti, 2001). The latter is based on the observed low-temperature release of Chass-S (Mathew and Marti, 2001), although in the analysis of Ott (1988) at least half of this primitive component was retained to higher temperature (see Section 3.2.1).

The only serious study of the gas inventory in vesicles from shergottites is that of Wiens (1988) who crushed a sample of EETA 79001 impact glass. Its presence in other shergottites, though, is indicated by results from crushing LEW 88516 (Mohapatra et al., 2009) and possibly by the rightward shift of shergottite data points in the $^{129}Xe/^{132}Xe$ versus $^{84}Kr/^{132}Xe$ plot relative to the Chassigny—MA mixing line (Fig. 3.3). In the Wiens (1988) experiment, Martian atmospheric gases were released, as expected, in the later heating steps, but crushing released noble gases with a very specific composition (EETV component; Section 3.2.4) characterized by extremely high $^{84}Kr/^{132}Xe$ (higher than either the Martian or terrestrial atmosphere) and low $^{129}Xe/^{132}Xe$, features in a sense opposite to those of "fractionated atmosphere." EETV was probably released from vesicles in the $10-100\,\mu m$ size range, with smaller vesicles unaffected by crushing (Wiens, 1988). At present, as back at the time of Wiens (1988), it is still not clear what the relationship is, or if there is any relationship at all, between the EETV component and the other Martian noble gas components. A possibility is that EETV constitutes a "deep-seated," (i.e., another interior) component already present in cracks and vugs into which the shock producing the impact glass injected molten material along with the atmospheric gas (Wiens, 1988). The importance of vesicles for the noble gas budget of Martian meteorites certainly deserves further attention.

While the presence of EETV gas seems a very reasonable explanation, another possible explanation suggested for the rightward shift of shergottite data points in the plot of Fig. 3.3 is terrestrial contamination (Bogard and Garrison, 1998). This would obviously explain a low $^{129}Xe/^{132}Xe$ ratio, but in order to match $^{84}Kr/^{132}Xe$ and $^{36}Ar/^{132}Xe$ it would require in several cases mixing of MA with terrestrial air that is elementally unfractionated (Figs. 3.2 and 3.3 in Bogard and Garrison 1998) or worse, even enriched in ^{36}Ar in case of one analysis of Zagami (Ott et al., 1988). Such a scenario seems highly unlikely, since generally contamination is a more serious problem for the heavier noble gases. An extreme case of this, often encountered in desert meteorites, has been studied by Mohapatra et al. (2002, 2009), who coined the term EFA (elementally fractionated air) to describe the phenomenon. While generally present in low-temperature extractions, EFA may persist up to release temperatures of 1000 °C.

On one hand, the presence of EFA is annoying, since it can mask the Martian noble gas features in the desert meteorites (Schwenzer et al., 2007b; Mohapatra et al. 2009). For one, given the low $^{129}Xe/^{132}Xe$ (0.9832 in air) and $^{84}Kr/^{132}Xe$ (~ 1) ratios, it may be mistaken for Chassigny-type gas (Tables 3.2 and 3.4). Its presence can also make the assessment of the fractionated Martian component difficult (Schwenzer and Ott, 2006). Adsorption or dissolution in water are often regarded as the first (or decisive) step in reaching highly fractionated ratios, but dissolution in water can account for a factor ~ 2 in Kr/Xe ratio only, while adsorption at low temperature will require an additional step of fixation. The extremely low Kr/Xe ratio of EFA thus seems to require a multistep fractionation process (Mohapatra et al., 2009).

On the other hand, understanding EFA may also help us understand more about elementally fractionated Martian noble gas components, since it may well be assumed that similar processes of adsorption and weathering were acting on Mars. There are secondary minerals of preterrestrial origin in the Martian meteorites (e.g., Drake et al., 1994; Swindle et al., 2000; Treiman, 2005 and references therein, Changela and Bridges, 2010; Hicks et al., 2014) that may have trapped noble gases. While iddingsite, contrary to earlier speculations (Drake et al., 1994), has not been found to be an important host phase in the nakhlites (Gilmour et al., 1999), noble gases can also be taken up by the weathered surfaces of primary minerals like olivine and pyroxene, or be adsorbed on fresh surfaces (Schwenzer et al., 2012). If this happened on Mars, the fixation by shock envisaged by Gilmour et al. (1999) of gases adsorbed on cold mineral surfaces may also have acted on gases acquired during weathering.

Naturally, environmental conditions (gas fugacity and temperature) on Mars will have differed from the places on Earth where EFA was acquired, and they changed over the course of Martian history. The Noachian-aged ALH 84001 and NWA 7034 meteorites probably witnessed a more "clement" Mars, warmer and more "Earth-like" than at later times, when the nakhlites and shergottites formed (Fassett and Head, 2011). Carbonate formation in ALH 84001 (Melwani-Daswami et al., 2014) and alteration vein formation in the nakhlites (Bridges and Schwenzer, 2012) seem to have happened under neutral to mildly alkaline conditions, but conditions may also have been, at times and locally at least, more acidic than typical for (most) terrestrial environments (e.g., Hurowitz et al., 2006; Tosca et al., 2008). Simulation experiments and study of terrestrial analogs as started by Schwenzer et al. (2007b,c,d) should be a useful tool to understand the effect of environment and weathering on noble gases. Finally, noble gas analyses of irradiated samples can help trace the flow and distribution of elements characteristic for aqueous (acidic) systems like the halogens chlorine, bromine, and iodine (e.g., Cartwright et al., 2013). This subject is being treated in detail Chapter 2 of this book.

Finally, while generally there is an agreement between MA derived from meteorite data and those obtained by SAM at the MSL by direct sampling of the current MA (in particular in the crucial $^{40}Ar/^{36}Ar$, $^{38}Ar/^{36}Ar$, and $^{129}Xe/^{132}Xe$ isotopic ratios), differences are apparent in the lightest Xe isotopes. Their enhancement reminds of a cosmogenic contribution (Conrad et al., 2016). However, it is not at all clear why the cosmogenic contribution in Xe (as well as the neutron capture contribution in Kr, for that matter) should be higher in the atmosphere proper than in MA trapped by shergottites. If it reflects differences in composition due to the fact that the atmosphere was sampled at different times, this would require a surprisingly large addition since the time MA was shock-emplaced into the shergottites. Along the same lines, somewhat disturbingly, the ratio of excess ^{124}Xe to excess ^{126}Xe is quite unlike that of common cosmogenic Xe: 1.0 ± 0.7 or 1.2 ± 0.7, respectively, depending on whether the atmospheric composition from Swindle et al. (1986) or Marti and Mathew (2015) is used as a baseline (Table 3.4) versus typically ~ 0.6 (e.g., Hohenberg et al., 1978, 1981). Still, the errors are large enough, so that Conrad et al. (choosing a ratio of 0.609) were able to match the spallation-corrected data within their uncertainties with the composition of the model fractionated solar wind. It is clear that more accurate data on these minor isotopes are required to substantiate (or refute) the presence of spallation Xe in the present MA.

3.4 SHORT SUMMARY

Noble gases have been instrumental in asserting a Martian origin for the shergottites, nakhlites, chassignites, and orthopyroxenite ALH 84001. The way was paved by the crucial discovery in impact glass of the shergottite EETA 79001 of noble gases (Bogard and Johnson, 1983) that appeared similar in composition to the atmosphere of Mars as determined by Viking (Owen et al., 1977). This has been beautifully confirmed by the new, high-quality data for the MA from the SAM instrument aboard the MSL rover Curiosity, as far as they are available by now (Atreya et al., 2013; Conrad et al., 2014, 2016; Franz et al., 2015). The similarities between the Martian and terrestrial atmosphere are remarkable, in terms of elemental abundance pattern as well as with regard to the isotopically strongly fractionated composition of Xe, although questions remain.

Martian meteorites also contain components of noble gases other than the atmospheric one, which are not as easily obtainable by instruments operating on the planet, and thus the meteorites provide additional information. These other components include in particular:

- Components that are considered as being of interior origin. The best characterized are those dominant in Chassigny (Chass-S, Chass-E). They are distinguished by extremely low Ar/Xe and Kr/Xe ratios and an isotopic composition of Xe resembling solar wind xenon. Shergottites show evidence for additional ^{40}Ar from an interior reservoir.
- Components found in the nakhlites on the one hand and the old Martian meteorite ALH 84001 on the other, which are suggestive of elementally fractionated atmosphere. It would be important to clarify whether these components do differ in detail as suggested by Mathew and Marti (2001).

More enigmatic is a component released by crushing of EETA 79001 (EETV), which is probably hosted by vesicles in the 10−100 μm range. Unfortunately, little attention has been paid to this component after its initial discovery by Wiens (1988).

Little doubt exists that the atmospheric gases in shergottite impact glass were introduced from the MA by shock. Less certain is the way the other components were acquired, but weathering may have played a role in case of the nakhlites. Perhaps this was aided by mild shock that helped in fixing the gas, since it is distributed among the major minerals, with weathering products per se adding only little to the budget. Future studies of how weathering affects the behavior of major minerals toward noble gases may shed more light on this question.

The most popular model for the origin of the current MA (Pepin, 1994, 2000; Pepin and Porcelli, 2002) is one in which hydrodynamic escape very early in Martian history led to essentially complete loss of noble gases save xenon, which partially remained and acquired its isotopically fractionated composition at this time. The lighter noble gases were replenished from the interior or by infall and, except for Kr, were further fractionated elementally and isotopically, most likely by sputtering/photochemical erosion. If Xe in the (elementally) fractionated component in >4 Ga old ALH 84001 is taken to represent atmospheric Xe trapped more than 4 Ga ago, a problem for this model arises from the observation that isotopically it appears unfractionated, solar wind−like (Mathew and Marti, 2001). It is crucial to confirm or refute this observation. If it holds up, alternative explanations for the elementally fractionated component in ALH 84001 should be considered, as well as other models for the origin and evolution of the MA. This may also bear on the origin and evolution of Earth's atmosphere, because terrestrial atmospheric Xe shows an almost identical fractionation pattern as Martian atmospheric Xe and thus a similar process/history is indicated.

REFERENCES

Agee, C.B., et al., 2013. Unique meteorite from early Amazonian Mars: water-rich basaltic breccia Northwest Africa 7034. Science 339, 780−785.

Atreya, S.K., et al., 2013. Primordial argon isotope fractionation in the atmosphere of Mars measured by the SAM instrument on *Curiosity* and implications for atmospheric loss. Geophys. Res. Lett. 40, 5605−5609.

Bart, G.D., Swindle, T.D., Olson, E.K., Treiman, A.H., 2001. Xenon and krypton in Nakhla mineral separates. Lunar Planet. Sci. 32, abstract #1363.

Becker, R.H., Pepin, R.O., 1984. The case for a martian origin of the shergottites: nitrogen and noble gases in EETA 79001. Earth Planet. Sci. Lett. 69, 225−242.

Bogard, D.D., Garrison, D.H., 1998. Relative abundances of argon, krypton, and xenon in the Martian atmosphere as measured in Martian meteorites. Geochim. Cosmochim. Acta 62, 1829−1835.

Bogard, D.D., Garrison, D.H., 1999. Argon-39-argon-40 "ages" and trapped argon in Martian shergottites, Chassigny, and Allan Hills 84001. Meteorit. Planet. Sci. 34, 451−473.

Bogard, D.D., Johnson, P., 1983. Martian gases in an Antarctic meteorite? Science 221, 651−654.

Bogard, D.D., Park, J., 2008. ^{39}Ar-^{40}Ar dating of the Zagami Martian shergottite and implications for magma origin of excess ^{40}Ar. Meteorit. Planet. Sci. 43, 1113−1126.

Bogard, D.D., Huneke, J.C., Burnett, D.S., Wasserburg, G.J., 1971. Xe and Kr analyses of silicate inclusions from iron meteorites. Geochim. Cosmochim. Acta 35, 1231—1254.

Bogard, D.D., Hörz, F., Johnson, P.H., 1986. Shock-implanted noble gases: an experimental study with implications for the origin of Martian gases in shergottite meteorites. Proc. 17th Lunar Planet Sci. Conf., J. Geophys. Res. 91, E99—E114.

Bogard, D.D., Clayton, R.N., Marti, K., Owen, T., Turner, G., 2001. Martian volatiles: isotopic composition, origin, and evolution. Space Sci. Rev. 96, 425—458.

Bridges, J.C., Schwenzer, S.P., 2012. The nakhlite hydrothermal brine on Mars. Earth Planet. Sci. Lett. 359-360, 117—123.

Busemann, H., Baur, H., Wieler, R., 2000. Primordial noble gases in "phase Q" in carbonaceous and ordinary chondrites studied by closed-system stepped etching. Meteorit. Planet. Sci 35, 949—973.

Busemann, H., Eugster, O., 2002. The trapped noble gas component in achondrites. Meteorit. Planet. Sci. 37, 1865—1891.

Busemann, H., Seiler, S., Wieler, R., Kuga, M., Maden, C., Irving, A.J., et al., 2015. Martian noble gases in recently found shergottites, nakhlites and breccia Northwest Africa 8114. 78th Annu. Meteoritical Soc. Meeting, abstract #5235.

Cartwright, J.A., Gilmour, J.D., Burgess, R., 2013. Martian fluid and Martian weathering signatures identified in Nakhla, NWA 998 and MIL 03346 by halogen and noble gas analysis. Geochim. Cosmochim. Acta 105, 255—293.

Cartwright, J.A., Ott, U., Herrmann, S., Agee, C.B., 2014. Modern atmospheric signatures in 4.4 Ga Martian meteorite NWA 7034. Earth Planet. Sci. Lett. 400, 77—87.

Changela, H.G., Bridges, J.C., 2010. Alteration assemblages in the nakhlites: Variation with depth on Mars. Meteorit. Planet. Sci. 45, 1847—1867.

Clayton, R.N., Mayeda, T.K., 1983. Oxygen isotopes in eucrites, shergottites, nakhlites, and chassignites. Earth Planet. Sci. Lett. 62, 1—6.

Conrad, P.G., Malespin, C., Franz, H., Trainer, M.G., Brunner, A., Manning, H., The MSL Science Team, et al., 2014. SAM measurements of krypton and xenon on Mars. Lunar Planet. Sci. 45, abstract #2366.

Conrad, P.G., Malespin, C.A., Franz, H.B., Pepin, R.O., Trainer, M.G., Schwenzer, S.P., et al., 2016. In situ measurement of atmospheric krypton and xenon on Mars with Mars Science Laboratory. Earth Planet. Sci. Lett. 454, 1—9.

Dauphas, N., 2003. The dual origin of the terrestrial atmosphere. Icarus 165, 326—339.

Drake, M.J., Swindle, T.D., Owen, T., Musselwhite, D.S., 1994. Fractionated martian atmosphere in the nakhlites? Meteoritics 29, 854—859.

Eberhardt, P., Eugster, O., Marti, K., 1965. A redetermination of the isotopic composition of atmospheric neon. Z. Naturforsch. 20a, 623—624.

Eugster, O., 2003. Cosmic-ray exposure ages of meteorites and lunar rocks and their significance. Chem. Erde — Geochem. 63, 3—30.

Eugster, O., Busemann, H., Lorenzetti, S., Terribilini, D., 2002. Ejection ages from krypton-81-krypton-83 dating and pre-atmospheric sizes of martian meteorites. Meteorit. Planet. Sci. 37, 1345—1360.

Fassett, C.I., Head, J.W., 2011. Sequence and timing of conditions on early Mars. Icarus 211, 1204—1214.

Franz, H.B., Trainer, M.G., Wong, M.H., Mahaffy, P.R., Atreya, S.K., Manning, H.L.K., et al., 2015. Reevaluated martian atmospheric mixing ratios from the mass spectrometer on the Curiosity rover. Planet. Space Sci. 109-110, 154—158.

Fritz, J., Artemieva, N., Greshake, A., 2005. Ejection of Martian meteorites. Meteorit. Planet. Sci. 40, 1393—1411.

Garrison, D.H., Bogard, D.D., 1998. Isotopic composition of trapped and cosmogenic noble gases in several Martian meteorites. Meteorit. Planet. Sci. 33, 721—736.

Gilmour, J.D., Whitby, J.A., Turner, G., 1998. Xenon isotopes in irradiated ALH84001: evidence for shock-induced trapping of ancient Martian atmosphere. Geochim. Cosmochim. Acta 62, 2555−2571.

Gilmour, J.D., Whitby, J.A., Turner, G., 1999. Martian atmospheric xenon contents of Nakhla mineral separates: implications for the origin of elemental mass fractionation. Earth Planet. Sci. Lett. 166, 139−147.

Gilmour, J.D., Whitby, J.A., Turner, G., 2000. Extraterrestrial xenon components in Nakhla. Lunar Planet. Sci. 31, abstract #1513.

Gilmour, J.D., Whitby, J.A., Turner, G., 2001. Disentangling xenon components in Nakhla: Martian atmosphere, spallation and Martian interior. Geochim. Cosmochim. Acta 65, 343−354.

Gooding, J.L., Muenow, D.W., 1986. Martian volatiles in shergottite EETA 79001: New evidence from oxidized sulfur and sulfur-rich aluminosilicates. Geochim. Cosmochim. Acta 50, 1049−1059.

Haack, H., Grau, T., Bischoff, A., Horstmann, M., Wasson, J., Sørensen, A., et al., 2012. Maribo—a new CM fall from Denmark. Meteorit. Planet. Sci. 47, 30−50.

Heber, V.S., Wieler, R., Baur, H., Olinger, C., Friedmann, T.A., Burnett, D.S., 2009. Noble gas composition of the solar wind as collected by the Genesis mission. Geochim. Cosmochim. Acta 73, 7414−7432.

Heber, V.S., Baur, H., Bochsler, P., McKeegan, K.D., Neugebauer, M., Reisenfeld, D.B., et al., 2012. Isotopic mass fractionation of solar wind: evidence from fast and slow solar wind collected by the *Genesis* mission. Astrophys. J. 759, 121.

Hébrard, E., Marty, B., 2014. Coupled noble gas-hydrocarbon evolution of the early Earth atmosphere upon solar UV irradiation. Earth Planet. Sci. Lett. 385, 40−48.

Hicks, L.J., Bridges, J.C., Gurman, S.J., 2014. Ferric saponite and serpentine in the nakhlite martian meteorites. Geochim. Cosmochim. Acta 136, 194−210.

Hohenberg C.M., Marti K., Podosek F.A., Reedy R.C. and Shirck J.R., 1978. Comparison between observed and predicted cosmogenic noble gases in lunar samples. In: Proc. Lunar Planet. Sci. 9th, pp. 2311−2344.

Hohenberg, C.M., Hudson, B., Kennedy, B.M., Podosek, F.A., 1981. Xenon spallation systematics in Angra dos Reis. Geochim. Cosmochim. Acta 45, 1909−1915.

Hudson G.B., Kennedy B.M., Podosek F.A. and Hohenberg C.M., 1989. The early solar system abundance of 244Pu as inferred from the St. Severin chondrite. In: Proc. 19th Lunar Planet. Sci. Conf., pp. 547−557.

Humayun, M., Nemchin, A., Zanda, B., Hewins, R.H., Grange, M., Kennedy, A., et al., 2013. Origin and age of the earliest Martian crust from meteorite NWA 7533. Nature 503, 513−516.

Hunten, D.M., Pepin, R.O., Walker, J.C.G., 1987. Mass fractionation in hydrodynamic escape. Icarus 69, 532−549.

Hurowitz, J.A., McLennan, S.M., Tosca, N.J., Arvidson, R.E., Michalski, J.R., Ming, D.W., et al., 2006. In situ and experimental evidence for acidic weathering of rocks and soils on Mars. J. Geophys. Res. E111, E02S19.

Hutchins, K.S., Jakosky, B.M., 1996. Evolution of martian atmospheric argon: Implications for sources of volatiles. J. Geophys. Res. 101, 14933−14949.

Hutchins, K.S., Jakosky, B.M., Luhmann, J.G., 1997. Impact of a paleomagnetic field on sputtering of martian atmospheric argon and neon. J. Geophys. Res. 102, 9183−9189.

Jakosky, B.M., Pepin, R.O., Johnson, R.E., Fox, J.L., 1994. Mars atmospheric loss and isotopic fractionation by solar-wind-induced sputtering and photochemical escape. Icarus 111, 271−288.

Lee, J.-Y., Marti, K., Severinghaus, J.P., Kawamura, K., Yoo, H.-S., Lee, J.B., et al., 2006. A redetermination of the isotopic abundances of atmospheric Ar. Geochim. Cosmochim. Acta 70, 4507−4512.

Llorca, J., Roszjar, J., Cartwright, J.A., Bischoff, A., Ott, U., Pack, A., et al., 2013. The Ksar Ghilane 002 shergottite—the 100th registered Martian meteorite fragment. Meteorit. Planet. Sci. 48, 493−513.

Lodders, K., 1998. A survey of shergottite, nakhlite and Chassigny meteorites whole-rock compositions. Meteorit. Planet. Sci. 33, A183−A190.

Lodders, K., Palme, H., Gail, H.P., 2009. Abundances of the elements in the solar system. In: Trümper, J.E. (Ed.), Landolt-Börnstein, New Series, vol. VI/4B. Springer-Verlag, Berlin, Heidelberg, New York, pp. 560−630. Chap. 4. 4.

MacArthur, J.L., Bridges, J.C., Hicks, L.J., Gurman, S.J., 2015. The thermal and alteration history of NWA 8114 Martian regolith. Lunar Planet. Sci. 46, abstract #2295.

Mahaffy, P.R., et al., 2013. Abundance and isotopic composition of gases in the Martian atmosphere from the Curiosity rover. Science 341, 263−266.

Marti, K., Mathew, K.J., 2015. Xenon in the protoplanetary disk (PPD-Xe). Astrophys. J. Lett. 806, L30.

Marti, K., Kim, J.S., Thakur, A.N., McCoy, T.J., Keil, K., 1995. Signatures of the Martian atmosphere in glass of the Zagami meteorite. Science 267, 1981−1984.

Marty, B., Heber, V.S., Grimberg, A., Wieler, R., Barrat, J.-A., 2006. Noble gases in the Martian meteorite Northwest Africa 2737: a new chassignite signature. Meteorit. Planet. Sci. 41, 739−748.

Marty, B., Altwegg, K., Balsiger, H., Bar-Nun, A., Bekaert, D.V., Berthelier, J.-J., et al., 2017. Xenon isotopes in Comet 67P/Churyumov-Gerasimenko show comets contributed to Earth's atmosphere. Science, 356, 1069−1072.

Mathew, K.J., Marti, K., 2001. Early evolution of Martian volatiles: Nitrogen and noble gas components in ALH84001 and Chassigny. J. Geophys. Res. E 106, 1401−1422.

Mathew, K.J., Kim, J.S., Marti, K., 1998. Martian atmospheric and indigenous components of xenon and nitrogen in the Shergotty, Nakhla, and Chassigny group meteorites. Meteorit. Planet. Sci. 33, 655−664.

McSween Jr., H.Y., Stolper, E.M., 1980. Basaltic meteorites. Sci. Am. 242 (6), 54−63.

Melwani Daswani, M., Schwenzer, S.P., Reed, M.H., Wright, I.P., Grady, M.M., 2014. Carbonate precipitation driven by clay leachates on early Mars. Lunar Planet. Sci. 45, abstract #1280.

Meshik, A., Mabry, J., Hohenberg, C., Marrocchi, Y., Pravdivtseva, O., Burnett, D., et al., 2007. Constraints on neon and argon isotopic fractionation in solar wind. Science 318, 433−435.

Meshik, A., Hohenberg, C., Pravdivtseva, O., Burnett, D., 2014. Heavy noble gases in solar wind delivered by Genesis mission. Geochim. Cosmochim. Acta 127, 326−347.

Meshik, A., Pravdivtseva, O., Hohenberg, C., Burnett, D., 2015. Refined composition of solar wind xenon delivered by Genesis: implication for primitive terrestrial xenon. Lunar Planet 46, abstract #2640.

Meteoritical Bulletin Database, 2015. <http://www.lpi.usra.edu/meteor/> (accessed 18.08.15.).

Miura, Y.N., Nagao, K., Sugiura, N., Sagawa, H., Matsubara, K., 1995. Orthopyroxenite ALH84001 and shergottite ALH77005: Additional evidence for a martian origin from noble gases. Geochim. Cosmochim. Acta 59, 2105−2113.

Mohapatra, R.K., Schwenzer, S.P., Ott, U., 2002. Krypton and xenon in Martian meteorites from hot deserts—the low temperature component. Lunar Planet. Sci. 33, abstract #1532.

Mohapatra, R.K., Schwenzer, S.P., Herrmann, S., Murty, S.V.S., Ott, U., Gilmour, J.D., 2009. Noble gases and nitrogen in Martian meteorites Dar al Gani 476, Sayh al Uhaymir 005 and Lewis Cliff 88516: EFA and extra neon. Geochim. Cosmochim. Acta 73, 1505−1522.

Mousis, O., Chassefière, E., Lasue, J., Chevrier, V., Elwood Madden, M.E., Lakhlifi, A., et al., 2013. Volatile trapping in martian clathrates. Space Sci. Rev. 174, 213−250.

Murty, S.V.S., Mohapatra, R.K., 1997. Nitrogen and heavy noble gases in ALH 84001: Signatures of ancient Martian atmosphere. Geochim. Cosmochim. Acta 61, 5417−5428.

Musselwhite, D.S., Swindle, T.D., 2001. Is release of Martian atmosphere from polar clathrate the cause of the nakhlite and ALH84001 Ar/Kr/Xe ratios? Icarus 154, 207−215.

Musselwhite, D.S., Drake, M.J., Swindle, T.D., 1991. Early outgassing of Mars supported by differential water solubility of iodine and xenon. Nature 352, 697−699.

Nier, A.O., Hanson, W.B., Seiff, A., McElroy, M.B., Spencer, N.W., Duckett, R.J., et al., 1976. Composition and structure of the Martian atmosphere: Preliminary results from Viking 1. Science 193, 786−788.

Notesco, G., Laufer, D., Bar-Nun, A., Owen, T., 1999. An experimental study of the isotopic enrichment in Ar, Kr, and Xe when trapped in water ice. Icarus 142, 298−300.

Notesco, G., Bar-Nun, A., Owen, T., 2003. Gas trapping in water ice at very low deposition rates and implications for comets. Icarus 162, 183−189.

Nyquist, L.E., Bogard, D.D., Shih, C.-Y., Greshake, A., Stöffler, D., Eugster, O., 2001. Ages and geologic histories of Martian meteorites. Space Sci. Rev. 96, 105−164.

Nyquist, L.E., Shih, C.-Y., Peng, Z.X., Agee, C., 2013. NWA 7034 Martian breccia: disturbed Rb-Sr systematics, preliminary ∼4.4 Ga Sm-Nd age. In: 76th Annu. Meteoritical Soc. Meeting, abstract #5318.

Nyquist, L.E., Park, J., Nagao, K., Haba, M.K., Mikouchi, T., Kusakabe, M., et al., 2015. "Normal planetary" Ne-Q in Chelyabinsk and Mars. In: 78th Annu. Meteoritical Soc. Meeting, abstract #5054.

Ott, U., 1988. Noble gases in SNC meteorites: Shergotty, Nakhla, Chassigny. Geochim. Cosmochim. Acta 52, 1937−1948.

Ott, U., 2014. Planetary and pre-solar noble gases in meteorites. Chemie der Erde − Geochem. 74, 519−544.

Ott, U., Begemann, F., 1985. Are all the 'martian' meteorites from Mars? Nature 317, 509−512.

Ott, U., Löhr, H.P., Begemann, F., 1988. New noble gas data for SNC meteorites: Zagami, Lafayette, and etched Nakhla (abstract). Meteoritics 23, 295−296.

Owen, T., Biemann, K., Rushneck, D.R., Biller, J.E., Howarth, D.W., Lafleur, A.L., 1977. The composition of the atmosphere at the surface of Mars. J. Geophys. Res. 82, 4635−4639.

Owen, T., Bar-Nun, A., Kleinfeld, I., 1992. Possible cometary origin of heavy noble gases in the atmospheres of Venus, Earth and Mars. Nature 358, 43−46. see also comment by. Ozima, M. Wada, N., as well as reply by Owen, T., Bar-Nun, A., 1993. Nature 361, 693−694.

Ozima, M., Podosek, F.A., 2002. Noble Gas Geochemistry. Cambridge University Press, Cambridge, p. 286.

Park, J., Nagao, K., 2006. New insights on Martian atmospheric neon from Martian meteorite, Dhofar 378. Lunar Planet. Sci. 37, abstract #1110.

Pellas, P., 1987. Les météorites "SNC" sont-elles des échantillons de la planète Mars?. Bulletin de la Société Géologique de France 3, 21−29.

Pepin, R.O., 1985. Evidence of Martian origins. Nature 317, 473−475.

Pepin, R.O., 1991. On the origin and early evolution of terrestrial planet atmospheres and meteoritic volatiles. Icarus 92, 2−79.

Pepin, R.O., 1994. Evolution of the Martian atmosphere. Icarus 111, 289−304.

Pepin, R.O., 2000. On the isotopic composition of primordial xenon in terrestrial planet atmospheres. Space Sci. Rev. 92, 371−395.

Pepin, R.O., Phinney, D., 1978. Components of xenon in the solar system. Unpublished preprint, quoted in Pepin (1991).

Pepin, R.O., Porcelli, D., 2002. Origin of noble gases in the terrestrial planets. In: Porcelli, D., Ballentine, C.J., Wieler, R. (Eds.), Noble Gases in Geochemistry and Cosmochemistry. Mineralogical Society of America, Washington, DC, pp. 191−246.

Pepin, R.O., Schlutter, D.J., Becker, R.H., Reisenfeld, D.B., 2012. Helium, neon, and argon composition of the solar wind as recorded in gold and other Genesis collector materials. Geochim. Cosmochim. Acta 89, 62−80.

Rao, M.N., Nyquist, L.E., Bogard, D.D., Garrison, D.H., Sutton, S.R., Michel, R., et al., 2011. Isotopic evidence for a Martian regolith component in shergottite meteorites. J. Geophys. Res. 116, E08006.

Sanloup, C., Hemley, R.J., Mao, H.-k, 2002. Evidence for xenon silicates at high pressure and temperature. Geophys. Res. Lett. 29, 1883.

Scheidegger, Y., Baur, H., Brennwald, M.S., Fleitmann, D., Wieler, R., Kipfer, R., 2010. Accurate analysis of noble gas concentrations in small water samples and its application to fluid inclusions in stalagmites. Chem. Geol. 272, 31−39.

Schwenzer, S.P., 2004. Marsmeteorite: Edelgase in Mineralseparaten, Gesamtgesteinen und terrestrischen Karbonaten. (Ph.D. thesis). Johannes Gutenberg-Universität Mainz, p. 143.

Schwenzer, S.P., Ott, U., 2006. Evaluating Kr- and Xe-data in the nakhlites and ALHA 84001—does EFA hide EFM? Lunar Planet. Sci. 37, abstract #1614.

Schwenzer, S.P., Herrmann, S., Ott, U., 2006. Pyroxenes from Governador Valadares and Lafayette: a nitrogen and noble gas study. Lunar Planet. Sci. 37, abstract #1612.

Schwenzer, S.P., Herrmann, S., Mohapatra, R.K., Ott, U., 2007a. Noble gases in mineral separates from three shergottites: Shergotty, Zagami, and EETA79001. Meteorit. Planet. Sci. 42, 387–412.

Schwenzer, S.P., Colindes, M., Herrmann, S., Ott, U., 2007b. Cold desert's fingerprints: terrestrial nitrogen and noble gas signatures, which might be confused with (Martian) meteorites signatures. Lunar Planet. Sci. 38, abstract #1150.

Schwenzer, S.P., Herrmann, S., Ott, U., 2007c. Hot deserts' fingerprints in nitrogen and noble gas budgets of (Martian) meteorites—continued. Lunar Planet. Sci. 38, abstract #1143.

Schwenzer, S.P., Billmeier, U., Schmale, K., Bischoff, A., Ott, U., 2007d. Weathering El Hammadi (H5) in the laboratory—petrography and noble gases. In: 70th Annu. Meteoritical Soc. Meeting, abstract #5030.

Schwenzer, S.P., Fritz, J., Stöffler, D., Trieloff, M., Amini, M., Greshake, A., et al., 2008. Helium loss from Martian meteorites mainly induced by shock metamorphism: evidence from new data and a literature compilation. Meteorit. Planet. Sci. 43, 1841–1859.

Schwenzer, S.P., Herrmann, S., Ott, U., 2012. Noble gas adsorption with and without mechanical stress: not Martian signatures, but fractionated air. Meteorit. Planet. Sci. 47, 1049–1061.

Schwenzer, S.P., Greenwood, R.C., Kelley, S.P., Ott, U., Tindle, A.G., Haubold, R., et al., 2013. Quantifying noble gas contamination during terrestrial alteration in Martian meteorites from Antarctica. Meteorit. Planet. Sci. 48, 929–954.

Stephen, N.R., Ross, A.J., 2014. Examining the petrology of "Martian" meteorite NWA 7034: a polymict fragmental breccia. Lunar Planet. Sci 45, abstract #2924.

Swindle, T.D., 2002. Martian noble gases. In: Porcelli, D., Ballentine, C.J., Wieler, R. (Eds.), Noble gases in Geochemistry and Cosmochemistry. Mineralogical Society of America, Washington, DC, pp. 171–190.

Swindle, T.D., Jones, J.H., 1997. The xenon isotopic composition of the primordial Martian atmosphere: Contributions from solar and fission components. J. Geophys. Res. E102, 1671–1678.

Swindle, T.D., Caffee, M.W., Hohenberg, C.M., 1986. Xenon and other noble gases in shergottites. Geochim. Cosmochim. Acta 50, 1001–1015.

Swindle, T.D., Grier, J.A., Burkland, M.K., 1995. Noble gases in orthopyroxenite ALH84001: A different kind of martian meteorite with an atmospheric signature. Geochim. Cosmochim. Acta 59, 793–801.

Swindle, T.D., Treiman, A.H., Lindstrom, D.J., Burkland, M.K., Cohen, B.A., Grier, J.A., et al., 2000. Noble gases in iddingsite from the Lafayette meteorite: Evidence for liquid water on Mars in the last few hundred million years. Meteorit. Planet. Sci. 35, 107–115.

Swindle, T.D., Thomas, C., Mousis, O., Lunine, J.I., 2009. Incorporation of argon, krypton and xenon into clathrates on Mars. Icarus 203, 66–70.

Tosca, N.J., Knoll, A.H., McLennan, S.M., 2008. Water activity and the challenge for life on early Mars. Science 320, 1204–1207.

Treiman, A.H., 1998. The history of Allan Hills 84001 revised: Multiple shock events. Meteorit. Planet. Sci. 33, 753–764.

Treiman, A.H., 2005. The nakhlite meteorites: augite-rich igneous rocks from Mars. Chem. Erde-Geochem. 65, 203–270.

Turner, G., Knott, S.F., Ash, R.D., Gilmour, J.D., 1997. Ar-Ar chronology of the Martian meteorite ALH84001: Evidence for the timing of the early bombardment of Mars. Geochim. Cosmochim. Acta 61, 3835–3850.

Vogel, N., Heber, V.S., Baur, H., Burnett, D.S., Wieler, R., 2011. Argon, krypton, and xenon in the bulk solar wind as collected by the Genesis mission. Geochim. Cosmochim. Acta 75, 3057–3071.

Walton, E.L., Kelley, S.P., Herd, C.D.K., 2008. Isotopic and petrographic evidence for young Martian basalts. Geochim. Cosmochim. Acta 72, 5819–5837.

Wasson, J.T., Wetherill, G.W., 1979. Dynamical, chemical and isotopic evidence regarding the formation location of asteroids and meteorites. In: Gehrels, T. (Ed.), Asteroids. University of Arizona Press, Tucson, AZ, pp. 926–974.

Wieler, R., 2002. Noble gases in the solar system. In: Porcelli, D., Ballentine, C.J., Wieler, R. (Eds.), Noble Gases in Geochemistry and Cosmochemistry. Mineralogical Society of America, Washington, DC, pp. 21–70.

Wiens, R.C., 1988. Noble gases released by vacuum crushing of EETA 79001 glass. Earth Planet. Sci. Lett. 91, 55–65.

Wiens, R.C., Pepin, R.O., 1988. Laboratory shock emplacement of noble gases, nitrogen, and carbon dioxide into basalt, and implications for trapped gases in shergottite EETA 79001. Geochim. Cosmochim. Acta 52, 295–307.

Wiens, R.C., Becker, R.H., Pepin, R.O., 1986. The case for a martian origin of the shergottites, II. Trapped and indigenous gas components in EETA 79001 glass. Earth Planet. Sci. Lett. 77, 149–158.

Wood, C.A., Ashwal, L.D., 1981. SNC meteorites: Igneous rocks from Mars? Proc. Lunar Planet. Sci. Conf. 12, 1359–1375.

Zahnle, K., Pollack, J.B., Kasting, J.F., 1990. Xenon fractionation in porous planetesimals. Geochim. Cosmochim. Acta 54, 2577–2586.

Zhu, L., Liu, H., Pickard, C.J., Zou, G., Ma, Y., 2014. Reactions of xenon with iron and nickel are predicted in the Earth's inner core. Nat. Chem. 6, 644–648.

HYDROGEN RESERVOIRS IN MARS AS REVEALED BY MARTIAN METEORITES

4

Tomohiro Usui

Tokyo Institute of Technology, Tokyo, Japan

4.1 INTRODUCTION

Mars exploration missions provide compelling evidence for the presence of liquid water during the earliest geologic eras (pre-Noachian and Noachian: ~ 3.7 to 4.5 Ga) of Mars. Observations of widespread fluvial landforms, dense valley networks, evaporites (e.g., gypsum), and hydrous minerals (e.g., clays) that are commonly formed by aqueous processes imply that Mars had an active hydrological cycle with lakes and possibly oceans (Chapters 7 and 8; Carr, 2006; Ehlmann and Edwards, 2014); note that recent climate models suggest that such a watery surface condition should not require a warm climate with a dense atmosphere (e.g., < 260 K at 1 bar; Wordsworth, 2016). In contrast to the ancient watery environment, the present surface of Mars is relatively cold and dry. The recent desert-like surface conditions, however, do not necessarily indicate a lack of surface or near-surface water/ice. Massive deposits of ground ice and/or icy sediments have been proposed based on subsurface radar sounder observations (Mouginot et al., 2012). Furthermore, landforms that appear to be glacial commonly occur even in the equatorial regions; this has led to the hypothesis that, similar to Earth, Mars had recent ice ages in which water-ice would have globally covered the surface (Head et al., 2003, 2005). Hence, accurate knowledge about the evolution of physical state (ice, liquid, and vapor) and global inventory of water is crucial to our understanding of the transition of climate and near-surface environments and habitability on Mars.

The evolution of water reservoirs and their distributions in the atmosphere, hydrosphere (surface water and groundwater), and cryosphere (ground ice and polar ice caps) can be traced by measurement of hydrogen isotope ratios (D/H: deuterium/hydrogen). Hydrogen is a major component of water (H_2O) and its isotopes fractionate during atmospheric escape and hydrological cycling between the atmosphere, surface water, and the polar ice caps. Telescopic studies report that the global mean D/H ratio of the Martian atmosphere is ~ 6 times the terrestrial values ($\delta D = \sim 5000‰$) (Table 4.1); $\delta D = [(D/H)_{sample}/(D/H)_{reference} - 1] \times 1000$, where the reference is standard mean ocean water (SMOW). The high atmospheric D/H ratio results from the preferential loss of hydrogen relative to heavier deuterium from the atmosphere throughout Martian history (Jakosky and Phillips, 2001; Kurokawa et al., 2014; Lammer et al., 2013). While telescopic

Volatiles in the Martian Crust. DOI: http://dx.doi.org/10.1016/B978-0-12-804191-8.00004-0

Table 4.1 Hydrogen Isotopic Compositions of Martian Water Reservoirs

Reservoir	δD (‰, SMOW)[a]	Method/Sample	Reference
Atmosphere			
At Gale crater	4950 ± 1080	SAM[b]/atmospheric air	Webster et al. (2013)
Global average	~5000		Owen et al. (1988)
Seasonal and regional variation	~2000–9000	Telescope/ atmospheric air	Villanueva et al. (2015), Aoki et al. (2015), Novak et al. (2011)
Crustal water			
At present	3870–7010	SAM/soil at Gale	Leshin et al. (2013)
Recent past (< 1.3 Ga)	~4600–6000	SIMS[c]/SNC meteorites	Chen et al. (2015), Hu et al. (2014), Usui et al. (2012, 2015), Greenwood et al. (2008), Boctor et al. (2003), Watson et al. (1994)
Early Mars (~3–4 Ga)	2,000 ± 200 500–3000	SAM/mudstone at Gale SIMS/ALH 84001	Mahaffy et al. (2015) Greenwood et al. (2008), Boctor et al. (2003), Sugiura and Hoshino (2000)
Subsurface reservoir	1000–2000	SIMS/shergottites	Usui et al. (2015)
Water in deep interior			
Shergottite source (enriched)	500–1000	SIMS/Zagami, Shergotty	Watson et al. (1994)
Shergottite source (depleted)	900 ± 250	SIMS/QUE 94201	Leshin (2000)
Shergottite source (depleted)	< 275	SIMS/Y-980459	Usui et al. (2012)
Nakhlite source	− 111 ± 62	SIMS/Nakhla	Hallis et al. (2012)
Chassignite source	95–225	SIMS/Chassigny	Boctor et al. (2003)

[a]*The D/H ratios are given relative to a D/H value for SMOW of 1.5576×10^{-4}.*
[b]*SAM (sample analysis at Mars) is a suite of instruments on the Curiosity rover.*
[c]*Secondary ion mass spectrometry.*

measurements help constrain the hydrogen isotopic composition of the present-day atmosphere, the historical variations of hydrogen reservoirs are potentially recorded in Martian meteorites that formed at different times over the past 0.2–4.4 Ga (Humayun et al., 2013; Lapen et al., 2010; Nyquist et al., 2001).

This chapter summarizes the history of hydrogen isotope studies of Martian meteorites since the first detection of nonterrestrial hydrogen from the meteorites. Contamination of terrestrial water has been a major issue throughout most of this history. This review, in particular, focuses on recent ion microprobe studies that have contributed to the identification of Martian hydrogen reservoirs and have constrained their evolution throughout the planet's history. NASA's Curiosity rover has also provided valuable insights on the water history based on in situ hydrogen isotopic measurements of rocks in Gale crater since 2012. The Curiosity data are briefly mentioned in this chapter but described in detail in Chapter 12, Volatile Detections in Gale Crater Sediment and Sedimentary Rock: Results from the Mars Science Laboratory's Sample Analysis at Mars Instrument.

4.2 LABORATORY ANALYSIS OF HYDROGEN ISOTOPES IN MARTIAN METEORITES

4.2.1 TOWARD THE DETECTION OF MARTIAN HYDROGEN

The study of Martian hydrogen isotopes began in the 1980s. Telescopic observations based on the Fourier transform spectroscopy first resolved Doppler-shifted infrared lines of HDO and H_2O in the Martian atmosphere, yielding the global mean atmospheric D/H ratio of 6 ± 3 (Owen et al., 1988). Five years before this discovery, Bogard and Johnson (1983) reported that gases implanted in shergottites during shock match with those measured for the Martian atmosphere, verifying the hypothesis that the SNC (shergottite, nakhlite, chassignite) meteorites were from Mars (McSween and Stolper, 1979). This result encouraged geochemists to investigate water (as hydrogen) in the SNC meteorites.

Two of the earliest works (Fallick et al., 1983; Yang and Epstein, 1985), however, did not detect unambiguous Martian hydrogen components. They analyzed gasses released by heating of bulk-rock SNC meteorites, but the measured D/H ratios were indistinguishable from terrestrial water. Later work by Kerridge (1988) first reported nonterrestrial D/H ratios: $\delta D = 878‰$ for Shergotty (shergottite) and 456‰ for Lafayette (nakhlite). Kerridge (1988) used larger amounts of samples (e.g., 2 g of Shergotty) than the previous studies and combusted each of them at two different temperatures (e.g., 800 °C and 1050 °C for Shergotty). Kerridge's experiments suggested a possibility that the stepwise heating of bulk meteorites under suitable temperature conditions could differentiate Martian hydrogen from contamination by terrestrial water.

The first systematic D/H measurements of bulk Martian meteorites were conducted by Leshin et al. (1996). This study heated eight meteorites [seven SNCs and Allan Hills (ALH) 84001] and measured the D/H ratios of the released gases in several temperature steps, extending down to room temperature in order to fully monitor the breakdown of hydrous phases. The D/H ratios progressively increase with temperature, suggesting that the water in the samples originated from at least two sources: a terrestrial component released largely at low temperature ($< \sim 200$ °C) and an extraterrestrial component released at high temperature ($> \sim 400$ °C). Leshin et al. (1996) concluded that the variation in δD values of the high-temperature hydrogen ($\sim 250‰ - 900‰$ for the nakhlites, $\sim 1200‰ - 2100‰$ for the shergottites, and $\sim 800‰$ for ALH 84001) could represent true Martian δD variations. However, these high-δD values are still lower than that of the Martian atmosphere ($\sim 5000‰$), whereas some of the samples contain secondary phases (e.g., carbonate) that evidently formed with aqueous crustal fluids that had exchanged with the Martian atmosphere. Therefore Leshin et al. (1996) also proposed another possibility that the δD variation of the high-temperature hydrogen could reflect varying contributions of the terrestrial endmember.

4.2.2 IN SITU HYDROGEN ISOTOPE ANALYSIS BY SECONDARY ION MASS SPECTROMETRY

Even careful stepwise heating experiments did not succeed in extracting the pure Martian atmospheric components from the bulk meteorite samples (Eiler et al., 2002; Kerridge, 1988; Leshin et al., 1996; Yung et al., 1988). Most Martian meteorites contain a variety of postmagmatic phases (e.g., carbonates, sulfates, clays, and iron oxides) that formed due to alteration and weathering processes both on Mars and Earth (Bridges et al., 2001); the gas released at each heating step could be sourced from more than one phase. Since each of the magmatic and post-magmatic phases in the Martian meteorites is expected to contain specific hydrogen components originating from Mars and/or Earth, D/H analysis of individual phases is required to examine the Martian hydrogen reservoirs. Recently, secondary ion mass spectrometry (SIMS, or simply referred to as "ion microprobe") that enables in situ isotope analysis with high spatial resolution (typically $< \sim 20$ μm) has been intensively applied to Martian meteorites.

A pioneering SIMS study by Watson et al. (1994) first reported a δD value (4400‰) equivalent to that of the Martian atmosphere. Watson et al. (1994) conducted in situ measurements of hydrogen isotopes in igneous hydrous phases of apatite, biotite, and kaersutite in SNC meteorites. These phases have variable δD values ranging from ~ 500‰ to 4400‰. This large δD variation was interpreted as indicating an overprinting of magmatic low D/H values by crustal fluids with D/H values possibly as high as that of the present Martian atmosphere. Consequently, Watson et al. (1994) proposed crustal fluids as a new hydrogen reservoir on Mars (Table 4.1); this crustal water reservoir had isotopically exchanged with atmospheric water over geologic time.

More comprehensive SIMS studies were conducted in the early 2000s in order to search for unidentified Martian hydrogen reservoirs (Boctor et al., 2003; Sugiura and Hoshino, 2000). These studies analyzed not only hydrous phases but also nominally anhydrous phases including carbonate, magmatic glass, maskelynite, high-pressure silica polymorph, whitlockite, pyroxene, and olivine, in an orthopyroxenite (ALH 84001), a dunite (Chassigny), and basalts (shergottites: EETA 79001, ALHA 77005, Dag 476, SaU 005, Shergotty, and Zagami). Although these analyses reported greater δD variations (-100‰ to 4100‰) than those of the previous studies (Leshin et al., 1996; Watson et al., 1994), all of these studies reached the same conclusion that there are at least two hydrogen sources in the meteorites: a high-δD (~ 4000‰) component representing Martian crustal fluids and a low-δD component either representing Martian magmatic water or reflecting terrestrial contamination.

The origin of the Martian low-δD component still remains controversial because the actual δD values for primary Martian magmatic fluids are difficult to determine. Magmatic fluid is potentially contained in igneous hydrous phases. Leshin (2000) focused on relatively large grains of apatites (> 100 μm) with high H_2O contents ($\sim 0.2-0.5$ wt%) in order to reduce the effect of contamination during the SIMS analyses. The apatite grains exhibit strong correlation with δD values and water contents (Fig. 4.1). This correlation was explained by two-component mixing with a low-δD endmember of 900‰ $\pm$ 250‰ and an atmospheric high-δD (4200‰) component. Leshin (2000) concluded that this low-δD endmember should represent the magmatic water reservoir for shergottites because it is evidently nonterrestrial. However, Greenwood et al. (2008) argued that this correlation (Fig. 4.1) may be attributable to the mixing of terrestrial water during the analyses, despite the use of large apatite grains with high water contents. Using a SIMS equipped with a CMOS image sensor, Greenwood et al. (2008) produced 2D maps of hydrogen and deuterium ions for apatites in shergottites and demonstrated that microfractures within apatite grains have significant terrestrial contamination (Fig. 4.2). The 2D ion images were used to find areas for spot analysis free of microfractures. Contrary to Leshin's results, no correlations between δD values and water contents were observed for microfracture-free apatites (Greenwood et al., 2008).

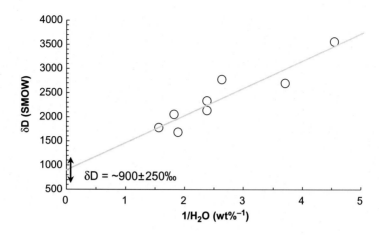

FIGURE 4.1

δD versus $1/H_2O$ mixing diagram for apatites in shergottite QUE 94001. Mixing of two components should produce a linear trend in the (δD vs $1/H_2O$) diagram assuming that the abundance of the minor isotope D is negligible relative to the major isotope H ($D/H = \sim 10^{-4}$). The intercept of the regression line indicates a δD value of the magmatic water component. *Modified after Leshin, 2000.*

Recent technical developments of ion microprobe analyses identified potential sources of contaminants and provided means to minimize the effect of contamination by terrestrial water (e.g., Greenwood et al., 2008; Hu et al., 2015; Usui et al., 2012). Sources of contaminants include air left in the vacuum system, oils (and/or water) used as lubricants during polishing, and epoxy (or acrylic) resin used as a mounting agent. Among them, resins were major unavoidable blank sources for Martian meteorites that were highly shocked (typically to > 20 GPa, Nyquist et al., 2001) and as a result have many microfractures; resins penetrate into these fractures and cannot be removed. While Greenwood et al. (2008) intended to avoid the microfractures by using the isotope imaging system (Fig. 4.2), Usui et al. (2015a, 2012) improved sample preparation using hydrogen-free indium metal instead of petrochemical resin to reduce the contamination. Recent SIMS studies have provided less-contaminated and well-evaluated D/H ratios of hydrogen sources in Martian meteorites. These datasets are accurate enough to identify separate Martian hydrogen reservoirs (Chen et al., 2015; Giesting et al., 2015; Greenwood et al., 2008; Hallis et al., 2012a, 2012b; Hu et al., 2014; Usui et al., 2012, 2015a). The origin and evolution of Martian hydrogen reservoirs are discussed based on these latest results in Section 4.3.

4.3 ORIGIN AND EVOLUTION OF HYDROGEN RESERVOIRS ON MARS

Three hydrogen reservoirs are now identified based on the analyses of Martian meteorites, telescopic observations, and Curiosity measurements: primordial water, atmospheric water, and crustal water (Table 4.1, Fig. 4.3). The primordial water is retained in the mantle and has a D/H ratio similar to those seen in Martian building blocks. The atmospheric water is interpreted to have

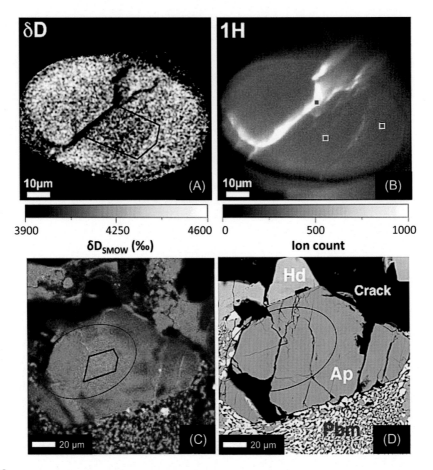

FIGURE 4.2

Ion and electron images of an apatite grain in shergottite Los Angeles (A) δD image. (B) ^{1}H image. Squares (2 μm across) indicate areas for SIMS spot analyses. (C) CL image with the area of (A) and (B) outlined with the black ellipse. The area of highest CL intensity is outlined with *black lines*. (D) Back-scattered electron image. *Ap*, apatite; *Hd*, hedenbergite; *Pbm*, pyroxferroite breakdown material. *Modified after Greenwood et al. (2008).*

progressively increased through the planet's history to reach the present-day mean value of ~5000‰. The crustal water reservoir is still enigmatic; this reservoir exhibits variable *D/H* ratios (~1000‰−7000‰), representing both local and historical variations.

4.3.1 PRIMORDIAL WATER

Mars is thought to represent a planetary embryo, which underwent differentiation during an early magma ocean stage, but did not experience later crust-mantle recycling or global melting events that significantly remixed mantle reservoirs or crustal components (e.g., Brandon et al., 2000;

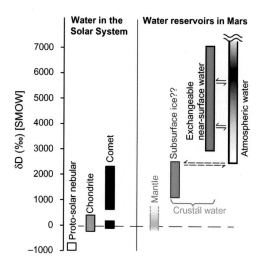

FIGURE 4.3

Hydrogen isotopic compositions of water in the Solar System and Martian water reservoirs. *Modified after Usui et al. (2015a).*

Dauphas and Pourmand, 2011; Debaille et al., 2007; Foley et al., 2005; Kleine et al., 2009). This suggests that water accreted during planetary formation should have been retained in the mantle. The *D/H* ratio of the Martian primordial water was estimated based on analyses of Martian meteorites that were thought to preserve geochemical signatures of their mantle sources.

A latest estimate for the primordial δD value ($<275‰$) was obtained from olivine-hosted melt inclusions in the most magnesian olivine-phyric shergottite Yamato (Y-) 980459 (Usui et al., 2012). This meteorite represents a parental melt from a geochemically depleted mantle source, and its melt inclusions are interpreted to possess undegassed water in the source mantle. The shergottite parental melt may not represent the *D/H* ratio of the Martian primordial mantle. Geochemical studies suggest that the depleted shergottite source was formed by a deep magma ocean at ~4.5 Ga, followed by an extraction of low-degree partial melt prior to the main shergottite magmatic events at ~0.2–0.6 Ga (e.g., Brandon et al., 2000; Dauphas and Pourmand, 2011; Debaille et al., 2007). These magmatic processes might have lost hydrogen and possibly fractionated (i.e., elevated) *D/H* ratio in the source mantle. Therefore Usui et al. (2012) concluded that the δD value (275‰) for the depleted shergottite mantle source yields the maximum estimate for the true primordial water that was originally derived from the Martian building blocks (Table 4.1).

Another δD estimate (900‰ ± 250‰) for the shergottite source was obtained based on the analyses of apatites from another shergottite Queen Alexandra Range (QUE) 94201 (Leshin, 2000). This value is distinctly higher than the maximum δD estimate (275‰) by Usui et al. (2012). However, the higher δD estimate of QUE 94201 was obtained by the extrapolation analysis of apatites (Fig. 4.1). Aforementioned in Section 4.2, the δD values of apatites may be disturbed by terrestrial water (Greenwood et al., 2008), which may change the slope of the regression line. Since QUE 94201 is one of the few meteorites representing magmatic liquids from the geochemically depleted source (McSween et al., 1996), these meteorites should be revisited by modern techniques.

Radiogenic isotope studies suggest that the shergottite source mantle is distinct from the chassignite−nakhlite source (e.g., Brandon et al., 2000). Chassigny has solar Xe isotopic ratios that are distinct from those in the Martian atmosphere and are therefore inferred to be a mantle signature (Swindle and Jones, 1997). Ion microprobe analyses of olivine-hosted melt inclusions in Chassigny showed low-δD values (95‰−225‰, Boctor et al., 2003); note that these values would need to be revisited because Chassigny melt inclusions were not evaluated for the presence of possibly contaminating cracks or epoxy. Recent analyses of apatites in Nakhla provided a δD value (-111‰ ± 62‰) of water in the nakhlite source (Hallis et al., 2012a). The δD values of chassignite (95‰−225‰) and nakhlite (-111‰ ± 62‰) sources are indistinguishable from the δD value of the shergottite source (<275‰) (Table 4.1), although their source mantles are chemically distinct regarding the Re−Os, Hf−W, and Sm−Nd systems (e.g., Brandon et al., 2000). Collectively, the recent hydrogen isotope studies of the SNC suite indicated that Martian primordial water has an Earth-like and carbonaceous chondrite-like D/H ratio (-111‰ to 275‰) (Table 4.1, Fig. 4.3; Alexander et al., 2012).

The origins of water in the terrestrial planets are still debated. Three extreme cases are envisioned: (1) wet accretion of chondrite-like materials, (2) dry accretion followed by the addition of volatile-rich materials (e.g., comets), or (3) adsorption of proto-solar nebular during the accretion (e.g., Albarede, 2009; Drake, 2005; Genda and Ikoma, 2008; Marty, 2012). The near-chondritic δD value for the Martian primordial water is not consistent with either a cometary origin from Oort cloud ($\delta D = 605$‰−1625‰) (Balsiger et al., 1995; Bockelée-Morvan et al., 1998; Eberhardt et al., 1995; Meier et al., 1998) or the proto-solar nebular origin ($\delta D = -870$‰) (Fig. 4.3; Geiss and Gloeckler, 1998). Alternatively, many carbonaceous chondrite-like materials match the source signature of primordial Martian water (cf., CM: ~ -230‰ to 340‰; CR: ~ 600‰ to 750‰; CI: ~ 80‰; CV: 14‰; CO: 50‰) (Alexander et al., 2012). Another possibility for the source of water for Mars is a Jupiter-family such as 103P/Hartley ice, which has a chondritic D/H ratio (33.6‰) (Hartogh et al., 2011). However, terrestrial planets would not have accreted only cometary ice but entire (i.e., bulk) comets that include deuterium-rich organic materials. Because the bulk hydrogen isotopic composition of a comet is likely to be significantly more deuterium-rich than its water, even comets such as Hartley-2 may not be a viable source of Martian water (Alexander et al., 2012). Furthermore, distinctly higher nitrogen isotope ratios (~ 700‰−1000‰) of Oort cloud and Jupiter-family comets than those of most carbonaceous chondrites and terrestrial mantles (~ 0‰) also discount the possibility of cometary origin for volatiles (Marty, 2012). Therefore Mars and Earth (and possibly the other terrestrial planets within the inner Solar System) could have accreted water from sources with similar chondritic D/H ratios (Hallis et al., 2012a; Usui et al., 2012).

4.3.2 ATMOSPHERIC WATER

The hydrogen isotopic composition of the "present-day" atmospheric water is the most reliable dataset among the Martian hydrogen reservoirs due to measurements by ground-based and space telescopes (Aoki et al., 2015; Krasnopolsky et al., 1998; Novak et al., 2011; Owen et al., 1988; Villanueva et al., 2015). These measurements yield a global mean D/H ratio of ~ 5000‰. Moreover, recent telescopic measurements have revealed a strong regional and seasonal variation in the hydrogen isotopic composition of atmospheric water, typically ranging from low δD (~ 2000‰) near the polar regions in the winter hemisphere to high δD (~ 6000‰−9000‰) near the equatorial regions in the spring hemisphere (Fig. 4.4; Aoki et al., 2015; Villanueva

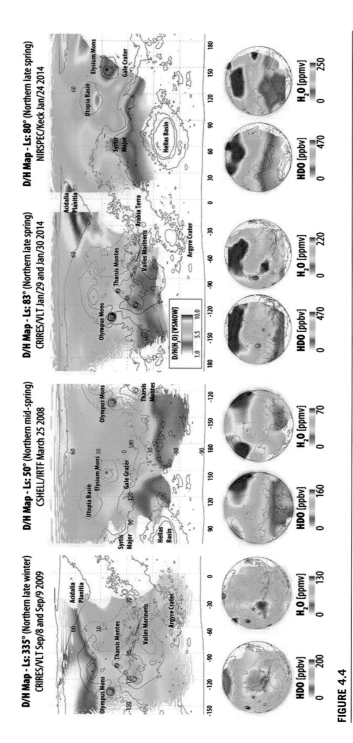

FIGURE 4.4

Maps of HDO and H_2O, and their ratio relative to VSMOW on Mars at four different seasons. *Modified after Villanueva et al. (2015).*

et al., 2015). The D/H maps (Fig. 4.4) are further corroborated by the in situ D/H measurement by Curiosity at Gale crater (4950‰ ± 1080‰) in late 2012 (see Chapter 12, Volatile Detections in Gale Crater Sediment and Sedimentary Rock: Results from the Mars Science Laboratory's Sample Analysis at Mars Instrument for details about the Curiosity measurements). This temporal and spatial D/H variation is explained by the interplay between local reservoirs (e.g., ice cap) with different D/H ratios and water-ice abundances (Fisher et al., 2008; Fisher, 2007).

The atmospheric D/H values in the recent past are approximated by that of the polar ice, because the polar ice including polar-layered deposits (PLD) dominate the inventory of surface hydrogen reservoirs: cf., total water volume in the polar ice = 20−30 mGEL (global equivalent layer) (Kurokawa et al., 2014) and the precipitable water column of the present-day atmosphere ≲ 50 μm (Fisher, 2007). Major fractions of polar ice should have been repeatedly vaporized and redeposited due to the significant obliquity variations over the past 10 Myr (0°−60°, Laskar and Robutel, 1993; Head et al., 2003, 2005). Consequently, all near-surface and polar ices should repeatedly interact with the atmosphere and thus share a common D/H ratio. Based on the D/H maps (Fig. 4.4) and simulation results from Martian General Circulation Model (MGCM) (Montmessin et al., 2005), the D/H ratio of the bulk polar ice was estimated to be ∼7000‰, representing an atmospheric D/H ratio in the recent past (Villanueva et al., 2015).

Hydrogen isotopic compositions of the ancient atmosphere are far more difficult to determine. The atmosphere should have exchanged with the topmost surficial water (or ice) reservoirs (Fisher et al., 2008; Fisher, 2007; Kurokawa et al., 2016; Montmessin et al., 2005). Thus the D/H evolution of the Martian atmosphere can be traced by the isotopic evolution of crustal fluids, which is discussed below.

4.3.3 CRUSTAL WATER

Martian meteorites are all igneous rocks (except for polymict breccia NWA 7034/7533) that were either extruded onto the surface as lava flows or solidified as magma bodies in the shallow crust (McSween and McLennan, 2013). Thus they potentially contain not only their original magmatic water but also surficial water components that could have been incorporated during and/or after the magma solidification. Since magmatic water has relatively low-δD values (e.g., < 275‰ for shergottite and −111‰ for nakhlite, Hallis et al., 2012a; Usui et al., 2012), the high-δD endmember components obtained from the meteorites are interpreted as representing the surficial water. SIMS analyses of apatites in shergottites, Shergotty and Los Angeles with young crystallization ages of ∼0.2 Ga, yielded a δD value of ∼4600‰ for a crustal water reservoir in the recent past (Greenwood et al., 2008). Moreover, a D/H zoning profile of a hydrated glass inclusion in shergottite GRV 020090 with a crystallization age of ∼0.2 Ga provided evidence of crustal fluid with a δD value of 6034‰ ± 72‰ (Fig. 4.5; Hu et al., 2014). These values (∼4600‰ and 6034‰) are consistent with in situ analyses of aeolian fines by Curiosity (3870‰−7010‰, Leshin et al., 2013) and are indistinguishable from the δD range of the atmosphere. These observations suggest that the near-surface water should have been isotopically exchanged with the atmospheric water since at least ∼0.2 Ga (Fig. 4.3).

In addition to the "exchangeable" surface water reservoir, Usui et al. (2015a) proposed an "isolated" water reservoir with an intermediate-δD range of ∼1000‰−2000‰, based on SIMS analyses of impact glasses in shergottites; these impact glasses are known to have trapped surficial and atmospheric components (Bogard and Johnson, 1983; Rao et al., 2011). This

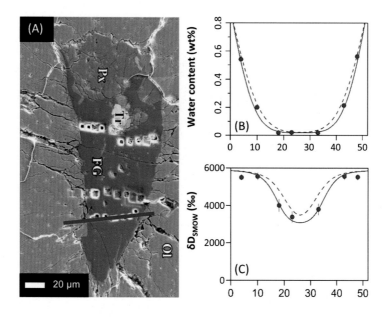

FIGURE 4.5

(A) SEM image of olivine-hosted glass inclusion in shergottite GRV 020090. Zoning profiles of (B) water contents and (C) δD values in the melt inclusion are also shown. The position of the profiles is shown as a *red line* in (A). *Modified after Hu et al. (2014).*

intermediate-δD range is consistent with that of Hesperian-aged clay minerals analyzed by Curiosity (2000‰ ± 200‰, Mahaffy et al., 2015) but is distinct from the low-δD primordial and the high-δD atmospheric water reservoirs. Usui et al. (2015a) implied that the intermediate-δD reservoir represents either hydrated crust and/or ground ice interbedded within sediments, which has existed relatively intact over geologic time (Fig. 4.6). The hydrated crustal materials and/or ground ice could have possibly acquired its intermediate-δD composition from the ancient surface water before the rise of the atmospheric D/H ratio to the present level of ~5000‰ (Fig. 4.7, Usui et al., 2015b).

The hydrogen isotopic composition of ancient surface water has been poorly constrained. Carbonates in ALH 84001 are the most promising containers of the ancient water component because they formed from fluid that was closely associated with the Noachian atmosphere (Borg and Drake, 2005; Niles et al., 2013). Previous SIMS analyses of ALH 84001 carbonates showed a large δD variation ranging from ~500‰ to 2000‰ (Boctor et al., 2003; Sugiura and Hoshino, 2000), yet these datasets would need to be revisited because the carbonates were not evaluated for the presence of possibly contaminating cracks or epoxy. On the other hand, an analysis of magmatic apatite in ALH 84001 provided a δD value of ~3000‰ (Greenwood et al., 2008). This value is apparently lower than those of the present-day atmosphere and the exchangeable surface water (~4000‰−7000‰) (Table 4.1). The δD increase from ~3000‰ to ~4000‰−7000‰ may indicate

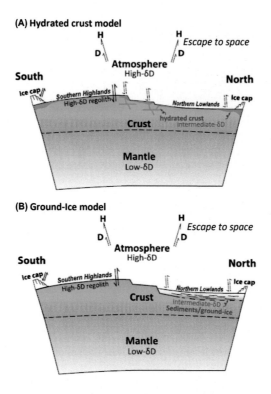

FIGURE 4.6

Schematic cross sections illustrating locations of the Martian water reservoirs based on the models of (A) hydrated crust and (B) ground ice. The intermediate-δD water/ice reservoir occurs as (A) hydrous phases such as clays and amphiboles formed by surface weathering and/or subsurface hydrothermal groundwater circulation or (B) ground-ice deposits interbedded within sediments in the northern lowlands. These models require limited interaction between the buried water/ice reservoirs and the atmosphere (shown as *dashed arrows*). *Modified after Usui et al. (2015a).*

the progressive atmospheric loss during the past 4 Ga (Fig. 4.7), although it is still uncertain whether the magmatic apatite represents the pure Noachian atmosphere.

Based on the *D/H* evolution for the exchangeable surface water, the loss of water due to atmospheric escape was estimated (Fig. 4.8). A box model calculation indicated that the water loss before 4 Ga (>41−99 mGEL) was more significant than in the rest of Martian history (>10−53 mGEL) (Kurokawa et al., 2014). These estimates were recently updated based on the new estimate for the representative atmospheric δD value of ~7000‰, yielding ~40% increases from the previous estimates (Villanueva et al., 2015). The amount of water loss estimated by the *D/H* evolution are significantly smaller than the size of putative paleo-oceans estimated by shoreline demarcation studies (~100−1000 mGEL; e.g., Fig. 4.8; Carr and Head, 2003; Di Achille and

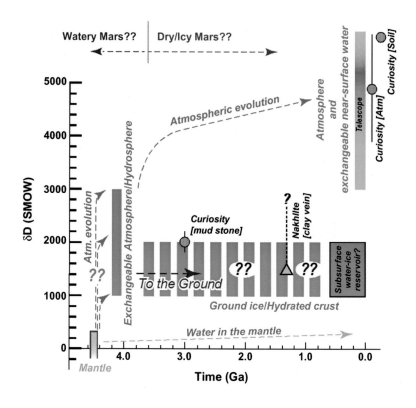

FIGURE 4.7

Evolution of hydrogen isotopic compositions of Martian water reservoirs. See text for discussion. *Modified after Usui et al. (2015a).*

Hynek, 2010; Head et al., 1999; Ormö et al., 2004). Granted the existence of these putative paleo-oceans, the difference in water volume suggests the existence of undetected subsurface water/ice reservoirs (Fig. 4.6; Usui et al., 2015a) and corroborates the hypothesis that a buried cryosphere accounts for a large part of the initial water budget of Mars (Clifford and Parker, 2001).

4.4 FUTURE PERSPECTIVE ON METEORITE STUDY

Hydrogen isotopes have the potential to be a powerful geochemical tracer for the evolution of water in Martian atmosphere, hydrosphere, and cryosphere. To maximize the potential scientific merits, accurate knowledge about the *D/H* evolution of individual Martian water reservoirs is crucial. As described earlier, *D/H* ratios of ancient water reservoirs are poorly constrained. This is mainly because the SNC meteorites formed relatively recently (< 1.3 Ga, Nyquist et al., 2009) and because there have been limited *D/H* datasets for the 4-Ga-old ALH 84001 meteorite (Greenwood et al., 2008) and polymict

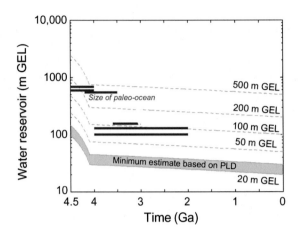

FIGURE 4.8

Evolution of water reservoirs for different amounts of present water reservoirs (*gray dotted lines*: 20, 30, 50, 100, 200, and 500 mGEL) and geological estimates on the size of paleo-oceans (*dark-blue horizontal bar*). The *light-blue* area indicates the evolution of surface water reservoir calculated based on the minimum present water reservoir (20−30 mGEL) estimated from PLD. *Modified after Kurokawa et al. (2014).*

breccia NWA 7034/7533, which contains a Noachian-aged clast (Agee et al., 2013). Along with the scarcity of ancient meteorite samples, technical issues have not been fully resolved. Each hydrogen isotopic composition of a phase in a meteorite reflects a mixture of more than two hydrogen sources (e.g., magmatic and surficial waters), which cannot be completely differentiated even by high-spatial resolution in situ SIMS analysis. Alternatively, the mixing ratio between the hydrogen components in the target phase should be independently determined on the basis of other geochemical tracers such as carbon and oxygen isotopes. Furthermore, the disturbance of hydrogen isotopic composition in the aftermath of impact-induced shock was recently reported, which includes both loss of original water and later uptake of an atmospheric high-*D/H* component (Giesting et al., 2015). Such shock effects should be carefully evaluated based on petrographic and mineralogical observations (e.g., occurrence of high-pressure minerals, zoning profiles of *D/H* ratio, and volatile abundances). Future investigations by well-designed, systematic volatile isotope studies on the Martian meteorites including NWA 7034/7533 and ALH 84001 will help identify the hydrogen isotopic compositions of the ancient water reservoirs on Mars.

REFERENCES

Agee, C.B., Wilson, N.V., McCubbin, F.M., Ziegler, K., Polyak, V.J., Sharp, Z.D., et al., 2013. Unique meteorite from early Amazonian Mars: water-rich basaltic breccia Northwest Africa 7034. Science 339, 780−785.
Albarede, F., 2009. Volatile accretion history of the terrestrial planets and dynamic implications. Nature 461, 1227−1233.

Alexander, C.M.O.D., Bowden, R., Fogel, M.L., Howard, K.T., Herd, C.D.K., Nittler, L.R., 2012. The provenances of asteroids, and their contributions to the volatile inventories of the terrestrial planets. Science 337, 721−723.

Aoki, S., Nakagawa, H., Sagawa, H., Giuranna, M., Sindoni, G., Aronica, A., et al., 2015. Seasonal variation of the HDO/H_2O ratio in the atmosphere of Mars at the middle of northern spring and beginning of northern summer. Icarus 260, 7−22.

Balsiger, H., Altwegg, K., Geiss, J., 1995. *D/H* and $^{18}O/^{16}O$ ratio in the hydronium ion and in neutral water from in situ ion measurements in comet Halley. J. Geophys. Res. 100, 5827−5834.

Bockelée-Morvan, D., Gautier, D., Lis, D.C., Young, K., Keene, J., Phillips, T., et al., 1998. Deuterated water in comet C/1996 B2 (Hyakutake) and its implications for the origin of comets. Icarus 133, 147−162.

Boctor, N.Z., Alexander, C.M.O.D., Wang, J., Hauri, E., 2003. The sources of water in Martian meteorites: clues from hydrogen isotopes. Geochim. Cosmochim. Acta 67, 3971−3989.

Bogard, D.D., Johnson, P., 1983. Martian gases in an Antarctic meteorite? Science 221, 651−654.

Borg, L., Drake, M.J., 2005. A review of meteorite evidence for the timing of magmatism and of surface or near-surface liquid water on Mars. J. Geophys. Res. 110, Available from: http://dx.doi.org/10.1029/2005JE002402.

Brandon, A.D., Walker, R.J., Morgan, J.W., Goles, G.G., 2000. Re−Os isotopic evidence for early differentiation of the Martian mantle. Geochim. Cosmochim. Acta 64, 4083−4095.

Bridges, J.C., Catling, D., Saxton, J., Swindle, T., Lyon, I., Grady, M., 2001. Alteration assemblages in Martian meteorites: implications for near-surface processes. Chronology and Evolution of Mars. Springer, New York, pp. 365−392.

Carr, M.H., 2006. The Surface of Mars. Cambridge University Press, Cambridge, p. 322.

Carr, M.H., Head, J.W., 2003. Oceans on Mars: an assessment of the observational evidence and possible fate. J. Geophys. Res.: Planets 108. Available from: http://dx.doi.org/10.1029/2002JE001963.

Chen, Y., Liu, Y., Guan, Y., Eiler, J.M., Ma, C., Rossman, G.R., et al., 2015. Evidence in Tissint for recent subsurface water on Mars. Earth Planet. Sci. Lett. 425, 55−63.

Clifford, S.M., Parker, T.J., 2001. The evolution of the Martian hydrosphere: implications for the fate of a primordial ocean and the current state of the northern plains. Icarus 154, 40−79.

Dauphas, N., Pourmand, A., 2011. Hf-W-Th evidence for rapid growth of Mars and its status as a planetary embryo. Nature 473, 489−492.

Debaille, V., Brandon, A.D., Yin, Q.Z., Jacobsen, B., 2007. Coupled $^{142}Nd−^{143}Nd$ evidence for a protracted magma ocean in Mars. Nature 450, 525−528.

Di Achille, G., Hynek, B.M., 2010. Ancient ocean on Mars supported by global distribution of deltas and valleys. Nat. Geosci. 3, 459−463.

Drake, M.J., 2005. Origin of water in the terrestrial planets. Meteorit. Planet. Sci. 40, 519−527.

Eberhardt, P., Reber, M., Krankowsky, D., Hodges, R., 1995. The *D/H* and $^{18}O/^{16}O$ ratios in water from comet P/Halley. Astron. Astrophys. 302, 301−316.

Ehlmann, B.L., Edwards, C.S., 2014. Mineralogy of the Martian surface. Annu. Rev. Earth Planet. Sci. 42, 291−315.

Eiler, J.M., Kitchen, N., Leshin, L., Strausberg, M., 2002. Hosts of hydrogen in Allan Hills 84001: evidence for hydrous Martian salts in the oldest Martian meteorite? Meteorit. Planet. Sci. 37, 395−405.

Fallick A., Hinton R., Mattey D., Norris S., Pillinger C., Swart P., et al., 1983. No unusual compositions of the stable isotopes of nitrogen, carbon and hydrogen in SNC meteorites. In: Lunar and Planetary Science Conference, pp. 183−184.

Fisher, D., Novak, R., Mumma, M.J., 2008. *D/H* ratio during the northern polar summer and what the Phoenix mission might measure. J. Geophys. Res. 113. Available from: http://dx.doi.org/10.1029/2007JE002972.

Fisher, D.A., 2007. Mars' water isotope (*D/H*) history in the strata of the North Polar Cap: inferences about the water cycle. Icarus 187, 430−441.

Foley, C.N., Wadhwa, M., Borg, L., Janney, P., Hines, R., Grove, T., 2005. The early differentiation history of Mars from 182 W−142 Nd isotope systematics in the SNC meteorites. Geochim. Cosmochim. Acta 69, 4557−4571.

Geiss, J., Gloeckler, G., 1998. Abundances of deuterium and helium-3 in the protosolar cloud. Space Sci. Rev. 84, 239−250.

Genda, H., Ikoma, M., 2008. Origin of the ocean on the Earth: early evolution of water *D/H* in a hydrogen-rich atmosphere. Icarus 194, 42−52.

Giesting, P.A., Schwenzer, S.P., Filiberto, J., Starkey, N.A., Franchi, I.A., Treiman, A.H., et al., 2015. Igneous and shock processes affecting chassignite amphibole evaluated using chlorine/water partitioning and hydrogen isotopes. Meteorit. Planet. Sci. 50, 433−460.

Greenwood, J.P., Itoh, S., Sakamoto, N., Vicenzi, E., Yurimoto, H., 2008. Hydrogen isotope evidence for loss of water from Mars through time. Geophys. Res. Lett. 35, L05203.

Hallis, L., Taylor, G., Nagashima, K., Huss, G., 2012a. Magmatic water in the Martian meteorite Nakhla. Earth Planet. Sci. Lett. 359, 84−92.

Hallis, L., Taylor, G., Nagashima, K., Huss, G., Needham, A., Grady, M., et al., 2012b. Hydrogen isotope analyses of alteration phases in the nakhlite Martian meteorites. Geochim. Cosmochim. Acta 97, 105−119.

Hartogh, P., Lis, D.C., Bockelee-Morvan, D., de Val-Borro, M., Biver, N., Kuppers, M., et al., 2011. Ocean-like water in the Jupiter-family comet 103P/Hartley 2. Nature 478, 218−220.

Head, J.W., Neukum, G., Jaumann, R., Hiesinger, H., Hauber, E., Carr, M., et al., 2005. Tropical to mid-latitude snow and ice accumulation, flow and glaciation on Mars. Nature 434, 346−351.

Head, J.W., Hiesinger, H., Ivanov, M.A., Kreslavsky, M.A., Pratt, S., Thomson, B.J., 1999. Possible ancient oceans on Mars: evidence from Mars Orbiter Laser Altimeter data. Science 286, 2134−2137.

Head, J.W., Mustard, J.F., Kreslavsky, M.A., Milliken, R.E., Marchant, D.R., 2003. Recent ice ages on Mars. Nature 426, 797−802.

Hu, S., Lin, Y., Zhang, J., Hao, J., Feng, L., Xu, L., et al., 2014. NanoSIMS analyses of apatite and melt inclusions in the GRV 020090 Martian meteorite: hydrogen isotope evidence for recent past underground hydrothermal activity on Mars. Geochim. Cosmochim. Acta 140, 321−333.

Hu, S., Lin, Y., Zhang, J., Hao, J., Yang, W., Deng, L., 2015. Measurements of water content and *D/H* ratio in apatite and silicate glasses using a NanoSIMS 50L. J. Anal. Atom. Spectrom. 30, 967−978.

Humayun, M., Nemchin, A., Zanda, B., Hewins, R., Grange, M., Kennedy, A., et al., 2013. Origin and age of the earliest Martian crust from meteorite NWA 7533. Nature 503, 513−516.

Jakosky, B.M., Phillips, R.J., 2001. Mars' volatile and climate history. Nature 412, 237−244.

Kerridge J., 1988. Deuterium in Shergotty and Lafayette (and on Mars?). In: Lunar and Planetary Science Conference, p. 599.

Kleine, T., Touboul, M., Bourdon, B., Nimmo, F., Mezger, K., Palme, H., et al., 2009. Hf-W chronology of the accretion and early evolution of asteroids and terrestrial planets. Geochim. Cosmochim. Acta 73, 5150−5188.

Krasnopolsky, V.A., Mumma, M.J., Gladstone, G.R., 1998. Detection of atomic deuterium in the upper atmosphere of Mars. Science 280, 1576−1580.

Kurokawa, H., Sato, M., Ushioda, M., Matsuyama, T., Moriwaki, R., Dohm, J.M., et al., 2014. Evolution of water reservoirs on Mars: constraints from hydrogen isotopes in Martian meteorites. Earth Planet. Sci. Lett. 394, 179−185.

Kurokawa, H., Usui, T., Sato, M., 2016. Interactive evolution of multiple water-ice reservoirs on Mars: insight from hydrogen isotope compositions. Geochem. J. 50, 67−79.

Lammer, H., Chassefière, E., Karatekin, Ö., Morschhauser, A., Niles, P.B., Mousis, O., et al., 2013. Outgassing history and escape of the Martian atmosphere and water inventory. Space Sci. Rev. 174, 113−154.

Lapen, T.J., Righter, M., Brandon, A.D., Debaille, V., Beard, B.L., Shafer, J.T., et al., 2010. A younger age for ALH84001 and its geochemical link to shergottite sources in Mars. Science 328, 347−351.

Laskar J. and Robutel P. 1993. The chaotic obliquity of the planets. Nature 362:608−612. Available from: http://www.nature.com/nature/journal/v361/n6413/abs/361608a0.html?foxtrotcallback=true.

Leshin, L., Mahaffy, P., Webster, C., Cabane, M., Coll, P., Conrad, P., et al., 2013. Volatile, isotope, and organic analysis of Martian fines with the Mars Curiosity rover. Science 341, 1238937.

Leshin, L.A., 2000. Insights into Martian water reservoirs from analyses of Martian meteorite QUE94201. Geophys. Res. Lett. 27, 2017−2020.

Leshin, L.A., Epstein, S., Stolper, E.M., 1996. Hydrogen isotope geochemistry of SNC meteorites. Geochim. Cosmochim. Acta 60, 2635−2650.

Mahaffy, P.R., Webster, C.R., Stern, J.C., Brunner, A.E., Atreya, S.K., Conrad, P.G., MSL Science Team, et al., 2015. The imprint of atmospheric evolution in the D/H of Hesperian clay minerals on Mars. Science 347, 412−414.

Marty, B., 2012. The origins and concentrations of water, carbon, nitrogen and noble gases on Earth. Earth Planet. Sci. Lett. 313−314, 56−66.

McSween, H.Y., McLennan, S.M., 2013. Mars. In: Holland, H.D., Turekian, K.K. (Eds.), Treatise on Geochemistry. Elsevier, Oxford, pp. 251−300.

McSween, H.Y., Stolper, E., 1979. Basaltic meteorites. Sci. Am. 242, 54−63.

McSween, H.Y., Eisenhour, D.D., Taylor, L.A., Wadhwa, M., Crozaz, G., 1996. QUE 94201 shergottite: crystallization of a Martian basaltic magma. Geochim. Cosmochim. Acta 60, 4563−4569.

Meier, R., Owen, T.C., Matthews, H.E., Jewitt, D.C., Bockelée-Morvan, D., Biver, N., et al., 1998. A determination of the HDO/H_2O ratio in comet C/1995 O1 (Hale-Bopp). Science 279, 842−844.

Montmessin, F., Fouchet, T., Forget, F., 2005. Modeling the annual cycle of HDO in the Martian atmosphere. J. Geophys. Res.: Planets 110. Available from: http://dx.doi.org/10.1029/2004JE002357.

Mouginot, J., Pommerol, A., Beck, P., Kofman, W., Clifford, S.M., 2012. Dielectric map of the Martian northern hemisphere and the nature of plain filling materials. Geophys. Res. Lett. 39, L02202.

Niles, P.B., Catling, D.C., Berger, G., Chassefière, E., Ehlmann, B.L., Michalski, J.R., et al., 2013. Geochemistry of carbonates on Mars: implications for climate history and nature of aqueous environments. Space Sci. Rev. 174, 301−328.

Novak, R.E., Mumma, M.J., Villanueva, G.L., 2011. Measurement of the isotopic signatures of water on Mars; implications for studying methane. Planet. Space Sci. 59, 163−168.

Nyquist, L.E., Bogard, D.D., Shih, C.-Y., Greshake, A., Stoffler, D., Eugster, O., 2001. Age and geologic histories of Martian meteorites. Space Sci. Rev. 96, 105−164.

Nyquist, L.E., Bogard, D.D., Shih, C.Y., Park, J., Reese, Y.D., Irving, A.J., 2009. Concordant Rb−Sr, Sm−Nd, and Ar−Ar ages for Northwest Africa 1460: a 346 Ma old basaltic shergottite related to "lherzo-litic" shergottites. Geochim. Cosmochim. Acta 73, 4288−4309.

Ormö, J., Dohm, J.M., Ferris, J.C., Lepinette, A., Fairén, A.G., 2004. Marine-target craters on Mars? An assessment study. Meteorit. Planet. Sci. 39, 333−346.

Owen, T., Maillard, J.P., de Bergh, C., Lutz, B.L., 1988. Deuterium on Mars: the abundance of HDO and the value of D/H. Science 240, 1767−1770.

Rao, M.N., Nyquist, L.E., Bogard, D.D., Garrison, D.H., Sutton, S.R., Michel, R., et al., 2011. Isotopic evidence for a Martian regolith component in shergottite meteorites. J. Geophys. Res.: Planets 116, E08006.

Sugiura, N., Hoshino, H., 2000. Hydrogen-isotopic compositions in Allan Hills 84001 and the evolution of the Martian atmosphere. Meteorit. Planet. Sci. 35, 373−380.

Swindle, T.D., Jones, J.H., 1997. The xenon isotopic composition of the primordial Martian atmosphere: contributions from solar and fission components. J. Geophys. Res. 102, 1671−1678.

Usui, T., Alexander, C.M.O.D., Wang, J., Simon, J.I., Jones, J.H., 2012. Origin of water and mantle−crust interactions on Mars inferred from hydrogen isotopes and volatile element abundances of olivine-hosted melt inclusions of primitive shergottites. Earth Planet. Sci. Lett. 357−358, 119−129.

Usui, T., Alexander, C.M.D., Wang, J., Simon, J.I., Jones, J.H., 2015a. Meteoritic evidence for a previously unrecognized hydrogen reservoir on Mars. Earth Planet. Sci. Lett. 410, 140−151.

Usui T., Simon J.I., Jones J.H., Kurokawa H., Sato M., Alexander C.M.O.D., et al., 2015b. Hydrogen isotopes record the history of the Martian hydrosphere and atmosphere. In: Lunar Planetary Science Conference, XLVI: Abstract #1593.

Villanueva, G.L., Mumma, M.J., Novak, R.E., Käufl, H.U., Hartogh, P., Encrenaz, T., et al., 2015. Strong water isotopic anomalies in the Martian atmosphere: probing current and ancient reservoirs. Science 348, 218−221.

Watson, L.L., Hutcheon, I.D., Epstein, S., Stolper, E., 1994. Water on Mars: clues from deuterium/hydrogen and water contents of hydrous phases in SNC meteorites. Science 265, 86−90.

Webster, R., et al., 2013. Isotope Ratios of H, C, and O in CO_2 and H_2O of the Martian Atmosphere. Science 341, 260. Available from: http://dx.doi.org/10.1126/science.1237961.

Wordsworth, R.D., 2016. The climate of early Mars. Annu. Rev. Earth Planet. Sci 44, 381−408.

Yang, J., Epstein, S., 1985. A study of stable isotopes in Shergotty meteorite. Lunar Planet. Sci. XVI 560, 25−26.

Yung, Y.L., Wen, J.-S., Pinto, J.P., Allen, M., Pierce, K.K., Paulson, S., 1988. HDO in the Martian atmosphere: implications for the abundance of crustal water. Icarus 146−159.

CARBONATES ON MARS

5

John C. Bridges[1], Leon J. Hicks[1], and Allan H. Treiman[2]
[1]*University of Leicester, Leicester, United Kingdom* [2]*Lunar and Planetary Institute, Houston, TX, United States*

5.1 INTRODUCTION

Carbonates have been the focus of much study in terrestrial sedimentary and hydrothermal systems because they record a history of fluid composition and pH, and temperatures during formation. The physicochemical characteristics and compositions of carbonate mineral deposits preserved throughout Earth's history have aided interpretations of geologic thermobarometry and provided insights about paleoclimatology and significant biogeochemical events on Earth, such as the evolution of photosynthesis (Essene, 1983; Gubler et al., 1967; Fairbridge, 1967).

The carbonation process, based on the reaction described by Xu et al. (2001), first involves atmospheric CO_2 dissolving in the water to produce weak carbonic acid (Eq. (5.1)).

$$2CO_2 + 2H_2O \rightarrow 2H_2CO_3 \qquad (5.1)$$

$$Fe_2SiO_2 + 2H_2CO_3 \rightarrow 2Fe^{2+} + SiO_2 + 2HCO_3^- + 2OH^- \qquad (5.2)$$

$$2Fe^{2+} + 2HCO_3^- + 2OH^- \rightarrow 2FeCO_3 + 2H_2O \qquad (5.3)$$

The associated increased acidity leads to the dissolution of minerals, such as Fe-bearing olivine, separating the divalent cations, for example, Fe^{2+}, as well as forming amorphous silica and the bicarbonate ions (Eq. (5.2)). The reaction between the bicarbonate and the divalent Fe^{2+} then precipitates the siderite carbonate under neutral pH conditions (Eq. (5.3)). As well as siderite ($FeCO_3$), the carbonation of basaltic composition igneous and sedimentary rocks commonly found on the surface of Mars, consisting of silicate minerals rich in Fe, Ca, and Mg, would also be expected to produce significant quantities of calcite ($CaCO_3$), magnesite ($MgCO_3$), and solid solutions between these end members.

Out of over 180 Martian meteorites including paired stones (LPI, 2017), Martian carbonates have only been identified in the nakhlite and ALH 84001 Martian meteorites (e.g., Bridges and Grady, 2000; Hicks et al., 2014; Mittlefehldt, 1994; Harvey and McSween, 1996). This is one of the reasons for the scientific significance of these meteorites and a focus of much research. Carbonates have also been identified at the Phoenix and Gusev landing sites (Morris et al., 2010; Boynton et al., 2009). The CRISM Near IR spectrometer data have led to the identification of carbonates, notably in Nili Fossae (Ehlmann et al., 2008) and Huygens crater, both within the ancient highlands (Wray et al., 2011). They are a record of different types of fluid activity near the Martian surface and water−atmosphere−rock interaction in potentially habitable environments at

Volatiles in the Martian Crust. DOI: http://dx.doi.org/10.1016/B978-0-12-804191-8.00005-2

different stages in Mars' evolution (Ehlmann and Edwards, 2014). Many rocks on the surface of Mars might be expected to contain significant amounts of carbonate where there has been interaction between the rocks and CO_2-rich waters as a result of a variety of processes including impact, igneous activity, weathering, and groundwater flow with diagenesis. Studying carbonate found on Mars can help reveal past environmental conditions, including the stable isotopic compositions and partial pressure pCO_2 of the atmosphere and the T, pH, compositions of fluids. CO_2 fixed as carbonate within the Martian crust is a key parameter in models of atmospheric evolution associated with the planet's evolution from "warm and wet" to today's cold and dry environment. In this paper, we use the current state of knowledge about Martian carbonates to consider atmospheric models. In particular, we consider whether "Thick" or "Thin" models (i.e., thousands or hundreds of mbar pCO_2) (Manning et al., 2006) are supported by the record of Martian carbonate.

We review the known Martian carbonate occurrences in order to show what we can learn about the variety of water−rock reactions and potentially habitable paleoenvironments on Mars. We also include new data on carbonates and associated phases from the nakhlite Martian meteorites as these provide a unique opportunity to study Martian carbonates in detail.

5.2 THE NAKHLITES

Lafayette, Nakhla, and Governador Valadares (GV) are members of the 9 classified plus 10 paired samples, of the SNC (shergottite, nakhlite, chassignite) nakhlite Martian meteorite group. These are basaltic clinopyroxenite rocks formed in a thick basic-ultrabasic lava flow or shallow sill intrusion (Treiman et al., 1993). The nakhlites are only mildly shocked, preserving the original crystallization ages, measured using isotopic data analyses, Rb−Sr, Sm−Nd (Shih et al., 1998), U−Pb, and $^{40}Ar/^{39}Ar$ (Nyquist et al., 2001), to have a common age of 1.3 Ga.

The nakhlites all share similar petrologic and isotopic characteristics, suggesting they all originated from a single magmatic flow formation or shallow sill intrusion. However, there are minor variations in the mesostasis, including its abundance and more specific features such as the size of plagioclase grains and the chemical compositions of the olivine and pyroxene rims and intercumulus melts. Based on these results, the nakhlites are considered to have each experienced a subtly different thermal history, having individually sampled different depths within the igneous body. Mikouchi et al. (2003, 2006, 2012) used mineral thermometry to suggest depths of formation. This nakhlite pile ranges from a crystallization depth of 1−2 m from the surface with MIL 090030/032/136 and NWA 5790. This is followed with increasing depth by NWA 817, MIL 03346, Y 000593 paired with Y 000749, GV along with Nakhla, and to the bottom with Lafayette and NWA 998 at a burial depth of > 30 m. The results are reported by Mikouchi et al. (2012, 2016).

A unique feature of the nakhlites is the presence of carbonate-bearing hydrothermal veining (Bridges and Grady, 2000; Changela and Bridges, 2011; Hicks et al., 2014). The secondary veins in places truncate the fusion crust in Nakhla (Gooding et al., 1991) and Lafayette (Treiman et al., 1993), establishing this veining material as Martian in origin. They likely formed ≤ 670 Ma based on K−Ar and Rb−Sr analyses (Swindle et al., 2000; Shih et al., 1998), decoupled from the original Amazonian era igneous event at ∼1.3 Gyr. Consisting predominantly of siderite, ferric saponite and Al-rich ferric serpentine, trace of ferric oxide, amorphous silicate gel of saponitic composition, halite, and Ca-sulfates (Bridges and Grady, 2000; Changela and Bridges, 2011; Bridges and Schwenzer, 2012; Hicks et al., 2014), the varying secondary mineralogy between the nakhlites makes them uniquely important in determining the fluid history of these rocks. Other less abundant secondary phases are traces of opal (Lee and Chatzitheodoridis, 2015).

To explain the hydrothermal system responsible for the secondary alteration, Changela and Bridges (2011) proposed a model of an impact-induced event with heated fluids percolating upward from buried H_2O-CO_2 ice and through the newly fractured nakhlite igneous mass. A quantitative determination of this fluid was proposed by Bridges and Schwenzer (2012). They concluded that partial dissolution of the surrounding ol-clinopyroxenites by CO_2-rich hydrothermal fluids, with initial temperatures of $150 \leq T \leq 200$ °C, pH 6–8, and a water:rock ratio (W/R) of ≤ 300. Due to exhaustion of the HCO_3^-, the early precipitates of Ca-siderite began to corrode as the fluids cooled to 50 °C, at pH 9 and a low W/R of 6, and dissolution of mesostasis began. At this stage, a crystalline ferric saponite precipitated within the olivine fractures, and ferric serpentine in mesostasis fractures. This was followed by a rapid cooling of the remaining fluid to form the amorphous gel in the center of veins, of similar composition to the ferric phyllosilicates. The other nakhlites also contain the hydrothermal assemblage, but predominantly this is the amorphous gel.

In order to test and develop the model presented by Bridges and Schwenzer (2012) and Changela and Bridges (2011), we have made a study of Fe-rich carbonates in the nakhlites, combining new data on Nakhla and Lafayette polished sections using SEM-EDX analyses, with that published from our previous work. In addition to the nakhlites, the orthopyroxene ALH 84001 Martian meteorite also contains Martian carbonates. These carbonates are 3.9–4.0 Ga old (Pb–Pb, Rb–Sr, Borg et al., 1999) and composed of calcite, dolomite–ankerite $Ca(Mg,Fe)(CO_3)_2$, and magnesite–siderite $(Mg,Fe)(CO_3)_2$ (Harvey and McSween, 1996).

5.2.1 NEW NAKHLITE SAMPLES AND METHODS

Nine nakhlites have been studied in polished sections and resin blocks: Lafayette (sample BM 1958, 775); Nakhla (sample BM 1911, 369); GV (sample BM 1975, M16); Y 000593 and Y 000749; MIL 03346; NWA 817; NWA 998; NWA 5790, which were studied by Changela and Bridges (2011) and Hicks et al. (2014), are reported here. An additional Nakhla sample, NWA 10153 and pairs, have new data reported here.

Carbonates in the veins were identified and characterized using back-scattered electron (BSE) imagery and energy dispersive X-ray (EDX) spectroscopy using a Phillips XL30 ESEM at the University of Leicester (UoL) Advanced Microscopy Centre. Normalized EDX spectra on new samples were measured, with an accelerating voltage of 20 kV and beam current of ~ 1.0 nA.

5.2.2 NAKHLITE CARBONATE TEXTURES AND COMPOSITIONS

Sideritic carbonates were found in Lafayette, Nakhla, and GV, but not in Y 000593/749, MIL 03346, NWA 817, NWA 998, and NWA 5790. Although a significant presence of calcite was also identified in NWA 998, NWA 817, and NWA 5790, near pure calcite is an ubiquitous feature of NWA meteorites of many types, chondrites, achondrites, Shergottite-Nakhlite-Chassignite (SNCs) and lunar, and so is clearly of terrestrial origin.

The Lafayette nakhlite meteorite contains sawtooth-shaped olivine fractures containing Ca-rich/Mg-poor siderite (often ≤ 100-μm rhombohedral grains) at the outer margins (Bridges and Grady, 2000). Changela and Bridges (2011) also identified grains of Mn-rich siderite. The siderite is partially corroded and replaced by crystalline ferric saponite, typically arranged perpendicular to fracture walls, with an Fe-oxide phase followed by an amorphous gel of similar saponite chemical composition occupying the center of the vein. Additionally, Lafayette also features crystalline Al-rich ferric serpentine with 0.7-nm lattice spacings found within mesostasis (Changela and Bridges, 2011; Hicks et al., 2014). Lee et al. (2013) showed a preferential widening of the fractures along a [0 0 1] direction in olivine grains, likely also associated with the shock effects.

The siderite−phyllosilicate−Fe-oxide−gel-hydrated assemblage is only found in Lafayette, whereas Nakhla and GV contain predominantly siderite—gel, with gypsum and anhydrite in interstitial, mesostasis areas. In contrast to Lafayette, the siderite grains in Nakhla and GV have a Mg-rich/Ca-poor content, with varying concentrations of Mn in Nakhla, and an irregular texture on the submicron scale (~ 50 nm in size) as shown by Bright-Field-TEM micrograph analysis (Bridges and Grady, 2000; Changela and Bridges, 2011).

The carbonates are present on the walls of fractures within the olivines of the three nakhlites (Fig. 5.1). In Lafayette, it is apparent that the siderite was partially dissolved and replaced by crystalline saponite phyllosilicates (Hicks et al., 2014), while hydrothermal fluids continued to flow, eventually depositing amorphous gel in the central regions of the fractures (Fig. 5.1A). The olivine fractures of Nakhla and GV also have amorphous gel in the central regions, but do not contain any crystalline phyllosilicates (see Fig. 5.1A and D). A few larger fractures of more than 30 μm width, often cutting through both olivine and pyroxene, in Nakhla (Fig. 5.1A) and Lafayette, contain siderite surrounded by the saponite phyllosilicates and/or amorphous gel.

In Nakhla and GV, small patches of sideritic carbonates were also found in this study within fractures of the mesostasis (see Fig. 5.1C), alongside amorphous gel of similar chemical composition to the ferric serpentine previously identified by Hicks et al. (2014), as well as silica and phosphate features. No carbonates were found in the mesostasis fractures of Lafayette.

Table 5.1 shows representative nakhlite carbonate compositions and the surrounding mineral in which the carbonates formed. All of the measurements reveal a siderite composition, with FeO contents ranging 22−55 wt%, but there is a notable variation with a Ca-rich composition for siderite in the Lafayette olivine fractures (CaO ≤ 17 wt%) and Ca-poor throughout GV and Nakhla (CaO < 4 wt%). The CaO content of the olivines in all of the nakhlites is negligible. Similarly, olivines in Lafayette have a MgO content of ≤ 14 wt%, whereas our analysis of siderite in those olivine fractures found only a trace MgO content, but Nakhla has a varied range (MgO $\approx 2-20$ wt%) with variations in composition unrelated to the surrounding minerals.

The sideritic composition of all of the carbonates, measured in Lafayette, GV, and Nakhla, is also apparent from the ternary plots in Figs. 5.2 and 5.3B−D. The ternary plot shows the variation in $MgCO_3$ and $CaCO_3$ content between Lafayette carbonates and the carbonates in GV and Nakhla, with a decreasing $CaCO_3$ content between the nakhlites as Lafayette $>$ GV $>$ Nakhla. The ternary plots also show that carbonate compositions lack any obvious carbonate and host-mineral relationship, or between olivine-hosted and mesostasis carbonate. The siderite associated with Lafayette olivine is $Cc_{0.18-0.31}Sd_{0.37-0.73}Rh_{0.07-0.35}$; GV olivine $Mg_{0.15-0.16}Cc_{0.03}Sd_{0.80}Rh_{0.02}$; GV mesostasis $Mg_{0.13-0.16}Cc_{0.05-0.06}Sd_{0.78-0.79}Rh_{0.01-0.02}$; Nakhla olivine $Mg_{0.06-0.23}Cc_{0.02-0.03}Sd_{0.73-0.82}Rh_{0.02-0.11}$; and Nakhla mesostasis $Mg_{0.06-0.32}Cc_{0.0-0.04}Sd_{0.62-0.82}Rh_{0.0-0.14}$, all mol%.

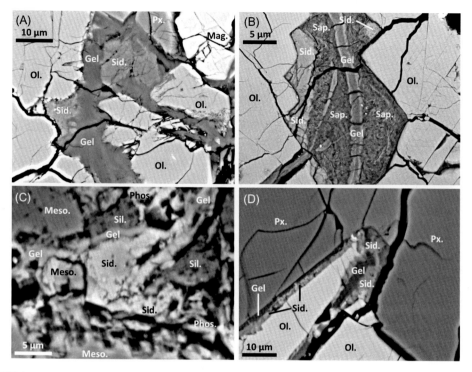

FIGURE 5.1

BSE images of the carbonates observed in the nakhlites. (A) A Nakhla vein between olivine (Ol) and pyroxene (Px) and magnetite (Mag) with siderite (Sid) forming on the walls of the fracture, and being surrounded by the silicate gel (Gel). (B) A 15–20-μm-wide vein formation in a Lafayette olivine, consisting of siderite formed on the walls of the fracture, with saponite phyllosilicates (Sap) within and a central formation of silicate gel. (C) A 15-μm-wide fracture within the mesostasis (Meso) and silica (Sil) of Nakhla, containing formations of siderite and phosphates (Phos) with silicate gel. (D) GV fracture, between olivine and pyroxene, filled with a central silicate gel and siderite forming on the walls of the fracture.

5.2.3 FORMATION OF THE NAKHLITE CARBONATES

The compositional ranges of nakhlite siderites in the olivines and mesostases strongly suggest that they are not solely dependent on adjacent minerals and cannot be an isochemical replacement of one phase. It has been suggested that that the siderite formed through carbonation of adjacent olivine (Lee et al., 2015). However, we suggest the siderite-bearing secondary assemblage formed from hydrothermal fluids flowing through the fractured nakhlite host rocks, introducing CO_2-rich waters and leading to a dissolution of the nakhlite mixture. The composition of fluids that formed the secondary mineral assemblage throughout the different nakhlites was modeled by Bridges and Schwenzer (2012). They concluded that partial dissolution of the surrounding olivine clinopyr-oxenites by CO_2-rich hydrothermal fluids, with temperatures of $150 \leq T \leq 200$ °C, pH 6–8,

Table 5.1 Carbonate Compositions Measured in Lafayette, GV, and Nakhla, ALH 84001 and Comache

Nakhlite veining	Lafayette Ol.	Lafayette Ol./Px.	GV Ol.	GV Px./Meso.	GV Meso.	Nakhla Ol.	Nakhla Ol./Px.	Nakhla Px.	Nakhla Px./Meso.	Nakhla Meso.	Lafayette[a]	GV[a]	Nakhla[a]	Nakhla[a]	ALH 84001[b]	Comanche outcrop[c]
SiO_2		2.1	4.9	2.0	0.2	1.6		2.5	1.9	1.5	tr.	0.5	2.2	0.3	0.1	
Al_2O_3						0.1			6.4					0.2		
FeO	31.3	38.1	43.6	41.2	45.5	45.5	39.8	37.6	38.7	42.8	36.9	48.9	39.3	24.5	21.9	19.2
MgO			6.2	7.1	7.1	5.6	13.3	13.4	5.2	10.8	0.3	3.2	13.8	7.0	24.2	26.2
MnO	13.0	3.3	1.1	1.2	0.8	3.1	1.0	1.3	2.6		4.1	1.3	1.1	19.9	0.8	1.4
CaO	14.9	15.7	1.5	3.9	2.6	0.9	1.6	1.6	1.4	1.3	18.0	5.5	2.3	0.2	6.7	6.6
Na_2O	1.1			0.8	1.6	1.5	2.1		0.5	1.3	0.9	1.1	0.3	1.7		
K_2O				0.2	0.2	tr.			0.2		tr.	0.2		0.4	0.3	
P_2O_5						0.5			1.0				0.1		tr.	
SO_3									0.3				0.1	2.1		
Cr_2O_3					0.1							0.3	0.1	0.2	tr.	
Cl				0.4	0.7						0.2	0.4	0.3	1.0		
CO_2[d]	39.7	40.8	42.6	43.2	41.1	41.0	42.2	43.5	41.6	42.2	39.6	38.8	40.6	42.5	46.0	46.4
Total	100.0	100.0	100.0	100.0	100.0	100.0	100.0	100.0	100.0	100.0	100.0	100.0	100.0	100.0	100.0	99.8
mol%																
Magnesite			14.8	16.6	15.7	12.8	28.8	29.9	13.6	24.0	0.8	9.0	29.4	16.9	58.0	62.0
Calcite	27.1	29.6	3.1	7.7	5.0	1.8	3.0	3.1	3.1	2.5	35.8	11.1	4.2	0.4	11.5	11.0
Siderite	51.4	64.8	80.0	73.6	77.9	79.9	66.5	64.7	78.1	73.5	57.0	77.8	64.7	45.4	29.4	25.0
Rhodochrosite	21.5	5.7	2.1	2.2	1.4	5.4	1.6	2.3	5.2		6.5	2.0	1.7	37.2	1.1	2.0
mol%[e] (n)	(n = 11)	(n = 9)	(n = 2)	(n = 5)	(n = 2)	(n = 3)	(n = 5)	(n = 19)	(n = 19)	(n = 13)	(n = 20)	(n = 12)	(n = 41)			
Magnesite	0.0–0.0	0.0–0.0	14.8–15.7	13.3–23.2	12.7–15.7	5.7–22.7	23.9–31.8	16.1–42.5	3.7–37.2	5.9–31.9	0.1–1.6	9.0–29.2	2.0–40.9			
Calcite	17.8–31.3	25.9–34.4	2.6–3.1	3.4–7.7	5.0–6.0	1.5–2.5	1.8–3.4	0.0–4.3	0.4–5.0	0.0–4.3	21.6–36.8	3.6–11.1	0.1–5.7			
Siderite	36.5–72.8	60.4–70.2	79.7–80.0	71.7–77.8	77.9–79.2	72.6–81.8	64.8–71.2	52.6–81.4	43.6–87.5	62.2–82.2	27.4–67.0	64.3–77.8	23.2–87.0			
Rhodochrosite	6.5–34.8	3.9–6.7	2.0–2.1	1.6–2.2	1.4–2.2	2.1–11.0	0.0–1.9	0.0–2.4	0.0–34.5	0.0–13.5	4.2–35.3	1.1–2.1	1.0–39.9			

Carbonates found within grains: olivine (Ol.); pyroxene (Px.); and mesostasis (Meso.). tr. = trace.
[a]*Bridges and Grady (2000);*
[b]*Mittlefehldt (1994);*
[c]*Morris et al. (2010).*
[d]*CO_2 calculated by difference.*
[e]*mol% range for all measurements made.*

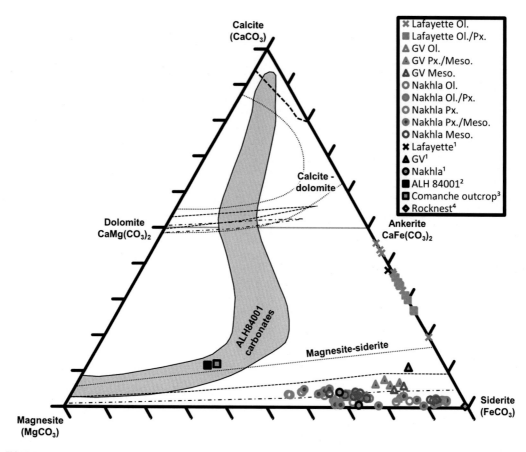

FIGURE 5.2

Ternary plot showing variations between calcite (CaCO₃), siderite (FeCO₃), and magnesite (MgCO₃), for the carbonates found in Lafayette (*crosses*), GV (*triangles*), and Nakhla (*circles*). The carbonates for each nakhlite have also been defined by the mineral type in which the fracture has been observed in or associated with, including olivine (blue), pyroxene (green), and mesostasis (red). The majority show a siderite-rich composition, but a clear separation of the magnesite-poor carbonates in Lafayette from the calcite-poor carbonates of Nakhla and GV, with no obvious grouping relevant to the surrounding mineral type. Additional points include Lafayette, GV, and two Nakhla samples referenced in Bridges and Grady (2000), Martian meteorite ALH 84001 (Mittlefehldt, 1994), the Martian surface Comanche outcrop carbonates referenced in Morris et al. (2010), and the Rocknest eolian deposits referenced in Sutter et al. (2015). The gray area shows the field of carbonate (mol%) composition in ALH 84001, based on the ternary plot presented in Bridges et al. (2001) of carbonate data from Harvey and McSween (1996). The lines represent calculated stability fields for carbonate one-phase solid solutions within these areas of the ternary system, at temperatures of 400 (*dash-dot*), 550 (*dashed*), and 700 °C (*dotted*), as presented by Anovitz and Essene (1987).

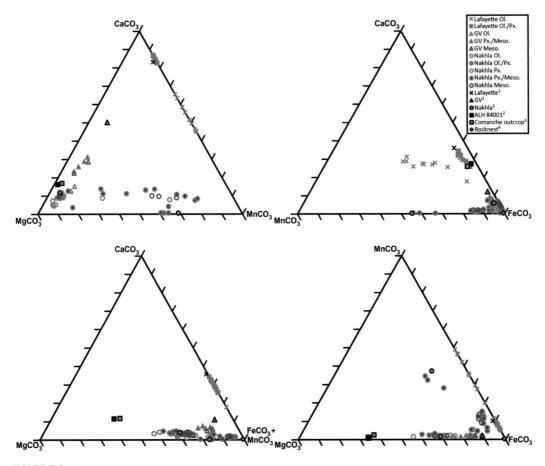

FIGURE 5.3

Four ternary plots showing variations between calcite (CaCO$_3$), siderite (FeCO$_3$), rhodochrosite (MnCO$_3$), and magnesite (MgCO$_3$), for the carbonates found in Lafayette (*crosses*), GV (*triangles*), and Nakhla (*circles*). Additional points include Lafayette, GV, and two Nakhla samples referenced in Bridges and Grady (2000), Martian meteorite ALH 84001 (Mittlefehldt, 1994), the Martian surface Comanche outcrop carbonates referenced in Morris et al. (2010), and the Rocknest eolian deposits referenced in Sutter et al. (2015). The carbonates for each nakhlite have also been defined by the mineral type in which the fracture has been observed in or associated with, including olivine (blue), pyroxene (green), and mesostasis (red). Three of the ternary plots (B–D) clearly show the majority of the carbonates to be siderites, with mostly FeCO$_3$-rich content. Ternary plot (A) shows the distinction between Ca, Mn, and Mg content: Lafayette siderite is Mg-poor, but Ca-rich (especially for siderites associated with pyroxene); Nakhla and GV are both Ca-poor and Mg-rich, with some variation between Mg-rich and Mn-rich for the Nakhla carbonates, but no apparent grouping based on such content and the surrounding mineral in which the carbonates are associated. Bridges and Grady (2000) had also presented ternary plots (B) and (D), with similar results.

and a water:rock ratio (*W/R*) of ≤ 300, produced the initial siderite. The bulk dissolved composition—throughout the complete hydrothermal event—was approximately 70% olivine, 20% bulk, 10% mesostasis (Bridges and Schwenzer, 2012). The heat source was likely impact-induced, and this was also responsible for the brittle fracturing seen throughout the nakhlites. An igneous heat source is unlikely as the radiometric ages of nakhlite alteration and igneous crystallization ages are decoupled by hundreds of millions of years (Treiman, 2005a). After the resultant melting of near surface H_2O-CO_2 ice, the fluid percolated upward through the nakhlite parent rocks, but it was not a long duration convecting hydrothermal system. Probably due to an exhaustion of HCO_3^-, the Ca-siderite began to corrode as the fluids cooled to $\sim 50 °C$, at pH 9 and a low *W/R* of 6, with dissolution of the pyroxene and mesostasis components of the nakhlites beginning. At this stage, crystalline ferric saponite phyllosilicates precipitated within the olivine fractures (or ferric serpentine in the mesostasis fractures). No evidence has been identified to suggest a second thermal or fluid event. The fluid cooled rapidly and formed the central vein filling of amorphous gel, which has a similar composition to the phyllosilicates (Hicks et al., 2014). The other nakhlites also contain the hydrothermal assemblage, but mainly the amorphous gel component. The final product of the fluid was evaporative precipitation of halite and sulfates within interstitial areas, and this is best preserved in Nakhla (e.g., Gooding et al., 1991; Bridges and Grady, 2000).

Fig. 5.2 is the $CaCO_3-MgCO_3-FeCO_3$ ternary diagram with relevant phase equilibria from experimental studies (Anovitz and Essene, 1987). The stability fields within the ternary reduce in size with decreasing temperature from 700 °C, through 550 °C, to 400 °C. Carbonates outside of these areas indicate metastable compositions and rapid crystallization (e.g., Goldsmith, 1983). This suggests that the Lafayette Ca-rich siderite, and the coexisting Ca-poor siderite, are metastable and crystallized rapidly. Recent experimental results on nakhlite mineral mixture analogs designed to follow the Bridges and Schwenzer (2012) modeling suggest that similar clay-like veins and carbonate in olivine can form within months (Schwenzer et al., 2017).

The initial fluid migrated through the nakhlite pile, from Lafayette through GV to Nakhla and beyond to the other nakhlite parent rocks. This is shown by the composition of the carbonate, becoming less Ca rich, and more Mg rich. However, after Nakhla, the fluid exhausted its HCO_3^- content. The composition of clay and gel also fractionated as the fluid migrated through the nakhlites away from the heat and fluid sources. This trend is a decrease in $100Mg/(Mg + Fe)$, and an increase in Fe/Si (Hicks et al., 2014; Changela and Bridges, 2011).

Stable isotope analyses clearly show that the nakhlite carbonate has preserved atmospheric signatures. In particular, stepped combustion mass spectrometry of carbonate within nakhlites is +10‰ to +55‰ (Carr et al., 1985; Wright et al., 1992; Jull et al., 1995), with these heavy isotopic values consistent with CO_2 originating within an atmosphere that had undergone substantial loss of ^{12}C (e.g., Jakosky and Phillips (2001)). They are similar to the current Mars atmosphere measured by the SAM instrument on Mars Science Laboratory (MSL): with 5 $\delta^{13}C$ analyses clustering around +50‰ (Webster et al., 2013). A minor abundance of ^{13}C-enriched siderite in Y 000593 was reported by Grady et al. (2007) using stepped combustion and mass spectrometry; however, a clear mineralogical composition of carbonate has not been obtained in this meteorite.

5.2.4 **TERRESTRIAL CALCITE**

Carbonates found in four other nakhlites, NWA 998, NWA 5790, NWA 817, and Y 000749, are usually pure calcite ($CaCO_3$) in composition, unlike the magnesite−siderite carbonates found in Lafayette, Nakhla, and GV, and are typical of desert weathering carbonate composition. These calcite features have not been seen to be truncated by the fusion crust, unlike those of Lafayette and Nakhla (Gooding et al., 1991; Treiman et al., 1993), and sometimes observed to exploit preexisting Martian veins, thus suggesting a terrestrial origin in these meteorite finds (Hicks et al., 2014). Terrestrial sulfates have also been identified cutting through the fusion crust in Y 000749 (Changela and Bridges, 2011).

5.2.5 **RECENTLY DISCOVERED NAKHLITES**

Mineral analyses reported by Mikouchi et al. (2012), for example high mesostasis abundance, suggest NWA 5790 is located at the top of the nakhlite igneous pile. Jambon et al. (2016) also agreed that NWA 5790 is from the topmost of an igneous flow, but suggested it was from a separate nakhlite igneous body of its own due to a lack of secondary minerals belonging to the deepest formation in the stack, below the other nakhlites. However, NWA 5790 located below all the other nakhlites is based on the assumption that the alteration fluids percolated down from above, which as summarized earlier we believe to be the opposite of the fluid flow. Alternatively, NWA 5790 may be entirely separate from the nakhlites, as suggested from the 7.3-Ma ejection age reported by Wieler et al. (2016), significantly below the ∼11.4-Ma ejection ages of other nakhlites (Nyquist et al., 2001), which would also explain the lack of hydrothermal alteration. The pure calcite features observed in NWA 5790 are considered to be terrestrial.

The paired nakhlites NWA 10659 and NWA 10153 are similar to Nakhla, GV, and Y 000593 in their augite core chemical composition and olivine zoning (Mikouchi et al., 2016). However, due to a lack of symplectic exsolution in the olivines, and Ca−Fe−Mg zoning unlike any other nakhlite, Mikouchi et al. suggested that NWA 10153 might have formed in a separate flow or sill from the other nakhlites. Alternatively, Mikouchi et al. suggested that the nakhlite cumulus pile was somewhat more heterogenous than previously recognized. The mesostasis and chemical composition values measured for NWA 10153 by Mikouchi et al. (2016) and Irving et al. (2015), along with its pair NWA 10659 (Hicks et al., 2016), compared to similar such measurements in the other nakhlites, suggest that the burial depth for NWA 10153/10659 may be ∼1−7 m, between NWA 817 and MIL 03346 above and Y 00593 and Nakhla below (see Mikouchi et al., 2016). NWA 10659 also features Fe-silicate alteration material similar to the other nakhlites (Irving et al., 2015; Hicks et al., 2016). However, no Martian carbonates have been identified.

5.3 **ALH 84001**

Martian meteorite Allan Hills 84001 (ALH 84001) is unique, the only Martian orthopyroxenite cumulate (Mittlefehldt, 1994), and the oldest known Martian meteorite to date at ∼4.1 Ga old (Lapen et al., 2010). The most distinctive features in ALH 84001 are its strikingly zoned carbonate masses (Fig. 5.4), which occur along cracks, as intergranular fillings, and as hemispheres. These carbonate globules are hosts to putative evidence of ancient Martian life (McKay et al., 1996), a

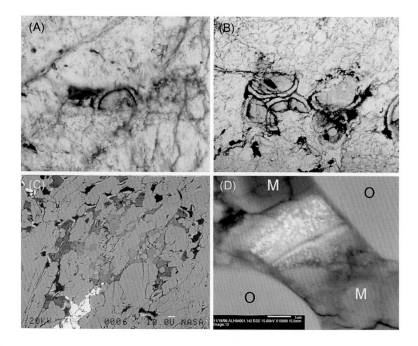

FIGURE 5.4

Carbonate masses in ALH 84001. (A) Disk-shaped carbonate mass, plane-polarized light in thin section (0.07); field of view 2.5 mm. Thin-section surface cuts carbonate disk along a fracture trace. The internal color bands represent the composition range of Fig. 5.2. (B) Hemispherical carbonate masses. Thin section (0.34), plane-polarized light, field of view 2.5 mm. Image shows portions of six hemispherical masses, each with the same compositional ranges and internal layering as in Figs. 5.2 and 5.4A. These masses are surrounded by orthopyroxene grains of $\sim 25\ \mu m$ dimension, and some maskelynite (white material, center right), from Treiman (1995). (C) Spherical carbonate mass in interstices among orthopyroxenes in a granular band; backscattered electron image, thin section (0.142), field of view $\sim 150\ \mu m$. Medium gray is orthopyroxene, bright white is chromite. Carbonate globule shows the same internal layering as carbonates in other settings: central sideritic carbonate as mid-tone to dark; magnetite as light tone (*arrow*); magnesite as dark tone; and again magnetite as light tone (*arrow*), from Treiman (1995). (D) High-resolution BSE image of carbonate mass enclosing orthopyroxene grains (thin section 0.142, see Fig. 5.4B). Scale bar is 1 μm. "O" is orthopyroxene; "M" is magnesite. Bright-tone grains are magnetite of the inner magnetite layer. Original chemical zoning in the globule was at < 0.1-μm scales, as shown by the dark band (low-Fe) within the magnetite-rich layer, from Treiman (2003).

hypothesis that has drawn abundant contrary evidence (Anders, 1996; Bradley et al., 1998; Treiman, 2003; Treiman and Essene, 2011).

Approximately 97% of ALH 84001 is orthopyroxene ($En_{70}Wo_{03}$) in anhedral to subhedral crystals up to ~ 6 mm across (Berkley and Boynton, 1992; Mittlefehldt, 1994). These large crystals are cut by fine-grained "granular bands" composed nearly completely of orthopyroxene, and all show evidence of shock deformation and strain (Mittlefehldt, 1994; Treiman, 1995, 1998). The other silicate phases in ALH 84001 include plagioclase-composition glasses (maskelynite and quenched melt), olivine inclusions (Fo_{65}) in orthopyroxene (Harvey and McSween, 1994, 1996; Shearer et al., 1999), augite ($En_{45}Wo_{42}$) (Mittlefehldt, 1994; Treiman, 1995), pigeonite ($En_{68}Wo_6$) (Treiman, 1995), and glasses of silica and sanidine compositions (Greenwood and McSween, 2001; McKay et al., 1996). Oxide minerals present include Fe^{3+}-rich chromite (~ 1 vol.%) (Berkley and Boynton, 1992; Mittlefehldt, 1994), and trace amounts of rutile (TiO_2) (Steele et al., 2007), periclase (MgO) (Barber and Scott, 2001, 2002), and magnetite (Fe_3O_4) (Barber and Scott, 2001; Bradley et al., 1996; McKay et al., 1996; Thomas-Keprta et al., 2000, 2002, 2009; Treiman, 2003; Treiman and Essene, 2011). The periclase and magnetite are spatially associated with the carbonates, and formed at least partially during their impact decarbonation (Treiman and Essene, 2011). Small proportions of graphitic poorly crystalline carbon is present in the carbonates (Steele et al., 2007), which could be original or products of impact decarbonation (Treiman and Essene, 2011).

ALH 84001 contains nearly no OH-bearing silicate minerals. Traces of preterrestrial, very fine-grained (< 100 nm length) phlogopite-like mica occur inside the carbonates (Brearley, 1998, 2000), some of which could have precipitated with the carbonates. Other phlogopite-like grains appear to crosscut the carbonates (Brearley, 2000). Smectite clays have been reported along grain boundaries (Wentworth et al., 1998; Thomas-Keprta et al., 2000), and these could reasonably be products of Antarctic weathering.

ALH 84001 contains a small proportion of phosphate minerals, primarily chlorapatite (Boctor et al., 2003; Cooney et al., 1998, 1999; Mittlefehldt, 1994) with lesser merrillite (Cooney et al., 1999; Greenwood et al., 2003). It also contains a small proportion of pyrite, FeS_2 (Greenwood et al., 2000; Mittlefehldt, 1994); trace proportions of pyrrhotite and greigite are present in the carbonates (Bradley et al., 1998).

ALH 84001 experienced several intense shock events (Treiman, 1995, 1998), which have (among other effects): faulted and broken the carbonate globules, decomposed the most Fe-rich carbonates (yielding magnetite and periclase), melted some original plagioclase, and crushed out some original porosity (Mittlefehldt, 1994; Treiman, 1998, 2003, 2005b; Brearley, 2003). What can be learned about Martian waters and aqueous processes from the ALH 84001 carbonate is always through this shock-generated haze.

5.3.1 CARBONATES

ALH 84001 contains a few percent (volume) of carbonate minerals, heterogeneously distributed in several distinct textural settings. They have been subject to intense scrutiny and interpretation, in part because of their putative connection to Martian life, and in part because they are the oldest known relics of aqueous activity on Mars. The carbonates were deposited at ~ 3.9 Ga (Borg et al., 1999), ~ 200 Ma after formation of the host pyroxenite (Lapen et al., 2010).

Most famous of the carbonates are disk-shaped masses along fractures in pyroxene and plagioclase (Fig. 5.4A); less common are coherent hemispherical globules (Fig. 5.4B) and globules enclosing finely granular pyroxene (Fig. 5.4C). Still less common are planar "slab" deposits, Figs. 5.1B and 5.2A of Corrigan and Harvey (2004) and unzoned masses (Eiler et al., 2002). The latter are interpreted as having been melted by shock, and not considered further here. But for the latter variety, all of the carbonate varieties show strong concentric or planar chemical zoning. Compositions range continuously from calcite (rarely observed) in the cores or interiors, through dolomite—ankerite to siderite—magnesite, and to nearly pure magnesite (Table 5.1, Fig. 5.2). Elementally, the zoning mostly follows this monotonic path; spatially, however, most occurrences

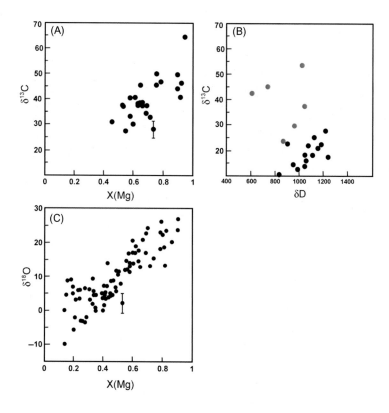

FIGURE 5.5

ALH 84001 carbonates: Correlations among stable isotopic and elemental compositions. (A) Carbon isotope ratios ($\delta^{13}C$, per mil relative to PDB) and X(Mg), molar Mg/(Mg + Fe + Ca). Uncertainty bar is typical for each point. (B) Carbon isotope ratios versus hydrogen isotope ratios (δD, per mil relative to VSMOW). *Black points* are from main carbonate globules; *gray points* from exterior magnesite-rich zones. Uncertainties not given. (C) Oxygen isotope ratios ($\delta^{18}O$, per mil relative to VSMOW). Uncertainty bar is typical for each point. *Data from Saxton et al. (1998), Eiler et al. (2002), Niles et al. (2005), and Usui et al. (2016).*

show some oscillatory zoning that is clearest as alternating bright and dark layers in backscattered electron imaging (Fe richer and poorer). This oscillatory zoning is apparent at the finest scales; Fig. 5.1d of Treiman (2003) shows such zoning at the 0.1-μm scale. The most iron-rich layers in the carbonates show a distinct peppering of iron-rich spots, which represent submicron grains of magnetite; these magnetite grains have been interpreted (in part) as magnetosomes from ancient Martian bacteria (McKay et al., 1996; Thomas-Keprta et al., 2000, 2009) and, as we adopt here, as products of shock-induced breakdown of iron-rich carbonate (Golden et al., 2004; Treiman, 2003; Treiman and Essene, 2011).

The stable isotopic compositions of the ALH 84001 carbonates also vary widely, for the most part in concert with their elemental compositions (Fig. 5.5). The carbonates' $^{18}O/^{16}O$ ratios, as

$\delta^{18}O$, vary monotonically and positively with Mg/Fe (Holland et al., 2005; Saxton et al., 1998; Usui et al., 2016), as do their $^{13}C/^{12}C$ ratios, as $\delta^{13}C$, Fig. 5.5 (Niles et al., 2005; Usui et al., 2016). Both of these isotopic ratios are high, which suggests significant (but variable) incorporation of C and O that were processed through the Martian atmosphere. The abundances of ^{17}O in the carbonates, as $\Delta^{17}O$, is characteristic of the Martian atmosphere rather than its interior, at $+0.55‰- +0.75‰$; carbonates with higher $\delta^{13}C$ (and hence Mg/Fe and δD, see below) have the lower $\Delta^{17}O$ values (Farquhar et al., 1998; Shaheen et al., 2015). Some late calcite grains, though, are distinct and may have a different origin. The carbonates are inferred to be the principal reservoir for hydrogen in ALH 84001 (Eiler et al., 2002; Leshin et al., 1996), and their hydrogen isotope ratios vary widely. The D/H ratios, as δD, of spots in the central portions of ALH 84001 carbonate globules vary linearly and monotonically with $\delta^{18}O$ and $\delta^{13}C$, and thus with Mg/Fe (Usui et al., 2016). The δD values, $+800‰- +1400‰$, are consistent with the inferred composition of Mars' surface water and not its mantle nor atmospheric water (Usui et al., 2015). The outermost magnesite-rich layer follows a different trend, suggesting a different origin. Although the carbonates do not contain measurable sulfur, they are associated with sulfide minerals (pyrite, pyrrhotite), and their S isotope ratios show nonmass-dependent fractionations that are attributed to photochemical processing in Mars' atmosphere (Franz et al., 2014; Greenwood et al., 1997, 2000).

5.3.2 FORMATION MECHANISM

Understanding of how ALH 84001 formed is severely limited because we lack any geological context for its carbonates, and even for the host pyroxenite. Understanding of the nakhlites and their aqueous alterations is much clearer, partly because they show minimal effects from shock, and because enough distinct nakhlites are known to allow reasonable guesses about their geology. With ALH 84001, we have a unique rock with unique alteration materials, from a time when our understanding of Martian geology and climate is limited. It is thus understandable that there have been many hypotheses proposed for the origin of the carbonates in ALH 84001, and even yet no strong consensus.

Early hypotheses on the origins of the ALH 84001 carbonates invoked high temperatures, but they are no longer tenable. A high formation temperature was first suggested, because the Ca contents of the Fe−Mg rich carbonates seemed consistent with equilibration at ∼700 °C (Harvey and McSween, 1996; Mittlefehldt, 1994), but this seemed inconsistent with the carbonates' fine-scale elemental zoning (Valley et al., 1997). An origin as impact melt was advocated (Scott et al., 1997, 1998), but the elemental zoning (core to rim) seemed inconsistent with igneous fractionation. High and rapidly decreasing temperature was invoked to explain the carbonates' varying oxygen isotope ratios (Leshin et al., 1998).

However, hypotheses invoking high temperature were rendered obsolete by Halevy et al. (2011), who showed that the "clumped" carbon and oxygen isotope ratios of the carbonates required a formation temperature near 18 °C. Halevy's result is entirely consistent with the presence and retention of fine-scale elemental and isotopic zoning in the carbonates (Eiler et al., 2002; Kent et al., 2001; Müller et al., 2012; Romanek et al., 1994; Treiman, 1995; Valley et al., 1997), and with laboratory experiments that produced similar carbonate globules at low temperatures (Golden et al., 2000, 2001; Niles et al., 2004).

Within this constraint of low temperature formation, hypotheses for the formation of ALH 84001 carbonates have invoked aqueous solutions, and have had to address two fundamental questions. First, how could Martian aqueous solutions arise, such that were capable of depositing Ca−Fe−Mg carbonates? And second, what physical or chemical events caused those solutions to deposit carbonates such as are present in ALH 84001? Especially important for the latter are the variations in stable isotope compositions among the carbonates.

The aqueous solution that deposited the ALH 84001 carbonates must have carried (of course) significant proportions of Ca, Mg, Fe, and carbonate. Silica is commonly associated with the carbonates, intimately in some locations, so the fluids must also have carried it. There is broad agreement that a solution of this sort could arise by low-temperature weathering of mafic rock like ALH 84001 itself (Halevy et al., 2011; Niles et al., 2009; van Berk et al., 2011; Warren, 1998) or clays formed by alteration of mafic rock (Melwani Daswani et al., 2016). There is also broad agreement that fluid/rock ratios must have been low, and that the altering solutions were mildly acidic. That acidity could have arisen through equilibration with atmospheric CO_2 at 0.1−1 bar pressure (Niles et al., 2009), which is significantly above the current level but possible on early Mars. It is likewise possible (though rarely invoked) that the fluids were acidic from oxidation of igneous sulfide minerals (Burns and Fisher, 1990; Fernández-Remolar et al., 2004). Finally, the stable isotopic compositions of the carbonates (specifically their H, O, C, and S) leave little doubt that their parental solutions were processed through the Martian atmosphere or had reacted chemically with the atmosphere (Holland et al., 2005; Saxton et al., 1998; Usui et al., 2016).

Similarly, there is a general (although not universal) agreement about the general physicochemical setting of carbonate formation. The depositional environment must permit the carbonates' wide variations in stable isotopic compositions (Fig. 5.5). In the absence of multiple isotopic reservoirs (e.g., fluids) or huge temperature changes (Halevy et al., 2011), this requires fractionations between phases of different bonding characteristics, specifically between aqueous fluid and gas. Nearly all current hypotheses for the ALH 84001 carbonates invoke evaporation as the primary process that forced their elemental and isotopic variations (Catling, 1999; Halevy et al., 2011; McSween and Harvey, 1998; Melwani Daswani et al., 2016; Warren, 1998). Alternately, similar chemical variations may be forced by degassing of CO_2 from solutions, from perhaps ~ 10 to 0.1 bars (Niles et al., 2009; van Berk et al., 2011). The latter certainly would have involved some evaporation of H_2O, but its effect on H isotope ratios has not been explored. The huge challenge for hypotheses of this sort is to quantitatively model the simultaneous variations in stable isotope ratios (O, C, H) and elemental abundances (Ca, Mg, Fe, Mn) in the ALH 84001 carbonates. Such a comprehensive model is not yet available.

On the other hand, the chemical and isotopic variations of the ALH 84001 carbonates could represent mixing of fluids, a process that (in general) is capable of precipitating carbonate minerals (Bischoff et al., 1993; Buchardt et al., 2001) and has been invoked without specifics by several researchers (Corrigan and Harvey, 2004; Farquhar et al., 1998; Farquhar and Thiemens, 2000; Gleason et al., 1997; Saxton et al., 1998). The strongest evidence for multiple fluids in the genesis of the ALH 84001 carbonates comes from the few available determinations of $\Delta^{17}O$. Materials in Martian meteorites show a significant range in $\Delta^{17}O$, a ratio that is essentially unchanged in processes without photochemistry (e.g., evaporation and degassing). High-temperature silicate minerals in Martian meteorites all have $\Delta^{17}O = +0.317‰$ (Ali et al., 2016). In ALH 84001, carbonate analyses show a range of $\Delta^{17}O$ from $\sim +0.55‰$ to $\sim +0.75‰$, with the high and low values distinctly outside analytical uncertainties (Farquhar et al., 1998; Shaheen et al., 2015); the high $\Delta^{17}O$ values are attributed to atmospheric contribution. The presence of distinctly different $\Delta^{17}O$ values in the ALH 84001 carbonate globules suggests that several fluids were involved, each with a slightly different mix of atmospheric and internal oxygen. Oxygen isotope analyses of secondary materials in the nakhlite meteorites show that a range of such mixes has happened, with $\Delta^{17}O \approx +1.0‰$ for Nakhla and $\approx +0.7‰$ for Lafayette (Farquhar and Thiemens, 2000). For ALH 84001, the simplest mixing hypothesis might involve two fluids (see Fig. 5.5): one with low δD, low $\delta^{13}C$, low $\delta^{18}O$, high $\Delta^{17}O$, and rich in Ca and Fe; and a second with high δD, high $\delta^{13}C$, high $\delta^{18}O$, not so high $\Delta^{17}O$, and rich in Mg. This speculation would be testable with precision $\Delta^{17}O$ analyses coordinated with elemental chemical analyses.

5.4 CARBONATES DETECTED ON THE SURFACE OF MARS

5.4.1 LANDER ANALYSES

Morris et al. (2010) reported carbonates identified by the NASA Spirit Mars exploration rover in the Comanche outcrops of the Columbia Hills in Gusev crater (Fig. 5.6). This carbonate identification has been given added significance by the current interest in Columbia Hills as a landing site for Mars2020 and possible sample return (Golombek et al., 2016; Fergason et al., 2017). The Comanche outcrop is composed of erosional remnants with a granular surface, where the surrounding rocks are olivine-bearing basaltic rocks of Hesperian age. The spectrum measured by Spirit's Mössbauer spectrometer observed Fe^{2+} doublets characteristic of olivine and Fe−Mg-rich carbonate in the outcrops. The Alpha Particle X-ray Spectrometer (APXS) measured a low-bulk CaO concentration, consistent with MB data. From these data, a forsteritic olivine ($Fo_{0.72}Fa_{0.28}$) and Fe−Mg-rich carbonate ($Mg_{0.62}Cc_{0.11}Sd_{0.25}Rh_{0.02}$) assemblage was calculated. Spirit's Miniature Thermal Emission Spectrometer (Mini-TES) instrument found this mineral assemblage to be relatively uniform across the outcrop, based on eight separate spectral measurements. The carbonate component is 26% of the total outcrop assemblage, which is significantly greater than the $\sim 1\%$ found in the form of veins and fracture-filling deposits in the nakhlites and ALH 84001, but as can be seen in Fig. 5.2. The Comanche outcrop carbonates (square) are similar in composition to those carbonates of ALH 84001 (gray field).

The Phoenix lander (Fig. 5.6) has also identified $CaCO_3$ in the Martian polar (68.2° N 234.3° E) soil in a scooped sample named Wicked Witch. The carbonates were detected using the on-board Thermal and Evolved-Gas Analyzer (TEGA) (Boynton et al., 2009). Endothermic phase transition peaks at 725 °C − 820 °C and 860 °C − 980 °C suggest a calcium-rich carbonate mineral phase, and an estimated $CaCO_3$ concentration of 4.5 wt%, and Mg or Fe carbonates may be responsible for releases at lower temperatures of 400 °C − 680 °C. Measuring the pH and concentrations of various ions in the soil, the results from the Wet Chemistry Laboratory (WCL) are also consistent with the TEGA results (Boynton et al., 2009).

The soil at the Phoenix landing site has subsurface ice just centimeters below and surface frost in winter (Smith et al., 2009). When unfrozen, the atmospheric CO_2 diffused through this localized water reservoir reacts with the basaltic soil material, along with the dissolution of Ca^{2+}, carbonates would precipitate as a cement between the soil particles. Alternatively, due to the 20-km proximity

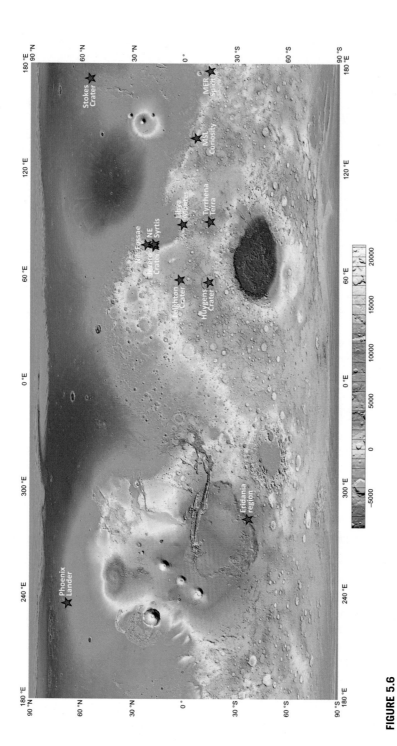

FIGURE 5.6

MOLA map of localities (*red stars*) on Mars where carbonate has been identified either in situ (Gale crater, Gusev, Phoenix) or from NIR remote observation at eight localities, particularly around the Syrtis-Terra Tyrrhena region in the Ancient Highlands. References listed in the text.

of the Phoenix lander to the 500-Ma Heimdall crater, it is also possible that the calcium carbonate formations and other aqueous alterations are due to localized impact-induced hydrothermal conditions and excavation (Boynton et al., 2009; Heet et al., 2009).

In Gale crater (6.6° S 137.4° E), Fe-rich carbonates ($\sim 0.7-1.0$ wt% $Fe_{0.99}X_{0.01}CO_3$) have been measured in a sample scoop of eolian deposits, probably having formed from the interaction of Fe^{2+}-rich olivine with the CO_2 atmosphere and transient water (Sutter et al., 2015). This measurement was achieved using the Sample Analysis at Mars (SAM) instrument on the MSL Curiosity rover, heating the sample in the range 25 °C − 715 °C. CO_2 release attributed to traces of Fe-rich carbonates peaked at 466 °C − 487 °C (Sutter et al., 2015). The SAM instrument was also used to measure cosmogenic noble gases ^{3}He, ^{21}Ne, and ^{36}Ar to estimate a time of 78 ± 30 Ma for which the surface rocks of the ~ 150 km wide crater have been exposed to cosmic rays (Farley et al., 2014). However, the carbonates at Rocknest were eolian deposits; thus, a definitive age given for the Fe-rich carbonates remains uncertain.

The absence, or at most trace abundance, of carbonates in the consolidated sediments of Gale crater is notable. Ming et al. (2014) used SAM pyrolysis data of Sheepbed Member mudstone to identify possible traces of carbonate, but so far X-ray diffraction with CheMin (e.g., Vaniman et al., 2014) has not identified carbonate. Bridges et al. (2015) and Schwenzer et al. (2016) suggested that the diagenetic fluids in the Sheepbed mudstone were CO_2-poor, neutral to alkaline groundwater, in order to explain the absence of carbonate. The diagenetic fluid at Yellowknife Bay, if originally in contact with acid and/or CO_2 sources from the atmosphere, would have been neutralized through reaction with the basaltic sediments and lost the CO_2 content prior to the diagenetic reactions recorded in the MSL analyses (Schwenzer et al., 2016).

5.4.2 SPACEBORNE OBSERVATIONS

Several regions on the surface of Mars have revealed the presence of carbonates based on spaceborne observations (Fig. 5.6). The Compact Reconnaissance Imaging Spectrometer for Mars (CRISM), on board the Mars Reconnaissance Orbiter (MRO) spacecraft, was designed for hyperspectral mapping of vibrational absorptions in the thermal infrared (TIR) wavelengths $0.4-4.0\,\mu m$ across 545 channels, and a resolution of 18 m per pixel. This allows CRISM to map carbonates, as well as to identify other secondary minerals including nontronite and other smectites, chlorite, prehnite, serpentine, kaolinite, K-mica, and hydrated silica (Murchie et al., 2007; Ehlmann et al., 2009). Mg-rich carbonates include spectral absorptions at 2.3 and $2.5\,\mu m$, due to C−O stretching and bending fundamental vibrations (Ehlmann et al., 2008, 2009; Gaffey, 1987), whereas absorptions identified at 2.33 and $2.53\,\mu m$ are consistent with Ca-rich siderite carbonates (Ehlmann et al., 2008; Wray et al. 2011). Other spectral absorption features can occur along with carbonates such as the presence of OH and water in hydrated minerals, notably clays with characteristic absorptions at 1.9 and $2.3\,\mu m$. The majority of CRISM-based carbonate observations are within or associated with impact craters. Their origin, whether through in situ, impact-induced hydrothermal activity or excavation of older, altered crust is often uncertain and the subject of ongoing debate, much of it now focused by the characterization and selection of potential landing sites. Of particular importance in CRISM-based identification of carbonate is the region to the west and south of Isidis, for example Jezero crater (76° E, 18° N), NE Syrtis (18° N, 77° E), Nili Fossae (23° N, 77° E), within which a regional carbonate unit or units are exposed. This region—Syrtis, Nili Fossae, and Terra

Tyrrhena—is the focus of current landing site selection and sample return studies (e.g., Goudge et al., 2017a, b; Bramble et al., 2017). It is the only regional carbonate unit on Mars (Niles et al., 2013), with thicknesses of approximately tens of meters (e.g., Bramble et al., 2017). CRISM spectra suggest the presence of siderite and/or ankerite, with absorptions centered near $1.91-1.92$, $2.31-2.32$, and $2.52-2.54$ µm, along with the possible presence of Fe-rich smectites, in the uplifted Noachian bedrock of craters in the Noachian Tyrrhena Terra region 90° E, 14.8° S (e.g., Carrozzo et al., 2013). In Jezero crater, Mg-carbonate is associated with basin fill and fluvio-deltaic deposits (Goudge et al., 2017a, b).

Impact excavation of carbonates from depths of ~ 6 km was suggested by Michalski and Niles (2010) to explain the identification of siderite and/or calcite—along with phyllosilicate—within the central uplift peak of Leighton crater (57° E, 3° N) within Tyrrhena Terra. It is possible these rock units were ancient sediments that may have extended over a much larger area, and were buried and metamorphosed by the Syrtis Major volcanic deposits. Possible evidence for Syrtis magma$-$carbonate interaction is suggested by the identification of carbonate decomposition products such as portlandite, and a high abundance of high-Ca pyroxene in the Syrtis Major lavas, analyzed from Mars Global Surveyor (MGS) TES data (Glotch and Rogers, 2013).

Ehlmann et al. (2008, 2009) reported magnesite carbonates exposed over a < 10-km^2 region of Nili Fossae (22.6° N, 76.8° E) and two nearby areas, west of Isidis Basin (Fig. 5.6). The carbonates are present in fractured Noachian or early-Hesperian rocks, stratigraphically similar to olivine-bearing rock units, overlying layers of Fe$-$Mg smectite bearing rock, with an unaltered mafic cap unit and Al-phyllosilicates deposited on top (Mustard et al., 2008; Ehlmann et al., 2008). Ehlmann et al. (2008) also reported some smaller exposures of carbonate-bearing rocks in the Libya Montes region (1.4° N, 88.2° E) of the ancient highland terrains.

Craters in the equatorial region, between Isidis Basin, Huygens Basin, Syrtis Major, and toward the Hellas Basin, have also been studied by Bultel et al. (2015). They report CRISM observations, detecting alteration minerals such as Fe/Mg chlorites, Fe/Mg smectites, possible serpentines, and carbonates—either hydrated or mixed with hydrated phases, which are present in and around the various craters of the studied area, many of which are Hesperian and Amazonian in age. Bultel et al. (2015) suggested that hydration and carbonation around the Amazonian impact craters may have been impact induced, but also discussed contradictory evidence such as the necessary quantity of subsurface water-ice being unlikely during Amazonian times in these equatorial regions (Madeleine et al., 2009). Other suggestions contradictory to the impact-induced hydrothermal hypotheses include the expectation that higher temperature phase minerals should be more abundant in the larger craters, plus the carbonates and alteration minerals are present in the rim, the ejecta, and the central peak of the craters, whereas evidence for impact-induced hydrothermal activity is expected to be limited to the floor and central peak (Barnhart and Nimmo, 2011). Thus Bultel et al. (2015) assumed that most, if not all, of the alteration minerals in the studied craters have been exhumed to the surface from the > 3.5-Ga Noachian basement.

Also in this region, Wray et al. (2011) reported Mg-rich carbonates in the ejecta and interior of several craters within the 460-km diameter Huygens, Noachian multiring crater (55.6° E, 14.0° S). Wray et al. suggested that these had been excavated from rocks that were initially buried ~ 5 km below the Martian surface, where the carbonate-bearing rocks were jointed and layered with dikes and/or large veins. Elsewhere on Mars, Mg-rich carbonates observed in crater interiors include those of the 3.8-Ga old McLaughlin crater. These carbonates may have formed as a result of sporadic groundwater activity during the Noachian to early Hesperian (Michalski et al., 2013). The central uplift of the Amazonian Stokes crater (55.6° N, 171.2° E) contains Fe-carbonate (Turner et al., 2016) in addition to Fe/Mg/Al phyllosilicates and chlorite (Carter et al., 2010). Carbonate deposits, of mixed Ca, Mg, Fe composition, are also reported by Gilmore et al. (2014) in the gullies of craters in the Noachian Eridania region (281.8° E, 37.1° S).

An estimate of the quantity of carbonates contained in Martian dust was reported by Bandfield et al. (2003), using data obtained by the MGS-TES instrument. Emissivity measured at wavenumbers $> 1300 \text{ cm}^{-1}$, including the emissivity minima at 1350 and 1580 cm^{-1} and maxima near 1480 and 1630 cm^{-1}, is consistent with coarse silicate surfaces with a concentration of $< 5\%$ carbonate material in the Martian dust. This carbonate dust is interpreted to be dominated by magnesite. However, subsequent analyses, for example, by MSL and CRISM have not confirmed crustal abundances in eolian deposits at this high level of abundance.

5.5 PAST ATMOSPHERIC $p\text{CO}_2$ AND ENVIRONMENTAL CONDITIONS

Chemical weathering is most rapid when atmospheric CO_2 pressures are large. Thus, carbonate abundance is one of the key factors in constraining past $p\text{CO}_2$. Manning et al. (2006) modeled past Martian $p\text{CO}_2$ as a function of atmospheric, ice cap, adsorbed CO_2 in the regolith, and carbonate-rich reservoirs. We can calculate the equivalent $p\text{CO}_2$ trapped as carbonate in order to constrain the thick and thin models for the ancient atmosphere described by Manning et al. (2006). If we assume a crustal density of 3000 kg/m^3, 1000 m of crust, with 1 wt% $CaCO_3$, would contain 2.2×10^{18} kg of carbonate ($= 2.2 \times 10^{19}$ mol CO_2). Every 1 wt% of $CaCO_3$, if assumed to be evenly distributed within the top 1000 m of Martian crust, is equivalent to 37,000 Pa (370 mbar) of trapped $p\text{CO}_2$ (Bridges et al., 2001). The Martian crustal abundances, up to 4.5 wt% at the Phoenix site—and higher at Spirit-Comache—are equivalent to at least 170,000 Pa $p\text{CO}_2$ (1.7 bar) if they are representative of the Martian crust. The lesser abundances detected at Gale crater suggest that it is an upper limit to a crustal average; however, it is clear that the Mars crust has an inventory of at least hundreds of mbar CO_2.

The thick atmospheric model of Manning et al. (2006) includes a 450-mbar carbonate reservoir, while thin models have 200 mbar. The recent surface and meteorite carbonate analyses summarized here suggest that a thick atmospheric model is the more accurate scenario. This predicts an early free inventory of 5 bar and that 300–700 mbar of free CO_2 remained by the late-Noachian. Thick models also imply a relatively large current CO_2 ice cap. Such models are consistent with recent MAVEN measurements of atmospheric loss rates (Jakosky et al., 2017). The sublimation of a massive cap at a high obliquity would create a climate swing with the likelihood of greenhouse warming effects on the Martian surface as atmospheric pressure exceeds the triple point of water.

5.6 SYNTHESIS

Carbonates are an expected weathering product for a Martian environment containing water and Fe−Mg silicate minerals, with an atmosphere largely composed of CO_2. The ages of carbonate vary from Noachian (e.g., CRISM identification in Nili Fossae) and ALH 84001 carbonate, to Amazonian (the nakhlite siderite and a relatively small number of Amazonian craters). Similarly, carbonate has been precipitated in a variety of different fluid environments: impact or magmatic induced hydrothermal, excavation of ancient carbonate-bearing rocks, groundwater diagenesis. However, it appears from spaceborne, lander, and meteorite-based observations of Mars that rather than extensive carbonate deposits similar to those on Earth, there are multiple carbonate formations in localized environments. The one exception to this is the Syrtis-Terra Tyrrhena region to the west and south of the Isidis Basin (Fig. 5.6) where ∼tens meter thick carbonates are exposed in craters. The exact processes that led to the crystallization of such carbonates are, in the absence of "ground truth" from landers, necessarily uncertain. Possible processes include weathering, fluid evaporation, serpentinization, hydrothermal activity, and sedimentary/lacustrine deposits associated with deltas at Jezero crater (e.g., Bramble et al., 2017; Goudge et al., 2017a, 2017b). However, by analogy with the nakhlites it is likely that whichever model is correct, it involves extensive alteration of olivine. For instance, the carbonates detected in the Noachian terrain of Nili Fossae probably formed from hydrothermal or near surface fluid alteration of olivine-rich igneous rocks, with impact or volcanic-induced heat sources (Ehlmann et al., 2008; 2009). ALH 84001 contains a range of magnesite to calcite compositions, and thus are similar in composition and age to the early-Noachian to Early Hesperian magnesite carbonates observed at Nili Fossae (Ehlmann et al., 2009) and Huygens crater (Wray et al., 2011), or the Noachian Fe-calcites at Leighton crater (Michalski and Niles, 2010). The carbonates measured in the Columbia Hills by the Spirit MER are also exactly similar in composition to some of those in ALH 84001 (Table 5.1), for example 62% magnesite, although much younger in age (Hesperian) than those of ALH 84001.

To explain the general lack of massive carbonate deposits, some researchers have suggested that a warmer, wetter early Mars was not sustained by a CO_2-rich atmosphere but instead relied on methane or ammonia (e.g., Kasting, 1997). Other processes such as warming through CO_2 clouds may also have facilitated the necessary global warming (Forget and Pierrehumbert, 1997). The formation of meteorite carbonate and evidence from landers suggest that the water−rock reactions could largely have been subsurface, such as diagenesis or hydrothermal, away from the CO_2 atmosphere, and reducing the production of carbonate deposits. Also, the environment may sometimes have been too acidic for the formation of massive carbonate deposits, or perhaps they have since been partially destroyed by large-scale acidic weathering from later Hesperian times (Ehlmann et al., 2009). The trace abundance of carbonates in the mudstones of Gale crater suggest that carbonate precipitation can be hindered through buffering of CO_2-mineral reactions to neutralize the acidity of fluids (e.g., Schwenzer et al., 2016 and Tosca et al., 2004).

However, recalculating the crustal carbonate abundances that do exist (e.g., at current landing sites and in some SNC meteorites as pCO_2) does suggest that "Thick" atmospheric models (Manning et al., 2006), that is with equivalent bars of CO_2, are applicable for ancient Mars. Loss of CO_2 by carbonate sequestration would stop at the triple point of water 6.1 mbar (e.g., Kahn, 1985), and thus in recent Martian history, relatively little CO_2 has been lost or trapped from the atmosphere in this way. The exact timing of major losses from the Mars atmosphere is not certain, with Webster et al. (2013) using the comparison between their results from the SAM experiment on MSL with Martian meteorites to infer that the major loss from an early atmosphere with bars of pCO_2 occurred after 4 billion years ago. This is consistent with the presence of extensive fluvial and deltaic deposits in Noachian to early Hesperian highland terranes, for example the thick mudstone deposits within Gale crater (Grotzinger et al., 2014), and might even have facilitated the existence of a northern ocean (e.g., Perron et al., 2007; Clifford and Parker, 2001). This evidence, together with the presence of mass independent O-isotopic anomalies in ALH 84001 and the nakhlites (e.g., Shaheen et al., 2015; Farquhar and Thiemens, 2000), suggests that the combined effects of sequestration, impact-induced and thermal and nonthermal loss processes are recorded by Martian carbonate.

However, some carbonate precipitation has also occurred in more recent Martian epochs. The nakhlite carbonates are ≤ 670 Ma (Swindle et al., 2000; Shih et al., 1998) in age, forming during the mid- to late-Amazonian epoch, and the carbonate identified by the Phoenix Polar lander is also likely to be of Amazonian age. In general, spaceborne detections of carbonates or any mineralogic record of water−rock interactions in the Amazonian are much sparser (Ehlmann and Edwards, 2014; Turner et al., 2016). If carbonate formations occurred in brief localized events, such as an impact event as we suggest for the nakhlite carbonates, then, in addition to the effects of a thinner atmosphere, it is to be expected that carbonate in the Amazonian will be sparse rather than in thick or wide-scale deposits.

5.7 CONCLUSIONS

Carbonate is present within the Martian crust due to water-ice−rock−CO_2 reactions resulting from a diverse range of processes—impact heating and excavation, hydrothermal fluids, and other less well-defined atmospheric-mineral reaction processes at the surface of Mars. Massive, sedimentary deposits of carbonate do not appear to exist except in Syrtis-Terra Tyrrhena where tens meter thick outcropping carbonate units are present. A wide range of Ca−Mg−Fe−Mn carbonate compositions can be inferred from NIR analysis by orbiters, analyses by the Spirit rover (Comache outcrop), Phoenix lander soil, and MSL eolian samples.

However, to date, detailed mineralogical analyses of compositions and textures of Martian carbonate have only been possible in a few of the SNC meteorites—the nakhlites and ALH 84001. Most carbonate found in Martian meteorites is of terrestrial origin. In published literature and new analyses and imagery presented here, Martian carbonates have been found in three of the nine nakhlite Martian meteorites—Lafayette, Nakhla, and GV—but not in any other nakhlite including recently discovered ones. The nakhlite carbonates are dominantly $FeCO_3$, with varying Mg, Mn, and Ca contents. The variation in composition is the result of HCO_3^- bearing fluids, which partially dissolved an olivine-rich, bulk nakhlite mixture. Other nakhlites are devoid of carbonates due to an early exhaustion of HCO_3^- in the hydrothermal fluids migrating through fractures in Lafayette, GV, and Nakhla in a single fluid event that—based on experimental work with analog mineral assemblages—probably lasted of the order of months, starting at $\leq 150\ °C$ and forming metastable assemblages. As the nakhlite secondary fluid evolved, some of the earlier sideritic carbonate was partially dissolved and replaced by ferric saponite and gel. The heat source may have been impact, and the remaining nakhlite meteorites were likely more distant from the source of melted $H_2O−CO_2$ ice. The ALH 84001 orthopyroxenite contains carbonate rosettes with zoned Fe−Ca-rich cores to Mg-rich rims, which probably formed at low temperatures such as $< 20\ °C$ (Halevy et al., 2011), with some shock-induced recrystallization.

Carbonate abundances on the Mars surface and in meteorites imply a crustal inventory equivalent to at least ~ 400 mbar $p\mathrm{CO_2}$. This, in turn, is consistent with models for the ancient atmosphere, which require a thick atmosphere of several bars $p\mathrm{CO_2}$. The nakhlite and ALH 84001 carbonates show signs of high $^{13}\mathrm{C}/^{12}\mathrm{C}$ ratios, for example $\delta^{13}\mathrm{C}$ $+41‰-50‰$. That range of values is similar to measurements of the current Mars atmosphere made by MSL of approximately $\delta^{13}\mathrm{C}$ $+50‰$. Variable D/H anomalies and also mass independent fractionation leading to positive $\Delta^{17}\mathrm{O}$-isotope anomalies are also seen within the Martian carbonate. Martian carbonates have thus recorded the effects of an ancient (e.g., early Hesperian and earlier) thick $\mathrm{CO_2}$ atmosphere with subsequent atmospheric loss processes.

ACKNOWLEDGMENTS

We would like to thank NIPR, Japan for the loan of sections of Y 000593 and Y 000749, the Natural History Museum, United Kingdom, for the loan of Lafayette, Governador Valadares, and Nakhla sections. We would like to thank Mohit Melwani-Daswani for useful discussions. JCB and LJH acknowledge funding from STFC. LPI Contribution #2064.

REFERENCES

Ali, A., Jabeen, I., Gregory, D., Verish, R., Banerjee, N.R., 2016. New triple oxygen isotope data of bulk and separated fractions from SNC meteorites: evidence for mantle homogeneity of Mars. Meteorit. Planet. Sci. 51, 981−995.

Anders, E., 1996. Evaluating the evidence for past life on Mars. Science 274, 2119−2121.

Anovitz, L.M., Essene, E.J., 1987. Phase equilibria in the system $\mathrm{CaCO_3-MgCO_3-FeCO_3}$. J. Petrol. 28, 389−414.

Bandfield, J.L., Glotch, T.D., Christensen, P.R., 2003. Spectroscopic identification of carbonate minerals in the Martian dust. Science 301, 1084−1087.

Barber, D.J., Scott, E.R.D., 2001. Transmission electron microscopy of carbonates and associated minerals in ALH 84001: impact-induced deformation and carbonate decomposition. Meteorit. Planet. Sci. 36, A13−A14.

Barber D.J., Scott E.R.D., 2002. Origins of potential Martian magneto-fossils in Allan Hills 84001 (abstract #6693). 27th General Assembly of the European Geophysical Society.

Barnhart, C.J., Nimmo, F., 2011. Role of impact excavation in distributing clays over Noachian surfaces. J. Geophys. Res. 116, E01009.

Berkley, J.L., Boynton, N.J., 1992. Minor/major element variation within and among diogenite and howardite orthopyroxenite groups. Meteorit. Planet. Sci. 27, 387−394.

Bischoff, J.L., Stine, S., Rosenbauer, R.J., Fitzpatrick, J.A., Stafford, T.W., 1993. Ikaite precipitation by mixing of shoreline springs and lake water, Mono Lake, California, USA. Geochim. Cosmochim. Acta 57, 3855−3865.

Boctor, N.Z., Alexander, C.M.O.'D., Wang, J., Hauri, E., 2003. The sources of water in Martian meteorites: clues from hydrogen isotopes. Geochim. Cosmochim. Acta 67, 3971−3989.

Borg, L.E., Connelly, J.N., Nyquist, L.E., Shih, C.Y., Wiesmann, H., Reese, Y., 1999. The age of the carbonates in Martian meteorite ALH84001. Science 286, 90−94.

Boynton, W.V., Ming, D.W., Kounaves, S.P., Young, S.M.M., Arvidson, R.E., Hecht, M.H., et al., 2009. Evidence for calcium carbonate at the Mars Phoenix landing site. Science 325, 61–64.

Bradley, J.P., Harvey, R.P., McSween Jr., H.Y., 1996. Magnetite whiskers and platelets in the ALH84001 Martian meteorite: evidence of vapor phase growth. Geochim. Cosmochim. Acta 60, 5149–5155.

Bradley, J., McSween, H., Harvey, R., 1998. Epitaxial growth of nanophase magnetite in Martian meteorite Allan Hills 84001: implications for biogenic mineralization. Meteorit. Planet. Sci. 33, 765–773.

Bramble M.S., Mustard J.F., Cannon K.M. 2017. Testing carbonate formation mechanisms at Northeast Syrtis Major using manual and automated hyperspectral analyses (abstract #2815). Lunar and Planetary Science Conference XLVIII.

Brearley A.J. 1998. Rare potassium-bearing mica in Allan Hills 84001: additional constraints on carbonate formation. Martian Meteorites: Where Do We Stand and Where Are We Going? Abstracts from a Workshop, p. 6.

Brearley A.J. 2000. Hydrous phases in ALH84001: further evidence for preterrestrial alteration and a shock-induced thermal overprint (abstract #1203). Lunar and Planetary Science Conference XXXI.

Brearley, A.J., 2003. Magnetite in ALH 84001: an origin by shock-induced thermal decomposition of iron carbonate. Meteorit. Planet. Sci. 38, 849–870.

Bridges, J.C., Grady, M.M., 2000. Evaporite mineral assemblages in the nakhlite (Martian) meteorites. Earth Planet. Sci. Lett. 176, 267–279.

Bridges, J.C., Schwenzer, S.P., 2012. The nakhlite hydrothermal brine on Mars. Earth Planet. Sci. Lett. 359–360, 117–123.

Bridges, J.C., Catling, D.C., Saxton, J.M., Swindle, T.D., Lyon, I.C., Grady, M.M., 2001. Alteration assemblages in Martian meteorites: implications for near-surface processes. Space Sci. Rev. 96, 365–392.

Bridges, J.C., Schwenzer, S.P., Leveille, R., Westall, F., Wiens, R.C., Mangold, N., et al., 2015. Diagenesis and clay mineral formation at Gale Crater, Mars. J. Geophys. Res. 120, 1–19.

Buchardt, B., Israelson, C., Seaman, P., Stockmann, G., 2001. Ikaite tufa towers in Ikka Fjord, southwest Greenland: their formation by mixing of seawater and alkaline spring water. J. Sediment. Res. 71, 176–189.

Bultel, B., Quantin-Nataf, C., Andréani, M., Clénet, H., Lozac'h, L., 2015. Deep alteration between Hellas and Isidis Basins. Icarus 260, 141–160.

Burns, R.G., Fisher, D.S., 1990. Iron-sulfur mineralogy of Mars: magmatic evolution and chemical weathering products. J. Geophys. Res.: Solid Earth 95, 14415–14421.

Carr, R.H., Grady, M.M., Wright, I.P., Pillinger, C.T., 1985. Martian atmospheric carbon dioxide and weathering products in SNC meteorites. Nature 314, 248–250.

Carrozzo F.G., Bellucci G., Altieri F., D'Aversa E., 2013. Detection of carbonate bearing-rocks in craters uplifts of Tyrrhena Terra, Mars (abstract #2241). Lunar and Planetary Science Conference XLIV.

Carter, J., Poulet, F., Bibring, J.-P., Murchie, S., 2010. Detection of hydrated silicates in crustal outcrops in the Northern Plains on Mars. Science 328, 1682–1686.

Catling, D.C., 1999. A chemical model for evaporites on early Mars: possible sedimentary tracers of the early climate and implications for exploration. J. Geophys. Res.: Planets 104, 16453–16469.

Changela, H.G., Bridges, J.C., 2011. Alteration assemblages in the nakhlites: variation with depth on Mars. Meteorit. Planet. Sci. 45, 1847–1867.

Clifford, S.M., Parker, T.J., 2001. The evolution of the Martian hydrosphere: implications for the fate of a primordial ocean and the current state of the Northern Plains. Icarus 154, 40–79.

Cooney T., Scott E.R., Krot A.N., Sharma S., Yamaguchi A. 1998. Confocal Raman microprobe and IR reflectance study of minerals in the Martian meteorite ALH84001 (abstract #1332). Lunar and Planetary Science Conference XXIX.

Cooney, T., Scott, E.R., Krot, A.N., Sharma, S., Yamaguchi, A., 1999. Vibrational spectroscopic study of minerals in the Martian meteorite ALH84001. Am. Mineral. 84, 1569–1576.

Corrigan, C.M., Harvey, R.P., 2004. Multi-generational carbonate assemblages in Martian meteorite Allan Hills 84001: implications for nucleation, growth, and alteration. Meteorit. Planet. Sci. 39, 17–30.

Ehlmann, B.L., Edwards, C.S., 2014. Mineralogy of the Martian surface. Annu. Rev. Earth Planet. Sci. 42, 291–315.

Ehlmann, B.L., Mustard, J.F., Murchie, S.L., Poulet, F., Bishop, J.L., Brown, A.J., et al., 2008. Orbital identification of carbonate-bearing rocks on Mars. Science 322, 1828–1832.

Ehlmann, B.L., Mustard, J.F., Swayze, G.A., Clark, R.N., Bishop, J.L., Poulet, F., et al., 2009. Identification of hydrated silicate minerals on Mars using MRO-CRISM: geologic context near Nili Fossae and implications for aqueous alteration. J. Geophys. Res. 114, E00D08.

Eiler, J.M., Valley, J.W., Graham, C.M., Fournelle, J., 2002. Two populations of carbonate in ALH84001: geochemical evidence for discrimination and genesis. Geochim. Cosmochim. Acta 66, 1285–1303.

Essene, E.J., 1983. Solid solutions and solvi among metamorphic carbonates with applications to geologic thermobarometry. In: Reeded, R.J. (Ed.), Carbonates: Mineralogy and Chemistry [Reviews in Mineralogy], Volume 11. Mineralogical Society of America, Michigan, USA, pp. 77–96.

Fairbridge, R.W., 1967. Carbonate rocks and paleoclimatology in the biogeochemical history of the planet. In: Chilingar, G.V., Bissell, H.J., Fairbridge, R.W. (Eds.), Carbonate Rocks: Origin, Occurrence and Classification [Developments in Sedimentology 9A]. Elsevier Publishing Company, Netherlands, pp. 399–434.

Farley, K.A., Malespin, C., Mahaffy, P., Grotzinger, J.P., Vasconcelos, P.M., Milliken, R.E., et al., 2014. In situ radiometric and exposure age dating of the Martian surface, and MSL Science Team. Science 343, 1247166.

Farquhar, J., Thiemens, M.H., 2000. Oxygen cycle of the Martian atmosphere-regolith system: $\Delta^{17}O$ of secondary phases in Nakhla and Lafayette. J. Geophys. Res.: Planets 105, 11991–11997.

Farquhar, J., Thiemens, M., Jackson, T., 1998. Atmosphere−sulphur interactions on Mars: $\Delta^{17}O$ measurements of carbonate from ALH84001. Science 280, 369.

Fergason R.L., Hare T.M., Kirk R.L., Piqueux S., Galuzska D.M., Golombek M.P., et al., 2017. Mars 2020 landing site evaluation: slope and physical property assessment (abstract #2163). Lunar and Planetary Science Conference XLVIII.

Fernández-Remolar, D., Gómez-Elvira, J., Gómez, F., Sebastian, E., Martıin, J., Manfredi, J., et al., 2004. The Tinto River, an extreme acidic environment under control of iron, as an analog of the Terra Meridiani hematite site of Mars. Planet. Space Sci. 52, 239–248.

Forget, F., Pierrehumbert, R.T., 1997. Warming early Mars with carbon dioxide clouds that scatter infrared radiation. Science 278, 1273–1276.

Franz, H.B., Kim, S.-T., Farquhar, J., Day, J.M., Economos, R.C., McKeegan, K.D., et al., 2014. Isotopic links between atmospheric chemistry and the deep sulphur cycle on Mars. Nature 508, 364–368.

Gaffey, S.J., 1987. Spectral reflectance of carbonate minerals in the visible and near infrared (0.35−2.55 μm): anhydrous carbonate minerals. J. Geophys. Res. 92, 1429–1440.

Gilmore M.S., Golder K.B., Korn L., Aaron L.M., 2014. Carbonate associated with gullies in the Eridania Region of Mars (abstract #1388). Eighth International Conference on Mars.

Gleason, J.D., Kring, D.A., Hill, D.H., Boynton, W.V., 1997. Petrography and bulk chemistry of Martian orthopyroxenite ALH84001: implications for the origin of secondary carbonates. Geochim. Cosmochim. Acta 61, 3503–3512.

Glotch, T.D., Rogers, A.D., 2013. Evidence for magma−carbonate interaction beneath Syrtis Major, Mars. J. Geophys. Res.: Planets 118, 126–137.

Golden, D., Ming, D., Schwandt, C., Morris, R., Yang, S., Lofgren, G., 2000. An experimental study on kinetically-driven precipitation of calcium−magnesium−iron carbonates from solution: implications for the low-temperature formation of carbonates in Martian meteorite Allan Hills 84001. Meteorit. Planet. Sci. 35, 457–466.

Golden, D., Ming, D.W., Schwandt, C.S., Lauer, H.V., Socki, R.A., Morris, R.V., et al., 2001. A simple inorganic process for formation of carbonates, magnetite, and sulfides in Martian meteorite ALH84001. Am. Mineral. 86, 370–375.

Golden, D., Ming, D., Morris, R., Brearley, A., Lauer, H., Treiman, A., et al., 2004. Evidence for exclusively inorganic formation of magnetite in Martian meteorite ALH84001. Am. Mineral. 89, 681–695.

Goldsmith, J.R., 1983. Phase relations of rhombohedral carbonates. In: Reeded, R.J. (Ed.), Carbonates: Mineralogy and Chemistry [Reviews in Mineralogy], Volume 11. Mineralogical Society of America, Ann Arbor, MI, pp. 49–76.

Golombek M.P., Grant J.A., Farley K.A., Williford K., Chen A., Otero R.E., et al., 2016. Downselection of landing sites proposed for the Mars 2020 rover mission (abstract #2324). Lunar and Planetary Science Conference XLVII.

Gooding, J.L., Wentworth, S.J., Zolensky, M.E., 1991. Aqueous alteration of the Nakhla meteorite. Meteoritics 326, 135–143.

Goudge, T.A., Milliken, R.E., Head, J.W., Mustard, J.F., Fassett, C.I., 2017a. Sedimentological evidence for a deltaic origin of the western fan deposit in Jezero crater, Mars and implications for future exploration. Earth Planet. Sci. Lett. 458, 357–365.

Goudge T.A., Ehlmann B.L., Fassett C.I., Head J.W., Mustard J.F., Mangold N., et al. 2017b. Jezero crater, Mars as a compelling site for future in situ exploration (abstract #1197). Lunar and Planetary Science Conference XLVIII.

Grady M.M., Anand M., Gilmore M.A., Watson J.S., Wright I.P., 2007. Alteration of the nakhlite lava pile: was water on the surface, seeping down, or at depth, percolating up? Evidence (such as it is) from carbonates (abstract #1826). Lunar and Planetary Science Conference XXXVIII.

Greenwood, J.P., McSween Jr., H.Y., 2001. Petrogenesis of Allan Hills 84001: constraints from impact-melted feldspathic and silica glasses. Meteorit. Planet. Sci. 36, 43–61.

Greenwood, J.P., Riciputi, L.R., McSween, H.Y., 1997. Sulfide isotopic compositions in shergottites and ALH84001, and possible implications for life on Mars. Geochim. Cosmochim. Acta 61, 4449–4453.

Greenwood, J.P., Mojzsis, S.J., Coath, C.D., 2000. Sulfur isotopic compositions of individual sulfides in Martian meteorites ALH84001 and Nakhla: implications for crust–regolith exchange on Mars. Earth Planet. Sci. Lett. 184, 23–35.

Greenwood, J.P., Blake, R.E., Coath, C.D., 2003. Ion microprobe measurements of $^{18}O/^{16}O$ ratios of phosphate minerals in the Martian meteorites ALH84001 and Los Angeles. Geochim. Cosmochim. Acta 67, 2289–2298.

Grotzinger, J.P., Sumner, D.Y., Kah, L.C., Stack, K., Gupta, S., Edgar, L., et al., 2014. A habitable fluvio-lacustrine environment at Yellowknife Bay, Gale Crater, Mars, and MSL Science Team. Science 343, 1242777

Gubler, Y., Bertrand, J.P., Mattavelli, L., Rizzini, A., Passega, R., 1967. Petrology and petrography of carbonate rocks. In: Chilingar, G.V., Bissell, H.J., Fairbridge, R.W. (Eds.), Carbonate Rocks: Origin, Occurrence and Classification [Developments in Sedimentology 9A]. Elsevier Publishing Company, Netherlands, pp. 51–86.

Halevy, I., Fischer, W.W., Eiler, J.M., 2011. Carbonates in the Martian meteorite Allan Hills 84001 formed at 18 ± 4 °C in a near-surface aqueous environment. Proc. Natl. Acad. Sci. 108, 16895–16899.

Harvey, R.P., McSween Jr., H.Y., 1994. Ancestor's bones and palimpsests: olivine in ALH 84001 and orthopyroxene in Chassigny. Meteoritics 29, 472.

Harvey, R.P., McSween, H.Y., 1996. A possible high-temperature origin for the carbonates in the Martian meteorite ALH84001. Nature 382, 49–51.

Heet T., Arvidson R.E., Mellon M. and The Phoenix Science Team, 2009. Regional geology and rock distributions of the Mars Phoenix landing site (abstract #1114). Lunar and Planetary Science Conference XL.

Hicks, L.J., Bridges, J.C., Gurman, S.J., 2014. Ferric saponite and serpentine in the nakhlite Martian meteorites. Geochim. Cosmochim. Acta 136, 194−210.

Hicks L.J., Bridges J.C., Greenwood R.C., Franchi I.A., 2016. NWA 10659: a clay-rich nakhlite pair of NWA 10153 (abstract #6421). 79th Annual Meeting of the Meteoritical Society.

Holland, G., Saxton, J.M., Lyon, I.C., Turner, G., 2005. Negative $\delta^{18}O$ values in Allan Hills 84001 carbonate: possible evidence for water precipitation on Mars. Geochim. Cosmochim. Acta 69, 1359−1369.

Irving A.J., Kuehner S.M., Ziegler K., Andreasen R., Righter M., Lapen T.J., et al., 2015. Chlorophaeite-bearing nakhlite Northwest Africa 10153: petrology, oxygen and hafnium isotopic composition, and implications for magmatic or crustal water on Mars (abstract #5251). 78th Annual Meeting of the Meteoritical Society.

Jakosky, B.M., Phillips, R.J., 2001. Mars' volatile and climate history. Nature 412, 237−244.

Jakosky, B.M., Slipski, M., Benna, M., Mahaffy, P., Elrod, M., Yelle, R., et al., 2017. Mars' atmospheric history derived from upper-atmosphere measurements of $^{38}Ar/^{36}Ar$. Science 355, 1408−1410.

Jambon, A., Sautter, V., Barrat, J., Gattacceca, J., Rochette, P., Boudouma, O., et al., 2016. Northwest Africa 5790: revisiting nakhlite petrogenesis. Geochim. Cosmochim. Acta 190, 191−212.

Jull, A.J.T., Eastoe, C.J., Xue, S., Herzog, G.F., 1995. Isotopic composition of carbonates in the SNC meteorites ALH84001 and Nakhla. Meteoritics 30, 311−318.

Kahn, R., 1985. The evolution of CO_2 on Mars. Icarus 62, 175−190.

Kasting, J.F., 1997. Warming early Earth and Mars. Science 276, 1213−1215.

Kent, A., Hutcheon, I., Ryerson, F., Phinney, D., 2001. The temperature of formation of carbonate in Martian meteorite ALH84001: constraints from cation diffusion. Geochim. Cosmochim. Acta 65, 311−321.

Lapen, T., Righter, M., Brandon, A., Debaille, V., Beard, B., Shafer, J., et al., 2010. A younger age for ALH84001 and its geochemical link to shergottite sources in Mars. Science 328, 347−351.

Lee M.R., Chatzitheodoridis E. 2015. Formation of berthierine in the Martian meteorite Nakhla be replacement of aluminosilicate glass (abstract #5219). 78th Annual Meeting of the Meteoritical Society.

Lee, M.R., Tomkinson, T., Mark, D.F., Stuart, F.M., Smith, C.L., 2013. Evidence for silicate dissolution on Mars from the Nakhla meteorite. Meteorit. Planet. Sci. 48, 224−240.

Lee, M.R., Tomkinson, T., Hallis, L.J., Mark, D.F., 2015. Formation of iddingsite veins in the Martian crust by centripetal replacement of olivine: evidence from the nakhlite meteorite Lafayette. Geochim. Cosmochim. Acta 154, 49−65.

Leshin, L.A., Epstein, S., Stolper, E.M., 1996. Hydrogen isotope geochemistry of SNC meteorites. Geochim. Cosmochim. Acta 60, 2635−2650.

Leshin, L.A., McKeegan, K.D., Carpenter, P.K., Harvey, R.P., 1998. Oxygen isotopic constraints on the genesis of carbonates from Martian meteorite ALH84001. Geochim. Cosmochim. Acta 62, 3−13.

LPI, 2017. The Meteoritical Bulletin Database. Available at: http://www.lpi.usra.edu/meteor/metbull.php.

Madeleine, J.-B., Forget, F., Head, J.W., Levrard, B., Montmessin, F., Millour, E., 2009. Amazonian northern mid-latitude glaciation on Mars: a proposed climate scenario. Icarus 203, 390−405.

Manning, C.V., McKay, C.P., Zahnle, K.J., 2006. Thick and thin models of the evolution of carbon dioxide on Mars. Icarus 180, 38−59.

McKay, D.S., Gibson, E.K., Thomas-Keprta, K.L., Vali, H., Romanek, C.S., Clemett, S.J., et al., 1996. Search for past life on Mars: possible relic biogenic activity in Martian meteorite ALH84001. Science 273, 924−930.

McSween, H.Y.J., Harvey, R.P., 1998. An evaporation model for formation of carbonates in the ALH84001 Martian meteorite. Int. Geol. Rev. 40, 774−783.

Melwani Daswani, M., Schwenzer, S.P., Reed, M.H., Wright, I.P., Grady, M.M., 2016. Alteration minerals, fluids, and gases on early Mars: predictions from 1-D flow geochemical modelling of mineral assemblages in meteorite ALH 84001. Meteorit. Planet. Sci. 51, 2154−2175.

Michalski, J.R., Niles, P.B., 2010. Deep crustal carbonate rocks exposed by meteor impact on Mars. Nat. Geosci. Lett. 3, 751−755.

Michalski, J.R., Cuadros, J., Niles, P.B., Parnell, J., Deanne Rogers, A., Wright, S.P., 2013. Groundwater activity on Mars and implications for a deep biosphere. Science 6, 133−138.

Mikouchi, T., Koizumi, E., Monkawa, A., Ueda, Y., Miyamoto, M., 2003. Mineralogy and petrology of Yamato 000593: comparison with other Martian nakhlite meteorites. Antarctic Meteorite Res. 16, 34−57.

Mikouchi T., Miyamoto M., Koizumi E., Makishima J., McKay G., 2006. Relative burial depths of nakhlites: an update (abstract #1865). Lunar and Planetary Science Conference XXXVII.

Mikouchi T., Makishima J., Kurihara T., Hoffmann V.H., Miyamoto M., 2012. Relative burial depth of nakhlites revisited (abstract #2363). Lunar and Planetary Science Conference XLIII.

Mikouchi T., Righter M., Ziegler K., Irving A.J., 2016. Petrology, mineralogy and oxygen isotope composition of the Northwest Africa 10153 nakhlite: a sample from a different flow from other nakhlites? (abstract #6369). 79th Annual Meeting of the Meteoritical Society.

Ming, D.W., Archer Jr., P.D., Glavin, D.P., Eigenbrode, J.L., Franz, H.B., MSL Science Team, et al., 2014. Volatile and organic compositions of sedimentary rocks in Yellowknife Bay, Gale Crater, Mars. Science 343, 1245267.

Mittlefehldt, D.W., 1994. ALH84001, a cumulate orthopyroxenite member of the Martian meteorite clan. Meteorit. Planet. Sci. 29, 214−221.

Morris, R.V., Ruff, S.W., Gellert, R., Ming, D.W., Arvidson, R.E., Clark, B.C., et al., 2010. Identification of carbonate-rich outcrops on Mars by the Spirit Rover. Science 329, 421−424.

Müller, T., Cherniak, D., Watson, E.B., 2012. Interdiffusion of divalent cations in carbonates: experimental measurements and implications for timescales of equilibration and retention of compositional signatures. Geochim. Cosmochim. Acta 84, 90−103.

Murchie, S., Arvidson, R., Bedini, P., Beisser, K., Bibring, J.-P., Bishop, J., et al., 2007. Compact Reconnaissance Imaging Spectrometer for Mars (CRISM) on Mars reconnaissance orbiter (MRO). J. Geophys. Res. 112, E05S03.

Mustard, J.F., Murchie, S.L., Pelkey, S.M., Ehlmann, B.L., Milliken, R.E., Grant, J.A., et al., 2008. Hydrated silicate minerals on Mars observed by the Mars Reconnaissance Orbiter CRISM instrument. Nature 454, 305−309.

Niles P.B., Leshin L., Socki R., Guan Y., Ming D., Gibson E., 2004. Cryogenic calcite—a morphologic and isotopic analog to the ALH84001 carbonates (abstract: #1459). Lunar and Planetary Science Conference XXXV.

Niles, P.B., Leshin, L., Guan, Y., 2005. Microscale carbon isotope variability in ALH84001 carbonates and a discussion of possible formation environments. Geochim. Cosmochim. Acta 69, 2931−2944.

Niles, P.B., Zolotov, M.Y., Leshin, L.A., 2009. Insights into the formation of Fe- and Mg-rich aqueous solutions on early Mars provided by the ALH 84001 carbonates. Earth Planet. Sci. Lett. 286, 122−130.

Niles, P.B., Catling, D.C., Berger, G., Chassefière, E., Ehlmann, B.L., Michalski, J.R., et al., 2013. Geochemistry of carbonates on Mars: implications for climate history and nature of aqueous environments. Space Sci. Rev. 174, 301−328.

Nyquist, L.E., Bogard, D.D., Shih, C.Y., Greshake, A., Stöffler, D., Eugster, O., 2001. Ages and geologic histories of Martian meteorites. Chronol. Evol. Mars 96, 105−164.

Perron, J.T., Mitrovica, J.X., Manga, M., Matsuyama, I., Richards, M.A., 2007. Evidence for an ancient Martian ocean in the topography of deformed shorelines. Nature 447, 840−843.

Romanek, C.S., Grady, M.M., Wright, I.P., Mittlefehldt, D.W., Socki, R.A., Pillinger, C.T., et al., 1994. Record of fluid−rock interactions on Mars from the meteorite ALH84001. Lett. Nat. 372, 655−657.

Saxton, J., Lyon, I., Turner, G., 1998. Correlated chemical and isotopic zoning in carbonates in the Martian meteorite ALH84001. Earth Planet. Sci. Lett. 160, 811−822.

Schwenzer, S.P., Bridges, J.C., Wiens, R.C., Conrad, P.G., Kelley, S.P., Leveille, R., et al., 2016. Fluids during diagenesis and sulfate vein formation in sediments at Gale Crater, Mars. Meteorit. Planet. Sci. 51, 2175−2202.

Schwenzer S.P., Bridges J.C., Miller M.A., Hicks L.J., Ott U., Filiberto J., et al., 2017. Diagenesis on Mars: insights into noble gas pathways and newly formed mineral assemblages from long term experiments (abstract: submitted). Lunar and Planetary Science Conference XXIX.

Scott, E.R., Yamaguchi, A., Krot, A.N., 1997. Petrological evidence for shock melting of carbonates in the Martian meteorite ALH84001. Nature 387, 377−379.

Scott, E.R., Krot, A.N., Yamaguchi, A., 1998. Carbonates in fractures of Martian meteorite Allan Hills 84001: petrologic evidence for impact origin. Meteorit. Planet. Sci. 33, 709−719.

Shaheen, R., Niles, P.B., Chong, K., Corrigan, C.M., Thiemens, M.H., 2015. Carbonate formation events in ALH84001 trace the evolution of the Martian atmosphere. Proc. Natl. Acad. Sci. 112, 336−341.

Shearer, C.K., Leshin, L.A., Adcock, C.T., 1999. Olivine in Martian meteorite ALH 84001: evidence for a high-temperature origin and implications for signs of life. Meteorit. Planet. Sci. 34, 331−339.

Shih C.Y., Nyquist L.E., Reese Y., Weismann H., 1998. The chronology of the nakhlite, Lafayette: Rb−Sr and Sm−Nd isotopic ages (abstract #1145). Lunar and Planetary Science Conference XXIX.

Smith, P.H., Tamppari, L.K., Arvidson, R.E., Bass, D., Blaney, D., Boynton, W.V., et al., 2009. H_2O at the Phoenix landing site. Science 325, 58−61.

Steele, A., Fries, M.D., Amundsen, H.E.F., Mysen, B.O., Fogel, M.L., Schweizer, M., et al., 2007. Comprehensive imaging and Raman spectroscopy of carbonate globules from Martian meteorite ALH 84001 and a terrestrial analogue from Svalbard. Meteorit. Planet. Sci. 42 (9), 1549−1566.

Sutter B., Heil R.E., Rampe E.B., Morris R.V., Ming D.W., Archer P.D., et al., 2015. Iron-rich carbonates as the potential source of evolved CO_2 detected by the Sample Analysis at Mars (SAM) instrument in Gale Crater (abstract #20150019435). American Geophysical Union, Fall Meeting.

Swindle, T.D., Treiman, A.H., Lindstrom, D.J., Burkland, M.K., Cohen, B.A., Grier, J.A., et al., 2000. Noble gases in iddingsite from the Lafayette meteorite: evidence for liquid water on Mars in the last few hundred million years. Meteorit. Planet. Sci. 35, 107−115.

Thomas-Keprta K.L., Wentworth S.J., McKay D.S., Gibson J.E.K., 2000. Field emission gun scanning electron microscopy (FEGSEM) and transmission electron (TEM) microscopy of phyllosilicates in Martian meteorites ALH84001, Nakhla, and Shergotty (abstract #1690). Lunar and Planetary Science Conference XXXI.

Thomas-Keprta, K.L., Clemett, S.J., Bazylinski, D.A., Kirschvink, J.L., McKay, D.S., Wentworth, S.J., et al., 2002. Magnetofossils from ancient Mars: a Robust Biosignature in the Martian Meteorite ALH84001. Appl. Environ. Microbiol. 68, 3663−3672.

Thomas-Keprta, K., Clemett, S., McKay, D., Gibson, E., Wentworth, S., 2009. Origins of magnetite nanocrystals in Martian meteorite ALH84001. Geochim. Cosmochim. Acta 73, 6631−6677.

Tosca, N.J., McLennan, S.M., Lindsley, D.H., Schoonen, M.H., 2004. Acid-sulfate weathering of synthetic Martianbasalt: the acid fog model revisited. J. Geophys. Res. 109, E05003. Available from: http://dx.doi.org/10.1029/2003JE002218.

Treiman, A.H., 1995. A petrographic history of Martian meteorite ALH84001: two shocks and an ancient age. Meteoritics 30, 294−302.

Treiman, A.H., 1998. The history of Allan Hills84001 revised: multiple shock events. Meteorit. Planet. Sci. 33, 753−764.

Treiman, A.H., 2003. Submicron magnetite grains and carbon compounds in Martian meteorite ALH84001: inorganic, abiotic formation by shock and thermal metamorphism. Astrobiology 3, 369−392.

Treiman, A.H., 2005a. The nakhlite meteorites: augite-rich igneous rocks from Mars. Chemie der Erde 65, 203−270.

Treiman A.H., 2005b. Olivine and carbonate globules in ALH84001: a terrestrial analog, and implications for water on Mars (abstract #1107). Lunar and Planetary Science Conference XXXVI.

Treiman, A.H., Essene, E.J., 2011. Chemical composition of magnetite in Martian meteorite ALH 84001: revised appraisal from thermochemistry of phases in Fe−Mg−C−O. Geochim. Cosmochim. Acta 75, 5324−5335.

Treiman, A.H., Barrett, R.A., Gooding, J.L., 1993. Preterrestrial aqueous alteration of the Lafayette (SNC) meteorite. Meteoritics 28, 86−97.

Turner, S.M.R., Bridges, J.C., Grebby, S., Ehlmann, B.L., 2016. Hydrothermal activity recorded in post Noachian-aged impact craters on Mars. J. Geophys. Res.: Planets 121, 608−625.

Usui, T., Alexander, C.M.D., Wang, J., Simon, J.I., Jones, J.H., 2015. Meteoritic evidence for a previously unrecognized hydrogen reservoir on Mars. Earth Planet. Sci. Lett. 410, 140−151.

Usui T., Alexander C., Wang J., Simon J., Jones J., 2016. Coordinated in situ NanoSIMS analyses of HCO isotopes in Allan Hills 84001 carbonates (abstract #1780). Lunar and Planetary Science Conference XLVII.

Valley, J.W., Eiler, J.M., Graham, C.M., Gibson, E.K., Romanek, C.S., Stolper, E.M., 1997. Low-temperature carbonate concretions in the Martian meteorite ALH84001: evidence from stable isotopes and mineralogy. Science 275, 1633−1638.

van Berk, W., Ilger, J.M., Fu, Y., Hansen, C., 2011. Decreasing CO_2 partial pressure triggered Mg−Fe−Ca carbonate formation in ancient Martian crust preserved in the ALH84001 Meteorite. Geofluids 11, 6−17.

Vaniman, D.T., Bish, D.L., Ming, D.W., Bristow, T.F., Morris, R.V., Blake, D.F., et al., 2014. Mineralogy of a mudstone on Mars. Science 343. Available from: http://dx.doi.org/10.1126/science1243480.

Warren, P.H., 1998. Petrologic evidence for low-temperature, possibly flood evaporitic origin of carbonates in the ALH84001 meteorite. J. Geophys. Res.: Planets 103, 16759−16773.

Webster C.R., Mahaffy P.R., Leshin L.A., Atreya S.K., Flesch G.J., Stern J., et al. and MSL Science Team, 2013. Mars atmospheric escape recorded by H, C and O isotope ratios in carbon dioxide and water measured by the SAM tunable laser spectrometer on the Curiosity Rover (abstract #1365). Lunar and Planetary Science Conference XLIV.

Wentworth S.J., Thomas-Keprta K.L., Mckay D.S., 1998. Alteration products and secondary minerals in Martian meteorite Allan Hills 84001. Martian Meteorites: Where Do We Stand and Where Are We Going? Abstracts from a Workshop, p. 59.

Wieler, R., Huber, L., Busemann, H., Seiler, S., Leya, I., Maden, C., et al., 2016. Noble gases in 18 Martian meteorites and angrite Northwest Africa 7812—exposure ages, trapped gases, and a re-evaluation of the evidence for solar cosmic ray-produced neon in shergottites and other achondrites. Meteorit. Planet. Sci. 51, 407−428.

Wray J.J., Murchie S.L., Ehlmann B.L., Milliken R.E., Bishop J.L., Seelos K.D., et al., 2011. Orbital evidence for iron or calcium carbonates on Mars (abstract #EPSC-DPS2011-1719). European Planetary Science Congress 2011.

Wright, I.P., Grady, M.M., Pillinger, C.T., 1992. Chassigny and the nakhlites: carbon-bearing components and their relationship to Martian environmental conditions. Geochim. Cosmochim. Acta 56, 817−826.

Xu T., Apps J., Pruess K., 2001. Analysis of mineral trapping for CO_2 disposal in deep aquifers. Lawrence Berkeley National Laboratory Report LBNL-46992, Permalink: http://escholarship.org/uc/item/59c8k6gb.

CHAPTER

SULFUR ON MARS FROM THE ATMOSPHERE TO THE CORE

6

Heather B. Franz[1], Penelope L. King[2], and Fabrice Gaillard[3]

[1]*NASA Goddard Space Flight Center, Greenbelt, MD, United States*
[2]*Australian National University, Canberra, ACT, Australia* [3]*CNRS-Université d'Orléans, Orléans, France*

6.1 INTRODUCTION

6.1.1 THE PREVALENCE OF SULFUR ON MARS

The widespread occurrence of sulfur on Mars was first noted by the Viking mission (Baird et al., 1976; Clark et al., 1976, 1977; Toulmin et al., 1977). Since that time, the ubiquitous nature of Martian sulfur has been confirmed through in situ measurements by subsequent rovers and landers, remote sensing observations by orbiting spacecraft, and analyses of Martian meteorites in terrestrial laboratories. These observations all point to a higher abundance of sulfur on the surface of Mars than on present-day Earth, leading to suggestions that the geochemical history of the Martian surface may have been dominated by the sulfur cycle, analogous to the manner in which the carbon cycle dominates surficial processes on Earth (Gaillard et al., 2013; Halevy et al., 2007; King et al., 2004; King and McLennan, 2010; McLennan and Grotzinger, 2008). The Martian sulfur cycle includes sulfur output from the deep interior to the surface and atmosphere through magmatism and volcanism, as well as assimilation of sulfur-bearing phases from the surface or shallow subsurface into rising magmas (Farquhar et al., 2000b; Franz et al., 2014a; Jones, 1989; King et al., 2004). Nonetheless, experimental constraints suggest that on Mars as on Earth, the abundance of sulfur in the core is a critical parameter in modeling the physical state of the core and its ability to sustain a global magnetic field (Stewart et al., 2007).

6.1.2 SULFUR GEOCHEMISTRY AND ISOTOPES

Sulfur is a versatile element that is found in a wide range of organic and inorganic compounds, in valence states from -2 to $+6$, and in solid, aqueous, and gas phases. It has four stable isotopes, ^{32}S, ^{33}S, ^{34}S, and ^{36}S, with abundances of approximately 95.04%, 0.75%, 4.20%, and 0.015%, respectively (Ding et al., 2001). Because these isotopes fractionate during processing of sulfur-bearing compounds, analysis of sulfur isotopes has emerged as a vital tool for understanding numerous processes, both geological and biological in nature. Most commonly, mass-dependent fractionation occurs, in which isotopes of a given element partition differentially between chemical

Volatiles in the Martian Crust. DOI: http://dx.doi.org/10.1016/B978-0-12-804191-8.00006-4

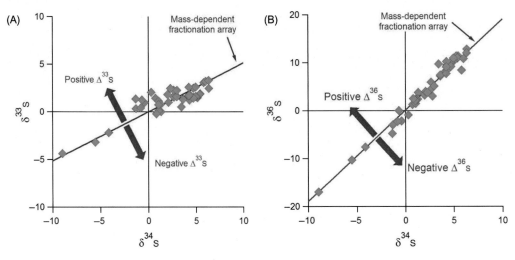

FIGURE 6.1

Illustration of mass-independent fractionation (MIF) of sulfur isotopes, (A) $\Delta^{33}S$ and (B) $\Delta^{36}S$. Points on the *red line* display mass-dependent fractionation, while points above or below the *red line* (i.e., possessing nonzero values of $\Delta^{33}S$ or $\Delta^{36}S$) display MIF. Data shown are from Archean samples of Farquhar et al. (2000a).

compounds or phases based on the small differences in their masses (Urey, 1947). In other cases, the partitioning of isotopes is controlled by additional factors and cannot be predicted simply from isotopic masses. For example, on Mars UV photochemistry is an important mechanism to partition isotopes of certain elements, including oxygen and sulfur, in ways that are not proportional to the difference in isotopic masses (Clayton and Mayeda, 1983; Farquhar et al., 2001; Thiemens, 1999). This type of behavior, commonly known as mass-independent fractionation (MIF), presents an additional means to track processes in elements with more than two stable isotopes through deviation of the minor isotopes from mass-dependent compositions. Anomalies in abundances of both ^{33}S and ^{36}S compared to the other isotopes produce distinctive patterns that differ with the processes responsible for the fractionation (Farquhar et al., 2000a, 2001; Gao and Thiemens, 1991, 1993a,b; Hulston and Thode, 1965; Ono et al., 2006). MIF is portrayed conceptually in Fig. 6.1.

Isotope ratios are usually reported in terms of delta (δ) values, representing per mil (‰) deviations from established reference standards. For sulfur, the original reference standard was Fe-sulfide from the Cañon Diablo iron meteorite, or Cañon Diablo Troilite (CDT) (Jensen and Nakai, 1962). Due to dwindling supply of material from the meteorite, as well as heterogeneity observed in isotopic composition of CDT samples (Beaudoin et al., 1994), the International Atomic Energy Agency (IAEA) established a modern sulfur isotope scale (Robinson, 1993). This scale, known as V-CDT, is closely linked to CDT but utilizes a set of synthetic Ag-sulfide mixtures spanning the range in sulfur isotopic compositions typically observed in natural samples as reference materials (Ding et al., 2001). The sulfur isotopic composition measured for a sample is then reported as the deviation from the reference composition for each ratio of minor isotope to ^{32}S:

$$\delta^{33}\text{S} = \left[\left(^{33}\text{S}/^{32}\text{S} \right)_{\text{meas}} / \left(^{33}\text{S}/^{32}\text{S} \right)_{\text{ref}} - 1 \right] \times 1000 \tag{6.1}$$

$$\delta^{34}\text{S} = \left[\left(^{34}\text{S}/^{32}\text{S} \right)_{\text{meas}} / \left(^{34}\text{S}/^{32}\text{S} \right)_{\text{ref}} - 1 \right] \times 1000 \tag{6.2}$$

$$\delta^{36}\text{S} = \left[\left(^{36}\text{S}/^{32}\text{S} \right)_{\text{meas}} / \left(^{36}\text{S}/^{32}\text{S} \right)_{\text{ref}} - 1 \right] \times 1000 \tag{6.3}$$

where the subscript "meas" denotes the measured ratio and "ref" denotes the isotopic ratio of the reference standard. "Capital delta" notation is used to report MIF:

$$\Delta^{33}\text{S} = \delta^{33}\text{S} - 1000 \times \left[\left(1 + \delta^{34}\text{S}/1000 \right)^{0.515} - 1 \right] \tag{6.4}$$

$$\Delta^{36}\text{S} = \delta^{36}\text{S} - 1000 \times \left[\left(1 + \delta^{34}\text{S}/1000 \right)^{1.90} - 1 \right] \tag{6.5}$$

where values of zero indicate mass-dependent behavior and nonzero values indicate MIF.

6.1.3 CLUES TO MARS' PAST

Studying sulfur compounds on Mars can provide clues to numerous facets of Martian history, which will be explored in the sections that follow. The abundance and redox state of sulfur-bearing minerals in the igneous rocks comprising our collection of Martian meteorites informs estimates of the planet's total inventory of sulfur and the oxidation state of magmas. These parameters are also linked to the quantity and identity of sulfurous gases that would have been released into the atmosphere due to volcanic activity. Since SO_2 in particular has the potential to act as a greenhouse gas, the atmospheric mixing ratio of sulfur, in conjunction with that of water, is critical in attempts to model the Martian climate and to address the question of whether the surface was ever sufficiently warm that liquid water could have existed even transiently.

The past climate and water activity also bear directly on the question of whether ancient Mars was habitable—in other words, whether it could have supported a biosphere with any resemblance to that of Earth. The terrestrial sulfur cycle is heavily dominated by biological processes, which are intimately tied to the sulfur isotopic compositions observed in all types of environments. The presence of sulfur-metabolizing microorganisms near the root of the phylogenetic tree (e.g., Edmonds et al., 1991; Gregersen et al., 2011; Shooner et al., 1996; Stetter, 2007; Woese et al., 1991) has sparked interest in whether such primitive metabolisms could also have arisen on Mars, with its rich abundance of sulfur. Sulfur isotope ratios also provide evidence suggesting that a UV photochemical cycle, distinct from that of Earth, operated in the Martian atmosphere in the past. Deciphering this isotopic fingerprint to reveal the nature of photochemical processes that produced it may yield clues to the composition of the atmosphere at the time, which affects climate scenarios as well as potential means for providing UV shielding to protect past Martian life.

6.2 MARTIAN SULFUR RESERVOIRS

6.2.1 SULFUR ABUNDANCE AND CORE FORMATION

It is informative first to assess Martian sulfur abundance in a Solar System context. Sulfur was the most abundant moderately volatile element in the solar nebula, present as H_2S that condensed out of the nebula by temperature-dependent reactions with solid phases (Ebel, 2011; Lodders et al., 2009). For example, highly refractory CaS condenses ~ 1200 K (Pasek et al., 2005), while Fe-sulfides form in the range $\sim 500-710$ K (Lauretta et al., 1997; Lodders et al., 2009; Pasek et al., 2005). Variations observed among classes of chondrites in volatile element abundances relative to solar composition reflect fractionation by gas–solid reaction processes during condensation. The CI chondrites have sulfur abundances similar to the solar value (2−10 wt%) (Palme and Jones, 2003), while other types of chondrites exhibit relative depletions (Lodders et al., 2009). Volatile element depletions are also observed in differentiated Solar System bodies, including the Earth, Moon, and eucrites (Palme et al., 1988). In comparison to Earth, analyses of Martian meteorites suggest that Mars contains more volatile and moderately volatile elements, except for chalcophile elements (Zn, Tl, In) (Dreibus and Wänke, 1985). The sulfur content of bulk Mars has been modeled in various ways using elemental volatility trends in comparison to chondritic meteorites, yielding a range of results as shown in Table 6.1.

In the case of both Earth and Mars, most sulfur is modeled as residing in the core. These models are based on partitioning of sulfur between Fe−Ni metal (the "core") and silicate melt (the "mantle") during core segregation (e.g., Kargel and Lewis, 1993), although Gaillard et al. (2013) noted the difficulties inherent in such modeling due to nonideal behavior of sulfur in Fe−S melts. The sulfur content of the core is a critical parameter in models of core solidification, since its presence depresses the melting temperature compared to pure Fe or Fe−Ni metal (Fei et al., 2000; Stewart et al., 2007). At least a portion of the core must be liquid to sustain a convectively driven magnetic field (Acuna et al., 1999; Connerney et al., 1999; Fei and Bertka, 2005; Schubert and Spohn, 1990; Stewart et al., 2007; Young and Schubert, 1974). Magnetic surveys of the Martian crust by the Mars Global Surveyor provide evidence to suggest that the young Mars had an active core dynamo for as long as several hundred million years (Acuna et al., 1999; Lillis et al., 2008; Solomon et al., 2005). This dynamo could have been driven by convection due to a compositional gradient in a liquid outer core as the inner core solidified or by thermal convection due to high heat flux out of a fully liquid core (Fei and Bertka, 2005; Stevenson, 2001). The sulfur abundance of the core thus strongly influences the ability of the planet to sustain a magnetic field, which has implications for the climate and surface radiation environment (Lillis et al., 2008).

Elemental abundance estimates for Mars have been generated through analyses of Martian meteorites combined with geophysical models, producing a range of results for the distribution of sulfur between Mars' mantle and core (Table 6.1). Estimates of mantle sulfur abundance based directly on sulfur abundances in meteorites have yielded a wide range of values, because the ultimate sulfur abundance and distribution among mineral phases in the meteorites reflect a multitude of factors (see Section 6.2.3). However, it is inferred from meteorite compositional analyses that the Martian mantle is roughly a factor of two richer in FeO than that of Earth (McSween, 1994; Wänke and Dreibus, 1994). As explained in Section 6.2.2, the amount of sulfur that can be dissolved in silicate melts is limited by its partitioning with other coexisting phases. The maximum sulfur content of a silicate melt, governed by the sulfur capacity, is expected to correlate with FeO content (Carroll and Rutherford, 1985; Gaillard et al., 2013; O'Neill and Mavrogenes, 2002; Wallace and Carmichael, 1992), so assuming that the Martian mantle was once molten (i.e., magma ocean) (Elkins-Tanton, 2008), the sulfur abundance of the Martian mantle is expected to be higher than that of the Earth's mantle (Gaillard and Scaillet, 2009; Wänke and Dreibus, 1994).

Table 6.1 S Contents of Different Reservoirs on Mars

Reservoir		Sulfur	References
Bulk Mars		2.2 wt%	Lodders and Fegley (1997)
Core		14.24 wt%	Dreibus and Wänke (1985)
		10.6 wt%	Lodders and Fegley (1997)
		16.2 wt%	Sanloup et al. (1999)
		16 ± 2 wt%	Rivoldini et al. (2011)
		10.6−14.9 wt%	Wang et al. (2013b)
		> 20 wt%	Khan and Connolly (2008)
		< 8 wt%	Shahar et al. (2015)
		5−10 wt%	Wang and Becker (2017)
Primitive mantle		700−1000 ppm	Ding et al. (2015), Gaillard and Scaillet (2009)
		700−2000 ppm	Gaillard et al. (2013) and references therein
		3000−7000 ppm	Righter et al. (2009)
		900 ppm	Lodders and Fegley (1997)
		360 ppm	Wang and Becker (2017)
Meteorites			
Olivine-phyric shergottites		408−2700 ppm	Aoudjehane et al. (2012), BasuSarbadhikari et al. (2009), Ding et al. (2015), Franz et al. (2014a), Gibson et al. (1985), Lodders (1998), Shirai and Ebihara (2004), Zipfel et al. (2000)
Basaltic shergottites		65−2865 ppm	Ding et al. (2015), Franz et al. (2014a), Gibson et al. (1985), Lodders (1998)
Nakhlites		34−1287	Ding et al. (2015), Farquhar et al. (2007b), Franz et al. (2014a), Gibson et al. (1985), Lodders (1998)
Chassignites		67−440	Franz et al. (2014a), Gibson et al. (1985), Lodders (1998)
Orthopyroxenite		103−110	Dreibus et al. (1994), Franz et al. (2014a)
Surface rocks measured in situ			
Viking 1 and 2		2.4−3.8 wt%	Clark (1993)
Mars Pathfinder		1.6−2.6 wt%	Bell et al. (2000)
MER-A Spirit	Soil	0.99−35.8	NASA Planetary Data System
	Brushed rocks	1.52−9.28	
	Abraded rocks	1.23−7.96 (12.88)	
MER-B	Soil	4.56−7.36 (16.46)	
Opportunity	Brushed rocks	2.33−25.2	
	Abraded rocks	(0.56) 17−28.62	
MSL Curiosity	Soil	1.85−2.81	
(< Sol 1182)	Igneous rocks	0.85−2.37	
APXS	Sed. rocks	0.36−5.36	
Phoenix Wet Chemistry Lab	Rosy Red	0.40 ± 0.12	Kounaves et al. (2010)
	Sorceress-2	0.48 ± 0.12	
GRS global average		1.76 wt%	McLennan et al. (2010)
Layered sulfate deposits		8−12 wt%	Gaillard et al. (2013)

Wänke and Dreibus (1988) applied general cosmochemical constraints to estimate the bulk composition of Mars by applying mass balance calculations to elemental abundances observed in chondrites. They demonstrated that two-component mixing models, originally developed to explain Earth's mantle composition (Ringwood, 1979; Wänke, 1981), could also be applied to Mars. The model of Wänke and Dreibus required a highly reduced component A, free of elements more volatile than Na, and an oxidized component B, containing all elements in CI abundances, in proportions A:B of approximately 60:40. Sulfur, present as FeS in component B, reacted with Fe−Ni metal from component A to form a sulfur-rich Fe−Ni alloy during core segregation, extracting chalcophile elements into the core according to their sulfide-silicate partition coefficients. The core composition was estimated as 77.8% Fe, 7.6% Ni, 0.4% Co, and 14.2% S (Wänke and Dreibus, 1988).

Other compositional models for Mars have incorporated oxygen isotopes to constrain accretion from mixtures of different types of chondrites (e.g., Lodders and Fegley, 1997). It was shown by Clayton et al. (1976) that the Earth, Moon, and chondritic meteorites may be classified into several distinct groups according to their oxygen isotope systematics, specifically by variations in the relationship between $^{18}O/^{16}O$ and $^{17}O/^{16}O$ that have been interpreted to reflect heterogeneities in solar nebula composition or localized effects of photochemical processes that occurred in the nebula (Clayton, 1993, 2002, 2003; Clayton and Mayeda, 1984; Clayton et al., 1991; Thiemens and Heidenreich, 1983). Analyses of Martian meteorites reveal that Mars also carries a characteristic mass-independent oxygen isotopic signature (Clayton and Mayeda, 1983, 1996), with silicates displaying $\Delta^{17}O$ of 0.32‰ on average compared to V-SMOW (Franchi et al., 1999). From mass balance calculations, Lodders and Fegley (1997) suggested that Mars accreted from about 85% H-, 11% CV-, and 4% CI-chondritic material, comprising ∼80% silicates with the remaining ∼20% in a metal-sulfide core containing ∼10.6% sulfur. This model produced an estimate of 2.2% for the bulk sulfur content of Mars and 900 ppm sulfur in the Martian mantle (Lodders and Fegley, 1997), approximately four times that of Earth (Allègre et al., 2001; Dreibus and Palme, 1996). The latter value overlaps with the range of 700−1000 ppm reported as upper limits for cold and hot Martian mantle models based on a recent analysis of meteorite bulk sulfur contents, combined with a mantle partial melting model (Ding et al., 2015).

Some more recent models for the bulk chemical composition of Mars have exploited improved knowledge of Martian geophysical data provided by radiometric tracking of spacecraft, such as moment of inertia (MOI), solar tidal deformation parameters, mass, and size, to constrain estimates of bulk composition. For example, a higher sulfur content in the core implies lower core density (Fig. 6.2), requiring a larger core radius to fit a given mass and producing a larger mean MOI factor, while a lower sulfur content would have the opposite effect (Wang et al., 2013b). Bertka and Fei (1998) performed high-pressure experiments with appropriate analog mixtures to evaluate the Wänke and Dreibus (1988) model in light of MOI measurements based on tracking of the Pathfinder rover (Folkner et al., 1997; Yoder and Standish, 1997). They concluded that accretion models based solely on CI chondrites cannot match the observed MOI of Mars and are insufficient to explain the variations in Fe/Si among the terrestrial planets, but that mixtures of carbonaceous chondrites with ordinary or enstatite

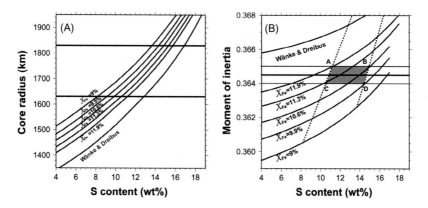

FIGURE 6.2

Calculated (A) core radius and (B) MOI factor as a function of sulfur content in the core for various Fe contents (X_{Fe}) in the mantle, assuming a Martian crust with thickness of 50 km and density of 3.0 g/cm^3. *Solid lines* in (A) and *dashed lines* in (B) denote the range of possible core radius, 1630 and 1830 km, inferred by Konopliv et al. (2011); *solid lines* in (B) are the observed value and error bars of the mean MOI factor. *Modified after Wang et al. (2013b) with permission.*

chondrites could produce bulk planet compositions consistent with geophysical constraints (Bertka and Fei, 1998). Sanloup et al. (1999), using an oxygen isotope approach similar to that of Lodders and Fegley (1997) but further constrained by Pathfinder MOI observations (Folkner et al., 1997), invoked a mixture of H- and EH-chondrites (Wasson and Kallemeyn, 1988) in a ratio of 55:45 to produce a core containing a maximum of 16.2% sulfur (Sanloup et al., 1999).

Wang et al. (2013b) used a method similar to that of Bertka and Fei (1998) with an updated Mars MOI based on measurements of Mars Reconnaissance Orbiter, Mars Global Surveyor, Mars Odyssey, Pathfinder, and Viking (Konopliv et al., 2011), but assumed a liquid Fe−S core (Yoder et al., 2003) as opposed to the solid core assumed by Bertka and Fei (1998). Wang et al. (2013b) used the mantle compositional model of Wänke and Dreibus (1988) and the temperature profile of Fei and Bertka (2005) to test the effects of core sulfur concentration on mantle−core density profiles, then calculated the MOI factors for each density profile and compared to observed MOI (Konopliv et al., 2011) to constrain Mars' composition. The range of compositions generated by their model indicates core sulfur content between 10.6 and 14.9 wt% (Wang et al., 2013b). Another study by Rivoldini et al. (2011) produced a model for the interior structure of Mars based on the thermoelastic properties of the mantle and core, constrained by the MOI and tidal deformation parameters. Those authors derived a sulfur abundance of 16 ± 2% for the core (Rivoldini et al., 2011).

Stewart et al. (2007) performed experiments with Fe−S and (Fe,Ni)−S mixtures at pressures corresponding to the core−mantle boundary and the center of Mars (~23 and 40 GPa, respectively) to investigate possible core crystallization scenarios. Their experiments included sulfur abundances of 10.6, 14.2, and 16.2 wt%, which encompass the range of core sulfur abundances determined by compositional studies, as well as the eutectic melt compositions at various core pressures (Table 6.1). By comparing liquidus intersections of the resulting pressure–temperature curves with temperature models for the Martian interior, they concluded that presently the Martian core is almost entirely, if not entirely, in a liquid state. Their model also examined potential mechanisms of core solidification that may ensue as Mars continues to cool. They identified two possible crystallization regimes for Martian core conditions, depending on the sulfur abundance. On the iron-rich side of the eutectic composition (10 and 14 wt% sulfur), iron-rich solids would nucleate in the outer region of the core and sink toward the center in a "snowing core" model. On the sulfur-rich side of the eutectic (14–16 wt% sulfur), Fe-sulfide phases would crystallize from the center of the core outward, in a "sulfide inner-core" model (Stewart et al., 2007).

Stewart et al. (2007) noted that since the Martian core is currently in a liquid state, the putative magnetic field that existed during the Noachian period must have been driven by a vigorous core convection different than that of the Earth. Some major change in the liquid core convection regime, other than core solidification, must have occurred to cause the cessation of Mars' early magnetic field. Furthermore, they suggested that when large-scale core crystallization ultimately begins on Mars, the resultant change in convection regime could reestablish a dynamo driving a second period of Mars' history characterized by a strong global magnetic field (Stewart et al., 2007). The crystallization regimes indicated by their model suggest that the sulfur abundance in the core will largely control the nature of this dynamo.

The highest existing estimate of Martian core sulfur content was obtained by Khan and Connolly (2008), who applied inversion of Martian geophysical parameters to produce an estimate of > 20 wt% sulfur in the core. Khan and Connolly (2008) also calculated a bulk planet Fe/Si ratio of ~ 1.2 and suggested that this indicates accretion of Mars from different Solar System materials than those that formed Earth, such as ordinary chondrites, which also have oxygen isotopic ratios similar to those in Martian meteorites.

Two recent studies have resulted in lower estimates for the sulfur content of the Martian core than those reported previously. Shahar et al. (2015) performed a series of experiments at high pressure and temperature designed to constrain the iron isotopic fractionation during core–mantle differentiation. Reconciliation of their experimental results with geophysical models and data for iron isotopic signatures in Martian meteorites requires an upper limit of ~ 8 wt% sulfur in the Martian core. Wang and Becker (2017) pointed out that the high FeO content of Martian meteorites would require a higher sulfur concentration in parent magmas than have generally been measured in meteorites. They examined chalcophile element concentrations in Martian meteorites and noted that limited variations in ratios of these elements in magmas with a wide range of volatility indicates little degassing loss of several volatile elements, including sulfur, during eruption. Based on limited variability in ratios of elements with different sulfide-silicate melt partition coefficients and negative correlations of chalcophile elements with MgO, they further suggested that sulfide-undersaturated magmas may have been prevalent throughout Mars' geological history. We should note here that under reduced conditions similar to those prevailing in the Martian interior during the earliest stages of magma ocean or even for more recent magmatic activities, sulfur is not a strongly volatile element; at $fO_2 < IW + 2$, sulfur tends to remain in the molten silicate (see Gaillard et al., 2013). That being said, through comparison of copper and sulfur contents in meteorites, Wang and Becker (2017) estimated a concentration of $360 \pm 120\,\mu g/g$ of sulfur in the Martian mantle and 5−10 wt% sulfur in the core. They state that such a low sulfur content in the interior of Mars is consistent with the low zinc content in the Martian mantle, from which an estimate of ~ 5 wt% sulfur in the core is derived (Wang and Becker, 2017).

Another recent study examining sulfur isotopic fractionation between metal and silicate melts at high pressure suggests that the $\delta^{34}S$ of mantle sulfur may yield information about the pressure of core−mantle differentiation (Labidi et al., 2016). Sulfur in the Martian mantle (Franz et al., 2014) and in chondrites (Gao and Thiemens, 1993a,b) has an isotopic composition indistinguishable from CDT, while sulfur in the Earth's mantle shows a small depletion in ^{34}S (Labidi et al., 2013). A series of high-pressure and temperature experiments with synthetic mixtures of metal and silicate melts revealed that the sulfur isotopic fractionation between metal and silicate phases increases linearly with aluminum and boron concentrations, which affect the structure of the melt through coordination number of bonds and lead to stronger sulfur bonds in metal alloys than in silicates (Labidi et al., 2016; Shahar et al., 2009). Because the coordination number is similarly affected by changes in pressure, the authors conclude that the occurrence of sulfur isotopic fractionation during formation of Earth's core and the absence of such fractionation on Mars suggests a lower average pressure during core−mantle equilibration on Mars compared to Earth (Labidi et al., 2016). This topic deserves further investigation.

6.2.2 SULFUR IN MELTS, VAPOR, SULFIDE, SULFATE, AND METAL

How much sulfur dissolves in silicate melts, commonly called sulfur solubility in silicate melts, is limited by the ability of the coexisting phase to concentrate sulfur. In the framework of the Martian cycling of sulfur and its secular evolution, the phases coexisting with a silicate melt of basaltic composition are metal, sulfide, sulfate, and vapor during the stages of core formation, mantle melting, magma fractionation, and magma degassing, respectively. Due to the tremendous range of oxidation states for sulfur (−2 to +6), the partitioning of sulfur between a silicate melt and any of the above-mentioned phases is critically controlled by oxygen fugacity, fO_2, which is a thermodynamic measure of the oxidation state of a multiphase system (Gaillard et al., 2015).

In silicate melts, sulfur essentially dissolves as S^{2-} and S^{6+} (Jugo et al., 2010). In metals, by definition, the oxidation state of sulfur is 0. In (liquid) Fe-sulfide, FeS, we generally assume a −2 oxidation state, although nonstoichiometric sulfide, FeS_x, is common (Kiseeva and Wood, 2013). The existence of a complete solid solution between molten Fe and FeS at high temperature complicates the matter, as it implies a change in the oxidation state of both Fe and S at a composition that remains unclear (Gaillard and Scaillet, 2009; Wang et al., 1991). This continuous solid solution between metal and sulfide has introduced confusion in the literature addressing core−mantle equilibration (Gaillard et al., 2013). Sulfate is a magmatic mineral that carries sulfur with the oxidation state +6 (Jugo, 2009). Finally, in the vapor phase, the commonly considered species, H_2S, S_2, and SO_2, carry sulfur with an oxidation state of −2, 0, and +4 (Gaillard et al., 2011). Variation of the prevalent oxidation state between phases—melt, metal, sulfide, and vapor—for a given fO_2 implies that the partitioning of sulfur between these phases is overwhelmed by fO_2. In turn, the transfer of sulfur between these phases is also expected to affect fO_2 (Gaillard et al., 2011).

The solubility of sulfur in silicate melts saturated in sulfide (solid or molten) has been widely studied by material scientists and geochemists (O'Neill and Mavrogenes, 2002, and references therein). The concept of sulfur capacity was introduced 70 years ago to quantify the following heterogeneous equilibrium:

$$S^{2-} + \tfrac{1}{2} O_2 = O^{2-} + \tfrac{1}{2} S_2. \tag{6.6}$$

Eq. (6.6) describes the partitioning of sulfur between gas and melt. The introduction of ionic species to a silicate melt in equilibrium, with a thermodynamic standard state difficult to define, led to the concept of sulfur capacity, representing a quantitative gauge of the ability of the silicate melt to dissolve sulfur. From Eq. (6.6), one can assume that the silicate melt is dominated by oxygen anions (i.e., activity of $O^{2-} = 1$), which defines the dependence of sulfur capacity, C_S, on oxygen fugacity, fO_2, and sulfur fugacity, fS_2:

$$\ln C_S^{silicate} = \ln X_{S2-}^{silicate} + \frac{1}{2} \ln fO_2 - \frac{1}{2} \ln fS_2. \tag{6.7}$$

This equilibrium is critical in the modeling of sulfur degassing and can also be related to another parameter critical for petrological modeling of sulfur: the sulfur content at sulfide saturation (SCSS) that corresponds to the following equilibrium:

$$S^{2-} + FeO = O^{2-} + FeS. \tag{6.8}$$

Eq. (6.8) implies that the SCSS can be calculated from the Gibbs free energy changes of Eq. (6.8), $\Delta G^{\circ}_{(6.8)}$, the activity of FeO in the silicate melt, $a_{(FeO)}$, and the sulfur capacity (Gaillard et al., 2013; O'Neill and Mavrogenes, 2002):

$$\ln (X_{S2-})^{ppm-basalt} = \Delta G^{\circ}_{(6.8)}/[RT] + \ln Cs - \ln a_{(FeO)}. \tag{6.9}$$

Several experimental studies have provided constraints on the dependence of C_S on the melt chemical composition (see O'Neill and Mavrogenes, 2002). In the framework of Martian magmatism, the critical parameter is the high FeO content of basalts, which is 2–3 times greater than that of basalts on Earth (Ding et al., 2014; Gaillard and Scaillet, 2009; Righter et al., 2009). Several empirical formulations of the SCSS have been produced (Ding et al., 2014; Righter et al., 2009) and applied to determine the maximum sulfur content of Martian basalts (see Section 6.2.3). It appears that Martian basalts, being richer in FeO than Earth basalts, can dissolve ~ 4000 ppm sulfur, or 3–4 times the sulfur content in terrestrial mid-ocean ridge basalts.

Less relevant to Martian magmatism, petrological modeling of the sulfur content at anhydrite saturation (SCAS) has been performed by Jugo (2009). The SCAS exceeds 1% S in basaltic magma, which is significantly greater than the SCSS even for the FeO-rich Martian basalts. However, this petrological parameter is relevant to fO_2 conditions $> \sim 1.5$–2 log units above the quartz-fayalite-magnetite (QFM) buffer, or at least 2 log units more oxidized than our current estimates of typical Martian basalt fO_2 (Wadhwa, 2001).

6.2.3 SULFUR IN THE PRIMITIVE MANTLE AND CRUST

The primitive crust of Mars most likely formed following core—mantle differentiation, although their coeval formation cannot be ruled out. A widely accepted model holds that the crust formed from crystallization of a magma ocean, accompanied by substantial outgassing (Elkins-Tanton, 2008; Elkins-Tanton et al., 2005). Estimates of the quantity of volatiles released during this stage and the pressure of the resultant proto-atmosphere are difficult to constrain (Elkins-Tanton, 2008; Erkaev et al., 2014; Hirschmann and Withers, 2008), and efforts to quantify volcanic production have primarily focused on activity during the past ~ 4 Ga (Baratoux et al., 2013; Grott et al., 2011; Hirschmann and Withers, 2008).

Most of our knowledge of Martian petrology and mineralogy has been acquired through the study of Martian meteorites. At the time of publication, 107 distinct (unpaired) meteorites of Martian origin have been recovered and classified on Earth (Irving, 2017; McSween, 1994; Meyer, 2015). They all derive from igneous parent rocks, with the single exception of an impact-produced breccia, NWA 7034, and its pairs (Agee et al., 2013; Humayun et al., 2013). Igneous rocks range in age from ~ 4.1 Ga for ALH 84001 (Lapen et al., 2010) to ~ 175 Ma for the youngest group of shergottites (Nyquist et al., 2001). Their primary sulfur mineralogy is dominated by Fe-sulfides, with minor sulfates, as discussed in more detail in Section 6.2.5. As mentioned earlier, the ultimate sulfur abundance and distribution among mineral phases in the meteorites reflect a complex interplay of factors, including mantle sulfur contents, degree of (partial) melting, pressure, temperature, iron, and volatile contents of the magma, sulfur degassing during ascent, water activity of the melt, sulfur speciation, fO_2, later weathering processes, and possible incorporation of crustal sulfur into magmas or impact glasses (Franz et al., 2014; Gaillard et al., 2013; Gaillard and Scaillet, 2009; Jones, 1989; Rao et al., 1999, 2011; Wallace and Carmichael, 1992; Walton et al., 2010). Accurate interpretation of meteorite data requires consideration of sulfur contents with respect to the abundances of other elements. Measurements of the bulk sulfur abundances of Martian meteorites have yielded a wide range of values (Table 6.1), sometimes even among multiple analyses of a single meteorite. These variations likely arise from a combination of sample heterogeneity and differences in measurement techniques between studies.

The most primitive Martian meteorites and arguably most informative about Mars' mantle composition are shergottites. The initial oxygen fugacity (fO_2) of shergottite magmas upon partial melting of the mantle source was typically in a reducing range, ~ 1 to 3 log units below the QFM buffer (Herd, 2003; Herd et al., 2002; Lorand et al., 2005; McSween et al., 1996; Smith and Hervig, 1979; Stolper and McSween, 1979; Wadhwa, 2001). Since the oxidation state and solubility of sulfur in a S-saturated melt is a function of fO_2, magmatic sulfur under these conditions would have either dissolved in the melt as S^{2-} (preserved in glass) or produced sulfide minerals (Jugo et al., 2005; Wallace and Carmichael, 1992). A limited number of shergottites record more oxidizing conditions with fO_2 near the QFM buffer (Herd, 2003). This is within the range where S^{6+}/total S becomes significant (Jugo et al., 2005) and may indicate that late-stage sulfates could have been stable (Gross et al., 2013).

Sulfides in the shergottites are commonly associated with Fe—Ti oxides (Lorand et al., 2005), reflecting the relationship between FeO content and sulfide saturation as the melts evolved. The behavior of sulfur in basaltic magma reflects the competing effects of pressure, temperature, composition, fO_2 and fS_2 (Haughton et al., 1974; O'Neill and Mavrogenes, 2002). Although each of these factors alone can result in sulfide saturation, their activities in concert control precipitation of sulfide minerals. Sulfur solubility decreases as a magma ascends from its mantle source region and cools along an adiabatic path. At conditions of emplacement, FeO content of the melt may decrease due to precipitation of Fe−Ti oxides, which may induce localized sulfide precipitation (Gaillard et al., 2013; Haughton et al., 1974; Kress and Carmichael, 1991; Li and Ripley, 2005, 2009; O'Neill and Mavrogenes, 2002; Righter et al., 2009; Wallace and Carmichael, 1992).

The sulfide abundances of shergottites have also been used in conjunction with compositional data of shergottites and chondritic meteorites to estimate the sulfur abundance of the Martian mantle (Table 6.1 and references therein) and the quantity of sulfur that may have been introduced to the surface through volcanic outgassing (Gaillard and Scaillet, 2009; Halevy et al., 2007; Righter et al., 2009). Calculations of this type may be best undertaken with meteorites that approximate melt compositions, as the sulfur abundances in meteorites with significant cumulate fractions may underrepresent the composition of the mantle (Ding et al., 2015). Another important consideration is that the meteorites may include sulfur from crustal or hydrothermal rather than magmatic sources (Franz et al., 2014). The maximum sulfur content of a Martian magma can be estimated by assuming that it was initially sulfide saturated (Gaillard et al., 2013; Righter et al., 2009). Gaillard et al. (2013) adopted a range of 700−2000 ppm for Martian mantle sulfur, although other studies have suggested that the lower end of this range from 700 to 1000 ppm is most realistic (Ding et al., 2015; Gaillard and Scaillet, 2009). Assuming that sulfur is perfectly incompatible during melting, more conservative estimates yield 3000 ± 500 ppm or more sulfur in Martian magmas upon mantle melting (Gaillard and Scaillet, 2009; Righter et al., 2009), largely exceeding the sulfur content of primary basalts on Earth (1000 ± 500 ppm). A first critical question recently raised by Ding et al. (2015) is whether the mantle melting that produced the shergottites on Mars left a mantle residue that remained sulfide saturated. Interestingly, Ding et al. (2015) argued based on geochemical modeling that it may not always be the rule, since sulfide can be exhausted during mantle melting. This implies that sulfur content < 3000 ppm in primary shergottites can be expected. The second critical question is whether all sulfur dissolved in the shergottite melts can be degassed into the atmosphere (see Section 6.2.4).

In contrast to the shergottites, the nakhlites have higher fO_2 values of ΔQFM 0 to $+1.5$ (e.g., Chevrier et al., 2011; Dyar, 2005; Szymanski et al., 2010). Such values allow for higher sulfur contents in any melt phase, and a larger proportion of S^{6+} in the melt that could saturate the melt in sulfates. See Section 6.2.2 for related discussion.

Although most existing constraints on the sulfur abundance of the Martian mantle have been derived from sulfur geochemistry of Martian meteorites, ratios of chalcophile elements may also be used to constrain sulfur behavior in melts. As discussed in Section 6.2.1, Wang and Becker (2017) estimated ~ 360 ppm sulfur in the Martian mantle using this method, significantly lower than most other estimates.

6.2.4 SULFUR FROM VOLCANIC OUTGASSING

As a magma ascends, a fraction of its sulfur escapes due to degassing (Fig. 6.3), with quantity and species of sulfur-bearing gases dependent on several factors, including the overlying pressure, temperature, melt fO_2, fS_2, melt composition, and the concentration and speciation of other volatile elements present (e.g., H, C, Cl) (Gaillard and Scaillet, 2014). Models of Martian volcanic

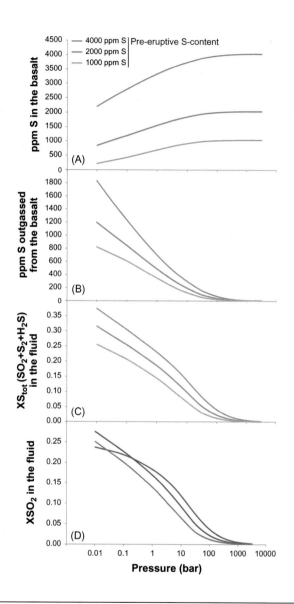

FIGURE 6.3

Degassing of sulfur from Martian basalts containing variable amounts of sulfur. In each plot, the preeruptive concentrations of water (500 ppm) and CO_2 (1600 ppm) are constant and taken from Usui et al. (2012), while fO_2 is taken as QFM -0.8, that is, the uppermost range of Martian fO_2 upon melting, which represents the most favorable case for sulfur degassing. (A) The equilibrium S content in basalt during degassing versus pressure; (B) the mass of S degassed as g/Mg basalt; (C) the mole fraction of total sulfur in the gas ($H_2S + S_2 + SO_2$); (D) the mole fraction of SO_2; comparison with (A) demonstrates that the fraction of SO_2 at equilibrium is independent of the initial, preeruptive, sulfur content in the undegassed melt.

degassing are discussed extensively in Gaillard and Scaillet (2009) and Gaillard et al. (2013). Besides the physical effect of pressure on gas solubility, sulfur degassing on Mars is particularly sensitive to fO_2 and magmatic water content, with both factors exerting positive effects of similar magnitude on quantity of sulfur released (Gaillard et al., 2013). The maximum sulfur degassing would be achieved by hydrated and oxidized melts (~0.4 wt% water and $fO_2 \sim$ QFM -0.5), which would release about 75% of their initial sulfur content into the atmosphere (Gaillard et al., 2013). Although H_2S would have dominated during degassing of an early magma ocean and at higher eruptive pressures (1−1000 bar), the dominant sulfur species at conditions expected for Martian volcanism during the past ~4 Ga (<0.1 bar) would be SO_2 (Gaillard et al., 2013; Zolotov, 2003). Release of SO_2 would also be favored by an increase in oxidation state of magmas due to assimilation of crustal material, a process that would have been facilitated by slow magma ascent in low gravity and high liquidus temperatures of magmas with low water content (Zolotov, 2003). An increase in fO_2 due to assimilation of oxidized crustal material (Franz et al., 2014; Herd et al., 2002; Jones, 1989; Wadhwa, 2001) would be offset by a decrease in fO_2 due to outgassing of SO_2 from the magma (Gaillard et al., 2013). A synthesis of the tradeoff between sulfur content in primary magma (during mantle melting) and the degassing process that transfers this magmatic sulfur into the atmosphere is summarized in Fig. 6.4. Here, different cases simulate the degassing of sulfur-rich and sulfur-poor magma. The conclusion is clear: the amount of sulfur degassing is affected by the amount of sulfur present in the primary melt, but a decrease of a factor of 4 in the sulfur melt content translates into a decrease by a factor of 2 in the quantity of sulfur degassed. Furthermore, the sulfur content and its speciation in the volcanic gases are only moderately affected by the sulfur content of the primary melt, with the highest mole fraction of SO_2 produced by the

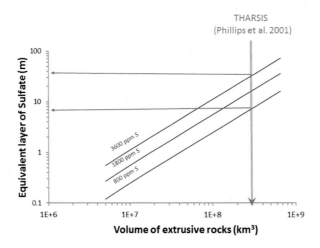

FIGURE 6.4

Model predictions for sulfate precipitated due to volcanic degassing, expressed as equivalent to a uniform layer around the planet. *Arrow* indicates the volume of extrusive rocks produced by the Tharsis province as estimated by Phillips et al. (2001). The Tharsis province is expected to have produced a uniform layer of sulfate covering the surface of Mars with depth from a few meters to a few tens of meters.

most sulfur-poor melts. This illustrates that degassing is the key factor in the transfer of sulfur from the mantle to the atmosphere.

Gaillard and Scaillet (2009) calculated the amount of sulfur that could have been introduced to the Martian surface during formation of the Tharsis volcanic province, assuming associated basalt production of 3×10^8 km^3 (Phillips et al., 2001). We revise their estimate here based on the amount of sulfur outgassed for sulfur-rich and sulfur-poor Martian melts as defined in Fig. 6.4. If all outgassed sulfur were ultimately deposited on the ground, we estimate that volcanic sulfur could have delivered a global equivalent layer of sulfates 7–35 m thick. Volcanic outgassing thus presents a feasible mechanism for generating the abundant sulfate minerals now observed and redistributed at the surface (Gaillard and Scaillet, 2009).

A substantial quantity of sulfur has likely been cycled through the Martian atmosphere due to volcanic outgassing. The timing of past activity is based on application of crater size-frequency distribution models to volcanic landforms (Hartmann and Neukum, 2001). Studies by Werner (2009) and Robbins et al. (2011) using images acquired with the Mars Express High Resolution Stereo Camera (HRSC) and the Mars Reconnaissance Orbiter Context Camera (CTX), respectively, concluded that volcanic activity in the Tharsis province ended between 100 and 200 million years ago. However, the unusually long and episodic lifetimes of volcanoes on Mars (Frankel, 1996) have led to suggestions that there may still be active regions on the planet today. A study of plain-style volcanoes with CTX indicated extended areas in Tharsis that have been resurfaced by basaltic lava flows in the past few tens of millions of years (Hauber et al., 2011). Neukum et al. (2004) examined HRSC images of the Tharsis and Elysium provinces and reported evidence for repeated activation and resurfacing of five volcanoes during the last 20% of Martian history. Citing phases of activity that occurred as recently as 2 million years ago, they suggested that Tharsis volcanoes may be currently active (Neukum et al., 2004). In addition, they reported evidence for glacial activity at the base of the Olympus Mons escarpment as recently as 4 million years ago (Neukum et al., 2004). Further HRSC images of Olympus Mons revealed evidence for very recent geological activity, $\leq 25-40$ Ma, on the volcano's eastern flank, including fluvial networks, emplacement of lava flows and dikes, and wrinkle ridges and troughs due to tectonic processes (Basilevsky et al., 2006).

6.2.5 SULFUR AT THE MARTIAN SURFACE

Our knowledge of sulfur abundance and mineralogy at the surface of Mars is based on a combination of data acquired through remote sensing from Earth and by spacecraft orbiting Mars, in situ measurements by instruments on rovers and landers, and studies of Martian meteorites. As indicated in Fig. 6.5 and discussed in the following sections, sulfur may be found on Mars in a range of oxidation states, dependent on reactions with iron, oxygen, hydrogen, carbon, and sulfur or interactions with radiation. Common sulfur oxidation states are: S^{2-} (e.g., in pyrrhotite, FeS_{1-x}), S^{1-} (e.g., pyrite FeS_2), S^{4+} ($SO_{2\,(g)}$, SO_2^{2-} bisulfine, SO_3^{2-} bisulfite) and S^{6+} ($SO_{3\,(g)}$, SO_4^{2-} sulfate, SO_2^{2+} sulfone), and S^+, S^{2+}, S^{3+}, and S^{5+} may also exist.

In the following sections, we briefly describe some of the measurement techniques that have provided existing data on sulfur chemistry and mineralogy from (A) the Martian surface, (B) Martian meteorites, and (C) orbiters.

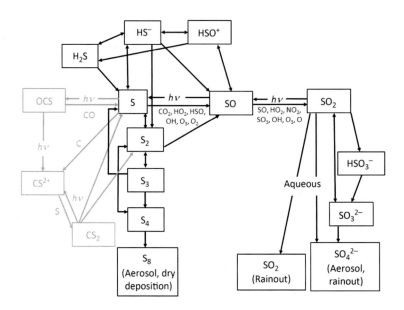

FIGURE 6.5

Schematic diagram of the primary atmospheric reactions involving sulfur species (after Johnson et al. 2009). Species are assumed to occur in gas phase with the exceptions of SO_2 and SO_4^{2-} rainout (aqueous phase) and S_8 and SO_4^{2-} aerosol deposition (solid phase). *Black lines and text* indicate species included in the model of Johnson et al. (2009), and *gray lines and text* indicate notional representation of reactions between S- and C-bearing species that were not considered in that model but may have played a significant role in the Martian atmosphere during periods of active volcanism. Notation "$h\nu$" generally refers to wavelengths in the ultraviolet (UV) range; discussion of specific wavelengths responsible for these reactions is beyond the scope of this chapter.

6.3 IN SITU OBSERVATIONS

6.3.1 MEASUREMENT TECHNIQUES

6.3.1.1 X-Ray Fluorescence and Alpha-Particle/Proton X-Ray Spectrometry

A range of sulfur contents has been measured in situ with rover and lander missions on Mars (Table 6.1; Fig. 6.6). X-ray fluorescence (XRF) analyses from the Viking missions revealed that Mars has sulfur-rich surface regolith or soil (Baird et al., 1976; Clark et al., 1976, 1982; Toulmin et al., 1977). We use the term "soil" in a general sense to imply loose, unconsolidated material at the planet's surface, as applied by previous authors (e.g., Blake et al., 2013; Morris et al., 2000; Vaniman et al., 2014). Subsequent rover missions confirmed high sulfur contents using two different alpha-particle/proton X-ray spectrometry (APXS) instruments: an alpha-*proton* X-ray spectrometer on the Pathfinder mission (Bell et al., 2000; Bruckner et al., 2001; Foley et al., 2001; McSween and Keil, 2000) and an alpha-particle X-ray spectrometer on each of the Mars exploration rovers (MER) Spirit and Opportunity (Gellert et al., 2006; McLennan and Grotzinger, 2008; Ming et al., 2006, 2008; Rieder et al., 2004) and the Mars Science Laboratory Curiosity rover (Berger et al., 2014, 2015;

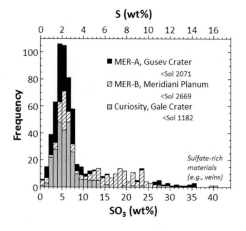

FIGURE 6.6

Histograms showing sulfur contents (as both S and SO_3 wt%) from APXS measurements of bulk rock samples using bins of 1 wt% (larger than the error for most measurements). Data are available in the Planetary Data System: MER-A, Gusev crater (< Sol 2071), MER-B, Meridiani Planum (< Sol 2669) and Curiosity, Gale crater (< Sol 1182), where "<" refers to sols prior to the given sol number.

Campbell et al., 2012; McLennan et al., 2014). The results of these studies are examined in more detail later.

Dust, by which we generally mean weathered grains usually less than a few micrometers in size that remain suspended in the atmosphere for prolonged periods of time, is ubiquitous on Mars' surface and in its atmosphere; therefore, dust or regolith is included as a coating on underlying rocks or sediments reported by XRF and many APXS analyses. APXS analysis of airfall dust on the observation tray of the Curiosity rover revealed the highest concentrations of SO_3 and Cl yet measured in Martian dust (8.3 wt% SO_3, or 3.3 wt% S, and 1.1 wt% Cl) (Berger et al., 2016). The surface of Mars is reported to have an average SO_3 content of 6.8 wt% (2.7 wt% S), with soil having an average SO_3 content of ~6.2 wt% (2.5 wt% S) (Taylor and McLennan, 2009; Yen et al., 2005). Our compilation in Fig. 6.6 suggests that the median value for most MER and MSL APXS analyses is ~5−6 wt%. Sulfur-rich veins and infill are common (Fig. 6.7) and these have higher sulfur contents (Fig. 6.6).

Using data from the Mars Pathfinder mission, Foley et al. (2003) calculated a soil-free rock composition using linear regression of rock and coating compositions to subtract a soil component (Foley et al., 2003). Similar approaches or ratios of raw data to global soil or dust compositions have been applied in the MER mission (e.g., Gellert et al., 2006; Yen et al., 2005). Alternately, authors have examined samples that have been brushed or abraded (RAT—rock abrasion tool) to minimize the contribution from surface material. Such studies show that the coatings on rocks contribute to the analysis of untouched (as is) rocks and that if rocks have not been brushed or abraded, the interpretation of their bulk sulfur contents is challenging. For example, Fig. 6.8 shows Cl versus SO_3 for rocks from the MER-A (Spirit rover) at Gusev crater, indicating that rocks that have been abraded have lower Cl and SO_3 than those that are brushed or untouched. These data indicate

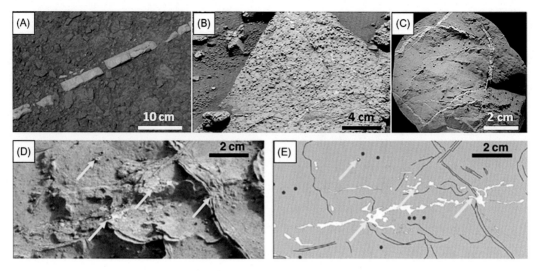

FIGURE 6.7

Compilation of images from the Martian surface showing sulfate-rich veins. (A) "Homestake" vein from the western rim of Endeavour crater. This image was collected from the panoramic camera (Pancam) on NASA's MER Opportunity using light with wavelengths centered at 601 nm (red), 535 nm (green), and 482 nm (blue). (B) Sheepbed target showing millimeter-sized nodules (dark) and also crosscutting white sulfate-bearing fractures. This image was collected as a Mastcam324 mosaic on NASA's Mars Science Laboratory mission. (C) Sheepbed target with sulfate veins (white) through nodule-rich mudstone. This image was collected as a ChemCam RMI image on NASA's Mars Science Laboratory mission. (D, E) Raised ridges (*red lines* in E are cements likely composed of akaganeite, magnetite, and/or smectite; Siebach et al. (2014), nodules (black in E) and Ca-sulfate veins in the Sheepbed target (white in E). *(A) Modified after Squyres et al. (2012), image credit NASA/JPL-Caltech/Cornell/ASU. (B and C) Modified after Grotzinger et al. (2014), image credit NASA/JPL-Caltech/MSSS. (D and E) Images are mosaics from MastCam, image credit NASA/JPL-Caltech/MSSS, interpretation from Siebach et al. (2014).*

mixing between the rock and either, or both, soil and Ca- and/or Mg-sulfate minerals (Ming et al., 2006). For instance, the samples with higher SO_3 content are inferred to contain sulfate minerals. Within the soil, sulfur is commonly correlated with Cl, Ca, Mg (Figs. 6.9−6.11) and/or Na and Br, as well as oxidized Fe^{3+} (Fig. 6.12; Morris et al., 2006).

6.3.1.2 Wet Chemistry Titration

The Phoenix lander included a Wet Chemical Laboratory (WCL) with four cells designed to analyze chemical properties of Martian soil samples (Kounaves et al., 2009). After mixing a soil sample with a leaching solution, sulfate detection was facilitated by titration with $BaCl_2$ powder, in which Ba^{2+} ions released by dissolution reacted with sulfate ions in solution. A Ba^{2+} ion-selective electrode allowed detection and quantification of sulfate present in the sample (Kounaves et al., 2010). The WCL carried three aliquots of $BaCl_2$ powder, each allowing detection of up to 5 wt% SO_3 (Kounaves et al., 2010).

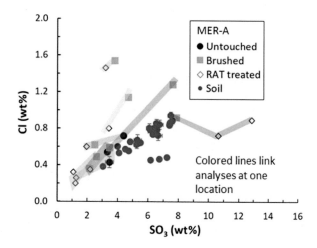

FIGURE 6.8

Cl versus SO_3 (wt%) for outcrop and float targets analyzed by MER-A, Gusev crater. Analyses at a single location are linked by *colored lines*. The pretreatment prior to analysis is indicated as untouched, brushed, and RAT treated (i.e., cleaned with an RAT). The precision error bars are within the size of the *larger symbols* and are small when present. We do not report the precision errors in subsequent figures because they are small and likely lower than the accuracy errors, which depend on detector performance, consolidation of the material being analyzed, distance, and angle from the target, which in combination are poorly constrained by robotic analyses on another planet.

6.3.1.3 Laser-Induced Breakdown Spectroscopy

Laser-Induced Breakdown Spectroscopy (LIBS) is one of the techniques used by ChemCam on the MSL Curiosity rover (Wiens et al., 2013). Sulfur is typically challenging to identify with LIBS because it tends to generate weak spectral lines (Dyar et al., 2011; Salle et al., 2004; Sobron et al., 2012; Vaniman et al., 2012). Nonetheless, sulfur has been detected by LIBS in veins, nodules, and bowls (subspherical mm-sized nodules); the sulfur-rich nature of some of these targets is confirmed through APXS analyses. For example, S lines have been detected in many targets at Yellowknife Bay, generally in combination with high Ca lines and H detection (Nachon et al., 2014). Sulfur has also been detected at Reddick Bight (#5 and #8) and Denault (#4) in the Shaler unit of Gale crater (Anderson et al., 2015).

6.3.1.4 Evolved Gas Analysis

Both the Thermal and Evolved Gas Analyzer (TEGA) instrument on the Phoenix lander (Boynton et al., 2001) and the Sample Analysis at Mars (SAM) instrument on the Curiosity rover (Mahaffy et al., 2012) include a pyrolysis oven coupled with a mass spectrometer that enables evolved gas analysis (EGA) of solid samples. In this technique, powdered rock or soil samples are heated in the pyrolysis oven, while evolved gases are sampled by the mass spectrometer. Sulfur-bearing gases may be identified by their characteristic mass spectra. EGA cannot definitively identify mineral phases, but comparison of the temperature at which a gas evolves with EGA libraries of known compounds can indicate possible mineral candidates. For example, under SAM operating

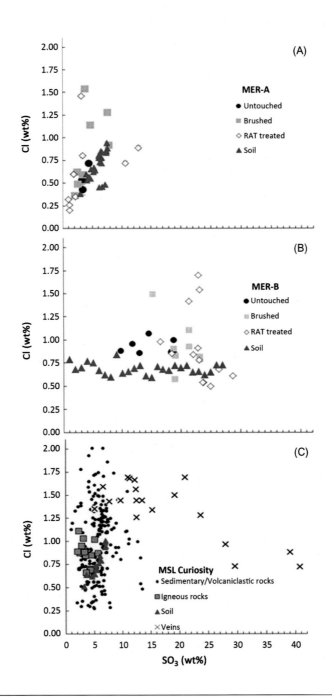

FIGURE 6.9

Cl versus SO_3 (wt%) for targets analyzed by (A) MER-A (Gusev crater), (B) MER-B (Meridiani Planum), and (C) Curiosity (Gale crater).

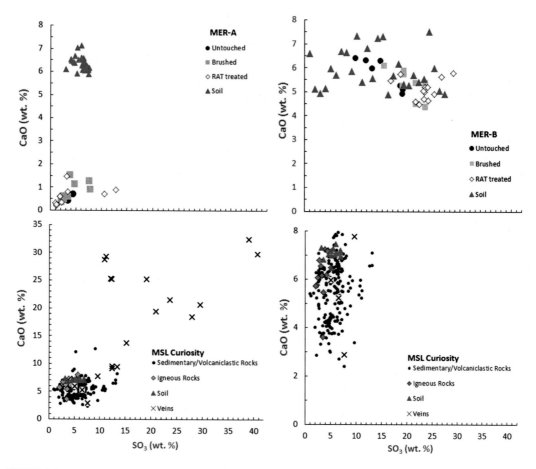

FIGURE 6.10

CaO versus SO₃ (wt%) for targets analyzed by (A) MER-A (Gusev crater), (B) MER-B (Meridiani Planum), and (C, D) Curiosity (Gale crater); (D) presents an expanded view of the graph in (C) to show data only for CaO mass fractions up to 8 wt%.

conditions, Fe-sulfates release SO_2 around 600 °C, while Mg- and Ca-sulfates typically do not release SO_2 until temperatures of 900 °C or higher (McAdam et al., 2014), although lower evolution temperatures are possible through catalysis by HCl derived from oxychlorine compounds in the samples (McAdam et al., 2016). See Section 6.3.2.5 and Fig. 6.16 for further details.

6.3.2 RESULTS FROM SURFACE MISSIONS

6.3.2.1 Viking Landers

Analyses from the Viking X-ray fluorescence spectrometer (XRFS) showed that the sulfur concentration in Martian fines is 10−100 times higher than that in common terrestrial and lunar rocks and

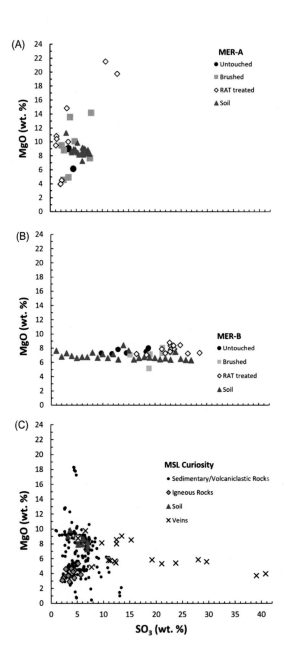

FIGURE 6.11

MgO versus SO$_3$ (wt%) for targets analyzed by (A) MER-A (Gusev crater), (B) MER-B (Meridiani Planum), and (C) Curiosity (Gale crater).

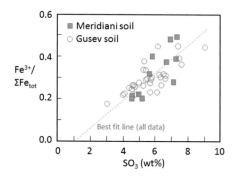

FIGURE 6.12

Fe^{3+}/Fe_{total} versus SO_3 (wt%) for Meridiani and Gusev soils based on Mössbauer and APXS analyses, respectively, with a best fit line through all data. *Modified after Morris et al. (2013).*

soils, but similar to that in many chondritic meteorites (Clark et al., 1976). Abundances of volatile sulfur compounds, including elemental S and organic sulfur, were found to be low based on Viking molecular analysis experiment results (Biemann et al., 1976). XRFS analyses indicated that salts inferred to be on the Martian surface are Mg−S-rich with significant Na and Cl (Baird et al., 1976; Clark et al., 1976; Clark and Van Hart, 1981) and that sulfur is predominantly present as S^{6+} (Toulmin et al., 1977). Relative abundances of sulfur and available cations suggested that Na-, Mg-, and Ca-sulfates and Fe-sulfides were the only reasonable possibilities for sulfur host minerals (Clark et al., 1976). Later refinement of these results showed that Ca, Fe, and Al were either uncorrelated or negatively correlated with S, and only Mg showed a positive correlation, although the precision was insufficient for strong conclusions to be drawn from these observations (Clark et al., 1982).

The two Viking landers analyzed a total of 21 samples (Clark et al., 1982). Bulk fines from both Viking lander sites were similar in composition. Sulfur contents of all samples ranged from 5.9 to 9.5 (+6/−2) wt% at Chryse Planitia and from 7.6 to 9 (+6/−2) wt% at Utopia Planitia (Clark et al., 1982). "Duricrust" fragments, possibly indicating Mg-sulfate, contained ∼50% higher sulfur concentrations than fines (Clark et al., 1977, 1982). Gooding (1978) proposed that with a scarcity of liquid water on Mars, weathering reactions may have been dominated by gas−solid interactions between atmospheric O_2, H_2O, and CO_2 and primary basaltic minerals. Such reactions would have transformed igneous pyrrhotite to sulfates (Gooding, 1978). Alternative formation mechanisms for surface sulfate include deposition of atmospheric aerosols (Settle, 1979) and low-temperature saline groundwater (Burns, 1988, 1993). Because sulfate salts are relatively soluble, if dissolved in water they would form high-salinity brines stable as liquid on the surface of Mars (e.g., on grain boundaries) at temperatures as low as ∼210 K with appropriate concentrations of water and Cl (Brass, 1980).

6.3.2.2 Pathfinder Rover

Pathfinder APXS analyses were more extensive than those of Viking and showed that Mg−S-rich materials are found in soils, cements, rocks, and dust in the atmosphere (Bell et al., 2000; Bruckner et al., 2001; Foley et al., 2001; McSween and Keil, 2000; Morris et al., 2000; Rieder, 1997). Morris et al. (2000) showed that Pathfinder rocks and soils are consistent with two-component mixtures of andesitic rock with low MgO and SO_3 concentrations and a global soil of basaltic composition with high MgO and SO_3.

Endmember models proposed to explain the S-rich nature of Mars' surface as seen by both Viking and Pathfinder include high-temperature volcanic gases ("acid fog") and hydrothermal fluids (Banin et al., 1997; Catling, 1999; Morris et al., 2000; Newsom et al., 1999). Newsom et al. (1999) proposed that hydrothermal systems on Mars could be described by two endmember fluid types: (1) neutral chloride fluids with high Cl and Na contents but low S/Cl ratios, and (2) acid-sulfate fluids with high S/Cl ratios and significant vapor transport, consistent with the low water abundance on Mars. A model combining these two types of fluids can account for the abundances of S, Cl, and other mobile elements in Martian soil (Newsom et al., 1999).

6.3.2.3 Mars Exploration Rovers

The MER rovers did not deploy a single instrument for definitive mineralogy, but the combined instrument suite provides appropriate information to identify mineral groups. Data from both Spirit and Opportunity indicate several types of sulfate minerals, but sulfides have not been detected. Sulfate minerals are inferred by correlating APXS-derived SO_3 and major cation contents (accounting for silicates and oxides), Mössbauer spectroscopy (definitive peak intensities and peak velocity splitting), visible and near-infrared (VNIR, Pancam) spectra, and/or thermal emission infrared (Mini-TES) spectra. In the near-infrared, sulfates are detected based on a range of features (e.g., polyhydrated sulfates with broad bands at 1.9–2.0 and 2.4 μm; monohydrated sulfates with a broad $\sim$2.1 μm band; Bishop et al., 2009; Gendrin et al., 2005; Murchie et al., 2009a). In the thermal infrared, sulfate minerals exhibit a range of diagnostic absorption features (Lane, 2007; Lane et al., 2007) with the main ν_3 stretch of SO_4^{2-} near $\sim$8.8–8.9 μm ($\sim$1120–1140 cm^{-1}) and a range of $\sim$8.2–9.3 μm ($\sim$1070–1220 cm^{-1}).

Meridiani Planum—Jarosite was the first sulfate mineral detected with the Opportunity rover, in iron oxide-rich sedimentary rocks at Meridian Planum (Fig. 6.13). Identification of jarosite was made using both Mössbauer spectrometry (Fig. 6.13B) in combination with high sulfur contents determined from APXS (Klingelhöfer et al., 2004; Morris et al., 2006) and Mini-TES spectra (Christensen et al., 2004c). Jarosite provides evidence for low-pH aqueous conditions or low water:rock ratios during deposition (King and McSween, 2005; Madden et al., 2004). This suggestion is supported by the laminar nature of the sedimentary layering and crystal casts suggesting subsequent dissolution and precipitation of concretions/nodules (Fig. 6.13C). Zolotov and Shock (2005) proposed that the assemblage of jarosite–goethite–hematite at Meridiani Planum was formed by hydrothermal pyrite precipitation followed by oxidation of pyrite to form jarosite and goethite. Hematite formation would have been promoted by later heating and dehydration of goethite (Zolotov and Shock, 2005).

The nearby Burns Formation in Endurance crater contains eolian sandstones (Fig. 6.14) with a suite of sulfate minerals that are thought to have formed through evaporation of brines and then modified diagenetically by periodic recharge (Grotzinger et al., 2005; McLennan et al., 2005). Chemical data (APXS) and spectral evidence indicate jarosite ($\sim$10% based on Mössbauer spectrometry), Mg-sulfate and Ca-sulfate ($\sim$18% and 10%, respectively, from Mini-TES), and possibly Na-bearing sulfates (Clark et al., 2005; McLennan and Grotzinger, 2008). The hydration states of the sulfates are unknown.

More recently, thick veins of Ca-sulfate (Fig. 6.7A) were identified at Endeavour crater by the Opportunity rover (Squyres et al., 2012). A strong correlation is observed between Ca and S using APXS and Pancam spectra, which are most consistent with the hydrated Ca-sulfate, gypsum.

Gusev Crater—The most S-rich soils identified at Gusev crater are the "Paso Robles class" soils (Fig. 6.15) that are inferred to contain Fe^{3+}-, Mg-, and Ca-sulfates. The Fe^{3+}-sulfates dominate, based on studies using Mössbauer, VNIR (with two different data-processing methods), Mini-TES, and APXS spectroscopies (Johnson et al., 2007; Lane et al., 2008). Ferricopiapite

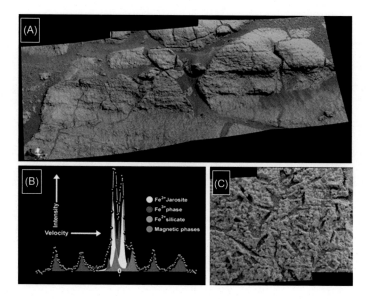

FIGURE 6.13

(A) Sulfate-rich sedimentary beds at El Capitan, Endurance crater, Meridiani Planum (NASA/JPL/Cornell/ASU). (B) Mössbauer spectrum of El Capitan (NASA/JPL/University of Mainz) reported in full in Klingelhöfer et al. (2004). Note that Mössbauer spectra are generally sensitive only to Fe-bearing phases. (C) Microscopic Imager mosaic of El Capitan outcrop with elongate holes inferred to result from the removal of salt minerals that formed in the sediment near deposition (McLennan et al. 2005).

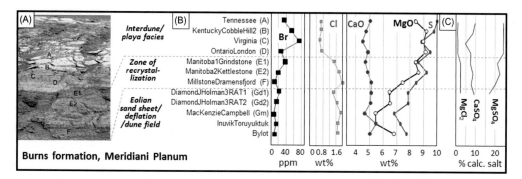

FIGURE 6.14

Burns formation, Meridiani Planum. (A) Outcrop showing the RAT-abraded analysis location (Grotzinger et al. 2005). (B) Variations in chemical elements as a function of the stratigraphy (data plotted from the Planetary Data System). (C) Calculated salt mineralogy of units A through Gd1 (McLennan et al. 2005).

$[Fe^{3+}_{4.6}(SO_4)_6(OH)_2 \cdot 20H_2O]$, (para)coquimbite $[Fe^{3+}_2(SO_4)_3 \cdot 9H_2O]$, and fibroferrite $[Fe^{3+}SO_4(OH) \cdot 5H_2O]$ are proposed based on at least four spectral methods, and parabutlerite $[Fe^{3+}(SO_4)(OH) \cdot 2H_2O]$ and rhomboclase $[(H_3O)Fe^{3+}(SO_4)_2 \cdot 3H_2O]$ are identified based on at least three methods (Johnson et al., 2007; Lane et al., 2008). Other possible minerals include

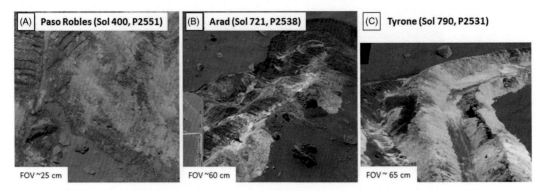

FIGURE 6.15

Approximately true-color images from Gusev crater with white Fe^{3+}-sulfate (blue = 432 nm, green = 535 nm, red = 754 nm) from NASA/JPL/Cornell University: (A) Paso Robles soil. (B) Arad soil. (C) Tyrone soil. These images were generated from a combination of filters ranging from 430 to 750 nm wavelength.

hydronium jarosite $[H_3O^+Fe^{3+}_6(SO_4)_4(OH)_{12}]$, bilinite $[Fe^{2+}Fe^{3+}_2(SO_4)_4 \cdot 22H_2O]$, butlerite $[Fe^{3+}(SO_4)(OH) \cdot 2H_2O]$, and metahohmannite $[Fe^{3+}_2(SO_4)_2O \cdot 4H_2O]$.

The Mini-TES instrument has also detected sulfates, most likely Ca-sulfates based on correlations between CaO and SO_3 (Ming et al., 2006), in the dust and in the classes of rocks called Clovis, Watchtower, and Wishstone at Gusev (Christensen et al., 2004b; Glotch et al., 2006; Ruff et al., 2006). The nearby Peace rocks are cemented by sulfates as evidenced from imaging, APXS chemistry, and Mini-TES spectra (Ruff et al., 2006; Squyres et al., 2006).

6.3.2.4 Phoenix Lander

The Phoenix WCL detected soluble sulfate equivalent to $\sim 1.3 \pm 0.5$ wt% SO_4 in the soil (Kounaves et al., 2010). Fe-sulfates were ruled out at this site because cyclic voltammetry and pH data did not indicate their presence. K and Na concentrations were low and Mg-sulfates (and some Ca-sulfates) were most favored based on geochemical models and measured cation abundances (Kounaves et al., 2010). EGA measurements by TEGA revealed no SO_2 release up to 1000 °C, suggesting that the sulfate was present as Ca-sulfate, most likely anhydrite, rather than Mg-sulfate (Golden et al., 2009).

6.3.2.5 Curiosity Rover

Unambiguous identification of sulfur-bearing minerals in situ on Mars has only been achieved using the CheMin X-ray diffraction instrument aboard the MSL Curiosity rover (Blake et al., 2012). Prior to this mission, sulfates were strongly suggested based on bulk chemistry (XRF and APXS) and spectral evidence (Mössbauer and near- to mid-infrared spectroscopy). Based on the CheMin analyses from Gale crater, anhydrite is the most common sulfate (Table 6.2). It is found in the Rocknest eolian sands (Blake et al., 2013). Anhydrite, bassanite, and pyrrhotite were found at the John Klein and Cumberland sites (Table 6.2) in the Sheepbed mudstone at Yellowknife Bay (Vaniman et al., 2014), as well as the Windjana drill site in the Dillinger sandstone at Kimberley (Treiman et al., 2016). Several phases are at the limits of CheMin detection, including pyrite at John Klein (Vaniman et al., 2014) and bassanite at Windjana (Rampe et al., 2015). Possible jarosite (near the detection limit) is the only sulfur-bearing

Table 6.2 Sulfur Minerals Detected by the Mars Science Laboratory CheMin in Gale Crater, Mars[a]

Rock Type	Sample	Sulfates			Sulfides		References
		Anhydrite	Bassanite	Jarosite	Pyrrhotite	Pyrite	
Modern sand deposit	Rocknest	1.1					Bish et al. (2013), Blake et al. (2013)
Sheepbed mudstone	John Klein	2.6	1.0		1.0		Vaniman et al. (2014)
	Cumberland	0.8	0.7		1.0	(0.3)	Vaniman et al. (2014)
Dillinger sandstone	Windjana	1.0			0.7		Treiman et al. (2016)
Pahrump Hills mudstone							
Concretions	Confidence Hills			(0.2)			Rampe et al. (2016)
Mineral pseudomorphs	Mojave2			~1			Rampe et al. (2016)
Laminations	Telegraph Peak			(~0.4)			Rampe et al. (2016)
Mudstone	Buckskin	1.8 ± 0.6					Morris et al. (2016)
Stimson unit (cross-bedded sandstone)	Big Sky	Yes					Yen et al. (2016)
Stimson unit in a fracture	Greenhorn	~7	~2				Yen et al. (2016)

[a]Quoted uncertainties are as reported in original publications. Parentheses around numbers in the table indicate tentative values near the CheMin detection limit.

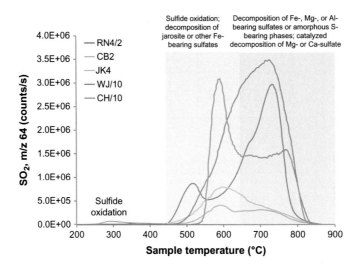

FIGURE 6.16

SO$_2$ released from thermal processing of Gale crater rocks and soils by SAM, shown as the signal for the major SO$_2$ isotopologue at m/z 64 as a function of sample temperature. The legend identifies samples of the Rocknest (RN) eolian deposit, Cumberland (CB) mudstone, John Klein (JK) mudstone, Windjana (WJ) sandstone, and Confidence Hills (CH) mudstone. A number after the sample identifier indicates a specific aliquot in repeated analyses of a given sample—for example, "RN4" refers to the fourth analysis of the Rocknest sample by SAM. *Modified after McAdam et al. (2016).*

phase at the Confidence Hills, Telegraph Peak, and Mojave2 drill sites in the Pahrump Hills member of the Murray Formation (Rampe et al., 2016) and may be present in trace abundance at Windjana (Treiman et al., 2016). The Buckskin drillhole in a mudstone member of the Murray Formation contains anhydrite (Morris et al., 2016). The Big Sky drill hole is located outside an altered fracture zone in the Stimson unit (a cross-bedded sandstone) and contains anhydrite, while the Greenhorn drillhole is within the fracture zone and yields anhydrite and bassanite (Yen et al., 2016). A sulfur-rich X-ray amorphous material—that is, with either a grain size too small to produce X-ray diffraction peaks in the detected energy range or without any crystalline structure such as a glass—is detected in all solid materials analyzed (e.g., Bish et al., 2013; Blake et al., 2013; Morris et al., 2015, 2016; Vaniman et al., 2014).

Although the SAM instrument on Curiosity is unable to provide definitive detection of specific minerals, SAM is a valuable complement to CheMin through its ability to observe sulfur-bearing phases in solid samples that may be present below the CheMin detection limit (Chapter 12; McAdam et al., 2014). SAM uses EGA protocols in which volatiles released through heating of solid samples in a pyrolysis oven are continuously sampled by SAM's quadrupole mass spectrometer. The temperature at which gases evolve provides information about possible mineral sources within the sample (Fig. 6.16). SAM has identified evidence supporting both oxidized and reduced forms of sulfur in Gale crater (Fig. 6.16), with possible sources including Fe-sulfides as well as Fe-, Mg-, and Ca-sulfates (McAdam et al., 2014; Ming et al., 2014). Sulfite salts (S^{4+}) such as Ca-sulfite are also consistent with SO$_2$ evolution temperatures in some SAM experiments (McAdam et al., 2014). Sulfites have been proposed as potential sulfur-bearing phases at the Martian surface (Halevy et al., 2007) but have not yet been definitively identified by mineralogical instruments. See Chapter 12, Volatile Detections in Gale Crater Sediment and Sedimentary Rock: Results from the Mars Science Laboratory's Sample Analysis at Mars Instrument in this volume for more detailed discussion of SAM results.

6.4 **SULFUR MINERALS IN MARTIAN METEORITES**

6.4.1 **SULFIDES IN MARTIAN METEORITES**

Sulfides constitute up to 1 wt% of the Martian meteorites and include troilite (FeS), pyrrhotite (FeS_{1-x}), pyrite (FeS_2), rare chalcopyrite ($CuFe^{2+}S_2$) and/or cubanite ($CuFe_2S_3$), pentlandite ((Fe^{2+},Ni)$_9S_8$), troilite—pentlandite—chalcopyrite intergrowths or pyrrhotite—pentlandite blebs (Fig. 6.17). In some cases, marcasite (FeS_2 polymorph) occurs as an alteration mineral. Most sulfides

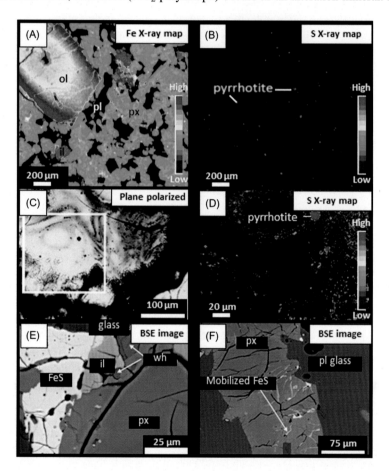

FIGURE 6.17

Iron sulfides in Martian meteorites. (A) Iron and (B) sulfur distribution from X-ray maps of the Martian meteorite NWA1110, where ol = olivine, pl = plagioclase, px = pyroxene, and il = ilmenite (King and McLennan 2010; courtesy of Paul Burger and Jim Papike, University of New Mexico). (C) Plane polarized light map of a large impact melt pocket in Tissint, showing schlierens and black sulfide globules (polished section #NHM-Vienna N 9404, glass "C"). (D) Sulfur X-ray map of the white area outlined in (C), showing the red sulfides. (E) Vesicular pyrrhotite in EETA 79001, shown in a BSE image from Walton et al. (2010). (F) Mobilized pyrrhotite infills cracks and fractures in pyroxene adjacent to plagioclase glass in EETA 79001, as shown in a BSE image from Walton et al. (2010).

are located near rims of pyroxene, in interstitial mesostasis, or in melt inclusions (Fig. 6.17A,B) indicating that sulfur is enriched in the late-stage magmatic liquids. Sulfides are also present as blebs in shock glasses or as veins in minerals associated with shock (Fig. 6.17C−F) and these are discussed separately later.

The shergottites range from the least evolved lherzolites (highest Mg#) through picritic basalt and basalt (lowest Mg#), where Mg# = 100Mg/(Mg + Fe) on a molecular basis. As the rocks become more evolved, the sulfides have lower pentlandite/pyrrhotite, and Ni content decreases in pyrrhotite (Balta et al., 2015; Grott et al., 2013; Lorand et al., 2005). These findings are interpreted to indicate that the sulfides are comagmatic and that sulfide melts segregated as the silicate melt differentiated (Grott et al., 2013; Lorand et al., 2005). In the LAR 06319 olivine-phyric shergottite, Ni and Co contents increase from troilite to pyrrhotite to pyrite, which provides evidence for pyrite formation from late-stage oxidation and/or sulfurization of pyrrhotite (Basu Sarbadhikari et al., 2009).

NWA 7533 and its pairs represent a singular example of a meteorite derived from Martian regolith breccia (Agee et al., 2013; Humayun et al., 2013; Lorand et al., 2015). NWA 7533 is also unique in containing accessory pyrite of up to 1 wt%, in contrast to the dominance of pyrrhotite in most Martian meteorites (Lorand et al., 2015). This pyrite has unusually high Ni contents of up to 4.5 wt%, which Lorand et al. (2015) interpret to indicate crystallization of pyrite from S-rich hydrothermal fluids that percolated through the breccia billions of years after breccia formation.

The nakhlites and chassignites, which have similar magmatic ages of ∼1.3 Ga (Nyquist et al., 2001), are depleted in sulfur relative to the shergottites (Burgess et al., 1989; Chevrier et al., 2011; Ding et al., 2015; Franz et al., 2014a; Gibson, 1983; Gibson et al., 1985; Lorand et al., 2012; Mikouchi et al., 2003, 2006). In the phase-specific extractions by Franz et al. (2014), sulfur was recovered predominantly as sulfate from these meteorites, despite observations of pyrrhotite or pyrite in petrographic sections of nakhlites (Imae and Ikeda, 2007; Imae et al., 2005; Mikouchi et al., 2003; Rochette et al., 2005) and Chassigny (Floran and Prinz, 1978). Some magmatic sulfides may have been oxidized to sulfate during weathering on Mars, in terrestrial environments prior to meteorite recovery, or even in curation. Observation of S-MIF signatures in magmatic pyrrhotite of nakhlites as well as some shergottites indicates that these sulfides formed from sulfur that had been processed in the Martian atmosphere and incorporated into magmas after surface deposition (Farquhar et al., 2000b; Franz et al., 2014: Sections 6.2.6 and 6.2.7). Pyrites in the ALH 84001 and Chassigny meteorites are interpreted as secondary alteration products (Franz et al., 2014a; Greenwood et al., 2000).

Several Martian meteorites contain regions with unusually high sulfur contents in shock glass pockets and veins (Fig. 6.17C−F). The Tissint meteorite has significant concentrations of sulfide blebs in shock glass (Fig. 6.17C,D), with up to 6000 ppm S (Aoudjehane et al., 2012). The Elephant Moraine A (EET) 79001 lithology C (EET 79001C) contains < 22 wt% SO_3 as small vesicular pyrrhotite blebs, veins of mobilized pyrrhotite in shocked minerals, and sulfur dissolved in shock glasses (Fig. 6.17E,F; Barrat et al., 2014; Schrader et al., 2011; Walton et al., 2010). Interestingly, shock glasses in Shergotty glasses (a meteorite "find") may have pyrrhotite blebs, but some S-bearing glasses with the highest SO_3 (∼3 wt%) contain Ca-sulfate crystals (Rao et al., 2008). In the case of all three meteorites, the sulfur in the shock glasses may be derived from preferential mobilization of Fe-sulfide from the target rock into the impact glass (Barrat et al., 2014; Walton et al., 2010), incorporation of secondary sulfates from the martian regolith into the impact melt (Rao et al., 2008, 2012), or, in the case of Tissint, desert weathering (Barrat et al., 2014).

The shock imparted by impact may also have resulted in sulfur devolatilization that converted pyrrhotite to troilite and metal alloys and may have eliminated Ca-sulfate in some meteorites (Lorand et al., 2012). For example, the chassignite NWA 2737 contains metal-saturated Fe−Ni sulfides with Fe/S of 1.00 ± 0.01, suggesting modification of primary magmatic pyrrhotites (Lorand et al., 2012). Changes in sulfide structures accompanying shock-induced sulfur degassing may also have resulted in loss of original magnetic properties in some meteorites (Lorand et al., 2012), analogous to the demagnetization of the Martian crust in regions with high impact crater densities (Louzada et al., 2007; Rochette et al., 2005).

6.4.2 SULFATES IN MARTIAN METEORITES

Sulfides are readily weathered by fluids (even at low fluid/rock ratios), and they may form alteration products such as pyrite (FeS_2—cubic) at higher temperatures, or marcasite (FeS_2—orthorhombic) and sulfate minerals at lower temperature (e.g., King and McSween, 2005). Sulfate minerals may also form at high temperature from volcanic gas (Gooding, 1978; Settle, 1979) or reactions of SO_2 gases with minerals and glasses (Henley et al., 2015).

The Martian meteorites "falls" (Shergotty, Zagami, Nakhla, Chassigny, and Tissint) are widely accepted to provide the best evidence for the primary sulfur mineralogy of the Martian crust because they are the least affected by terrestrial processes. Because Tissint is a recent fall, the other four falls have been studied much more extensively (Leshin and Vicenzi, 2006). The best characterized sulfate-bearing minerals from Mars are Ca-sulfates (Fig. 6.18A−C), Mg-sulfates (Fig. 6.18D), jarosite (Fig. 6.18E), and K-(Na)-Fe-(phosphate)-sulfates (Fig. 6.18F). Calcium- and Mg-sulfates have been observed in the interiors of some Martian meteorites (i.e., inside fusion crusts), suggesting that these secondary minerals formed on Mars (e.g., Bridges and Grady, 2000; Gooding et al., 1991; Treiman et al., 1993; Wentworth and Gooding, 1994). Iron-bearing sulfates (some with Na, K, or P) have been identified in Martian meteorites (Gooding et al., 1991; Wentworth and Gooding, 1994), including jarosite [$KFe_3(SO_4)_2(OH)_6$] trapped in a melt inclusion (Fig. 6.18E, McCubbin et al., 2009).

Mass spectrometry combined with thermal analysis of the interior portions of the Nakhla meteorite as well as shergottites ALH 77005 and EET 79001 indicate that a significant fraction of SO_2 is evolved at high temperatures ($> \sim 800$ °C), which is consistent with sulfates (Gooding et al., 1990) and the observation of anhydrite in Shergotty shock glass (Rao et al., 2008). We note that Walton et al. (2010) only found reduced sulfur in EET 79001 samples that they analyzed. Oxidized sulfur was detected in other aliquots of EET 79001 glass (Sutton et al., 2008), but beam damage cannot be ruled out (cf. Metrich et al., 2009). More investigations of these shock glasses should better resolve the oxidation state of the dissolved sulfur.

Sulfates are especially labile due to their high solubility and the sensitivity of phase transitions to relative humidity and temperature (e.g., Hogenboom et al., 1995; Vaniman et al., 2004; Wang et al., 2009, 2011); therefore, it is necessary to document textural features, age, and isotopic signatures, which can provide evidence in support of a Martian origin (e.g., Leshin and Vicenzi, 2006). Unfortunately, the original hydration state of most of these sulfates (except jarosite) is unknown. For this reason, particular care must be taken in analyzing many sulfate minerals (e.g., Hyde et al., 2011).

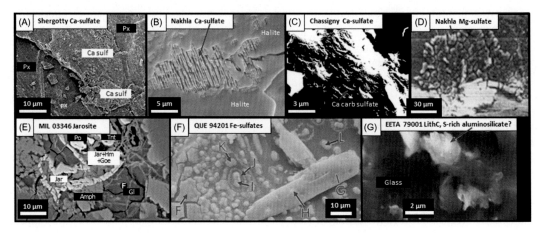

FIGURE 6.18

Sulfate minerals in the Martian meteorites. (A) Ca-sulfates in the Shergotty meteorite (Wentworth et al. 2000). (B) Ca-sulfate in the Nakhla meteorite covered in a thin layer of halite (Wentworth and McKay 1999). (C) Ca-sulfate in the Chassigny meteorite, associated with carbonate in some cases (Wentworth and Gooding 1994). (D) Mg-sulfate in the Nakhla meteorite (Gooding et al. 1991). (E) Jarosite hosted in a melt inclusion inside clinopyroxene from MIL 03346 (McCubbin et al. 2009). (F) Fe-sulfates (*red arrows*) in QUE 94201 (Wentworth and Gooding 1996). (G) A S-rich aluminosilicate or S-mineral on glass (Gooding and Muenow 1986).

In the case of the Martian meteorites, it is possible that primary sulfates are hydrated on Earth, which would result in ambiguous, disrupted textures (cf. Orgueil meteorite; Gounelle and Zolensky, 2001) because the hydrated forms of sulfate tend to have higher molar volumes.

6.4.3 OTHER SULFUR-BEARING MINERALS DETECTED IN MARTIAN METEORITES

Sulfur-rich aluminosilicates were observed in shergottite sample EET 79001 (Fig. 6.18G) by Gooding and Muenow (1986). The aluminosilicates were tentatively identified as amphiboles, but they contain up to 3 wt% SO_3, making it more likely that the beam volume overlapped glass underlying a sulfate mineral.

Alteration minerals rich in sulfur were identified in the Lafayette meteorite, a nakhlite (Treiman et al., 1993). Again, the exact host of the sulfur is unknown: sulfur may be adsorbed to phyllosilicates, present as a substituent in smectite or ferrihydrite, or perhaps present as gypsum. Similarly, sulfur is found in association with iddingsite (< 0.69 wt% SO_3) and carbonate (< 0.26 wt% SO_3) alteration minerals in Shergotty, although the exact host is unknown (Rao et al., 2008). More analyses, preferably with higher lateral resolution techniques such as transmission electron microscopy, would help determine the host phase for the sulfur.

6.5 REMOTE SENSING OBSERVATIONS

6.5.1 GAMMA RAY SPECTROMETRY

The Mars Odyssey Gamma Ray Spectrometer (GRS) detects element abundances in approximately the upper few decimeters of the Martian surface (Boynton et al., 2004). The average GRS sulfur abundance is 1.76 wt% ($SO_3 = 4.4$ wt%) (McLennan et al., 2010), although the values vary widely (Fig. 6.19A) due to the variable nature of the surface materials (e.g., dust vs bed rock). Hot spots in sulfur content (~ 3 wt%) are observed in the region surrounding Gale crater, an area near Nakong Vallis and the western Vallis Marineris catchment (Fig. 6.19A). As noted earlier, S-rich veins and infill have been detected at Gale crater with ~ 40 wt% SO_3 (Figs. 6.6, 6.9–6.11), and such materials would undoubtedly contribute to the detected GRS sulfur content.

The lower average sulfur abundance determined by the GRS, compared to average soils measured in situ, constrains the relative distributions of soil and igneous bedrock exposed at the surface. Overall, these findings are consistent with the observation that sulfur has been extensively transported, deposited, and remobilized across the Martian surface in dust, veins, and infill.

6.5.2 SPECTRAL DETECTION OF SULFATE MINERALS FROM EARTH

The first remote measurements suggesting sulfates were present on Mars were made of dust in the atmosphere using Earth-based telescopes (Bell et al., 1994; Blaney and McCord, 1995). These studies examined infrared spectra of Mars at wavelength ranges encompassing absorption features characteristic of certain minerals, including sulfates, carbonates, and hydrated silicates. Comparison with terrestrial infrared spectral libraries indicated the presence of sulfates, carbonates (or bicarbonates), and oxides on the Martian surface and/or in atmospheric dust, providing confirmation that the bulk sulfur chemistry measured by Viking was best explained by oxidized sulfur.

6.5.3 SPECTRAL DETECTION OF SULFATE MINERALS FROM ORBIT AROUND MARS

Infrared spectrometers deployed on orbiters around Mars provide information in both the thermal and visible to near-infrared parts of the electromagnetic spectrum. Note that spectra at the wavelength range employed by these instruments are typically sensitive to surficial composition to depths of a few tens of micrometers at most, as opposed to GRS instruments capable of detecting elemental abundances to depths of a few decimeters (Boynton et al., 2004). Inferred mineralogy of stratigraphic sections, such as Candor Mensa shown in Fig. 6.20, are thus based on the spectra of phases on exposed surfaces of the section and may reflect weathering that affected the uppermost layer of minerals but not the entire stratigraphic layer. Later, we summarize the primary results from each of the remote sensing instruments in sequence of the satellite's launch date.

6.5.3.1 Thermal Infrared Imaging Spectrometers

(a) *Thermal Emission Spectrometer*

The Mars Global Surveyor was launched in 1996 and contains the Thermal Emission Spectrometer (TES) that detects signals in the thermal IR part of the spectrum (Christensen et al., 2001). This instrument collects spectra from known 3×6 km areas over the

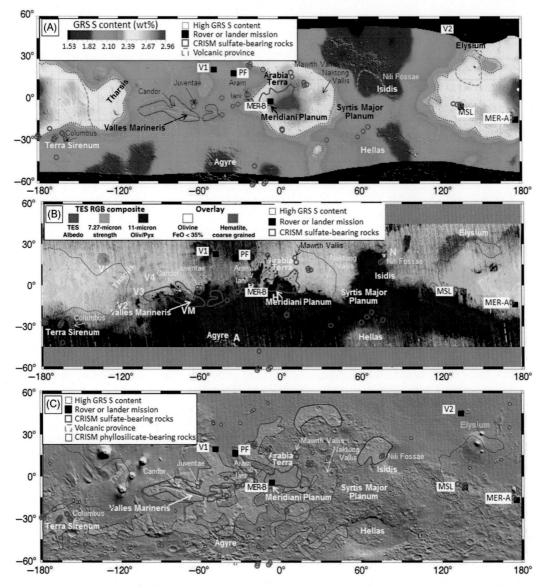

FIGURE 6.19

Sulfur or sulfate concentrations and locations on the surface of Mars. Viking (V-1, V-2), Pathfinder (PF), Spirit (MER-A), Opportunity (MER-B), Phoenix (PH), and Curiosity (MSL) landing sites are plotted. Localities discussed in the text are labeled. (A) S content (in stoichiometrically equivalent mass fractions) in the upper few decimeters of the Martian surface, as mapped by the Mars Odyssey Gamma Ray Spectrometer (GRS). The GRS map, from Karunatillake et al. (2014), is based on 10×10 degree bins using a boxcar filter with a smoothing radius of 25 degrees and is draped onto a Mars Orbiter Laser Altimeter shaded-relief map. Regions of very high hydrogen content are not shown (gray) due to the presence of abundant subsurface ice. (B) Thermal Emission Spectrometer (TES) red-green-blue (RGB) composite showing areas expected to be sulfate rich (7.27 micron) in green. This diagram is modified after Hoefen and Clark presented at http://speclab.cr.usgs.gov/mars.press. release.10.2000.html. (C) Diagram showing sulfate-bearing rocks and phyllosilicate-bearing rocks based on data from CRISM (Ehlmann and Edwards 2014). Volcanic provinces are from Michalski and Bleacher (2013).

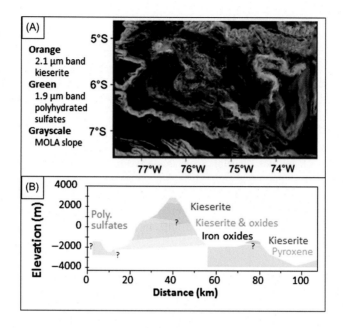

FIGURE 6.20

Interpretation of an area in Candor Mensor. (A) CRISM data with the 2.1 μm band diagnostic of kieserite (red to yellow) and 1.9 μm band related to polyhydrated sulfates (green to light green) plotted over MOLA slope map in grayscale. (B) Modification of a N-S cross section of Candor Mensa with layering postulated as subhorizontal and "?" indicating uncertain boundaries. *Modified from Mangold et al. (2008).*

~6–50 μm wavelength range. Data from TES confirmed that sulfates are present in the Martian dust (Bandfield et al., 2003; Christensen et al., 2004b). Also, the 7.27 μm band was tentatively assigned to sulfate with TES and the relative band intensity has been mapped as shown in Fig. 6.19 (USGS, 2000). Interestingly, this work shows that 7.27 μm band is not related to large volcanic provinces on Mars (Tharsis, Olympus Mons, Elysium), perhaps suggesting that any sulfur that was derived from volcanoes has been remobilized (Hoefen et al., 2003, 2000).

(b) *Thermal Emission Imaging System*

Mars Odyssey, launched in 2001, carries the Thermal Emission Imaging System (THEMIS), which has primarily been used to identify minerals and to determine thermal inertia (Christensen et al., 2004a). The instrument collects images over seven bands in the 6.7–14.8 μm wavelength range and 100 × 100 m areas. The THEMIS spectrometer should be capable of detecting sulfates, but an intensive survey of putative Martian paleolakes found no evidence for them, which suggested either high detection limits or that they are obscured by dust (Stockstill et al., 2005). In 2009, the Mars Odyssey moved to an earlier orbit time, to provide better detection of sulfates at warmer temperatures. The warmer, post-2009 THEMIS data do indicate sulfate minerals (Baldridge et al., 2013), although not always in the same location that has been identified by the Compact Reconnaissance Imaging Spectrometer for Mars (CRISM, below), possibly due to the spectral and lateral resolution of THEMIS, uncertainty in the diurnal temperature, and scattering effects from induration and particle size variations at the surface (Baldridge et al., 2013).

6.5.3.2 Visible/Near-Infrared Imaging Spectrometers

In the last decade or so, visible/near-infrared spectrometers have collected a large volume of information on the location of sulfates on Mars. Sulfates have been identified in the Valles Marineris canyon system, Meridiani Planum, Terra Meridiani, Mawrth Valles, Gale crater, Aram Chaos (Figs. 6.19C, 6.20, 6.21), and in the northern circumpolar dune fields and deposits (Fig. 6.21E). The majority of sulfate-rich rocks are coincident with areas rich in phyllosilicate-bearing rocks (Fig. 6.21E; Bibring et al., 2007, 2005, 2006; Ehlmann and Edwards 2014) and, in common with

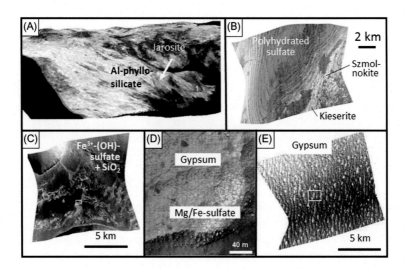

FIGURE 6.21

Sulfates in different settings on the Martian surface inferred from CRISM-derived spectra. (A) Jarosite in Mawrth Vallis from a CRISM composite draped over topography with $20 \times$ vertical exaggeration from HighRise showing jarosite-bearing unit (yellow) and Al-phyllosilicate unit (light blue) from Farrand et al. (2009). (B) Fe-sulfates at Juventae Chasma showing szomolnokite (yellow) and kieserite (orange) with polyhydrated sulfates in blue (Bishop et al., 2009). (C) Fe^{3+}-OH-sulfate (e.g., jarosite) in pink in the plains W of Juventae Chasma with sulfate beds in pink and silica beds—light blue. R = 2.53, G = 1.51, B = 1.08 from Murchie et al. (2009b). (D) Gypsum and Mg/Fe-sulfate-rich units in the Columbus crater near the crater wall showing albedo and textural contrast between the different spectral (mineralogic) units in the ring based on HiRISE image (ESP_013960_1510) (Wray et al., 2011). (E) Gypsum (dark color) in dunes from the northern plains (Murchie et al., 2009b). R = 2.53 μm, G = 1.51 μm, B = 1.08 μm.

the TES data, the sulfate-rich areas are adjacent to volcanic provinces rather than closely associated with them (Michalski and Bleacher, 2013).

(a) *Observatoire pour la Minéralogie, l'Eau, les Glaces et l'Activité*

The Mars Express satellite was launched in 2003 and it includes the Observatoire pour la Minéralogie, l'Eau, les Glaces et l'Activité (OMEGA) spectrometer. This instrument collects spectral images with resolution of 200–2000 m/pixel over the 0.4–5 μm wavelength range. OMEGA has been used to identify extensive deposits of kieserite, gypsum, and an unidentified "polyhydrated sulfate" (Bibring et al., 2005; Gendrin et al., 2005; Roach et al., 2010) in Valles Marineris and surrounds. Also, elevated sulfate concentrations were identified in Aureum and Iani Chaos (Gendrin et al., 2005; Noe Dobrea et al., 2006a, 2006b). In Candor Chasma (Fig. 6.20), kieserite occurs on steep, heavily eroded scarps overlying polyhydrated sulfates that formed on less eroded slopes. Iron oxides are typically associated with the sulfate-rich scarps and at the base of layered deposit scarps (Fig. 6.20B; Bibring et al., 2007; Mangold et al., 2008). At Meridiani Planum and locations elsewhere, Hesperian sedimentary rocks have been identified with hematite and both polyhydrated and monohydrated sulfates, overlying clay-bearing, highly cratered Noachian units (Arvidson et al., 2005, 2006; Bibring et al., 2007; Gendrin et al., 2005; Roach et al., 2010; Wiseman et al., 2010). Jarosite and polyhydrated sulfate minerals are identified on the NE flank of Syrtis Major (Fig. 6.19C) and are interpreted to indicate hydrothermally altered basaltic flows (Mustard and Ehlmann, 2010), and jarosite and alunite have been proposed in the Terra Sirenum basins (Swayze et al., 2008; Wray et al., 2011). An unusual $Fe^{3+}SO_4(OH)$ is proposed to occur at Aram Chaos (Lichtenberg et al., 2010).

(b) *Compact Reconnaissance Imaging Spectrometer for Mars (CRISM)*

The Mars Reconnaissance Orbiter satellite, launched in 2005, contains the CRISM instrument (Murchie et al., 2009b). This instrument collects spectral images with resolution of 18–200 m/pixel over the 0.4–4 μm wavelength range. Murchie et al. (2009b) summarize an extensive inventory of sulfate minerals in different settings as follows: (1) *Meridiani-type layered deposits* host monohydrated or polyhydrated sulfates on top of cratered terrain, with common hematite-rich layers (e.g., Terra Meridiani and Aram Chaos; Fig. 6.19). (2) *Valles-type layered deposits* are dominated by polyhydrated sulfates overlying monohydrated sulfates and commonly form mounds or deeply eroded plateaus or infill chasmata. Some of these deposits include ferric oxides, oxyhydroxides, or hydrated sulfates (e.g., Valles Marineris, Juvente Chasma) (Bishop et al., 2009). (3) *Intracrater clay–sulfate deposits* have sulfate layers (e.g., gypsum, polyhydrated, and monohydrated Mg/Fe-sulfates, jarosite, alunite, and szomolnokite) that are interbedded with phyllosilicate layers and exposed on crater walls (e.g., craters in Terra Sirenum) (Wray et al., 2011, 2009). The sulfates make up about 10% of the minerals present and are considered to have formed in lakes fed by groundwater, although the alunite may have a hydrothermal origin. (4) *Gypsum plains* are made of sandy material including gypsum and they are reworked into dunes, probably with other hydrated minerals (e.g., north polar erg and layered deposits) (Fishbaugh et al., 2007). (5) *Siliceous layered deposits* consist of jarosite associated with amorphous silica layers (e.g., plains surrounding Valles Marineris, Mawrth Vallis) (Farrand et al., 2009). The Nili Fossae region (Fig. 6.19C) also contains a range of sulfate minerals (Ehlmann et al., 2009).

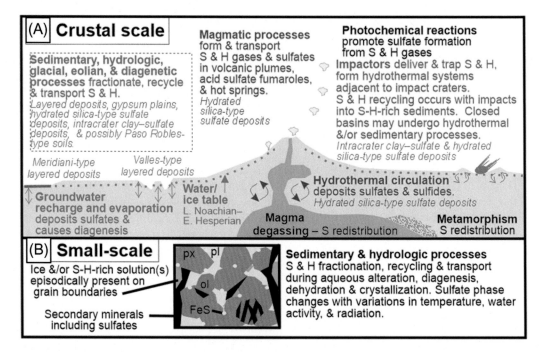

FIGURE 6.22

Mechanisms for distributing sulfur (and also H) on Mars' near surface on both (A) the crustal scale; and (B) the small scale. *From King and McLennan (2010).*

6.5.4 SURFACE PROCESSES FOR SULFUR

Although impactors may have provided some sulfur to the Martian surface, most sulfur is ultimately derived from the Martian mantle and transported to the surface at high temperatures through magmatic and hydrothermal processes (e.g., magmatic degassing, acid sulfate fumaroles, hot springs; King and McLennan, 2010). Redistribution at high temperatures can also occur via impact and possibly metamorphic processes (Fig. 6.22).

Geomorphic evidence indicates that sulfates are extensively remobilized through the sedimentary processes with water at low temperature on Mars' (near-)surface (Fig. 6.22; King and McLennan, 2010; McLennan and Grotzinger, 2008; Ming et al., 2008). Mechanical (or physical) plus chemical transport of sulfur occurs through sedimentary, hydrologic, glacial, eolian, impact and diagenetic processes with sulfur in solids (sulfate, sulfide, amorphous S-bearing material, and possibly sulfur), gas phase (e.g., species in Fig. 6.5), or in solution. In these processes, sulfate minerals may undergo changes in grain size, density sorting, deliquescence, dissolution, evaporation, reaction, and crystallization. These surface processes produce the "basin type" assemblages of sulfates, including Ca- and Mg-sulfates and possibly Na- and Fe-sulfates.

Geochemical evidence from the laboratory, Earth, and theoretical models indicates that sulfate minerals on Mars may vary depending on the starting composition of any solution, temperature, the other phases present, whether the system remains open or closed, adsorption and desorption reactions, and cation−exchange reactions. This area of research has received a huge amount of attention that is well beyond the scope of this review, and the reader is referred to the following key publications (Altheide et al., 2009; Arvidson et al., 2010; Bridges et al., 2001; Bridges and Grady, 2000; Bruckner et al., 2008; Burns, 1987, 1988; Burns and Fisher, 1990; Catling, 1999; Chevrier and Mathé, 2007; Chevrier and Altheide, 2008; Clark et al., 2005; Dehouck et al., 2012; King et al., 2004; King and McSween, 2005; Liu and Wang, 2015; Marion et al., 2008; McLennan et al., 2005; McLennan and Grotzinger, 2008; Ming et al., 2006, 2008; Rampe et al., 2016; Tosca and McLennan, 2006; Tosca et al., 2005; Wang et al., 2006a, 2009, 2011, 2013a, 2016; Wang and Ling, 2011; Wang and Zhou, 2014; Wilson and Bish, 2011; Xu and Parise, 2009; Xu et al., 2008).

The terrestrial sulfur cycle is dominated by biological processes that introduce large isotopic fractionations between sulfur phases (Canfield, 2001a). Due to the early emergence of sulfur-metabolizing microbes on Earth (Shen et al., 2001; Wacey et al., 2011) and the high abundance of sulfur on Mars, sulfur isotopes provide a key target in the search for Martian biomarkers.

6.5.4.1 Importance of sulfates for hydrogen on Mars

Most sulfates contain H_2O and/or OH^- and may have total H_2O contents as high as 62 wt% (Wang et al., 2009), rendering them potentially some of the most water-rich minerals on Mars. Therefore, there is a strong link between water and sulfur transport on Mars (Fig. 6.22). Furthermore, sulfates can substantially decrease the freezing temperature of any solutions (Brass, 1980; Fairen et al., 2009; King et al., 2004), which allows liquid brines and films of liquid on grain boundaries to be stable to much lower temperatures than liquid water ($< -60\ °C$).

To understand the transfer of water on Mars, it is necessary to account for dehydration/hydration reactions among sulfates. Such reactions are common (Cloutis et al., 2007; Janchen et al., 2005), and hydration can be associated with large increases in molar volume (Peterson and Wang, 2006). Sulfate hydration reactions, such as those due to interlayer H_2O-loss from clays (Bish et al., 2003), may produce either higher hydrate sulfates or amorphous gel-like materials (e.g., Chipera and Vaniman, 2007). Also, sulfates may dissolve if sufficient water is present (e.g., Elwood Madden et al., 2004 and Fig. 6.13C), or they may undergo cation exchange with clay minerals, even in relatively dry environments (Wilson and Bish, 2011). Sulfate stability depends on temperature, relative humidity, composition, and, for the Fe-bearing sulfates, the partial pressure of oxygen and pH (Brass, 1980; Chipera and Vaniman, 2007; Chipera et al., 2007; Chou and Seal, 2007; Fairen et al., 2009; Fialips et al., 2005; Hogenboom et al., 1995; King and McSween, 2005; Peterson and Grant, 2005; Peterson et al., 2008; Peterson and Wang, 2006; Tosca et al., 2005; Vaniman et al., 2004; Vaniman and Chipera, 2006; Wang et al., 2006a, 2006b, 2007, 2009). Because sulfates have such sensitive phase relations, they may be used to place constraints on environmental variables. In summary, sulfates play a key role in the formation of geomorphic and textural features on Mars, including some catastrophic outflow features and gullies (Chevrier et al., 2009), secondary pores/veins in sedimentary rocks, physical properties in the soils (e.g., clods), and formation of small-scale fracture/vein systems.

6.5.4.2 Age of the sulfates on Mars

The absolute age of sulfate deposition is a topic of debate, and it is probable that sulfate formed *both* (1) from the late Noachian to Early Amazonian (e.g., recent summary by Ehlmann and Edwards, 2014) and (2) during episodic events that remobilized sulfur (King et al., 2004; McLennan and

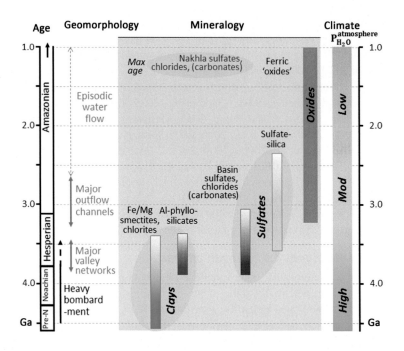

FIGURE 6.23

Generalized stratigraphy on Mars based on CRISM data (Bibring et al., 2006; Ehlmann and Edwards, 2014). Major events modifying the surface are also shown. Note that "oxides" include anhydrous oxides (e.g., magnetite, hematite, maghemite), oxyhydroxides (goethite), hydroxides (ferrihydrite), and hydrated iron oxide minerals.

Grotzinger, 2008). Most studies of the age of sulfate deposition are based on remote sensing linked with geomorphological evidence and crater counting to determine the age of a surface. As indicated in Section 6.2.5, sulfate minerals are closely associated with areas enriched in phyllosilicate minerals (Fig. 6.19C). Bibring et al. (2006) proposed that sulfates record a significant change in the past climate on Mars, specifically that the oldest Noachian terrains contained phyllosilicate minerals formed by pervasive aqueous alteration and that in the late Noachian to Hesperian, sulfates formed in an acidic environment. This environmental change is linked to the end of the heavy bombardment and the formation of major valley networks (Fig. 6.23) that transported saline solutions to basins where sulfates and chlorides precipitated (e.g., King et al., 2004; Bibring et al., 2006; Murchie et al., 2009b). This hypothesis agrees with subsequent work showing that sulfates occur most abundantly in areas of groundwater upwelling (Andrews-Hanna et al., 2007), including Terra Sirenum, Meridiani Planum, and Aram Chaos (Murchie et al., 2009a; Roach et al., 2010; Wray et al., 2011). After about 3 Ga, sulfates predominantly coexist with silica, suggesting that they are most likely linked to volcanic and impact processes (Ehlmann and Edwards, 2014; Murchie et al., 2009b; Newsom et al., 1999). From the Amazonian onward, ferric oxides (hematite, magnetite, maghemite), oxyhydroxides (goethite), and hydroxides (ferrihydrite) dominate the surface of Mars, which is interpreted to indicate the occurrence of a drying process, coincident with geomorphic evidence for only episodic water flow.

Despite remote sensing data indicating drying, a number of Martian meteorites contain sulfate and/or sulfide with mass-independent ^{33}S anomalies (i.e., nonzero values of Δ^{33}S, Section 6.1.2), indicating the presence of sulfur that has been processed in Mars' atmosphere (Farquhar et al., 2000b, 2007b; Franz et al., 2014; Section 6.5.4.1). The largest Δ^{33}S anomalies have been observed in the nakhlite meteorites (magmatic age ~ 1.3 Ga), which contain both sulfate and magmatic sulfide minerals with negative Δ^{33}S (Farquhar et al., 2007b; Franz et al., 2014). Thus, 1.3 Ga provides a maximum age constraint for fluids that exchanged sulfur with the atmosphere and then deposited sulfur-bearing minerals onto the surface prior to their assimilation into nakhlite parent magmas to form isotopically anomalous sulfides (Franz et al., 2014). However, some secondary sulfates in the nakhlites may have formed during later alteration events, suggesting episodic periods of water activity in the vicinity of the nakhlite parent rock as recently as $\sim 600-650$ Ma (Misawa et al., 2003; Swindle et al., 2000). In addition, the Lafayette and Chassigny meteorites, which both have magmatic ages of ~ 1.3 Ga, as well as shergottites dating to as recently as ~ 175 Ma, contain sulfates and/or sulfides with positive Δ^{33}S anomalies (Farquhar et al., 2000b; Franz et al., 2014). This observation extends the possible window for sulfur exchange between the Martian atmosphere and surface fluids to at least ~ 175 Ma (Franz et al., 2014).

6.5.5 ATMOSPHERIC SULFUR

6.5.5.1 Atmospheric abundance of sulfur

Telescopic surveys focused on atmospheric trace gases have found no evidence for currently active sources releasing sulfur into the atmosphere, with the most recent observations establishing upper limits of 0.3 ppb (Encrenaz et al., 2011; Krasnopolsky, 2012) to 1.1 ppb (Khayat et al., 2015) SO_2, 0.7 ppb SO, and 1.3 ppb H_2S (Khayat et al., 2015). Of these gases, SO_2 represents the best candidate in the search for active volcanism on Mars, because in addition to its expected dominance during magmatic degassing under current surface conditions, it has a photochemical lifetime of 600 days (Wong et al., 2005), compared to 9 days for H_2S and 4.6 hours for SO (Wong and Atreya, 2003; Wong et al., 2005). Given the global mixing timescale for the Martian atmosphere of 0.5 year (Krasnopolsky et al., 2004), detection of H_2S or SO would probably require contemporaneous emission from multiple sources on Mars (Khayat et al., 2015) or one very large source.

6.5.5.2 Photochemical processing and isotopic fractionation of sulfur

Evidence for atmospheric processing of sulfur during Mars' past is found in the form of MIF isotopic signatures in sulfur-bearing phases of some Martian meteorites (Farquhar et al., 2000b, 2007b; Franz et al., 2014a). These meteorites reveal the fingerprint of a S-MIF signature unique among Solar System bodies and indicate that the photochemical mechanism for fractionation of sulfur isotopes on Mars was different from that of Earth (Franz et al., 2014a). Specifically, S-MIF signatures on Earth show a distinctive covariation between Δ^{33}S and Δ^{36}S (Farquhar et al., 2000a), while S-MIF signatures on Mars are characterized by variations in Δ^{33}S with near zero values of Δ^{36}S (Farquhar et al., 2007b; Franz et al., 2014a). It is not yet clear whether this is due to differences in chemical composition or optical depth of the planets' atmospheres that affected UV shielding conditions for sulfur-bearing gases such as SO_2 and thus the wavelengths at which reactions occurred or to a difference in dominant sulfur-bearing species in the Martian atmosphere compared to Earth

(Franz et al., 2014a). If Martian outgassed sulfur comprised a mixture of oxidized and reduced species, variations in the SO_2:H_2S ratio, for example, would be expected to produce variability in the S-MIF signature over time, as observed in Archean terrestrial rocks (Halevy et al., 2010). Due to differences in the isotopic fractionation produced during H_2S and SO_2 photochemistry (Farquhar et al., 2000b), the net S-MIF signature would reflect a combination of these processes (Halevy et al., 2010). The continuity in S-MIF signature observed in meteorites of ages ranging from ~4.1 to ~175 Ma suggests that the dominant photochemical mechanism operating in the Martian atmosphere remained constant throughout this time period, supporting continuity in the sulfur speciation of atmospheric gases over time (Franz et al., 2014a). Although sulfur in Mars' proto-atmosphere may have been dominated by H_2S or S_2, modeling by Gaillard et al. (2013) indicated that volcanic emissions over the last 4—4.5 Ga should have contained mostly SO_2 due to the relatively low atmospheric pressure compared to Earth, suggesting SO_2 as the most likely reactant for S-MIF production (Franz et al., 2014a). Laboratory experiments to simulate potential processes on Mars that may have produced the S-MIF signature observed in Martian meteorites have proven difficult due to low product yields that compromise measurements of $\Delta^{36}S$ in experiments investigating optically thin conditions (Franz et al., 2013). However, ab initio modeling of SO_2 photolysis suggests that primary absorption effects in an optically thin SO_2 atmospheric column may offer a feasible scenario for explaining the Martian signature (Franz et al., 2013). Additional laboratory

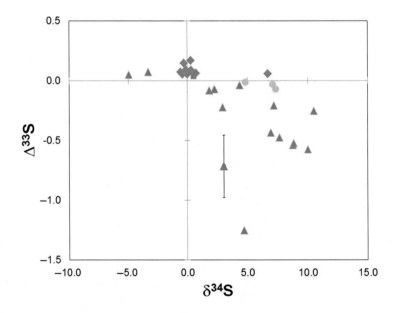

FIGURE 6.24

Plot of $\Delta^{33}S$ and $\delta^{34}S$ for isotopically anomalous sulfur extracted from Martian meteorites (Farquhar et al., 2000b, 2007b; Franz et al., 2014). ALH 84001—*gold circles*; nakhlites—*green triangles*; Chassigny—*pink square*; shergottites—*blue diamonds*. The single point with the large error bar represents average results of ion microprobe analyses of MIL 03346 pyrrhotite (Franz et al., 2014). All other points were obtained through bulk extraction protocols and have error bars smaller than the symbols.

experiments with SO_2 and other sulfur-bearing species are needed to understand Martian sulfur photochemistry more fully.

Fig. 6.24 depicts $\Delta^{33}S$ and $\delta^{34}S$ for Martian meteorites found to date to carry S-MIF signatures (Farquhar et al., 2000b, 2007b; Franz et al., 2014a). If these signatures were generated by a single common mechanism in the atmosphere, the presence of both positive and negative $\Delta^{33}S$ signals among the meteorites indicates that different photochemical products must have followed different preservation pathways leading to surface deposition. Otherwise, recycling reactions in an oxidizing atmosphere would have favored ultimate homogenization of all products as sulfate and removal of the S-MIF signal (Pavlov and Kasting, 2002). S-MIF signatures are present in both oxidized and reduced phases of the meteorites, including igneous sulfides that provide evidence for assimilation of atmospherically processed sulfur into magmas (Farquhar et al., 2000a, 2007a; Franz et al. 2014a).

Interestingly, regardless of oxidation state, the $\Delta^{33}S$ in sulfur phases of ALH 84001 and most nakhlites is negative, while all shergottites carrying S-MIF so far have shown positive $\Delta^{33}S$. The reason for this difference is uncertain, but the presence of sulfate with positive $\Delta^{33}S$ in Lafayette, in contrast to the other nakhlites analyzed to date, as well as Chassigny, with similar crystallization age to the nakhlites, may argue against a simple temporal explanation for the shift (Franz et al., 2014a). It is possible that the distinct signs observed in $\Delta^{33}S$ provide evidence tracing preservation of separate photochemical products in the meteorites in different proportions. The schematic diagram in Fig. 6.5, following that of Johnson et al. (2009), shows atmospheric reactions relevant to sulfur cycling and deposition. Based on their modeling results, Johnson et al. (2009) concluded that under weakly reducing conditions that may have existed on ancient Mars, sulfur could have been deposited in a variety of oxidation states, from elemental sulfur or sulfate aerosols or through rain-out of SO_2. The specific chemical pathways and dominant species deposited at the surface were found to be highly sensitive to other model parameters, especially atmospheric water abundance. Modeling of the Archean terrestrial sulfur cycle by Halevy et al. (2010) indicated sensitivity of the S-MIF signature to the partial pressure of CO_2, primarily through its effect on surface temperature and the acidity of water reservoirs involved in S-MIF preservation. Although there were significant differences between the ancient terrestrial and Martian atmospheres that prohibit application of terrestrial models directly to Mars, it is possible that episodic variations in local or global CO_2 abundance as a result of volcanic activity may have affected Martian S-MIF preservation.

Ueno et al. (2009) proposed another potential mechanism for affecting the sign of $\Delta^{33}S$ preserved in Archean sulfate, which may also be relevant to Mars. Through ab initio modeling of SO_2 photolysis in the presence of various overlying UV absorbers, they determined that carbonyl sulfide (OCS) could build up in a reducing atmosphere and produce a UV shielding effect that influenced the nature of S-MIF signal preserved in sulfates (Ueno et al., 2009). This topic deserves further attention of modeling and experimental efforts.

6.6 SULFUR AND THE MARTIAN CLIMATE

The process for depositing atmospheric sulfur onto the surface is intimately tied to the Martian climate through the greenhouse effect exerted by sulfurous gases. The existence of geomorphological features (Carr and Head, 2010) and hydrated minerals (Bibring et al., 2005, 2006; Ehlmann et al.,

2011; Murchie et al., 2009b; Poulet et al., 2005) indicating past activity of liquid water has led to extensive debate about the nature of Mars' early climate and whether these features necessarily imply warm, wet conditions in the past or if they were carved by episodic outflows of groundwater on a cold planet (Baker, 1978; Carr, 1983; Gulick and Baker, 1989; Peale et al., 1975; Squyres, 1984; Wallace and Sagan, 1979). Questions pertinent to warm, wet scenarios include whether such conditions prevailed persistently (Pollack et al., 1987; Postawko and Kuhn, 1986; Ramirez et al., 2013; Sagan and Mullen, 1972; Toon et al., 1980) or for discrete intervals of time (Halevy and Head, 2014; Johnson et al., 2008, 2009; King et al., 2004; Squyres and Kasting, 1994), and whether they were achieved globally or in isolated geographical regions (Mischna et al., 2013). Attempts to explain warm, wet conditions through a CO_2 atmosphere alone (Kasting, 1991) have proven difficult, especially given a "faint young sun" with reduced luminosity early in Solar System history (Gough, 1981; Sagan and Chyba, 1997). Therefore, various trace gases have been proposed as contributors to a greenhouse effect that warmed surface temperatures above the freezing point of water or brines (Halevy et al., 2007; Hirschmann and Withers, 2008; Johnson et al., 2008, 2009; Mischna et al., 2013; Postawko and Kuhn, 1986; Ramirez et al., 2013; von Paris et al., 2013).

Both SO_2 and H_2S figure prominently in the list of candidate Martian greenhouse gases, due to their absorption in the infrared (Haberle, 1998), as well as their potential presence in volcanic gases, as discussed earlier. Attempts to model the effects of these gases on the Martian climate have produced a range of conflicting conclusions regarding their warming capacity. Early efforts using a 1-D annually averaged model suggested that 1000 ppmv of SO_2 could generate up to 8−10 K of warming in a dry 500 mb CO_2 atmosphere, insufficient to reach a global average temperature above the 273 K needed for stability of liquid water, but possibly high enough to enable the existence of brines (Postawko and Kuhn, 1986). Yung et al. (1997), applying a similar model, suggested that SO_2 could have significantly warmed the middle atmosphere of early Mars, inhibiting CO_2 condensation that would produce a cooling effect (Kasting, 1991). Their calculations showed that the addition of 0.1 ppmv of SO_2 to a 2 bar CO_2 atmosphere would produce ~ 10 K of warming in the middle atmosphere, raising surface temperatures to ~ 276 K and ensuring that the upper atmosphere remained above the condensation temperature of CO_2. In addition, they suggested that such abundance of SO_2 in the atmosphere could have served as an effective UV shield for an early Martian biosphere (Yung et al., 1997).

Halevy et al. (2007) proposed that in an oxidant-limited system, outgassing of H_2S and SO_2 in approximately equal proportions during emplacement of the Tharsis province would have warmed the surface sufficiently to allow liquid water to exist. They implemented a model of early Martian carbon and sulfur cycles featuring an ocean covering $\sim 30\%$ of the planet. Volcanic emissions of CO_2, SO_2, and H_2S were countered by formation of carbon and sulfur-bearing minerals. Dissolution of SO_2 (S^{4+}) into surface waters would have promoted acidic surface conditions, favoring precipitation of Ca-sulfite and inhibiting formation of Ca-carbonate. Exhaustion of the oxidant supply would have allowed buildup of SO_2 to levels of a few ppm to several hundred ppm near the surface. A climate feedback involving SO_2 would result, in which an increase in supply of volcanic SO_2 caused atmospheric pSO_2 and thus temperature to rise, raising rates of weathering reactions and SO_2 removal by sulfite precipitation. The large reservoir of dissolved S^{4+} would buffer atmospheric SO_2 and stabilize surface temperatures. The partial pressure of CO_2 would have increased gradually up to a critical point at which the combined radiative effect of CO_2 and SO_2 allowed liquid water at the surface. Later alteration processes may have oxidized sulfite to sulfate after cessation of volcanic activity and a return to cold, more oxidized conditions (Halevy et al., 2007).

More recent modeling work has focused on SO_2 based on its longer photochemical lifetime (Wong and Atreya, 2003; Wong et al., 2005), pathways favoring atmospheric oxidation of H_2S to SO_2, and estimates that SO_2 was actually the dominant sulfur species outgassed under conditions prevalent during most of Mars' history (Gaillard et al., 2013; Gaillard and Scaillet, 2009). Johnson et al. (2008) explored scenarios for pulsed degassing of sulfur volatiles into atmospheres with background of 50 and 500 mb CO_2 and H_2S and SO_2 mixing ratios from 10^{-3} to 10^{-6}. Using the Mars Weather Research and Forecasting (MarsWRF) general circulation model, they investigated the greenhouse effect produced by this range of sulfur abundances with variable water mixing ratios. Their results suggested that the pulses of sulfur gases alone could have generated surface warming of up to 25 K above a CO_2 atmosphere. With the addition of a water vapor feedback, the surface temperature could have reached a point where brines and possibly transient liquid water could exist (Johnson et al., 2008).

Johnson et al. (2009) further explored scenarios for an ancient Martian atmosphere with pressures of 500 mb CO_2 and 100 mb or 6 mb N_2 through adaptation of a 1-D photochemical model previously developed for sulfur cycling on the Archean Earth (Pavlov and Kasting, 2002). Water vapor was controlled by temperature within the troposphere such that it remained beneath saturation pressure, and precipitation constraints were based on extremely arid conditions of the Atacama Desert in Chile, which was taken as a suitable analog for Mars. The lifetime of SO_2 in the atmosphere was controlled by the relative efficiencies of removal as S_8 or sulfate aerosols or through SO_2 rainout. Results of their simulations showed that SO_2 could persist for much longer time periods than previously assumed under weakly reducing conditions, reaching e-folding times on the order of hundreds of years for SO_2 mixing ratios of $1-100$ ppmv, with rainout as the dominant removal mechanism (Johnson et al., 2009).

Tian et al. (2010) questioned the results of Halevy et al. (2007) and Johnson et al. (2008, 2009), suggesting that the greenhouse warming produced by $1-100$ ppmv SO_2 in atmospheres with $0.5-3$ bar CO_2 and baseline surface temperatures of 235 K would have been more than offset by a cooling effect due to Rayleigh scattering by elemental sulfur and sulfate aerosols. In contrast to the work of Johnson et al. (2009), Tian et al. (2010) assumed zero dry deposition or rainout for SO_2, and terrestrial rates of rainout for other gases and aerosols. The lack of rainout as a removal mechanism for SO_2 in their model would increase its lifetime in the atmosphere and allow more buildup of aerosols, enhancing the radiative cooling effect (Tian et al., 2010). Tian et al. (2010) concluded that any transient warming effects would prevail only for months, not years.

Mischna et al. (2013) followed up on the work of Johnson et al. (2008) using the same MarsWRF GCM with improvements in the radiation code. They point out that while the current Martian climate is controlled primarily by changes in obliquity, with minimal effect from volcanically generated greenhouse gases, the reverse was likely true during the Noachian when the Tharsis province developed. During an extensive transitional period between these two extremes, obliquity may have served as a switch that initiated transitions between cold/dry and warm/wet periods (Mischna et al., 2013). Mischna et al. (2013) applied their modified GCM model to a range of obliquity conditions, but with the same atmospheric conditions assumed by Johnson et al. (2008). Their results showed that the presence of water vapor and SO_2 in the atmosphere did not produce warm periods globally, but rather that localized environments experienced warm conditions for brief time periods. Notably, the water cycle was influenced by obliquity cycles such that when obliquity decreased, atmospheric water vapor abundance dropped sharply and the atmosphere became incapable of sustaining warm temperatures anywhere but in limited equatorial regions and only for very short time periods. Similar to the finding of Johnson et al. (2008), however, a trace gas was necessary to trigger sufficient warmth initially to allow widespread sublimation from surface water reservoirs. The greenhouse effect of water alone was moderate and generally insufficient to maintain warm temperatures through the diurnal cycle. Mischna et al. (2013) also note that they had to reduce planetary albedo in their model to obtain perpetually warm temperatures, which could be reached by the presence of a dark liquid water ocean or widespread occurrence of low-albedo rock at the surface in earlier periods. Since they neglected the formation of water clouds and sulfur aerosols in their model, which would have had a cooling effect, Mischna et al. (2013) acknowledged that their results represent an optimistic estimate for the plausibility of liquid water on Mars.

Halevy and Head (2014) investigated the climatic effects of brief, intense volcanic eruptions on Mars using a microphysical aerosol model. Their study focused on post-Tharsis volcanism that produced extensive basaltic plains covering over 30% of the planet's surface. Evidence suggests that the basalts were formed through a series of brief eruptions with extremely high effusion rates that were separated by quiescent periods lasting thousands of years, analogous to those responsible for generating continental flood basalts on Earth (Head et al., 2002, 2006; Self et al., 2006; Tanaka et al., 2014). Effusion rates during these eruptive episodes are estimated to be more than 1000 times higher than those associated with the earlier formation of the Tharsis province (Wilson et al., 2001). Based on the higher sulfur content of Martian magmas compared to terrestrial magmas and the much greater areal extent of the basaltic plains on Mars compared to continental flood basalts, Halevy and Head (2014) estimate that transient sulfur outgassing rates up to a few thousand times the global terrestrial average were likely during emplacement of the Martian plains. The microphysical model employed for their study tracked populations of both homogeneous $H_2O-H_2SO_4$ aerosols and H_2SO_4-coated dust. They found that higher SO_2 abundances produced during volcanic episodes generated net positive radiative forcing up to 27 W/m^2, despite formation of H_2SO_4-bearing aerosols (Halevy and Head, 2014). Halevy and Head (2014) attributed this to the physical properties of some aerosols comprising moderately absorptive dust cores with reflective H_2SO_4 coating of variable thickness, as well as the background load of atmospheric dust, which offset the cooling effect due to scattering. Punctuated episodes characterized by high SO_2 emissions allowed higher transient concentrations of SO_2 than would be possible with lower, long-term outgassing rates and generated peak daily temperatures at low latitudes exceeding 273 K for several months of the year (Halevy and Head, 2014). The duration of warm episodes was limited to tens or hundreds of years following volcanic events, with cold, dry conditions prevailing during intervening quiescent periods (Halevy and Head, 2014). Melted ice and snow at low latitudes during warm periods may have mobilized soluble sulfate and transported it to low-lying regions (Halevy and Head, 2014).

Kerber et al. (2015) revisited the question of SO_2 greenhouse with the Laboratoire de Météorologie Dynamique Generic GCM and investigated sensitivity of results to SO_2 and H_2O mixing ratios, background atmospheric pressure, and the abundance, size, and composition of aerosols. They concluded that the presence of aerosols of realistic physical dimensions even in minute quantities has a pronounced cooling effect, rendering SO_2 incapable of producing sustained greenhouse conditions on early Mars (Kerber et al., 2015). Their model predicted local temperatures above 273 K only for transient periods with CO_2 pressures above 500 mb, extremely high SO_2 mixing ratios above ~ 100 ppmv, and an absence of aerosols. Less SO_2 may have been required at higher pressures, which may have produced an increase in temperature, but would have also decreased the total abundance of sulfur gases and raised the ratio of H_2S, a less efficient greenhouse gas, to SO_2 (Gaillard and Scaillet, 2009; Kerber et al., 2015).

Kerber et al. (2015) highlighted several areas of uncertainty that plague modeling efforts. Results of climate models are very sensitive to pressure, but the pressure of the early Martian atmosphere is poorly constrained, with values from 100 mb (Postawko and Kuhn, 1986) to 4 bars (Tian et al., 2010) appearing in the literature. Both pressure and the redox state of the atmosphere affect the identities of sulfur-bearing species outgassed by volcanoes (Gaillard and Scaillet, 2009), and the redox state further affects reaction networks in the atmosphere (Kerber et al., 2015). The outgassing rate of sulfur is also critical in determining the strength of any SO_2 greenhouse effect and the rate of aerosol formation (Kerber et al., 2015), which also depends on the partial pressures of other gases such as CO_2 and H_2O. As we have already discussed, the effusion and outgassing rates may have differed between volcanic events associated with the formation of the Tharsis province and the basaltic plains that were emplaced later (Halevy and Head, 2014). Details of Mars' past hydrologic cycle, including effects of high weathering rates of the mid-Noachian, the valley networks that arose in the late Noachian and early Hesperian, and the existence of pure water as opposed to brines, should also be considered in the evaluation of SO_2 greenhouse potential (Kerber et al., 2015).

6.7 FUTURE DIRECTIONS

Current research seeks to synthesize information about Martian sulfur from all available sources to enable better understanding of the processes that have mobilized sulfur between different reservoirs and chemical phases throughout Mars' history. In particular, future work is needed to account for multiple brine compositions possible on Mars and the processes by which sulfate has been remobilized and redeposited (e.g., King and McSween, 2005; Tosca et al., 2005).

Sulfates have high potential to preserve biomarkers and paleoenvironmental conditions since they may be rapidly precipitated from a brine (e.g., Foster et al., 2010; Rouchy and Monty, 1981) and impurities in sulfate crystals can form a barrier to radiation (Aubrey et al., 2006; Hughes and Lawley, 2003). Furthermore, laboratory experiments at near-equatorial Martian surface temperature and relative humidity suggest that coexistence with smectites may extend the stability field of hydrated Mg-sulfate minerals and improve the prospects of biomarker preservation within sulfate crystals (Wilson and Bish, 2012). Analog studies on Earth show that biological material is commonly trapped in fluid inclusions or along grain boundaries (Foster et al., 2010). Future work is needed on examining how halophilic bacteria, Archaea, and their genetic lineages respond to conditions like those on Mars.

In addition to the potential for preservation of organic compounds as biomarkers within sulfate minerals, sulfur isotope ratios may function directly as biomarkers (e.g., Canfield, 2001a; Thode et al., 1953) and thus are of prime interest from an astrobiological perspective. Sulfur is one of the six elements essential to all known life forms on Earth and is incorporated into a variety of biochemically relevant compounds, including amino acids, enzymes, and proteins (Clark, 1981). Its role in biomarker generation is most closely associated with energy production, in which sulfur is cycled through redox reactions that typically produce sulfide isotopically depleted compared to sulfur of higher valence states (Canfield, 2001a). For example, dissimilatory sulfate reduction introduces fractionations of 30–50‰ between sulfide and sulfate (Canfield, 2001b; Kemp and

Thode, 1968). The difference in $\delta^{34}S$ of sulfide and sulfate may grow to even larger values through successive redox reactions by consortia of sulfur-metabolizing microbes (Canfield, 2001a).

Sulfur isotope ratios provide biomarkers for some of the earliest life to emerge on Earth, with evidence for sulfate reduction dating back to ~ 3.47 Ga (Shen et al., 2001) and sulfide oxidation to ~ 3.4 Ga (Wacey et al., 2011). In extant life on Earth, we observe the versatility of sulfur metabolisms and their pervasiveness in a range of extreme environments, such as deep subseafloor basalt (e.g., Lever et al., 2013; Orsi et al., 2016; Turchyn et al., 2016), geothermal springs and hydrothermal vents (e.g., Elsgaard et al., 1994; Frank et al., 2015; Kamyshny et al., 2014; Lin et al., 2016; Topcuoglu et al., 2016), perennially cold habitats (e.g., Knoblauch and Jorgensen, 1999; Konneke et al., 2013; Lamarche-Gagnon et al., 2015; Sagemann et al., 1998; Sattley and Madigan, 2010), and highly saline environments (e.g., Benison and Bowen, 2013; Blum et al., 2012; Johnson et al., 2015; Lamarche-Gagnon et al., 2015; Stam et al., 2010).

For all of these reasons, the possibility that sulfur-metabolizing microbes may have emerged on Mars at some point in history is of great interest to astrobiology. Assuming similarities between terrestrial and Martian biota, sulfur isotopic biomarkers would be the most likely evidence of these life forms. The sulfur isotopic composition of sediments at Gale crater is a key target of investigations by the Curiosity rover's SAM instrument for this reason (Franz et al., 2017). Identification of sulfur isotope biosignatures from in situ measurements such as these is complicated by other fractionation mechanisms, such as atmospheric photochemistry (see Section 6.2.7), that could produce similar values of $\delta^{34}S$. Analyses of Martian sedimentary materials returned to Earth could provide additional means to distinguish between different sources of fractionation, such as the presence or absence of S-MIF signatures accompanying variations in $\delta^{34}S$.

For Mars sample return, improved methods are needed to prepare samples while avoiding contamination and modification of the mineralogy or isotopic composition. Dependent on environmental conditions, sulfates may dissolve or react with water, exchange with clays, decompose in the presence of heat, react with organic material in epoxy (e.g., some sulfates), or decompose under vacuum. Delicate sulfates require preparation in specific conditions (e.g., under kerosene, H-free ethanol, etc.) and may require specific analysis steps (e.g., Fe-sulfates; Hyde et al., 2011). Understanding sulfate phase changes is critical for preparing transport devices for Mars sample return (King et al., 2008; MacPherson and II, 2005; McLennan et al., 2012; Vaniman et al., 2008).

By facilitating isotopic (S-MIF) analyses of sedimentary materials at a high level of precision currently achievable only in terrestrial laboratories, Mars sample return also promises an opportunity for significant growth in understanding Martian sulfur photochemistry and interactions between the atmosphere and surface. Continued progress in improving technology for in situ measurements of sulfur isotope ratios, such as tunable laser spectrometry, may also allow future surface missions to contribute to this effort. These results in turn will help to tighten constraints on past Martian atmospheric composition and climate.

REFERENCES

Acuna, M.H., et al., 1999. Global distribution of crustal magnetization discovered by the Mars Global Surveyor MAG/ER experiment. Science 284, 790−793.

Agee, C.B., et al., 2013. Unique meteorite from Early Amazonian Mars: water-rich basaltic breccia Northwest Africa 7034. Science 339, 780−785.

Allègre, C., et al., 2001. Chemical composition of the Earth and the volatility control on planetary genetics. Earth Planet. Sci. Lett. 185, 49−69.

Altheide, T.S., et al., 2009. Experimental investigation of the stability and evaporation of sulfate and chloride brines on Mars. Earth Planet. Sci. Lett. 282, 69−78.

Anderson, R., et al., 2015. ChemCam results from the Shaler outcrop in Gale crater, Mars. Icarus 249, 2−21.

Andrews-Hanna, J.C., et al., 2007. Meridiani Planum and the global hydrology of Mars. Nature 446, 163−166.

Aoudjehane, H.C., et al., 2012. Tissint Martian meteorite: a fresh look at the interior, surface, and atmosphere of Mars. Science 338, 785−788.

Arvidson, R.E., et al., 2005. Spectral reflectance and morphologic correlations in eastern Terra Meridiani, Mars. Icarus 307, 1591−1594.

Arvidson, R.E., et al., 2006. Nature and origin of the hematite-bearing plains of Terra Meridiani based on analyses of orbital and Mars Exploration Rover data sets. J. Geophys. Res. 111, E12S08. Available from: https://doi.org/10.1029/2006JE002728.

Arvidson, R.E., et al., 2010. Spirit Mars rover mission: overview and selected results from the northern Home Plate Winter Haven to the side of Scamander crater. J. Geophys. Res. 115, CiteID E00F03. Available from: https://doi.org/10.1029/2010je003633.

Aubrey, A., et al., 2006. Sulfate minerals and organic compounds on Mars. Geology 34, 357.

Baird, A.K., et al., 1976. Mineralogic and petrologic implications of Viking geochemical results from Mars: interim report. Science 194, 1288−1293.

Baker, V.R., 1978. The Spokane flood controversy and the Martian outflow channels. Science 202, 1249−1256.

Baldridge, A.M., et al., 2013. Searching at the right time of day: evidence for aqueous minerals in Columbus crater with TES and THEMIS data. J. Geophys. Res. Planets 118, 179−189.

Balta, J.B., et al., 2015. Petrology and geochemistry of new Antarctic shergottites: LAR 12011, LAR 12095, and LAR 12240. In: Lunar Planet. Sci., XLVI. The Woodlands, TX. Abstract #2294.

Bandfield, J.L., et al., 2003. Spectroscopic identification of carbonate minerals in the Martian dust. Science 301, 1084−1087.

Banin, A., et al., 1997. Acidic volatiles and the Mars soil. J. Geophys. Res. 102, 13341−13356.

Baratoux, D., et al., 2013. The petrological expression of early Mars volcanism. J. Geophys. Res. Planets 118, 59−64.

Barrat, J.A., et al., 2014. No Martian soil component in shergottite meteorites. Geochim. Cosmochim. Acta 125, 23−33.

Basilevsky, A.T., et al., 2006. Geologically recent tectonic, volcanic and fluvial activity on the eastern flank of the Olympus Mons volcano, Mars. Geophys. Res. Lett. 33, L13201. Available from: https://doi.org/10.1029/2006GL026396.

Basu Sarbadhikari, A., et al., 2009. Petrogenesis of olivine-phyric shergottite Larkman Nunatak 06319: implications for enriched components in Martian basalts. Geochim. Cosmochim. Acta 73, 2190−2214.

Beaudoin, G., et al., 1994. Variations in the sulfur isotope composition of troilite from the Cañon Diablo iron meteorite. Geochim. Cosmochim. Acta 58, 4253−4255.

Bell, J.F., et al., 1994. Spectroscopy of Mars from 2.04 to 2.44 μm during the 1993 opposition: absolute calibration and atmospheric VS mineralogic origin of narrow absorption features. Icarus 111, 106−123.

Bell, J.F., et al., 2000. Mineralogic and compositional properties of Martian soil and dust: results from Mars Pathfinder. J. Geophys. Res. 105, 1721−1756.

Benison, K.C., Bowen, B.B., 2013. Extreme sulfur-cycling in acid brine lake environments of Western Australia. Chem. Geol. 351, 154−167.

Berger, J.A., et al., 2014. MSL-APXS titanium observation tray measurements: laboratory experiments and results for the Rocknest fines at the Curiosity field site in Gale Crater, Mars. J. Geophys. Res. Planets 119, 1046−1060.

Berger, J.A., et al., 2015. Germanium enrichments in sedimentary rocks in Gale Crater, Mars: constraining the timing of alteration and character of the protolith. In: Lunar Planet. Sci., XLVI. The Woodlands, TX. Abstract #1832.

Berger, J.A., et al., 2016. A global Mars dust composition refined by the alpha-particle X-ray spectrometer in Gale Crater. Geophys. Res. Lett. 43, 67−75.

Bertka, C.M., Fei, Y., 1998. Implications of Mars Pathfinder data for the accretion history of the terrestrial planets. Science 281, 1838−1840.

Bibring, J.P., et al., 2005. Mars surface diversity as revealed by the OMEGA/Mars Express observations. Science 307, 1576−1581.

Bibring, J.P., et al., 2006. Global mineralogical and aqueous Mars history derived from OMEGA/Mars Express data. Science 312, 400−404.

Bibring, J.P., et al., 2007. Coupled ferric oxides and sulfates on the Martian surface. Science 317, 1206−1210.

Biemann, K., et al., 1976. Search for organic and volatile inorganic compounds in two surface samples from the Chryse Planitia region of Mars. Science 194, 72−76.

Bish, D.L., et al., 2003. Stability of hydrous minerals on the Martian surface. Icarus 164, 96−103.

Bish, D.L., et al., 2013. X-ray diffraction results from Mars Science Laboratory: mineralogy of Rocknest at Gale Crater. Science 341. Available from: http://dx.doi.org/10.1126/science.1238932.

Bishop, J.L., et al., 2009. Mineralogy of Juventae Chasma: sulfates in the light-toned mounds, mafic minerals in the bedrock, and hydrated silica and hydroxylated ferric sulfate on the plateau. J. Geophys. Res. 114, E00D09. Available from: https://doi.org/10.1029/2009JE003352.

Blake, D., et al., 2012. Characterization and calibration of the CheMin mineralogical instrument on Mars Science Laboratory. Space Sci. Rev. 170, 341−399.

Blake, D.F., et al., 2013. Curiosity at Gale Crater, Mars: characterization and analysis of the Rocknest sand shadow. Science 341. Available from: https://doi.org/10.1126/science.1239505.

Blaney, D.L., McCord, T.B., 1995. Indications of sulfate minerals in the Martian soil from Earth-based spectroscopy. J. Geophys. Res. 100, 14433−14442.

Blum, J.S., et al., 2012. Desulfohalophilus alkaliarsenatis gen. nov., sp., nov., an extremely halophilic sulfate- and arsenate-respiring bacterium from Searles Lake, California. Extremophiles 16, 727−742.

Boynton, W.V., et al., 2001. Thermal and Evolved Gas Analyzer: part of the Mars volatile and climate surveyor integrated payload. J. Geophys. Res. 106, 17683−17698.

Boynton, W.V., et al., 2004. The Mars Odyssey Gamma-Ray Spectrometer instrument suite. Space Sci. Rev. 110, 37−83.

Brass, G.W., 1980. Stability of brines on Mars. Icarus 42, 20−28.

Bridges, J.C., Grady, M.M., 2000. Evaporite mineral assemblages in the nakhlite (Martian) meteorites. Earth Planet. Sci. Lett. 176, 267−279.

Bridges, J.C., et al., 2001. Alteration assemblages in Martian meteorites: implications for near-surface processes. Space Sci. Rev. 96, 365−392.

Brückner, J., et al., 2001. Revised data of the Mars Pathfinder alpha proton X-ray spectrometer: geochemical behavior of major and minor elements. In: Lunar Planet. Sci., XXXII. Houston, TX, p. 1293.

Brückner, J., et al., 2008. Mars Exploration Rovers: chemical composition by the APXS. In: Bell III, J.F. (Ed.), The Martian Surface: Composition, Mineralogy, and Physical Properties. Cambridge University Press, Cambridge, pp. 58−102.

Burgess, R., et al., 1989. Distribution of sulphides and oxidised sulphur components in SNC meteorites. Earth Planet. Sci. Lett. 93, 314−320.

Burns, R.G., 1987. Ferric sulfates on Mars. J. Geophys. Res. 92, 570−674.

Burns, R.G., 1988. Gossans on Mars. Lunar Planet Sci. XVIII, 713–721.

Burns, R.G., 1993. Rates and mechanisms of chemical weathering of ferromagnesium silicate minerals on Mars. Geochim. Cosmochim. Acta 57, 4555–4574.

Burns, R.G., Fisher, D.S., 1990. Chemical evolution and oxidative weathering of magmatic iron sulfides on Mars. Lunar Planet Sci. XXI, 145.

Campbell, J.L., et al., 2012. Calibration of the Mars Science Laboratory alpha particle X-ray spectrometer. Space Sci. Rev. 170, 319–340.

Canfield, D.E., 2001a. Biogeochemistry of sulfur isotopes. Rev. Mineral. 43, 607–636.

Canfield, D.E., 2001b. Isotope fractionation by natural populations of sulfate-reducing bacteria. Geochim. Cosmochim. Acta 65, 1117–1124.

Carr, M.H., 1983. Stability of streams and lakes on Mars. Icarus 56, 476–495.

Carr, M.H., Head, J.W., 2010. Geologic history of Mars. Earth Planet. Sci. Lett. 294, 185–203.

Carroll, M.R., Rutherford, M.J., 1985. Sulfide and sulfate saturation in hydrous silicate melts. In: 15th Lunar and Planetary Science Conference, Part 2, Houston, TX, pp. C601–C612.

Catling, D.C., 1999. A chemical model for evaporites on early Mars: possible sedimentary tracers of early climate and implications for exploration. J. Geophys. Res. 104, 16453–16469.

Chevrier, V.F., et al., 2009. Viscosity of liquid ferric sulfate solutions and application to the formation of gullies on Mars. J. Geophys. Res. 114, CiteID E06001. Available from: https://doi.org/10.1029/2009JE003376.

Chevrier, V.F., et al., 2011. Sulfide petrology of four nakhlites: Northwest Africa 817, Northwest Africa 998, Nakhla, and Governador Valadares. Met. Planct. Sci. 46, 769–784.

Chevrier, V.F., Altheide, T.S., 2008. Low temperature aqueous ferric sulfate solutions on the surface of Mars. Geophys. Res. Lett. 35, CiteID L22101. Available from: https://doi.org/10.1029/2008GL035489.

Chevrier, V.F., Mathé, P.E., 2007. Mineralogy and evolution of the surface of Mars: a review. Planet. Space Sci. 55, 289–314.

Chipera, S.J., Vaniman, D.T., 2007. Experimental stability of magnesium sulfate hydrates that may be present on Mars. Geochim. Cosmochim. Acta 71, 241–250.

Chipera, S.J., et al., 2007. The effect of temperature and water on ferric-sulfates. In: Lunar planet. Sci., XXXVIII. p. 1409.

Chou, I.M., Seal, R.R., 2007. Magnesium and calcium sulfate stabilities and the water budget of Mars. J. Geophys. Res. 112, E11004. Available from: https://doi.org/10.1029/2007JE002898.

Christensen, P.R., et al., 2001. Mars Global Surveyor Thermal Emission Spectrometer experiment: investigation description and surface science results. J. Geophys. Res. 106, 23823–23871.

Christensen, P.R., et al., 2004a. The Thermal Emission Imaging System (THEMiS) for the Mars 2001 Odyssey Mission. Space Sci. Rev. 110, 85–130.

Christensen, P.R., et al., 2004b. Initial results from the Mini-TES experiment in Gusev Crater from the Spirit Rover. Science 305, 837–842.

Christensen, P.R., et al., 2004c. Mineralogy at Meridiani Planum from the Mini-TES experiment on the opportunity rover. Science 306, 1733–1739.

Clark, B.C., 1981. Sulfur: fountainhead of life in the universe? In: Billingham, J. (Ed.), Conference on Life in the Universe. MIT Press.

Clark, B.C., 1993. Geochemical components in Martian soil. Geochim. Cosmochim. Acta 57, 4575–4581.

Clark, B.C., Van Hart, D.C., 1981. The salt on Mars. Icarus 45, 370–378.

Clark, B.C., et al., 1976. Inorganic analyses of Martian surface samples at the Viking landing sites. Science 194, 1283–1288.

Clark, B.C., et al., 1977. The Viking X-ray fluorescence experiment: analytical methods and early results. J. Geophys. Res. 82, 4577–4594.

Clark, B.C., et al., 1982. Chemical composition of Martian fines. J. Geophys. Res. 87, 10059–10067.

Clark, B.C., et al., 2005. Chemistry and mineralogy of outcrops at Meridiani Planum. Earth Planet. Sci. Lett. 240, 73–94.

Clayton, R.N., et al., 1976. A classification of meteorites based on oxygen isotopes. Earth Planet. Sci. Lett. 30, 10–18.

Clayton, R.N., et al., 1991. Oxygen isotope studies of ordinary chondrites. Geochim. Cosmochim. Acta 55, 2317–2337.

Clayton, R.N., 1993. Oxygen isotopes in meteorites. Ann. Rev. Earth Planet. Sci. 21, 15–149.

Clayton, R.N., 2002. Self-shielding in the solar nebula. Nature 415, 860–861.

Clayton, R.N., 2003. Oxygen isotopes in the solar system. Space Sci. Rev. 106, 19–32.

Clayton, R.N., Mayeda, T.K., 1983. Oxygen isotopes in eucrites, shergottites, nakhlites, and chassignites. Earth Planet. Sci. Lett. 62, 1–6.

Clayton, R.N., Mayeda, T.K., 1984. The oxygen isotope record in Murchison and other carbonaceous chondrites. Earth Planet. Sci. Lett. 67, 151–161.

Clayton, R.N., Mayeda, T.K., 1996. Oxygen isotope studies of achondrites. Geochim. Cosmochim. Acta 60, 1999–2017.

Cloutis, E.A., et al., 2007. Stability of hydrated minerals on Mars. Geophys. Res. Lett. 34. Available from: https://doi.org/10.1029/2007GL031267.

Connerney, J.E.P., et al., 1999. Magnetic lineations in the ancient crust of Mars. Science 284, 794–798.

Dehouck, E., et al., 2012. Evaluating the role of sulfide-weathering in the formation of sulfates or carbonates on Mars. Geochim. Cosmochim. Acta 90, 47–63.

Ding, S., et al., 2014. Sulfur concentration of Martian basalts at sulfide saturation at high pressures and temperatures—implications for deep sulfur cycle on Mars. Geochim. Cosmochim. Acta 131, 227–246.

Ding, S., et al., 2015. New bulk sulfur measurements of Martian meteorites and modeling the fate of sulfur during melting and crystallization—implications for sulfur transfer from Martian mantle to crust-atmosphere system. Earth Planet. Sci. Lett. 409, 157–167.

Ding, T., et al., 2001. Calibrated sulfur isotope abundance ratios of three IAEA sulfur isotope reference materials and V-CDT with a reassessment of the atomic weight of sulfur. Geochim. Cosmochim. Acta 65, 2433–2437.

Dreibus, G., Palme, H., 1996. Cosmochemical constraints on the sulfur content in the Earth's core. Geochim. Cosmochim. Acta 60, 1125–1130.

Dreibus, G., Wänke, H., 1985. Mars, a volatile-rich planet. Meteoritics 20, 367–381.

Dreibus, G., et al., 1994. Chemical and mineral composition of ALH 84001: a Martian orthopyroxenite. Meteoritics 29, 461.

Dyar, M.D., 2005. MIL03346, the most oxidized Martian meteorite: a first look at spectroscopy, petrography, and mineral chemistry. J. Geophys. Res. 110, 10.1029/2005JE002426.

Dyar, M.D., et al., 2011. Strategies for Mars remote laser-induced breakdown spectroscopy analysis of sulfur in geological samples. Spectrochim. Acta Part B 66, 39–56.

Ebel, D.S., 2011. Sulfur in extraterrestrial bodies and the deep Earth. In: Behrens, H., Webster, J.D. (Eds.), Reviews in Mineralogy and Geochemistry. Minerological Society of America, pp. 315–336.

Edmonds, C.G., et al., 1991. Posttranscriptional modification of transfer-RNA in thermophilic archaea (archae-abacteria). J. Bacteriol. 173, 3138–3148.

Ehlmann, B.L., Edwards, C.S., 2014. Mineralogy of the Martian surface. Ann. Rev. Earth Planet. Sci. 42, 291–315.

Ehlmann, B.L., et al., 2009. Identification of hydrated silicate minerals on Mars using MRO-CRISM: geologic context near Nili Fossae and implications for aqueous alteration. J. Geophys. Res. 114. Available from: http://dx.doi.org/10.1029/2009JE003339.

Ehlmann, B.L., et al., 2011. Subsurface water and clay mineral formation during the early history of Mars. Nature 479, 53–60.

Elkins-Tanton, L., 2008. Linked magma ocean solidification and atmospheric growth for Earth and Mars. Earth Planet. Sci. Lett. 271, 181−191.

Elkins-Tanton, L.T., et al., 2005. Possible formation of ancient crust on Mars through magma ocean processes. J. Geophys. Res. 110. Available from: https://doi.org/10.1029/2005JE002480.

Elsgaard, L., et al., 1994. Microbial sulfate reduction in deep-sea sediments at the Guaymas Basin—hydrothermal vent area—influence of temperature and substrates. Geochim. Cosmochim. Acta 58, 3335−3343.

Elwood Madden, M.E., et al., 2004. Jarosite as an indicator of water-limited chemical weathering on Mars. Nature 431, 821−823.

Encrenaz, T., et al., 2011. A stringent upper limit to SO_2 in the Martian atmosphere. Astron. Astrophys. 530, A37.

Erkaev, N.V., et al., 2014. Escape of the Martian protoatmosphere and initial water inventory. Planet. Space Sci. 98, 106−119.

Fairen, A.G., et al., 2009. Stability against freezing of aqueous solutions on early Mars. Nature 459, 401−404.

Farquhar, J., et al., 2000a. Atmospheric influence of Earth's earliest sulfur cycle. Science 289, 756−758.

Farquhar, J., et al., 2000b. Evidence of atmospheric sulphur in the Martian regolith from sulphur isotopes in meteorites. Nature 404, 50−52.

Farquhar, J., et al., 2001. Observation of wavelength-sensitive mass-independent sulfur isotope effects during SO_2 photolysis: implications for the early atmosphere. J. Geophys. Res. 106, 32829−32839.

Farquhar, J., et al., 2007a. Implications of conservation of mass effects on mass-dependent isotope fractionations: influence of network structure on sulfur isotope phase space of dissimilatory sulfate reduction. Geochim. Cosmochim. Acta 71, 5862−5875.

Farquhar, J., et al., 2007b. Implications from sulfur isotopes of the Nakhla meteorite for the origin of sulfate on Mars. Earth Planet. Sci. Lett. 264, 1−8.

Farrand, W.H., et al., 2009. Discovery of jarosite within the Mawrth Vallis region of Mars: implications for the geologic history of the region. Icarus 204, 478−488.

Fei, Y., Bertka, C.M., 2005. The interior of Mars. Science 308, 1120−1121.

Fei, Y., et al., 2000. Structure type and bulk modulus of Fe_3S, a new iron-sulfur compound. Am. Mineral. 85, 1830−1833.

Fialips, C.I., et al., 2005. Hydration state of zeolites, clays, and hydrated salts under present-day Martian surface conditions: can hydrous minerals account for Mars Odyssey observations of near-equatorial water-equivalent hydrogen? Icarus 178, 74−83.

Fishbaugh, K.E., et al., 2007. On the origin of gypsum in the Mars north polar region. J. Geophys. Res. 112. Available from: https://doi.org/10.1029/2006JE002862.

Floran, R.J., Prinz, M., 1978. The Chassigny meteorite: a cumulate dunite with hydrous amphibole-bearing melt inclusions. Geochim. Cosmochim. Acta 42, 1213−1229.

Foley, C.N., et al., 2001. Chemistry of Mars Pathfinder samples determined by the APXS. In: Lunar Planet. Sci., XXXII. Houston, TX, p. 1979.

Foley, C.N., et al., 2003. Final chemical results from the Mars Pathfinder alpha proton X-ray spectrometer. J. Geophys. Res. 108. Available from: https://doi.org/10.1029/2002JE002019.

Folkner, W.M., et al., 1997. Interior structure and seasonal mass redistribution of Mars from ratio tracking of Mars Pathfinder. Science 278, 1749−1752.

Foster, I.S., et al., 2010. Characterization of halophiles in natural $MgSO_4$ salts and laboratory enrichment samples: astrobiological implications for Mars. Planet. Space Sci. 58, 599−615.

Franchi, I.A., et al., 1999. The oxygen-isotopic composition of Earth and Mars. Met. Planet. Sci. 34, 657−661.

Frank, K.L., et al., 2015. Key factors influencing rates of heterotrophic sulfate reduction in active seafloor hydrothermal massive sulfide deposits. Front. Microbiol. 6. Available from: https://doi.org/10.3389/fmicb.2015.01449.

Frankel, C., 1996. Volcanoes of the Solar System. Cambridge University Press, Cambridge.

Franz, H.B., et al., 2013. Mass-independent fractionation of sulfur isotopes during broadband SO_2 photolysis: comparison between ^{16}O- and ^{18}O-rich SO_2. Chem. Geol. 362, 56−65.

Franz, H.B., et al., 2014a. Isotopic links between atmospheric chemistry and the deep sulphur cycle on Mars. Nature 508, 364−368.

Franz, H.B., et al., 2017. Large sulfur isotope fractionations in Martian sediments at Gale crater. Nature Geosci. 10, 658−662.

Gaillard, F., Scaillet, B., 2009. The sulfur content of volcanic gases on Mars. Earth Planet. Sci. Lett. 279, 34−43.

Gaillard, F., Scaillet, B., 2014. Sulfur outgassing by volcanoes on Mars: what really matters. In: Workshop of Volatiles in the Martian Interior, Houston, TX, p. 1014.

Gaillard, F., et al., 2011. Atmospheric oxygenation caused by a change in volcanic degassing pressure. Nature 478, 229−232.

Gaillard, F., et al., 2013. Geochemical reservoirs and timing of sulfur cycling on Mars. Space Sci. Rev. 174, 251−300.

Gaillard, F., et al., 2015. The redox geodynamics linking basalts and their mantle sources through space and time. Chem. Geol. 418, 217−233.

Gao, X., Thiemens, M.H., 1991. Systematic study of sulfur isotopic composition in iron meteorites and the occurrence of excess ^{33}S and ^{36}S. Geochim. Cosmochim. Acta 55, 2671−2679.

Gao, X., Thiemens, M.H., 1993a. Isotopic composition and concentration of sulfur in carbonaceous chondrites. Geochim. Cosmochim. Acta 57, 3159−3169.

Gao, X., Thiemens, M.H., 1993b. Variations of the isotopic composition of sulfur in enstatite and ordinary chondrites. Geochim. Cosmochim. Acta 57, 3171−3176.

Gellert, R., et al., 2006. Alpha Particle X-Ray Spectrometer (APXS): results from Gusev crater and calibration report. J. Geophys. Res. 111. Available from: https://doi.org/10.1029/2005JE002555.

Gendrin, A., et al., 2005. Sulfates in Martian layered terrains: the OMEGA/Mars Express view. Science 307, 1587−1591.

Gibson Jr., E.K., 1983. Sulfur in achondritic meteorites. Proc. Lunar Planet. Sci. 14, 247−248.

Gibson, E.K., et al., 1985. Sulfur in achondritic meteorites. Meteoritics 20, 503−511.

Glotch, T.D., et al., 2006. Mineralogy of the light-toned outcrop at Meridiani Planum as seen by the Miniature Thermal Emission Spectrometer and implications for its formation. J. Geophys. Res. 111. Available from: https://doi.org/10.1029/2005JE002672.

Golden, D.C., et al., 2009. Sulfur mineralogy at the Mars Phoenix landing site. In: Lunar Planet Sci., XL. The Woodlands, TX, p. 2319.

Gooding, J.L., 1978. Chemical weathering on Mars. Icarus 33, 483−513.

Gooding, J.L., Muenow, D.W., 1986. Martian volatiles in shergottite EETA 79001: new evidence from oxidized sulfur and sulfur-rich aluminosilicates. Geochim. Cosmochim. Acta 50, 1049−1059.

Gooding, J.L., et al., 1990. Volatile compounds in shergottite and nakhlite meteorites. Meteoritics 25, 281−289.

Gooding, J.L., et al., 1991. Aqueous alteration of the Nakhla meteorite. Meteoritics 26, 135−143.

Gough, D.O., 1981. Solar interior structure and luminosity variations. Solar Phys. 74, 21−34.

Gounelle, M., Zolensky, M.E., 2001. A terrestrial origin for sulfate veins in CI1 chondrites. Met. Planet. Sci. 36, 1321−1329.

Greenwood, J.P., et al., 2000. Sulfur isotopic compositions of individual sulfides in Martian meteorites ALH 84001 and Nakhla: implications for crust-regolith exchange on Mars. Earth Planet. Sci. Lett. 184, 23−35.

Gregersen, L.H., et al., 2011. Mechanisms and evolution of oxidative sulfur metabolism in green sulfur bacteria. Front. Microbiol. 2. Available from: http://dx.doi.org/10.3389/fmicb.2011.00116.

Gross, J., et al., 2013. Petrography, mineral chemistry, and crystallization history of olivine-phyric shergottite NWA 6234: a new melt composition. Met. Planet. Sci. 48, 854−871.

Grott, M., et al., 2011. Volcanic outgassing of CO_2 and H_2O on Mars. Earth Planet. Sci. Lett. 308, 391−400.

Grott, M., et al., 2013. Long-term evolution of the Martian crust-mantle system. Space Sci. Rev. 174, 49−111.

Grotzinger, J.P., et al., 2005. Stratigraphy and sedimentology of a dry to wet eolian depositional system, Burns formation, Meridiani Planum, Mars. Earth Planet. Sci. Lett. 240, 11−72.

Grotzinger, J.P., et al., 2014. A habitable fluvio-lacustrine environment at Yellowknife Bay, Gale Crater, Mars. Science 343. Available from: http://dx.doi.org/10.1126/science.1242777.

Gulick, V.C., Baker, V.R., 1989. Fluvial valleys and Martian paleoclimates. Nature 341, 514−516.

Haberle, R.M., 1998. Early Mars climate models. J. Geophys. Res. 103, 28467−28480.

Halevy, I., Head, J.W.I., 2014. Episodic warming of early Mars by punctuated volcanism. Nat. Geosci. 7, 865−868.

Halevy, I., et al., 2007. A sulfur dioxide climate feedback on early Mars. Science 318, 1903−1907.

Halevy, I., et al., 2010. Explaining the structure of the Archean mass-independent sulfur isotope record. Science 329, 204−207.

Hartmann, W.K., Neukum, G., 2001. Cratering chronology and the evolution of Mars. Space Sci. Rev. 96, 165−194.

Hauber, E., et al., 2011. Very recent and wide-spread basaltic volcanism on Mars. Geophys. Res. Lett. 38. Available from: https://doi.org/10.1029/2011GL047310.

Haughton, D.R., et al., 1974. Solubility of sulfur in magic magmas. Bull. Soc. Econ. Geol. 69, 451−467.

Head, J.W., et al., 2002. Northern lowlands of Mars: evidence for widespread volcanic flooding and tectonic deformation in the Hesperian period. J. Geophys. Res. Planets 107. Available from: https://doi.org/10.1029/2000JE001445.

Head, J.W., et al., 2006. The Huygens-Hellas giant dike system on Mars: implications for Late Noachian-Early Hesperian volcanic resurfacing and climatic evolution. Geology 34, 285−288.

Henley, R.W., et al., 2015. Porphyry copper deposit formation by sub-volcanic sulphur dioxide flux and chemisorption. Nat. Geosci. 8, 210−215.

Herd, C.D.K., 2003. The oxygen fugacity of olivine-phyric Martian basalts and the components within the mantle and crust of Mars. Met. Planet. Sci. 38, 1793−1805.

Herd, C.D.K., et al., 2002. Oxygen fugacity and geochemical variations in the Martian basalts: implications for Martian basalt petrogenesis and the oxidation state of the upper mantle of Mars. Geochim. Cosmochim. Acta 66, 2025−2036.

Hirschmann, M.M., Withers, A.C., 2008. Ventilation of CO_2 from a reduced mantle and consequences for the early Martian greenhouse. Earth Planet. Sci. Lett. 270, 147−155.

Hoefen, T.M., et al., 2000. Unique spectral features in Mars Global Surveyor thermal emission spectra: implications for surface mineralogy in Nili Fossae. Am. Astronom. Soc. 32, 1118.

Hoefen, T.M., et al., 2003. Discovery of olivine in the Nili Fossae region of Mars. Science 302, 627−630.

Hogenboom, D.L., et al., 1995. Magnesium sulfate-water to 400 MPa using a novel piezometer: densities, phase equilibria, and planetological implications. Icarus 115, 258−277.

Hughes, K.A., Lawley, B., 2003. A novel Antarctic microbial endolithic community within gypsum crusts. Environ. Microbiol. 5, 555−565.

Hulston, J.R., Thode, H.G., 1965. Variations in the S^{33}, S^{34}, and S^{36} contents of meteorites and their relation to chemical and nuclear effects. J. Geophys. Res. 70, 3475−3484.

Humayun, M., et al., 2013. Origin and age of the earliest Martian crust from meteorite NWA 7533. Nature 503, 513−516.

Hyde, B.C., et al., 2011. Methods to analyze metastable and microparticulate hydrated and hydrous iron sulfate minerals. Am. Mineral. 96, 1856−1869.

Imae, N., Ikeda, Y., 2007. Petrology of the Miller Range 03346 nakhlite in comparison with the Yamato-000593 nakhlite. Met. Planet. Sci. 42, 171−184.

Imae, N., et al., 2005. Petrology of the Yamato nakhlites. Met. Planet. Sci. 40, 1581−1598.

Irving, A.J., 2017. An up-to-date list of Martian meteorites. Available from: http://www.imca.cc/mars/Martian-meteorites-list.htm.

Janchen, J., et al., 2005. Experimental studies of the water sorption properties of Mars-relevant porous minerals and sulfates. In: Lunar Planet. Sci., XXXVI. p. 1263.

Jensen, M.L., Nakai, N., 1962. Sulfur isotope meteorite standards, results and recommendations, Biogeochemistry of sulfur isotopes. In: NSF Symposium, New Haven, CT.

Johnson, J.R., et al., 2007. Mineralogic constraints on sulfur-rich soils from Pancam spectra at Gusev crater, Mars. Geophys. Res. Lett. 34. Available from: http://dx.doi.org/10.1029/2007GL029894.

Johnson, S.S., et al., 2008. Sulfur-induced greenhouse warming on early Mars. J. Geophys. Res. 113, E08005.

Johnson, S.S., et al., 2009. Fate of SO_2 in the ancient Martian atmosphere: implications for transient greenhouse warming. J. Geophys. Res. 114. Available from: https://doi.org/10.1029/2008JE003313.

Johnson, S.S., et al., 2015. Insights from the metagenome of an acid salt lake: the role of biology in an extreme depositional environment. PLoS One 10. Available from: https://doi.org/10.1371/journal.pone.0122869.

Jones, J.H., 1989. Isotopic relationships among the shergottites, the nakhlites, and Chassigny. Proc. Lunar Planet. Sci. 19, 465–474.

Jugo, P.J., 2009. Sulfur content at sulfide saturation in oxidized magmas. Geology 37, 415–418.

Jugo, P.J., et al., 2005. Experimental data on the speciation of sulfur as a function of oxygen fugacity in basaltic melts. Geochim. Cosmochim. Acta 69, 497–503.

Jugo, P.J., et al., 2010. Sulfur K-edge XANES analysis of natural and synthetic basaltic glasses: implications for S speciation and S content as a function of oxygen fugacity. Geochim. Cosmochim. Acta 74, 5926–5938.

Kamyshny Jr., A., et al., 2014. Multiple sulfur isotopes fractionations associated with abiotic sulfur transformations in Yellowstone National Park geothermal springs. Geochem. Trans. 15, 7.

Kargel, J.S., Lewis, J.S., 1993. The composition and early evolution of Earth. Icarus 105, 1–25.

Karunatillake, S., et al., 2014. Sulfates hydrating bulk soil in the Martian low and middle latitudes. Geophys. Res. Lett. 41, 7987–7996.

Kasting, J.F., 1991. CO_2 condensation and the climate of early Mars. Icarus 94, 1–13.

Kemp, A.L.W., Thode, H.G., 1968. The mechanism of the bacterial reduction of sulphate and of sulphite from isotope fractionation studies. Geochim. Cosmochim. Acta 32, 71–91.

Kerber, L., et al., 2015. Sulfur in the early Martian atmosphere revisited: experiments with a 3-D global climate model. Icarus 261, 133–148.

Khan, A., Connolly, J.A.D., 2008. Constraining the composition and thermal state of Mars from inversion of geophysical data. J. Geophys. Res. 113. Available from: https://doi.org/10.1029/2007JE002996.

Khayat, A.S., et al., 2015. A search for SO_2, H_2S and SO above Tharsis and Syrtis volcanic districts on Mars using ground-based high-resolution submillimeter spectroscopy. Icarus 253, 130–141.

King, P.L., McLennan, S.M., 2010. Sulfur on Mars. Elements 6, 107–112.

King, P.L., McSween, H.Y., 2005. Effects of H_2O, pH, and oxidation state on the stability of Fe minerals on Mars. J. Geophys. Res. 110. Available from: https://doi.org/10.1029/2005JE002482.

King, P.L., et al., 2004. The composition and evolution of primordial solutions on Mars, with application to other planetary bodies. Geochim. Cosmochim. Acta 68, 4993–5008.

King, P.L., et al., 2008. Fe-sulfates on Mars: considerations for Martian environmental conditions, Mars sample return and hazards, ground truth from Mars: science payoff from a sample return mission. In: LPI, Albuquerque, pp. 46–47.

Kiseeva, E.S., Wood, B.J., 2013. A simple model for chalcophile element partitioning between sulphide and silicate liquids with geochemical applications. Earth Planet. Sci. Lett. 383, 68–81.

Klingelhöfer, G., et al., 2004. Jarosite and hematite at Meridiani Planum from Opportunity's Mössbauer Spectrometer. Science 306, 1740–1745.

Knoblauch, C., Jorgensen, B.B., 1999. Effect of temperature on sulphate reduction, growth rate and growth yield in five psychrophilic sulphate-reducing bacteria from Arctic sediments. Environ Microbiol. 1, 457–467.

Konneke, M., et al., 2013. Desulfoconvexum algidum gen. nov., sp. no., a psychrophilic sulfate-reducing bacterium isolated from a permanently cold marine sediment. Int. J. System. Evol. Micriobiol. 63, 959–964.

Konopliv, A.S., et al., 2011. Mars high resolution gravity fields from MRO, Mars seasonal gravity, and other dynamical parameters. Icarus 211, 401–428.

Kounaves, S.P., et al., 2009. The MECA Wet Chemistry Laboratory on the 2009 Phoenix Mars Scout Lander. J. Geophys. Res. 114. Available from: https://doi.org/10.1029/2008JE003084.

Kounaves, S.P., et al., 2010. Soluble sulfate in the Martian soil at the Phoenix landing site. Geophys. Res. Lett. 37. Available from: https://doi.org/10.1029/2010GL042613.

Krasnopolsky, V.A., 2012. Search for methane and upper limits to ethane and SO_2 on Mars. Icarus 217, 144–152.

Krasnopolsky, V.A., et al., 2004. Detection of methane in the Martian atmosphere: evidence for life? Icarus 172, 537–547.

Kress, V.C., Carmichael, I.S.E., 1991. The compressibility of silicate liquids containing Fe_2O_3 and the effect of composition, temperature, oxygen fugacity and pressure on their redox states. Contr. Min. Petr. 108, 82–92.

Labidi, J., et al., 2013. Non-chondritic sulphur isotope composition of the terrestrial mantle. Nature 501, 208–212.

Labidi, J., et al., 2016. Experimentally determined sulfur isotope fractionation between metal and silicate and implications for planetary differentiation. Geochim. Cosmochim. Acta 175, 181–194.

Lamarche-Gagnon, G., et al., 2015. Evidence of in situ microbial activity and sulphidogenesis in perenially sub-0 degrees C and hypersaline sediments of a high Arctic permafrost spring. Extremophiles 19, 1–15.

Lane, M.D., 2007. Mid-infrared emission spectroscopy of sulfate and sulfate-bearing minerals. Am. Mineral. 92, 1–18.

Lane, M.D., et al., 2007. Identifying the phosphate and ferric sulfate minerals in the Paso Robles soils (Gusev crater, Mars) using an integrated spectral approach. In: Lunar Planet. Sci., XXXVIII. p. 1338. League City, TX.

Lane, M.D., et al., 2008. Mineralogy of the Paso Robles soils on Mars. Am. Mineral. 93, 728–739.

Lapen, T.J., et al., 2010. A younger age for ALH84001 and its geochemical link to shergottite sources in Mars. Science 328, 347–351.

Lauretta, D.S., et al., 1997. Experimental simulations of sulfide formation in the solar nebula. Science 277, 358–360.

Leshin, L.A., Vicenzi, E., 2006. Aqueous processes recorded by Martian meteorites: analyzing Martian water on Earth. Elements 2, 157–162.

Lever, M.A., et al., 2013. Evidence for microbial carbon and sulfur cycling in deeply buried ridge flank basalt. Science 339, 1305–1308.

Li, C., Ripley, E.M., 2005. Empirical equations to predict the sulfur content of magic magmas at sulfide saturation and applications to magmatic sulfide deposits. Mineral. Depos. 40, 218–230.

Li, C., Ripley, E.M., 2009. Sulfur contents at sulfide-liquid or anhydrite saturation in silicate melts: empirical equations and example applications. Econ. Geol. 104, 405–412.

Lichtenberg, K.A., et al., 2010. Stratigraphy of hydrated sulfates in the sedimentary deposits of Aram Chaos. Mars. J. Geophys. Res. 115. Available from: https://doi.org/10.1029/2009JE003353.

Lillis, R.J., et al., 2008. Rapid decrease in Martian crustal magnetization in the Noachian era: implications for the dynamo and climate of early Mars. Geophys. Res. Lett. 35. Available from: https://doi.org/10.1029/2008GL034338.

Lin, T.J., et al., 2016. Linkages between mineralogy, fluid chemistry, and microbial communities within hydrothermal chimneys from the endeavour segment. Juan de Fuca Ridge. Geochem. Geophys. Geosys. 17, 300–323.

Liu, Y., Wang, A., 2015. Dehydration of Na-jarosite, ferricopoapite, and rhomboclase at temperatures of 50 deg C and 95 deg C: implications for Martian ferric sulfates. J. Raman Spectrosc. 46, 493–500.

Lodders, K., 1998. A survey of shergottite, nakhlite and chassigny meteorites whole-rock composition. Met. Planet. Sci. 33, A183–A190.

Lodders, K., Fegley Jr., B., 1997. An oxygen isotope model for the composition of Mars. Icarus 126, 373–394.

Lodders, K., et al., 2009. Abundances of the elements in the solar system. In: Trumper, J.E. (Ed.), Landolt-Bornstein, New Series, Astron. Astrophys. Springer-Verlag, Berlin, Heidelberg, New York.

Lorand, J.P., et al., 2005. Sulfide mineralogy and redox conditions in some shergottites. Met. Planet. Sci. 40, 1257–1272.

Lorand, J.P., et al., 2012. Metal-saturated sulfide assemblages in NWA 2737: evidence for impact-related sulfur devolatilization in Martian meteorites. Met. Planet. Sci. 47, 1830–1841.

Lorand, J.P., et al., 2015. Nickeliferous pyrite tracks pervasive hydrothermal alteration in Martian regolith breccia: a study in NWA7533. Met. Planet. Sci. Available from: http://dx.doi.org/10.1111/maps.12565.

Louzada, K.L., et al., 2007. Effect of shock on the magnetic properties of pyrrhotite, the Martian crust, and meteorites. Geophys. Res. Lett. 34. Available from: https://doi.org/10.1029/2006GL027685.

MacPherson, G.J., II, T.M.S.R.S.S.G., 2005. The first Mars surface-sample return mission: revised science considerations in light of the 2004 MER results. In: Appendix III of Science Priorities for Mars Sample Return, p. 68.

Madden, M.E., et al., 2004. Jarosite as an indicator of water-limited chemical weathering on Mars. Nature 431, 821–823.

Mahaffy, P.R., et al., 2012. The Sample Analysis at Mars investigation and instrument suite. Space Sci. Rev. 170, 401–478.

Mangold, N., et al., 2008. Spectral and geological study of the sulfate-rich region of West Candor Chasma, Mars. Icarus 194, 519–543.

Marion, G.M., et al., 2008. Modeling ferrous-ferric iron chemistry with application to Martian surface geochemistry. Geochim. Cosmochim. Acta 72, 242–266.

McAdam, A.C., et al., 2014. Sulfur-bearing phases detected by evolved gas analysis of the Rocknest Aeolian deposit, Gale Crater, Mars. J. Geophys. Res. Planets 119, 373–393.

McAdam, A.C., et al., 2016. Reactions involving calcium and magnesium sulfates as potential sources of sulfur dioxide during MSL evolved gas analysis. In: Lunar Planet Sci., XLVII. p. 2277. The Woodlands, TX.

McCubbin, F.M., et al., 2009. Hydrothermal jarosite and hematite in a pyroxene-hosted melt inclusion in Martian meteorite Miller Range (MIL) 03346: implications for magmatic-hydrothermal fluids on Mars. Geochim. Cosmochim. Acta 73, 4907–4917.

McLennan, S.M., Grotzinger, J.P., 2008. The Martian surface—composition, mineralogy, and physical properties. In: Bell, J.F.I. (Ed.), The Sedimentary Rock Cycle of Mars. Cambridge University Press, Cambridge, p. 541.

McLennan, S.M., et al., 2005. Provenance and diagenesis of the evaporite-bearing Burns formation, Meridiani Planum, Mars. Earth Planet. Sci. Lett. 240, 95–121.

McLennan, S.M., et al., 2010. Distribution of sulfur on the surface of Mars determined by the 2001 Mars Odyssey Gamma-Ray Spectrometer. In: Lunar Planet. Sci., XLI. p. 1533. The Woodlands, TX.

McLennan, S.M., et al., 2012. Planning for Mars returned sample science: final report of the MSR End-to-End International Science Analysis Group (E2E-iSAG). Astrobiology 12, 175–230.

McLennan, S.M., et al., 2014. Elemental geochemistry of sedimentary rocks at Yellowknife Bay, Gale Crater, Mars. Science 343. Available from: http://dx.doi.org/10.1126/science.1244734.

McSween, H.Y., 1994. What we have learned about Mars from SNC meteorites. Meteoritics 29, 757–779.

McSween, H.Y., Keil, K., 2000. Mixing relationships in the Martian regolith and the composition of globally homogeneous dust. Geochim. Cosmochim. Acta 64, 2155–2166.

McSween Jr., H.Y., et al., 1996. QUE 94201 shergottite: crystallization of a Martian basaltic magma. Geochim. Cosmochim. Acta 60, 4563–4569.

Metrich, N., et al., 2009. The oxidation state of sulfur in synthetic and natural glasses determined by X-ray absorption spectroscopy. Geochim. Cosmochim. Acta 73, 2382–2399.

Meyer, C., 2015. The Martian Meteorite Compendium. Available from: http://curator.jsc.nasa.gov/antmet/mmc/index.cfm.

Michalski, J.R., Bleacher, J.E., 2013. Supervolcanoes within an ancient volcanic province in Arabia Terra, Mars. Nature 502, 47–52.

Mikouchi, T., et al., 2003. Mineralogy and petrology of Yamato 000593: comparison with other Martian nakhlite meteorites. Ant. Met. Res. 16, 34–57.

Mikouchi, T., et al., 2006. Relative burial depths of nakhlites: an update. In: Lunar Planet. Sci., XXXVII. p. 1865. Houston, TX.

Ming, D.W., et al., 2006. Geochemical and mineralogical indicators for aqueous processes in the Columbia Hills of Gusev crater, Mars. J. Geophys. Res. 111. Available from: https://doi.org/10.1029/2005JE002560.

Ming, D.W., et al., 2008. Geochemical properties of rocks and soils in Gusev Crater, Mars: results of the alpha particle X-ray spectrometer from Cumberland Ridge to Home Plate. J. Geophys. Res. 113. Available from: https://doi.org/10.1029/2008JE003195.

Ming, D.W., et al., 2014. Volatile and organic compositions of sedimentary rocks in Yellowknife Bay, Gale Crater, Mars. Science 343. Available from: https://doi.org/10.1126/science.1245267.

Misawa, K., et al., 2003. Crystallization and alteration ages of the Antarctic nakhlite Yamato 00093. In: Lunar Planet. Sci., XXXIV. p. 1556.

Mischna, M.A., et al., 2013. Effects of obliquity and water vapor/trace gas greenhouses in the early Martian climate. J. Geophys. Res. Planets 118, 560–576.

Morris, R.V., et al., 2000. Mineralogy, composition, and alteration of Mars Pathfinder rocks and soils: evidence from multispectral, elemental, and magnetic data on terrestrial analogue, SNC meteorites, and Pathfinder samples. J. Geophys. Res. 105, 1757–1818.

Morris, R.V., et al., 2006. Mössbauer mineralogy of rock, soil, and dust at Meridiani Planum, Mars: Opportunity's journey across sulfate-rich outcrop, basaltic sand and dust, and hematite lag deposits. J. Geophys. Res. 111. Available from: https://doi.org/10.1029/2006JE002791.

Morris, R.V., et al., 2015. Update on the chemical composition of crystalline, smectite, and amorphous components for Rocknest soil and John Klein and Cumberland mudstone drill fines at Gale crater, Mars. In: Lunar Planet. Sci., XLVI. Abstract # 1832. The Woodlands, TX.

Morris, R.V., et al., 2016. Silicic volcanism on Mars evidenced by tridymite in high-SiO_2 sedimentary rock at Gale crater. Proc. Natl Acad. Sci. 113, 7071–7076.

Murchie, S., et al., 2009a. Evidence for the origin of layered deposits in Candor Chasma, Mars, from mineral composition and hydrologic modeling. J. Geophys. Res. 114. Available from: https://doi.org/10.1029/2009JE003343.

Murchie, S.L., et al., 2009b. A synthesis of Martian aqueous mineralogy after 1 Mars year of observations from the Mars Reconnaissance Orbiter. J. Geophys. Res. Planets 114. Available from: https://doi.org/10.1029/2009JE003342.

Mustard, J.F., Ehlmann, B.L., 2010. Intact stratigraphy traversing the phyllosilicate to sulfate eras at the Syrtis-Isidis contact, Mars. In: Lunar Planet. Sci., XLI. Abstract # 1533. The Woodlands, TX.

Nachon, M., et al., 2014. Calcium sulfate veins characterized by ChemCam/Curiosity at Gale crater, Mars. J. Geophys. Res. Planets 119, 1991−2016.

Neukum, G., et al., 2004. Recent and episodic volcanic and glacial activity on Mars revealed by the high resolution stereo camera. Nature 432, 971−979.

Newsom, H.E., et al., 1999. Mixed hydrothermal fluids and the origin of the Martian soil. J. Geophys. Res. 104, 8717−8728.

Noe Dobrea, E.Z., et al., 2006a. Analysis of a spectrally unique deposit in the dissected Noachian terrain of Mars. J. Geophys. Res. 111. Available from: https://doi.org/10.1029/2005JE002431.

Noe Dobrea, E.Z., et al., 2006b. OMEGA analysis of light-toned outcrops in the chaotic terrain of the eastern Valles Marineris region. In: Lunar Planet Sci., XXXVII. p. 2068. Houston, TX.

Nyquist, L.E., et al., 2001. Ages and geologic histories of Martian meteorites. Space Sci. Rev. 96, 105−164.

O'Neill, H.S.C., Mavrogenes, J., 2002. The sulfide saturation capacity and the sulfur content at sulfide saturation of silicate melts at 1400 deg C and 1 bar. J. Petrol. 43, 1049−1087.

Ono, S., et al., 2006. Mass-dependent fractionation of quadruple stable sulfur isotope system as a new tracer of sulfur biogeochemical cycles. Geochim. Cosmochim. Acta 70, 2238−2252.

Orsi, W.D., et al., 2016. Transcriptional analysis of sulfate reducing and chemolithosutrophic sulfur oxidizing bacteria in the deep subseafloor. Environ Microbiol. Rep. 8, 452−460.

Palme, H., Jones, A., 2003. Solar system abundances of the elements. Treatise on Geochemistry 41−61.

Palme, H., et al., 1988. Moderately volatile elements. In: Kerridge, J.F., et al., (Eds.), Meteorites and the Early Solar System. University of Arizona Press, Tuscan, AZ, pp. 436−461.

Pasek, M., et al., 2005. Sulfur chemistry with time-varying oxygen abundance during Solar System formation. Icarus 175, 1−14.

Pavlov, A.A., Kasting, J.F., 2002. Mass-independent fractionation of sulfur isotopes in Archean sediments: strong evidence for an anoxic Archean atmosphere. Astrobiology 2, 27−41.

Peale, S.J., et al., 1975. Origin of Martian channels: clathrates and water. Science 187, 273−274.

Peterson, R.C., Grant, A., 2005. Dehydration and crystallization reactions of secondary sulfate minerals found in mine waste: in situ powder diffraction experiments. Can. Mineral. 41, 1173−1181.

Peterson, R.C., Wang, R., 2006. Crystal molds on Mars: melting of a possible new mineral species to create chaotic terrain. Geology 34, 957−960.

Peterson, R.C., et al., 2008. Meridianite: a new mineral species observed on Earth and predicted to exist on Mars. Am. Mineral. 92, 1756−1759.

Phillips, R.J., et al., 2001. Ancient geodynamics and global-scale hydrology on Mars. Science 291, 2587−2591.

Pollack, J.B., et al., 1987. The case for a warm, wet climate on early Mars. Icarus 71, 203−224.

Postawko, S.E., Kuhn, W.R., 1986. Effect of the greenhouse gases (CO_2, H_2O, SO_2) on Martian paleoclimate. J. Geophys. Res. 91, 431−438.

Poulet, F., et al., 2005. Phyllosilicates on Mars and implications for early Martian climate. Nature 438, 623−627.

Ramirez, R.M., et al., 2013. Warming early Mars with CO_2 and H_2. Nat. Geosci. 7, 59−63.

Rampe, E.B., et al., 2015. Potential cement phases in sedimentary rocks drilled by Curiosity at Gale crater, Mars. In: Lunar Planet. Sci., XLVI. p. 2038. The Woodlands, TX.

Rampe, E.B., et al., 2016. Mineralogical and geochemical trends in a fluviolacustrine sequence in Gale crater, Mars. In: Goldschmidt Conference, Japan.

Rao, M.N., et al., 1999. Martian soil component in impact glasses in a Martian meteorite. Geophys. Res. Lett. 26, 3265−3268.

Rao, M.N., et al., 2008. The nature of Martian fluids based on mobile element studies in salt-assemblages from Martian meteorites. J. Geophys. Res. 113. Available from: https://doi.org/10.1029/2007JE002958.

Rao, M.N., et al., 2011. Isotopic evidence for a Martian regolith component in shergottite meteorites. J. Geophys. Res. Planets 116. Available from: https://doi.org/10.1029/2010JE003764.

Rao, M.N., et al., 2012. Laboratory shock experiments on basalt-iron sulfate mixes at $\sim$40-50 GPa and their relevance to the Martian regolith component present in shergottites. In: Lunar Planet. Sci., XLIII. p. 1659. The Woodlands, TX.

Rieder, R., 1997. The chemical composition of Martian soil and rocks returned by the mobile alpha proton X-ray spectrometer: preliminary results from the X-ray mode. Science 278, 1771–1774.

Rieder, R., et al., 2004. APXS on Mars: analyses of soils and rocks at Gusev Crater and Meridiani Planum. In: Lunar Planet. Sci., XXXV. p. 2172. League City, TX.

Righter, K., et al., 2009. Experimental evidence for sulfur-rich Martian magmas: implications for volcanism and surficial sulfur sources. Earth Planet. Sci. Lett. 288, 235–243.

Ringwood, A.E., 1979. On the Origin of the Earth and Moon. Springer Verlag, New York.

Rivoldini, A., et al., 2011. Geodesy constraints on the interior structure and composition of Mars. Icarus 213, 451–472.

Roach, L.H., et al., 2010. Diagenetic haematite and sulfate assemblages in Valles Marineris. Icarus 207, 659–674.

Robbins, S.J., et al., 2011. The volcanic history of Mars: high-resolution crater-based studies of the calderas of 20 volcanoes. Icarus 211, 1179–1203.

Robinson, B.W., 1993. Sulfur isotope standards, Reference and intercomparison materials for stable isotopes of light elements. In: IAEA Consultants Meeting, Vienna.

Rochette, P., et al., 2005. Matching Martian crustal magnetization and magnetic properties of Martian meteorites. Met. Planet. Sci. 40, 529–540.

Rouchy, J.-M., Monty, C., 1981. Stromatolites and cryptalgal alminites of Messinian gypsum of Cyprus. In: Monty, C. (Ed.), Phanerozoic Stromatolites. Springer, Berlin, pp. 155–178.

Ruff, S.W., et al., 2006. The rocks of Gusev crater as viewed by the Mini-TES instrument. J. Geophys. Res. 111. Available from: https://doi.org/10.1029/2006JE002747.

Sagan, C., Chyba, C., 1997. The early faint Sun paradox: organic shielding of ultraviolet-labile greenhouse gases. Science 276, 1217–1221.

Sagan, C., Mullen, G., 1972. Earth and Mars: evolution of atmospheres and surface temperatures. Astrophys. J. 360, 727–736.

Sagemann, J., et al., 1998. Temperature dependence and rates of sulfate reduction in cold sediments of Svalbard. Arctic Ocean. Geomicro. J. 15, 85–100.

Salle, B., et al., 2004. Laser-induced breakdown spectroscopy for Mars surface analysis: capabilities at stand-off distances and detection of chlorine and sulfur elements. Spectrochim. Acta Part B 59, 1413–1422.

Sanloup, C., et al., 1999. A simple chondritic model of Mars. Phys. Earth Planet. Int. 112, 43–54.

Sattley, W.M., Madigan, M.T., 2010. Temperature and nutrient induced responses of Lake Fryxell sulfate-reducing prokaryotes and description of Desulfovibrio lacusfryxellense, sp. nov., a pervasive, cold-active, sulfate-reducing bacterium from Lake Fryxell, Antarctica. Extremophiles 14, 357–366.

Schrader, C.M., et al., 2011. Ni, S, and Cl in EETA 79001 lithology C. In: Lunar Planet. Sci., XLII. p. 2814. The Woodlands, TX.

Schubert, G., Spohn, T., 1990. Thermal history of Mars and the sulfur content of its core. J. Geophys. Res. 95, 14095–14104.

Self, S., et al., 2006. Volatile fluxes during flood basalt eruptions and potential effects on the global environment: a Deccan perspective. Earth Planet. Sci. Lett. 248, 518–532.

Settle, M., 1979. Formation and deposition of volcanic sulfate aerosols on Mars. J. Geophys. Res. 84, 8343–8354.

Shahar, A., et al., 2009. Sulfur isotopic fractionation during the differentiation of Mars. Geochim. Cosmochim. Acta 73, A1201.

Shahar, A., et al., 2015. Sulfur-controlled iron isotope fractionation experiments of core formation in planetary bodies. Geochim. Cosmochim. Acta 150, 253–264.

Shen, Y., et al., 2001. Isotopic evidence for microbial sulphate reduction in the early Archaean era. Nature 410, 77–81.

Shirai, N., Ebihara, M., 2004. Chemical characteristics of an olivine-phyric shergottite, Yamato 980459. In: Lunar Planet. Sci., XXXV. p. 1511. Houston, TX.

Shooner, F., et al., 1996. Isolation, phenotypic characterization, and phylogenetic position of a novel, facultatively autotrophic, moderately thermophilic bacterium, Thiobacillus thermosulfatus sp. nov. Int. J. Syst. Bacteriol. 46, 409–415.

Siebach, K.L., et al., 2014. Subaqueous shrinkage cracks in the Sheepbed mudstone: implications for early fluid diagenesis, Gale crater, Mars. J. Geophys. Res. Planets 119, 1597–1613.

Smith, J.V., Hervig, R.L., 1979. Shergotty meteorite: mineralogy, petrography and minor elements. Meteoritics 14, 121–142.

Sobron, P., et al., 2012. Extraction of compositional and hydration information of sulfates from laser-induced plasma spectra recorded under Mars atmospheric conditions—implications for ChemCam investigations on Curiosity rover. Spectrochim. Acta Part B 68, 1–16.

Solomon, S.C., et al., 2005. New perspectives on ancient Mars. Science 307, 1214–1220.

Squyres, S.W., 1984. The history of water on Mars. Ann. Rev. Earth Planet. Sci. 12, 83–106.

Squyres, S.W., Kasting, J.F., 1994. Early Mars: how warm and wet? Science 265, 744–749.

Squyres, S.W., et al., 2006. Rocks of the Columbia Hills. J. Geophys. Res. 111. Available from: https://doi.org/10.1029/2005JE002562.

Squyres, S.W., et al., 2012. Ancient impact and aqueous processes at Endeavour Crater, Mars. Science 336, 570–576.

Stam, M.C., et al., 2010. Sulfate reducing activity and sulfur isotope fractionation by natural microbial communities in sediments of a hypersaline soda lake (Mono Lake, California). Chem. Geol. 278, 23–30.

Stetter, K.O., 2007. Hyperthermophiles in the history of life. In: Bock, G.R., Goode, J.A. (Eds.), Ciba Foundation Symposium 202 - Evolution of Hydrothermal Ecosystems on Earth (and Mars?). John Wiley & Sons, Chichester, UK, pp. 1–18.

Stevenson, D.J., 2001. Mars' core and magnetism. Nature 412, 214–219.

Stewart, A.J., et al., 2007. Mars: a new core-crystallization regime. Science 316, 1323–1325.

Stockstill, K.R., et al., 2005. Thermal Emission Spectrometer hyperspectral analyses of proposed paleolake basins on Mars: no evidence for in-place carbonates. J. Geophys. Res. 110. Available from: https://doi.org/10.1029/2004JE002353.

Stolper, E., McSween Jr., H.Y., 1979. Petrology and origin of the shergottite meteorites. Geochim. Cosmochim. Acta 43, 1475–1498.

Sutton, S.R., et al., 2008. Sulfur and iron speciation in gas-rich impact-melt glasses from basaltic shergottites determined by micro-XANES. In: Lunar Planet. Sci., XXXIX. p. 1961. Houston, TX.

Swayze, G.A., et al., 2008. Discovery of the acid-sulfate mineral alunite in Terra Sirenum, Mars, using MRO CRISM: possible evidence for acid-saline lacustrine deposits? In: AGU Fall Meeting, pp. P44A-04.

Swindle, T.D., et al., 2000. Noble gases in iddingsite from the Lafayette meteorite: evidence for liquid water on Mars in the last few hundred million years. Met. Planet. Sci. 35, 107–115.

Szymanski, A., et al., 2010. High oxidation state during formation of Martian nakhlites. Met. Planet. Sci. 45, 21–31.

Tanaka, K.L., et al., 2014. The digital global geologic map of Mars: chronostratigraphic ages, topographic and crater morphologic characteristics, and updated resurfacing history. Planet. Space Sci. 95, 11–24.

Taylor, J., McLennan, S.M., 2009. Aqueous alteration and Martian bulk chemical composition. In: AGU Fall Meeting, pp. P12A-07.

Thiemens, M.H., 1999. Mass-independent isotope effects in planetary atmospheres and the early solar system. Science 283, 341−345.

Thiemens, M.H., Heidenreich, J.E.I., 1983. The mass-independent fractionation of oxygen: a novel isotope effect and its possible cosmochemical implications. Science 219, 1073−1075.

Thode, H.G., et al., 1953. Sulphur isotope fractionation in nature and geological and biological time scales. Geochim. Cosmochim. Acta 3, 235−243.

Tian, F., et al., 2010. Photochemical and climate consequences of sulfur outgassing on early Mars. Earth Planet. Sci. Lett. 295, 412−418.

Toon, O.B., et al., 1980. The astronomical theory of climate change on Mars. Icarus 44, 552−607.

Topcuoglu, B.D., et al., 2016. Hydrogen limitation and syntrophic growth among natural assemblages of thermophilic methanogens at deep-sea hydrothermal vents. Front. Microbiol. 7, 1240. Available from: https://doi.org/10.3389/fmicb.2016.01240.

Tosca, N.J., McLennan, S.M., 2006. Chemical divides and evaporite assemblages on Mars. Earth Planet. Sci. Lett. 241, 21−31.

Tosca, N.J., et al., 2005. Geochemical modeling of evaporation processes on Mars: insight from the sedimentary record at Meridiani Planum. Earth Planet. Sci. Lett. 240, 122−148.

Toulmin III, P., et al., 1977. Geochemical and mineralogical interpretation of the Viking inorganic chemical results. J. Geophys. Res. 82, 4625−4634.

Treiman, A.H., et al., 1993. Preterrestrial aqueous alteration of the Lafayette (SNC) meteorite. Meteoritics 28, 86−97.

Treiman, A.H., et al., 2016. Mineralogy, provenance, and diagenesis of a potassic basaltic sandstone on Mars: CheMin X-ray diffraction of the Windjana sample (Kimberley area, Gale Crater). J. Geophys. Res. Planets 121, 75−106.

Turchyn, A.V., et al., 2016. Microbial sulfur metabolism evidenced from pore fluid isotope geochemistry at Site U1385. Global Planet. Change 141, 82−90.

Ueno, Y., et al., 2009. Geological sulfur isotopes indicate elevated OCS in the Archean atmosphere, solving faint young sun paradox. Proc. Natl Acad. Sci. 106, 14784−14789.

Urey, H.C., 1947. The thermodynamic properties of isotopic substances. J. Chem. Soc. 562−581.

USGS, 2000. New Evidence Suggests Mars Has Been Cold and Dry "Red Planet" Abundant with Green Minerals. Available from: http://speclab.cr.usgs.gov/mars.press.release.10.2000.html.

Usui, T., et al., 2012. Origin of water and mantle−crust interactions on Mars inferred from hydrogen isotopes and volatile element abundances of olivine-hosted melt inclusions of primitive shergottites. Earth Planet. Sci. Lett. 357-358, 119−129.

Vaniman, D.T., Chipera, S.J., 2006. Transformations of Mg- and Ca-sulfate hydrates in Mars regolith. Am. Mineral. 91, 1628−1642.

Vaniman, D.T., et al., 2004. Magnesium sulphate salts and the history of water on Mars. Nature 431, 663−665.

Vaniman, D.T., et al., 2008. Salt-hydrate stabilities and Mars sample return missions, Ground Truth from Mars: science Payoff from a Sample Return Mission. In: LPI, Albuquerque, pp. 103−104.

Vaniman, D.T., et al., 2012. Ceramic ChemCam calibration targets on Mars Science Laboratory. Space Sci. Rev. 170, 229−255.

Vaniman, D.T., et al., 2014. Mineralogy of a mudstone at Yellowknife Bay, Gale Crater, Mars. Science 343. Available from: https://doi.org/10.1126/science.1243480.

von Paris, P., et al., 2013. N_2-associated surface warming on early Mars. Planet. Space Sci. 82, 149−154.

Wacey, D., et al., 2011. Earliest microbially-mediated pyrite oxidation in ∼3.4 billion-year-old sediments. Earth Planet. Sci. Lett. 301, 393−402.

Wadhwa, M., 2001. Redox state of Mars' upper mantle and crust from Eu anomalies in shergottite pyroxenes. Science 291, 1527–1530.

Wallace, D., Sagan, C., 1979. Evaporation of ice in planetary atmospheres: ice-covered rivers on Mars. Icarus 39, 385–400.

Wallace, P., Carmichael, I.S.E., 1992. Sulfur in basaltic magmas. Geochim. Cosmochim. Acta 56, 1863–1874.

Walton, E.L., et al., 2010. Martian regolith in Elephant Moraine 79001 shock melts? Evidence from major element composition and sulfur speciation. Geochim. Cosmochim. Acta 74, 4829–4843.

Wang, A., Ling, Z.C., 2011. Ferric sulfates on Mars: a combined mission data analysis of salty soils at Gusev crater and laboratory experimental investigations. J. Geophys. Res. 116. Available from: https://doi.org/10.1029/2010JE003665.

Wang, A., Zhou, Y., 2014. Experimental comparison of the pathways and rates of the dehydration of Al-, Fe-, Mg-, and Ca-sulfates under Mars relevant conditions. Icarus 234, 162–173.

Wang, A., et al., 2006a. Sulfates on Mars: a systematic Raman spectroscopic study of hydration states of magnesium sulfates. Geochim. Cosmochim. Acta 70, 6118–6135.

Wang, A., et al., 2006b. Sulfate deposition in subsurface regolith in Gusev crater. Mars. J. Geophys. Res. 111. Available from: https://doi.org/10.1029/2005JE002513.

Wang, A., et al., 2009. Phase transition pathways of the hydrates of magnesium sulfates in the temperature range 50°C to 5°C: implication for sulfates on Mars. J. Geophys. Res. 114. Available from: https://doi.org/10.1029/2008JE003266.

Wang, A., et al., 2011. Stability of Mg-sulfates at −10°C and the rates of dehydration/rehydration processes under conditions relevant to Mars. J. Geophys. Res. 116. Available from: https://doi.org/10.1029/2011JE003818.

Wang, A., et al., 2013a. The preservation of subsurface sulfates with mid-to-high degree of hydration in equatorial regions on Mars. Icarus 226, 980–991.

Wang, A., et al., 2016. Setting constraints on the nature and origin of the two major hydrous sulfates on Mars: monohydrated and polyhydrated sulfates. J. Geophys. Res. Planets 121, 678–694.

Wang, C., et al., 1991. Phase equilibria of liquid Fe-S-C-ternary. ISIJ Intl. 11, 1292–1299.

Wang, T., et al., 2007. Dehydration of iron(II) sulfate heptahydrate. Thermochim. Acta 462, 89–93.

Wang, Y., et al., 2013b. Composition of Mars constrained using geophysical observations and mineral physics modeling. Phys. Earth Planet. Int. 224, 68–76.

Wang, Z., Becker, H., 2017. Chalcophile elements in Martian meteorites indicate low sulfur content in the Martian interior and a volatile element-depleted late veneer. Earth Planet. Sci. Lett. 463, 56–68.

Wänke, H., 1981. Constitution of terrestrial planets. Phil. Trans. R, Soc. Lond. A 303, 287–302.

Wänke, H., Dreibus, G., 1988. Chemical composition and accretion history of terrestrial planets. Phil. Trans. R, Soc. Lond. A 325, 545–557.

Wänke, H., Dreibus, G., 1994. Chemistry and accretion history of Mars. Phil. Trans.: Phys. Sci. Eng. 349, 285–293.

Wasson, J.T., Kallemeyn, G.W., 1988. Compositions of chondrites. Phil. Trans. R, Soc. Lond. A 325, 535–544.

Wentworth, S.J., Gooding, J.L., 1994. Carbonates and sulfates in the Chassigny meteorite: further evidence for aqueous chemistry on the SNC parent planet. Meteoritics 29, 860–863.

Wentworth, S.J., Gooding, J.L., 1996. Water-based alteration of the Martian meteorite, QUE 94201, by sulfate-dominated solutions. In: Lunar Planet. Sci., XXVII. p. 1421.

Wentworth, S.J., McKay, D.S., 1999. Weathering and secondary minerals in the Nakhla meteorite. In: LPSC XXX, Houston, TX, p. 1946.

Wentworth, S.J., et al., 2000. Weathering and secondary minerals in the Martian meteorite Shergotty. In: Lunar Planet. Sci., XXXI. p. 1888. Houston, TX.

Werner, S.C., 2009. The global Martian volcanic evolutionary history. Icarus 201, 44−68.

Wiens, R.C., et al., 2013. Pre-flight calibration and initial data processing for the ChemCam laser-induced breakdown spectroscopy instrument on the Mars Science Laboratory rover. Spectrochim. Acta Part B 82, 1−27.

Wilson, S.A., Bish, D.L., 2011. Formation of gypsum and bassanite by cation exchange reactions in the absence of free-liquid H_2O: implications for Mars. J. Geophys. Res. Planets 116. Available from: https://doi.org/10.1029/2011JE003853.

Wilson, S.A., Bish, D.L., 2012. Stability of Mg-sulfate minerals in the presence of smectites: possible mineralogical controls on H_2O cycling and biomarker preservation on Mars. Geochim. Cosmochim. Acta 96, 120−133.

Wilson, L., et al., 2001. Evidence for episodicity in the magma supply to the large Tharsis volcanoes. J. Geophys. Res. Planets 106, 1423−1433.

Wiseman, S.M., et al., 2010. Spectral and stratigraphic mapping of hydrated sulfate and phyllosilicate-bearing deposits in northern Sinus Meridiani, Mars. J. Geophys. Res. 115. Available from: https://doi.org/10.1029/2009JE003354.

Woese, C.R., et al., 1991. Archaeal phylogeny—reexamination of the phylogenetic position of Archaeoglobidus-fulgidus in light of certain composition-induced artifacts. Syst. Appl. Microbiol. 14, 364−371.

Wong, A.-S., Atreya, S.K., 2003. Chemical markers of possible hot spots on Mars. J. Geophys. Res. 108. Available from: https://doi.org/10.1029/2002JE002003.

Wong, A.S., et al., 2005. Correction to and updated reaction in "Chemical markers of possible hot spots on Mars". J. Geophys. Res. 110. Available from: https://doi.org/10.1029/2005JE002509.

Wray, J.J., et al., 2009. Diverse aqueous environments on ancient Mars revealed in the southern highlands. Geology 37, 1043−1046.

Wray, J.J., et al., 2011. Columbus crater and other possible groundwater-fed paleolakes of Terra Sirenum, Mars. J. Geophys. Res. 116. Available from: https://doi.org/10.1029/2010JE003694.

Xu, W., Parise, J.B., 2009. $(H3O)Fe(SO4)2$, A new phase formed by dehydrating rhomboclase. In: Lunar Planet. Sci., XL. Abstract # 1816. The Woodlands, TX.

Xu, W., et al., 2008. Relative humidity-induced production of ferricopiapite and rhomboclase from ferric sulfate anhydrate: X-ray diffraction studies under controlled conditions. In: Lunar Planet. Sci., XXXIX. p. 1391. League City, TX.

Yen, A.S., et al., 2005. An integrated view of the chemistry and mineralogy of Martian soils. Nature 436, 49−54.

Yen, A.S., et al., 2016. Cementation and aqueous alteration of a sandstone unit under acidic conditions in Gale crater, Mars. In: Lunar Planet. Sci., XLVII. p. 1649. The Woodlands, TX.

Yoder, C.F., Standish, E.M., 1997. Martian precession and rotation from Viking lander range data. J. Geophys. Res. 102, 4065−4080.

Yoder, C.F., et al., 2003. Fluid core size of Mars from detection of the solar tide. Science 300, 299−303.

Young, R.E., Schubert, G., 1974. Temperatures inside Mars: is the core liquid or solid? Geophys. Res. Lett. 1, 157−160.

Yung, Y.L., et al., 1997. CO_2 greenhouse in the early Martian atmosphere: SO_2 inhibits condensation. Icarus 130, 222−224.

Zipfel, J., et al., 2000. Petrology and chemistry of the new shergottite Dar al Gani 476. Met. Planet. Sci. 35, 95−106.

Zolotov, M.Y., 2003. Martian volcanic gases: are they terrestrial-like? In: Lunar Planet. Sci., XXXIV. p. 1795. Houston, TX.

Zolotov, M.Y., Shock, E.L., 2005. Formation of jarosite-bearing deposits through aqueous oxidation of pyrite at Meridiani Planum, Mars. Geophys. Res. Lett. 32. Available from: https://doi.org/10.1029/2005GL024253.

THE HYDROLOGY OF MARS INCLUDING A POTENTIAL CRYOSPHERE

7

Jérémie Lasue[1], Stephen M. Clifford[2], Susan J. Conway[3], Nicolas Mangold[3], and Frances E.G. Butcher[4]

[1]IRAP-OMP, CNRS-UPS, Toulouse, France [2]Lunar and Planetary Institute, Houston, TX, United States [3]LPG, Université de Nantes, Nantes, France [4]The Open University, Milton Keynes, United Kingdom

7.1 OUTSTANDING QUESTIONS ABOUT WATER ON MARS

The unifying central theme of Mars exploration over the past two decades has been "follow the water," as water is a key factor in our understanding of the evolution of the Martian surface and its potential habitability through time, and could provide an in situ resource for sustaining future human explorers. The size of the water inventory on Mars is usually expressed in terms of a Global Equivalent Layer (GEL) of liquid water, averaged over the planet's surface. The same convention is used throughout this chapter.

There is considerable geomorphological evidence for the sustained presence and flow of liquid water on Mars in the ancient past, including the valley networks, outflow channels, hundreds of paleolake basins and delta fans, and possible shorelines of a former northern ocean. Further evidence of the activity of water on early Mars is indicated by the diversity of aqueous alteration mineralogical products detected in outcrops. Finally, geomorphic evidence of ancient glaciations is preserved at the surface and can be linked to ice and snow deposition during ancient surface water cycles. Such features are discussed in detail in Section 7.2. Therefore, our current understanding is that water has played a major role in the geologic evolution of the Martian surface and may have provided both sporadic and relatively long-lasting habitable environments (e.g., Carr, 1986; SSES, 2003; CASSEP, 2008; Grotzinger et al., 2014).

As discussed by Hynek (2016), there appears to be a "great paradox" about the Martian climate as we understand it. Climate modelers have consistently pointed out the difficulty of warming up early Mars sufficiently for liquid water to carve the geomorphological features seen on its ancient surface. Other researchers have argued that the early climate does not have to be "warm and wet" to explain the observed geomorphic features and have proposed "cold and icy" early Mars models. In the "warm and wet" scenario, it is envisioned that Mars possessed a persistent and extensive water cycle similar to Earth, with precipitation, efficient erosion, transport and deposition, and formation of lakes and seas. In the cold and icy scenario, liquid water is envisioned to be episodically present, with the majority of the water stored as ice in ice sheets and glaciers. These end members have profoundly different implications for Mars' early history and habitability and are also discussed in Section 7.2.

Volatiles in the Martian Crust. DOI: http://dx.doi.org/10.1016/B978-0-12-804191-8.00007-6

The current Martian atmosphere is both cold and dry, with a mean annual surface temperature of ~ 210 K, an atmospheric water content of ~ 10 precipitable microns and an average surface pressure of about 6 mbar — conditions which do not permit the sustained presence of liquid water on the surface of the planet today (e.g., Smith, 2008). Instead, remaining surface and near-surface water is stored in ice reservoirs including the polar cap deposits, high-to-mid-latitude ice-rich mantle, and debris-covered mid-latitude glaciers in the present day. However, recently detected activity of surface features such as gullies and recurring slope lineae (RSL) could be evidence that limited flow of liquid water is possible under present-day conditions. Spacecraft investigations give a number of clues as to the possible pervasive presence of water beneath the surface to substantial depths, but the global inventory of water at depth remains debated. The current state of present water detections on Mars is discussed in Section 7.3.

Section 7.4 is dedicated to a discussion of how the water inventory of Mars may have evolved with time. We discuss how its origin can be assessed and what are the roles and extent of the various sinks or loss processes. Finally, we discuss the current characteristics and extent of the Martian cryosphere under reasonable hypotheses (Section 7.5), and suggest which investigations need to be performed in the future in order to test these assumptions (Section 7.6).

7.2 EVIDENCE FOR THE PAST PRESENCE OF WATER ON THE SURFACE OF MARS

Geologic evidence of a possible warm and wet early Mars is based on the identification and distribution of various water-related landforms whose occurrence is largely restricted to terrains dating from the Late Noachian/Early Hesperian (Carr, 1996; Fasset and Head, 2008a; Carr and Head, 2010). In short, these include the Martian valley networks (Craddock and Maxwell, 1993; Irwin et al., 2005a; Fassett and Head, 2008a; Luo and Stepinski, 2009; Hynek et al., 2010), river deltas (Malin and Edgett, 2003; Fassett and Head, 2005), and topographic basins with paleoshorelines and drainage channels (Irwin et al., 2005a; Fassett and Head, 2008b).

7.2.1 FLUVIAL VALLEYS

Networks of valleys are found throughout the planet's most heavily cratered, Noachian age (> 3.7 Gy) terrain (Sharp and Malin, 1975; Carr and Clow, 1981; Carr, 1996). They have generally been interpreted as having formed by surface runoff associated with snow deposition and melting and/or rainfall because of their resemblance to the terrestrial fluvial valley networks (Carr and Clow, 1981; Craddock and Howard, 2002; Mangold et al., 2004a; Howard et al., 2005; Irwin et al., 2005a; Ansan and Mangold, 2006). The largest valleys have widths of a few kilometers and depths of hundreds of meters (Williams and Phillips, 2001). Valley networks have typical drainage densities (length of fluvial drainages divided by the watershed area) of $0.1-1 \text{ km}^{-1}$, which are equivalent to terrestrial valley networks in drylands (e.g., Carr and Chuang, 1997; Stepinski and Stepinski, 2005; Hynek et al., 2010; Ansan et al., 2008).

Inferred discharge rates from channels within the valleys have been estimated to range from a few 100s to as much as 5000 m^3/s, values consistent with fluvial activity on Earth (Irwin et al., 2005b).

However, this calculation is valid for the last episode of activity only. In addition, these channels in valleys are only observed in a few locations, thus leaving uncertainty about the hydrology of the earliest water cycle. The amount of water required to form the valley networks and to animate their associated water cycle remains poorly constrained. Rough estimations provided from valley depths suggest a GEL of 100−1000 m (e.g., Carr, 1996). The distribution and density of the ancient valley networks is consistent with a sustained hydrologic cycle involving precipitation and surface runoff (e.g., Luo and Stepinski, 2009; Hynek et al., 2010).

The size of the valleys and amplitude of this activity appear to have declined progressively with time (Hoke and Hynek, 2009; Hoke et al., 2011), but significant fluvial activity is thought to have persisted into the Late Hesperian and Amazonian (Ansan and Mangold, 2013). As shown in Fig. 7.1, recent maps of the distribution of valley networks indicate that they were active, at least episodically, from the Late Noachian (∼4 Gy) to the Early Amazonian (∼2.8 Gy). Several dendritic fluvial valley networks are found on Late Hesperian terrains, suggesting that liquid water was still present at the surface, at least episodically, during this period (e.g., Mangold et al., 2004a; Bouley et al., 2010). Further, fluvial valleys, which are thought to have been active in the Early and Middle Amazonian, are observed, but their distribution is patchier. Notably, small fluvial valleys dissect Amazonian volcanoes such as Alba Patera and Hecates Tholus (Gulick and Baker, 1990; Hauber et al., 2005). In addition, several large post-Noachian impact craters are superposed by small, poorly dendritic valleys that locally exhibit interior channels, for instance at Cerulli,

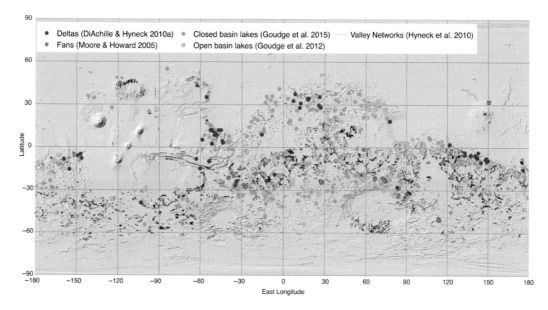

FIGURE 7.1

Map of the distribution of valley network, deltas, fans, and paleolakes on Mars. Basemap is Mars Orbiter Laser Altimeter (MOLA) gridded topography over MOLA hillshade (Smith et al., 2001). *Data from Hynek et al. (2010); Di Achille and Hynek (2010a); Goudge et al. (2012, 2015); Moore and Howard (2005).*

Hale, or Lyot craters (Mouginis-Mark, 1987; Dickson et al., 2009; Jones et al., 2011). Many of these young fluvial features are found at mid-latitudes (30–50° N and S), where ice is present close to the surface (see Section 7.3). Rather than being associated with a global hydrological cycle, it is likely that these "young" valleys developed through local melting of this subsurface ice due to impacts or other transient climatic processes (Morgan and Head, 2009; Fassett et al., 2010; Mangold, 2012; Parsons et al., 2013; Adeli et al., 2016).

However, the interpretation of an ancient Martian hydrologic cycle has been challenged by the fact that climate modelers have consistently pointed out the difficulty of warming up early Mars sufficiently for liquid water to carve such geomorphological features. This is in part due to the faint young Sun, which had 75% of its current luminosity around 4 Gy ago (Shaw, 2016) and the requirement for much higher early atmospheric pressures approaching 1 bar. Many different combinations of greenhouse gases (CO_2, H_2O, H_2, CH_4, SO_2, H_2S, etc.) have been tested with climate models and all require amounts of atmospheric CO_2 that vastly exceed the amounts that are consistent with the sedimentary carbonate measurements of Bristow et al. (2017) and rates of exospheric loss to space (Bibring et al., 2005). Some specific greenhouse gas combinations may work, for example, 1.3–4 bars of CO_2 and 5%–20% of H_2 (Ramirez et al., 2014) or 2%–10% of H_2 and CH_4 (Wordsworth, 2016; Wordsworth et al., 2017).

For atmospheric surface pressures > 0.5 bars, surface temperatures are determined not only by the surface insolation balance, but by altitude due to adiabatic cooling effects (e.g., Wordsworth, 2016). This creates cold-traps at high elevation and is the theoretical basis for the icy-highlands models of early Mars where snow and ice accumulate in the southern highlands (and other local topographic highs). The transient melting of such deposits may have served as the source of water for the formation of the valley networks (e.g., Segura et al., 2002; Wordsworth et al., 2013). The melt could be subglacial or supraglacial (Fastook et al., 2014; Wordsworth, 2016), perhaps facilitated by heating provided by local volcanoes or impact craters (Segura et al., 2002; Mangold, 2012). While such transient fluvial activity provides a good explanation for the patchy and apparently episodic valley formation of the Late Hesperian and Amazonian, whether the same processes could have led to the widespread dissection exhibited by the Noachian terrains is currently a subject of debate (Williams and Phillips, 2001; Som et al., 2009; Jaumann et al., 2010; Penido et al., 2013).

7.2.2 INVERTED CHANNELS, ALLUVIAL FANS, DELTAS, AND PALEOLAKES

Fluvial and fluvio-lacustrine (delta) deposits on Mars are more frequently identified from their morphology and topography than by their interior architecture. Fluvial deposits are of two types: alluvial fans forming at foothills with a conical shape and alluvial deposits forming in alluvial plains and visible from residual sediments often characterized by inverted channels, that is, channels that are preserved from erosion and inverted by differential erosion. Series of inverted channels are detected in the Arabia Terra region where they are interpreted as erosional relics of developed Noachian-aged valley networks (Davis et al., 2016). Spectacular sinuous inverted channels are observed in the Zephyria region where they represent Hesperian-aged relics of extensive fluvial activity (Burr et al., 2010).

Inverted channels exist on the surface of eroded alluvial fans or as relics of alluvial deposits. Alluvial fans on Mars are located preferentially in impact craters in the southern highlands, as

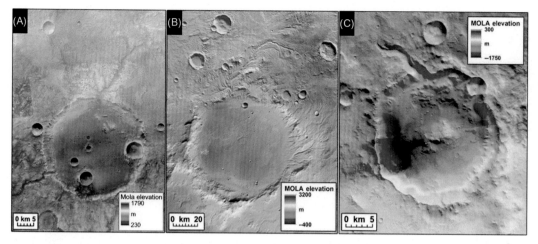

FIGURE 7.2

Different types of paleolake basins on Mars as classified by Goudge et al. (2016). Images show MOLA gridded topography overlaid on a mosaic of Context Camera (CTX) images. (A) Open-basin lake in heavily degraded crater ($-20.5°$ N, $87.0°$ E. CTX images B05_011498_1584, B10_013621_1594, G08_021480_1594, and G14_023748_1601 and High Resolution Stereo Camera (HRSC) nadir image h6523_0000. North is to the right). (B) Closed-basin lake in heavily degraded crater ($-18.8°$ N, $59.2°$ E. CTX images B21_017868_1583 and P16_007148_1628, and HRSC nadir images h8440_0000 and h0532_0000. North is to the right). (C) Closed-basin lake in less degraded crater ($-9.6°$ N, $144.1°$ E. CTX images P16_007158_1703 and B02_010230_1715. North is to the right).

illustrated in Fig. 7.2 (Moore and Howard, 2005; Kraal et al., 2008a; Morgan et al., 2013). Based on crater counts on alluvial fans and their host craters, alluvial fans were mostly formed during the Hesperian to Early Amazonian period (i.e., 3.4–2.5 Gy, Mangold et al., 2012; Grant et al., 2014). Strongly degraded Noachian craters do not usually host alluvial fans, but this may be caused by a preservation bias. The abundance of Hesperian alluvial fans further strengthens the argument that ongoing aqueous activity occurred during the Hesperian. Modeling of alluvial fans suggests episodic activity could last for less than 100,000 years (Armitage et al., 2011). There are spectacular small alluvial fans hosted in recent craters, such as Mojave (< 5 My old, Werner et al., 2014), which form an exception to this general trend and whose origin is still under debate (Williams and Malin, 2008). Alluvial fans have a conical shape and concave profile. Their average slopes vary from 0.5 degrees to 3–4 degrees, as observed on Earth for the same type of sedimentary body. These gentle slopes suggest that Martian alluvial fans are dominated by fluvial processes as opposed to fans formed by debris flows that tend to be steeper (> 5 degrees). A 10 km long alluvial fan has been sampled in situ at Gale crater by the Curiosity rover (Williams et al., 2013). This alluvial fan has a slope of ~ 1 degree and a thickness < 10 m (Palucis et al., 2014). The Curiosity rover has analyzed conglomerates that were deposited by shallow rivers, from water flows < 1 m of depth and velocity < 1 m/s (Williams et al., 2013).

Paleolakes are identified most frequently from the presence of preserved delta fans in topographic depressions, but also via the presence of sedimentary terraces, and evidence of

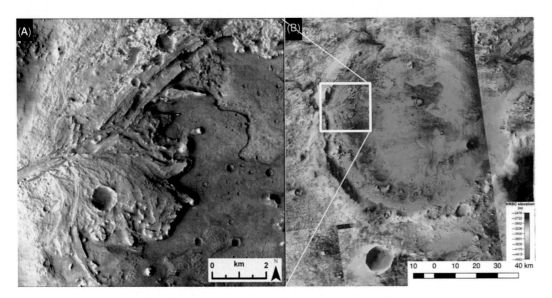

FIGURE 7.3

Map of Jezero crater, one of the candidate landing sites selected for the NASA Mars rover 2020. (A) Erosionally resistant channel deposits consistent with delta deposits; infrared spectral signatures of clays have been identified on the delta deposits. (B) Overlay of a CTX mosaic on a topographic map of Jezero crater from HRSC camera. *Adapted from Schon et al. (2012)* and *Hare et al. (2016), credit NASA/JPL/JHUAPL/MSSS/Brown University.*

paleoshorelines (e.g., De Hon, 1992; Newsom et al., 1996; Cabrol and Grin, 1999, 2001, 2003; for a review, see Cabrol and Grin, 2010 and references therein). Hundreds of paleolakes have now been identified, as illustrated in Fig. 7.2 (Goudge et al., 2015, 2016). Martian deltas are often Gilbert deltas, formed by an accumulation of fluvial sediments in a closed depression. They are identified mostly from their topography: convex shape, flat plain, and steep front, with some of them displaying locally inverted channels and/or clinoforms, that is, change in sedimentary rocks dip typical of the plain to front deposition of sediments. Notable examples are found in craters (10s to 100s of km in diameter), including Holden, Eberswalde, Gale, Gusev, Terby, and Jezero craters, the last example being illustrated in Fig. 7.3 (e.g., Malin and Edgett, 2000; Ori et al., 2000; Fassett and Head, 2005; Mangold and Ansan, 2006; Hauber et al., 2009; Morris et al., 2010; Ansan et al., 2011; Mangold et al., 2012; Goudge et al., 2017). They are also found in closed depressions, such as in Melas Chasma and Ismenius Cavus (Quantin et al., 2005; Metz et al., 2009; Dehouck et al., 2010). Eberswalde crater has the best preserved deltaic deposits, which comprise many inverted channels, meander belts, and local clinoforms (Malin and Edgett, 2003; Moore et al., 2003; Pondrelli et al., 2008). The thickness of deltas varies from several tens of meters to hundreds of meters up to 2 km at Terby crater.

The rate of (re)occurrence of paleolakes and the duration of the aqueous activity related to their formation are still a matter of debate. It is difficult to estimate the time required to form the sedimentary deposits associated with paleolakes. Initial estimates suggested that impact-generated hydrothermal lakes might survive for up to $\sim 10^4$ years (Newsom et al., 1996). However, many paleolakes seem to have formed as simple topographic depressions that served as catchment basins for local precipitation and surface runoff, indicating at least local hydrological cycles (e.g., Buhler et al., 2013). For instance, the delta in Eberswalde crater has a volume of 5 km^3, requiring an activity over only 100−100,000 years (Mangold et al., 2012; Irwin et al., 2015). In contrast, Terby crater has 2000 km^3 of preserved deposits suggesting activity over a more sustained time period (Ansan et al., 2011). Similar to alluvial fans, fresh-looking delta deposits are not linked to the most developed Noachian valley networks, but to smaller watersheds. Well-preserved late-stage deltas may not be representative of the most intense period of fluvial activity and therefore are dissociated from the known peak of fluvial activity in the Late Noachian (e.g., Fassett and Head, 2008a).

Although most paleolakes date from the Late Noachian/Early Hesperian, several studies indicate that some lakes may have formed as recently as the Late Hesperian (Cabrol et al., 2000; Ori et al., 2000; Mangold and Ansan, 2006; Mangold et al., 2012; Irwin et al., 2011). The finding that nearly all valley network-fed paleolakes have outlet drainage systems, while most single inlet paleolakes are closed-basin, indicates large surface run-off on early Mars with declining levels of fluvial activity over time (Fasset and Head, 2008b; Goudge et al., 2016). Paleolakes are also found associated with outflow channels (Irwin et al., 2002). Some of the smallest deltas often display steps that may have formed by ephemeral ponding (Kraal et al., 2008b).

In situ investigations of Gale crater by the Mars Science Laboratory (MSL) rover have confirmed that the crater's lowermost sedimentary landforms were deposited in fluvial, deltaic, and lacustrine environments. The lake system associated sedimentary deposits reach 75 m and would have required 10 ky to 10 My of intermittent activity to be emplaced suggesting an episodically wet climate that supplied moisture to the crater throughout much of the Hesperian (Grotzinger et al., 2015). At the same time, Bristow et al. (2017) report no carbonates at the level of detection of the Sample Analysis at Mars (SAM) instrument on MSL. This means that the atmospheric pressure at the time of deposition of the lower Gale crater strata (~ 3.5 Gy ago) must have been 10s to 100s of times lower than the CO_2 pressures necessary to produce a greenhouse effect sufficient to allow liquid water to flow on the surface. This paradoxical observation would seem to preclude an open lake system as the origin of the sedimentary deposits of Gale and will need to be explored by further measurements.

7.2.3 OUTFLOW CHANNELS

Evidence that Mars has experienced numerous episodes of catastrophic flooding is found at many locations across its surface, although principally concentrated along the dichotomy boundary. Outflow channels consist of broad scoured depressions, tens of kilometers wide, $\sim 10^2 - 10^3$ m deep, and hundreds to thousands of kilometers long, with streamlined islands along their beds. The example of Kasei Valles is shown in Fig. 7.4. Typically, they originate abruptly, from a fracture or region of collapsed and jumbled chaotic terrain, suggesting a formation by the catastrophic discharge of a large reservoir of subpermafrost groundwater (e.g., Manga, 2004; Carr, 2007 and references therein, Marra et al., 2014). Outflow channels are found on terrains dating from the Late Noachian to the Early Amazonian, with a peak activity during the Hesperian reflecting a later activity than the valley networks described earlier (Tanaka et al., 2014).

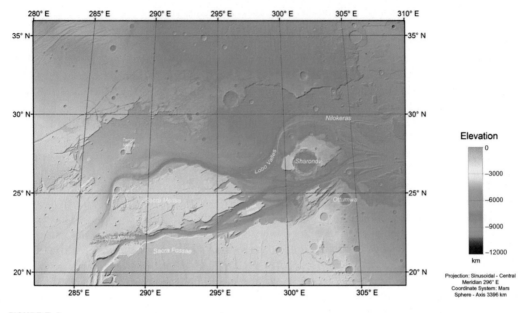

FIGURE 7.4

The Kasei Valles outflow channel system (HRSC FU Berlin/MOLA). This map shows the region of Kasei Valles, lying approximately between 21° and 28° North at 292.5° East. Connecting the Southern Echus Chasma and the plain Chryse Planitia to the east, Kasei Valles has a width of roughly 500 km and, if Echus Chasma is included, extends for approximately 2500 km. Both the North and South valley branches exhibit a depth of 2900 m.

Assuming this origin, discharge rates for the largest channels are estimated to have reached 10^7 m^3/s, more than 100 times greater than the present discharge of the Amazon river on Earth (e.g., Burr et al., 2002a, b; Manga, 2004; Williams and Malin, 2004). With such large discharge rates, the outflow channels may have formed in as little as several days (Manga, 2004; Andrews-Hanna and Phillips, 2007), although they may have persisted for months or years, if the discharge rates were smaller. The peak in outflow channel activity appears to have spanned the period from the Late Hesperian to Early Amazonian ($\sim$3–2 Gy, Tanaka, 1986); however, individual outflow channel ages cover a much greater range, including some as old as $\sim$3.4 Gy and as recent as $\sim$10 My at Athabasca Vallis (Hartmann and Neukum, 2001; Burr et al., 2002a, b; Chapman et al., 2010a, b). Recent high resolution mapping of the circum-Chryse outflow channels suggests significant activity persisting until the Mid-Amazonian (Rodriguez et al., 2014, 2015).

At the time of peak outflow channel activity ($\sim$3.4 to $\sim$2 Gy, e.g., Tanaka, 1986; Hartmann and Neukum, 2001), Carr (1986, 1996) and Baker et al. (1991) estimated that Mars possessed a planetary inventory of water equivalent to about 0.5–1 km GEL. As this peak activity postdates the period when the most efficient water loss processes were active (atmospheric escape and large impacts during the Pre- and Early-Noachian, >4.2 Gy ago), the majority of this water is expected to have been retained, at least in some form, to the present day.

However, Carr and Head (2015) have recently argued that the actual amount of water required to erode the channels may have been as small as ~ 40 m GEL, suggesting that this small inventory may have been recycled several times, based on the icy-highland hypothesis. Casanelli et al. (2015) considered the potential of the icy highlands model to recycle a small inventory of water through the basal melting of a high-altitude ice sheet or the underlying cryosphere. However, they found that groundwater recharge by this process was insufficient to explain the duration and magnitude of outflow channel activity—hence, if recharge occurred, it was either earlier in Martian geologic history (when the geothermal heat flow was higher), or by some other process.

While the outflow channels are widely regarded to have formed by fluvial processes, it has also been argued that they originate from thermal erosion by low viscosity lavas (Leverington, 2004, 2009, 2011). However, recent studies of the erosive efficiencies of lavas have generally concluded that they cannot explain erosion at the scale required for the formation of the channels (Dundas and Keszthelyi, 2014; Cataldo et al., 2015), whereas the Okavango outflow channels display terminal fans and fluvial bars inconsistent with lava flows (Mangold and Howard, 2013). Identification of the correct formation process and duration of activity required to explain the size of the outflow channels is paramount to constraining the global water inventory over time.

7.2.4 PAST OCEANS AND SEAS

The northern plains of Mars are younger and several kilometers lower in elevation than the southern highlands (Tanaka et al., 2014), making it a logical sink for sediments carried and deposited by the outflow channels and valley networks (e.g., Kreslavsky and Head, 2002). The possibility that an ocean may have once covered the northern plains of Mars is supported by the identification of a series of nested levels in high-resolution Viking images — interpreted as shorelines — which are located along the highland/lowland boundary (Parker et al., 1989; Baker et al., 1991; Parker et al., 1993, 2010). The presence of an ocean would also be consistent with interior layered deposits and possible shorelines in Hellas and Argyre (Moore and Edgett, 1993; Tanaka and Leonard, 1995; Tanaka, 2000; Tanaka et al., 2002; Wilson et al., 2007). The existence of an early ocean is also consistent with the observed distribution of deltas and valleys along the dichotomy boundary (Di Achille and Hynek, 2010b; DiBiase et al., 2013). The highest and oldest of these potential paleoshorelines has been called both the "Arabian Level" (Clifford and Parker, 2001) and "Contact 1" (Parker et al., 1989, 1993). Contact 1 is thought to date back to the Late Noachian (Clifford and Parker, 2001) and has a mean elevation of -1680 m with a standard deviation of 1.7 km, as determined from Mars Orbiter Laser Altimeter (MOLA) (Head et al., 1999). Contact 1 is not a good approximation of an equipotential surface (Head et al., 1999); however, when the elevations of Contact 1 are corrected for the long-wavelength topographic effects of true polar wander, the agreement is significantly improved (Perron et al., 2007). The lowest and most recent of the levels, known as "Contact 2," has a mean elevation of -3760 m and a standard deviation of 560 m (Head et al., 1999). It represents a much better fit to an equipotential surface, a fit that is also improved by consideration of the effects of true polar wander (Perron et al., 2007).

The former existence of an ocean remains a matter of intense debate. Arguments presented against include the paucity of well-defined coastal constructional landforms along the proposed shorelines (Ghatan and Zimbelman, 2006), as well as recent simulations that indicate that Martian surface conditions are unfavorable for the formation of shorelines due to weak winds and the estimated high rate of water and ice sublimation (Kraal et al., 2006; Banfield et al., 2015). Yet, a review of the latest high-resolution imagery and topographic data by Parker and Calef (2012) identifies further geomorphic evidence for the shorelines' existence. Additional support is provided by evidence of littoral resurfacing by run-up water and backwash channels from one or more Late Hesperian/Early Amazonian tsunamis (Rodriguez et al., 2016; Costard et al., 2017) — consistent with the location of Contact 2 and the presence of a northern ocean.

From a mineralogical perspective, Ehlmann et al. (2008) have argued that the small amounts of carbonate and hydrated minerals detected from orbit are inconsistent with a persistent aqueous environment in the northern plains, although hydrated minerals have been excavated by large impacts from beneath the plains (Carter et al., 2010). A possible explanation for the apparent low abundance of carbonate and hydrated minerals may be that their formation was inhibited by the potential combination of low temperature, high salinity, and acidic conditions that may have characterized the ocean environment (Chevrier et al., 2007; Fairén et al., 2011).

Finally, Mars Advanced Radar for Subsurface and Ionospheric Sounding (MARSIS) orbital radar surface reflectivity investigations suggest that the volume-averaged dielectric constant of the northern plains is quite low, at least to a depth of one wavelength ($\sim 60-80$ m). This can be explained by either a volume-averaged porosity of $\sim 35\%$ (consistent with the presence of high-porosity sediment) or by a volume-averaged ice content of $\sim 60\%$ (consistent with the frozen remnant of a former northern ocean, Mouginot et al., 2012). Whichever the cause, the agreement between the distribution of low dielectric material and the area encompassed by Contact 2 is intriguing. It is also interesting to note the higher concentration of double-layered ejecta craters in this region, which has been interpreted as evidence of impacts into a volatile-rich target (Costard, 1989; Barlow, 2005).

Estimates of the total volume of water required to form a northern ocean vary from ~ 160 m GEL (Carr and Head, 2003) up to 700 m GEL (Clifford and Parker, 2001; Fairén et al., 2003). Based on their analysis of the present near-surface inventory of water, and its likely sources and sinks over the course of Martian geological history, Carr and Head (2015) argued that only 24 m GEL was available as surface water to form a putative northern ocean and thus concluded that no ocean was present on Mars during the Noachian or any subsequent time.

7.2.5 CRUSTAL ALTERATION

The observed widespread occurrence of phyllosilicates in Noachian age terrains by the orbital spectrometers OMEGA (L'Observatoire pour la minéralogie, l'eau, les glaces et l'activité, Poulet et al., 2005; Bibring et al., 2006; Loizeau et al., 2007; Mangold et al., 2007) and CRISM (Compact Reconnaissance Imaging Spectrometer for Mars, Mustard et al., 2008; Murchie et al., 2009) strongly supports the hypothesis that early Mars was water rich. High-resolution images reveal the presence of mineral-filled fractures (Okubo and McEwen, 2007) and deep occurrence of phyllosilicate minerals in kilometer-scale vertical exposures of the crust indicating aqueous activity that has spanned Mars' history (Mustard et al., 2008; Murchie et al., 2009; Ehlmann et al., 2011). The processes and conditions that formed these minerals are consistent with the occurrence of aqueous alteration at a variety of depths, along with evidence for low- to moderate-temperature hydrothermal circulation of vapor and liquid water in the deep subsurface that may continue to the present day (Clifford, 1993; Bibring et al., 2006). In addition, ancient sediments especially the case of light-toned deposits that are exhumed by erosion (such as at Mawrth Vallis, Loizeau et al., 2007; McKeown et al., 2009; Bishop et al., 2013) or found as erosional windows at North Argyre, Terra Sirenum, or North Hellas (Milliken and Bish, 2010; Adeli et al., 2015; Salese et al., 2016) attest to extensive alteration and the presence of water at or close to the surface. These altered sediments may be similar to those observed by the Curiosity rover, in which abundant clay minerals have been detected (Vaniman et al., 2014; Grotzinger et al., 2015). Further discussion of the aqueous alteration of the Martian crust can be found in this book in Chapter 5, Carbonates on Mars, Chapter 6, Sulfur on Mars From the Atmosphere to the Core, and Chapter 8, Sequestration of Volatiles in the Martian Crust Through Hydrated Minerals: A Significant Planetary Reservoir of Water.

Fig. 7.5 summarizes the activity of water on the surface of Mars, over geologic time based on the different events and geomorphologic features' ages that are described in this section.

7.2.6 LAYERED EJECTA AND PEDESTAL CRATERS

Using Viking images, a class of impact craters with a continuous and lobate-shaped ejecta blanket was suggested to be associated with the presence of ground ice (Carr et al., 1977; Mouginis-Mark, 1979; Schultz and Gault, 1979; Costard, 1989; Baratoux et al., 2002; Barlow, 2004). This interpretation was later supported by the presence of similar ejecta shapes on some of Jupiter's icy satellites (e.g., Boyce et al., 2010). Lobate ejecta extend up to >3 times the crater radius, and are observed at mid- to high-latitudes. The volatilization of subsurface ground ice on impact is thought to cause fluidization of the ejected material, which forms a ground-hugging flow, resulting in a lobate planform shape (Barlow, 1994; Baratoux et al., 2002; Barnouin-Jha et al., 2005). However, lobate ejecta are also observed on Venus where they are instead associated with its thick atmosphere (Barnouin-Jha and Schultz, 1996; Barnouin-Jha, 1998), and similar processes are also proposed for Mars (Komatsu et al., 2007). There is currently no definitive conclusion with regard to the exact process that formed these ejecta on Mars.

A subset of lobate classification of ejecta is the double layer ejecta (DLE) craters, which have an inner layer with a smaller radius than, and superposed upon, the outer ejecta layer (Barlow et al., 2000; Barlow, 2005). Most DLE craters are found in both hemispheres (from about 25° to 60°), at a variety of elevations and terrain types and ages (Barlow, 2005). Morphologic studies of DLE have shown a distinguishing set of characteristics from other ejecta: two ejecta layers, well-defined radial grooves and ridges through their continuous ejecta blankets, and the apparent lack of secondary impact craters. Based on these observations, a double-step emplacement process requiring subsurface volatiles was suggested (Boyce and Mouginis-Mark, 2006). Another mechanism suggested for their formation relies on glacial snow and ice deposited on the surface of Mars at times of high obliquity (Senft and Stewart, 2008; Weiss and Head, 2013). These ejecta appear unlikely to be formed by atmospheric effects from ejecta curtain-induced vortices or a base surge (Weiss and Head, 2017). Overall, their global distribution and relationship with other types of crater ejecta indicate varying spatial and/or temporal subsurface abundances of water (Boyce and Mouginis-Mark, 2006; Jones and Osinski, 2015).

Pedestal craters are another specific type of lobate ejecta craters, observed in the northern plains. The ejecta of these craters is anomalously thick (>50 m) even distal from the crater center, and such a thickness cannot be explained by emplacement of the ejecta alone (Meresse et al., 2006; Black and Stewart, 2008; Kadish et al., 2009). These "perched" ejecta blankets are likely the result

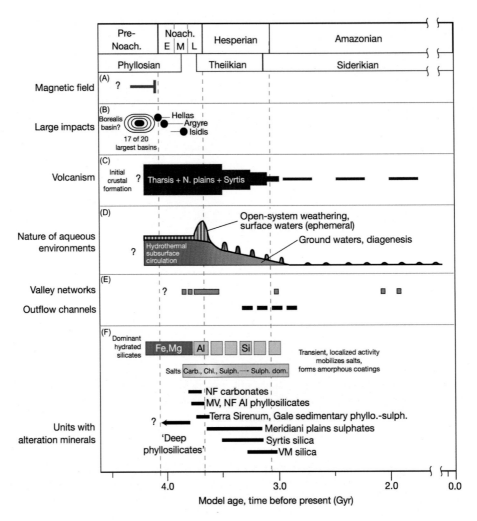

FIGURE 7.5

Hydrologic history of Mars based on estimated ages of water-related geomorphologic features (Ehlmann et al., 2011).

of a differential erosion of the material outside the ejecta while the material below the ejecta was protected from erosion. This observation is linked with the presence of volatile-rich deposits in the northern plains (see Section 7.3).

7.2.7 PAST GLACIATIONS

Piedmont-like fan forms bounded by extensive concentric ridges on the north-west flanks of the Tharsis volcanoes (Arsia, Pavonis, and Ascraeus Mons) have been interpreted as drop moraines and sublimation tills deposited by cold-based tropical ice sheets that retreated a few hundred My ago (e.g., Head and Marchant, 2003; Head et al., 2005; Shean et al., 2005). These ice sheets may have attained average thicknesses of $\sim$1.6–2.4 km over areas of 166,000 km^2, corresponding to a total ice volume of $\sim$3.2–4.7 $\times$ 10^6 km^3 (Shean et al., 2005; Fastook et al., 2008). General circulation model (GCM) simulations predict snowfall foci on the flanks of the Tharsis volcanoes at high ($>\sim$45°) axial obliquity (Head and Marchant, 2003; Head et al., 2005; Shean et al., 2005; Levrard et al., 2004, 2007; Forget et al., 2006; Madeleine et al., 2009). The locations of viscously deformed crater floor-filling materials, which resemble ice-rich mid-latitude concentric crater fill (CCF) deposits (see Section 7.3.1.3), are consistent with a second focus of snowfall predicted by these GCMs, south of the Schiaparelli impact basin (Shean, 2010).

Late Amazonian-aged ($\sim$40 My—1 Gy, Levy et al., 2007; Baker et al., 2010; Hartmann et al., 2014; Berman et al., 2015; Baker and Head, 2015) mid-latitude viscous flow features (VFF, discussed in detail in Section 7.3.1.3) have been widely interpreted as debris-covered glaciers, analogous to those in the Antarctic Dry Valleys on Earth (e.g., Head et al., 2010; Marchant and Head, 2007). Shallow Radar (SHARAD) data indicate that VFF comprise bulk water-ice ($>$90%) with a thin ($<$10 m) cover of surface debris, implying ice derivation from atmospheric precipitation during a period when surface ice was stable at these latitudes (Holt et al., 2008; Plaut et al., 2009). GCM simulations by Madeleine et al. (2009) demonstrate that such conditions could arise with a shift from high ($\sim$45°) to intermediate ($\sim$35°) obliquity. Preservation of remnant ice into the present "interglacial" climate is widely attributed to a ubiquitous blanket of dust and debris, which inhibits sublimation of the ice to the atmosphere (e.g., Marchant and Head, 2007; Head et al., 2010).

Under cold Amazonian climates, recent glaciations were likely cold-based and exchange of water with nonglacial reservoirs was dominated by sublimation to the atmosphere. However, limited evidence for meltwater production does exist. Proglacial valleys (e.g., Fassett et al., 2010; Adeli et al., 2016) and alluvial fans (Adeli et al., 2016) imply some degree of glacial melting during the Amazonian. Isolated examples of small valleys incised into VFF surfaces (Fassett et al., 2010) show that runoff of transient supraglacial meltwater produced by topographic focussing of solar insolation onto the surfaces of cold-based glaciers could explain such proglacial meltwater morphologies. More difficult to explain under cold-based regimes are landforms deposited by meltwater draining along the bed of a glacier, such as eskers. Eskers are sedimentary ridges deposited in ice-walled glacial meltwater conduits. Two candidate eskers have been identified in association with existing mid-latitude VFF on Mars, in Phlegra Montes (Gallagher and Balme, 2015; Butcher et al., 2017a) and Tempe Terra (Butcher et al., 2017b). These eskers are interpreted as deposits of past wet-based regimes induced by locally elevated geothermal heat at the beds of their parent VFF.

Networks of larger (kilometer-scale) ancient (3.48–3.6 Gy, Bernhardt et al., 2013; Kress and Head, 2015) eskers in the Argyre impact basin (Kargel and Strom, 1992; Banks et al., 2009; Bernhardt et al., 2013) and south polar Dorsa Argentea Formation (Howard, 1981; Kargel and Strom, 1992; Head and Pratt, 2001; Tanaka et al., 2014; Kress and Head, 2015; Butcher et al., 2016) indicate that ancient ice sheets accumulated in these areas and underwent extensive basal melting at the Noachian—Hesperian transition. The morphometry of these esker populations is consistent with formation under high meltwater discharges. Banks et al. (2009) calculate a minimum discharge of 1×10^4 m^3/s for formation of the Argyre eskers, within the range of 10^3–10^5 m^3/s calculated for the Dorsa Argentea eskers by Scanlon and Head (2015). Ice sheet reconstructions imply that the parent ice sheet of the Argyre eskers was up to ~ 2 km thick, and comprised $\sim 100,000-150,000$ km^3 of ice (Bernhardt et al., 2013).

GCM simulations show that, given estimated Noachian−Hesperian geothermal heat flux of $45-65$ mW/m^2, basal melting of an ice sheet in the Dorsa Argentea Formation would require atmospheric warming of $\sim 25-50$ °C from the present mean south polar temperature (-100 °C) (Fastook et al., 2012). Basal melting beneath thick polar ice sheets is not inconsistent with a "cold and icy" early climate scenario, as the necessary south polar warming would not raise mean annual temperatures at lower latitudes above the melting point (Fastook et al., 2012).

Under the "Late Noachian icy highlands" model for a cold and icy early Martian climate, higher atmospheric pressures on ancient Mars induced stronger vertical temperature gradients through the Martian atmosphere, generating cold-traps at high-elevation surfaces where glacial ice could accumulate (e.g., Wordsworth et al., 2013, 2015; Wordsworth, 2016; Fastook and Head, 2015). GCM simulations show correlations between snowfall and the locations of the ancient Martian valley networks in the highlands under surface pressures > 0.5 bar and intermediate-to-high planetary obliquity (Wordsworth et al., 2015). Ice sheet modeling predicts that, for the most likely Late Noachian water supply ($2-5$ times present GEL of water) and geothermal heat ($45-65$ mW/m^2) scenarios, thin ($\sim 10^2$ m), cold-based ice sheets could accumulate over large areas of the southern highlands (Fastook and Head, 2015). Fastook and Head (2015) show that, for a $5 \times$ present water supply scenario with ~ 55 mW/m^2 geothermal heat flux, a transient climate perturbation of $+18$ °C lasting 2000 years could release sufficient meltwater (~ 0.42 Mkm^3) from a highland ice sheet to fill open basin lakes associated with the ancient valley networks (Fastook et al., 2008). Transient warming of this magnitude is within the range predicted by Halevy and Head (2014) following strong volcanic eruptions (Fastook and Head, 2015). Therefore, it has been argued that a warm and wet early Martian climate is not a prerequisite for formation of the ancient valley networks or open-basin lakes (e.g., Wordsworth et al., 2013, 2015; Wordsworth, 2016; Fastook and Head, 2015). However, caution must be exercised when considering ancient glaciations under the framework of the Late Noachian icy highlands model, given the predicted absence of direct geomorphic signatures under cold-based glacial regimes (Fastook and Head, 2015). Direct geomorphic evidence for ancient (3.5 Gy) ice-sheet scale glaciation of Valles Marineris has been documented by Gourronc et al. (2014) but evidence for ancient glaciation of the southern highlands remains sparse. Therefore, at present, our understanding of these glaciations under this climate scenario is predominantly guided by predictions of global climate and ice sheet models.

7.3 REMOTE SENSING EVIDENCE FOR WATER ON MARS TODAY

The current observations indicate that even if water was abundant on Mars in the past, it is much less apparent on the surface today. Understanding the inventory of water on Mars and how it has evolved with time is a major crosscutting theme for the exploration of the planet (CASEM, 2007; CPSDS, 2011). In this section, we review current estimates of the present inventory of water on Mars based on the latest measurements.

7.3.1 VISIBLE RESERVOIRS OF WATER

7.3.1.1 Polar layered deposits

The polar layered deposits (PLDs) are the largest visible reservoirs of water on Mars (Fanale et al., 1986; Zent et al., 1986; Byrne, 2009b). Although some details of their growth and stability remain uncertain, atmospheric dust, raised by local and global dust storms, is considered to act as nucleation centers for the condensation of H_2O ice (Soderblom et al., 1973; Cutts, 1973; Pollack et al., 1979). Due to cold winter temperatures, CO_2 can condense on the polar surface and H_2O ice aerosols. In this way, it deposits a $\sim 1-2$ m thick cover poleward of $\sim 50°$, which contributes to the growth of the polar caps under favorable climatic conditions (Cutts, 1973; Pollack et al., 1979).

The deposition and entrainment of atmospheric dust in the polar ice results in the presence of numerous horizontal layers. The stratigraphic variations in thickness and albedo of these layers are thought to reflect climatic variations, modulated by the periodic changes in insolation caused by changes in obliquity and other orbital parameters (Ward, 1992; Touma and Wisdom, 1993; Laskar and Robutel, 1993; Laskar et al., 2004).

The North PLDs, which are ~ 1100 km in diameter, are among the youngest surfaces to be found on Mars — with no craters larger than 300 m within the $\sim 10^6$ km^2 covered by the deposits (Fig. 7.6A). This observation argues for an age of the exposed surface that is less than $\sim 10^5$ years. This result may reflect either the deposition of new material (Putzig et al., 2009; Smith et al., 2016), or the constant reworking of older material at the pole (Herkenhoff and Plaut, 2000). More recent assessments of the resurfacing suggest that the age of the NPLD may actually be orders of magnitude younger: 1.5–20 ky (Banks et al., 2010; Landis et al., 2016). However, it is thought that the layered polar basal unit (BU), on which the North PLD rests, may be much older — perhaps > 1 Gy (Tanaka et al., 2008).

Measurements made by MOLA (Zuber et al., 1998), the Mars Advanced Radar for Subsurface and Ionospheric Sounding (MARSIS, Picardi et al., 2005), and the SHAllow RADar (SHARAD, Phillips et al., 2008), indicate that the North PLD reach a maximum height of ~ 3 km above the surrounding plains (Fig. 7.6C). The volume of the North PLD, derived from several radar studies (Putzig et al., 2009; Grima et al., 2009; Selvans et al., 2010), is estimated to be $\sim 1.14 \times 10^6$ km^3, equivalent to ~ 7.5 m GEL of H_2O, assuming a dust fraction of a few percent. This dust fraction is consistent with the column-averaged permittivity of the PLD determined by Grima et al. (2009). The volume of the BU is $\sim 0.45 \pm 0.1 \times 10^6$ km^3. The volume fraction of dust in the BU is not well known, but is assumed to be greater than the PLD since its albedo is darker (Byrne and Murray, 2002; Fishbaugh and Head, 2005). If the dust fraction is as high as 15%–50%, then the volume of water stored in the BU is somewhere between ~ 1.6 and 2.6 m GEL.

In the South, the small residual ice cap (~ 400 km diameter) is superimposed on an extensive region of masked layered deposits (~ 1500 km in diameter) whose areal coverage is almost twice that occupied by the equivalent unit in the North (Fig. 7.6B). The exposure age of the southern PLD is older than that in the North, with ~ 36 craters, with diameters > 800 m visible on its surface and another 12 whose identification has yet to be confirmed (Koutnik et al., 2002). These results indicate a surface age of $\sim 30-100$ My, which is $\sim 2-3$ orders of magnitude older than in the North (Herkenhoff and Plaut, 2000; Koutnik et al., 2002; Kolb and Tanaka, 2006). The SPLDs are in turn surrounded by the Dorsa Argentea Formation ($\sim 1.5 \times 10^6$ km^2), which is thought to be a relic of a much larger expanse of PLD dating back to the Late Noachian/Early Hesperian (Ghatan and Head, 2004; Kress and Head, 2015; Butcher et al., 2016).

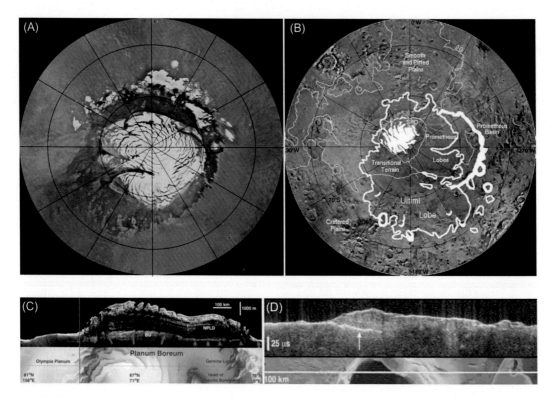

FIGURE 7.6

(A) Residual north polar cap and layered deposits (Viking MDIM colored relief map). (B) Residual south polar cap and layered deposits shown in *thick white lines* (Viking MDIM colored relief map). Adapted From Schenk and Moore (2000). (C) SHARAD radar cross section of the NPLD (Philips et al., 2008). (D) MARSIS radar cross section of the SPLD (Plaut et al., 2007). Note that, for (D), the apparent dip of the basal interface is an artifact of the lack of correction for the dielectric properties of ice. With this correction, the basal interface of the SPLD would be flat.

Based on an analysis of MARSIS data, Plaut et al. (2007) found that the South PLD has a maximum thickness of ~ 3.7 km (Fig. 7.6D) and a total volume of 1.6×10^6 km^3. This corresponds to a water volume of ~ 9.5 m GEL, assuming a dust fraction of $\sim 10\%-15\%$. This dust fraction is consistent with the slightly greater radar attenuation characteristics of the SPLD versus the NPLD (Grima et al., 2009).

While the thicknesses of the North and South PLDs are similar, the differences in their surface age, areal extent, and mass distribution are substantial, and are probably related to the ~ 6 km elevation difference between the south polar highlands and the northern plains (Smith et al., 1999). Taken together, the amount of H$_2$O stored in the North and South PLD is estimated to be ~ 22 m GEL (Carr and Head, 2015).

7.3.1.2 Latitude-dependent mantle

At times of high ($>45°$) obliquity, the sublimation of polar ice results in its redistribution to mid-latitudes, creating episodic "ice ages" (e.g., Head et al., 2003; Forget et al., 2006). From orbital image analysis, Soderblom et al. (1973) noted that, poleward of 30° latitude in both hemispheres, the Martian surface is covered by a mantle of debris that has undergone differential erosion — leaving it well preserved in some areas, but with extensive pits, knobs, and remnant mesas in others. Soderblom et al. (1973) inferred that these mantles consisted of eolian debris, derived from the PLDs, that was redistributed to lower latitudes as the result of climatically varying inso-lation and winds. This mantle was more fully resolved and characterized by the Mars Orbiter Camera (MOC, resolution 1.5–12 m/pix, Mustard et al., 2001). Evidence of variations in topo-graphic roughness coincides with geomorphological evidence for this Latitude-Dependent Mantle (LDM) (Kreslavsky and Head, 1999, 2000; Neumann et al., 2003). The thickness of the LDM is estimated to range from few meters to 10s of meters (Schon et al., 2009; Zanetti et al., 2010; Conway and Balme, 2014). The absence of any large superposed impact craters indicates that this unit has been very recently emplaced with inferred ages ranging from 0.1 to 5 My (Kostama et al., 2006; Willmes et al., 2012).

As shown in Fig. 7.7, detailed mapping of the geographic distribution of the LDM indicates that it is concentrated in two latitudinal bands: 30 to 70° N and 25 to 65° S (Mustard et al., 2001). It is further associated with polygonal patterned ground from thermal contraction cracking that is

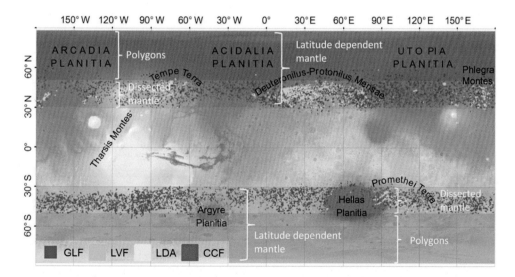

FIGURE 7.7

The distribution of lineated valley fill (LVF), lobate debris aprons (LDA), concentric crater fill (CCF) (Levy et al., 2014a), and glacier-like forms (GLF) (Souness et al., 2012), based on Levy et al. (2014a). Features detection is limited to <50° latitudes north and south. Note that LVF, LDA, and CCF are mapped with a 1 pixel outline to enhance clarity, which exaggerates their apparent area. GLFs are plotted as points. Basemap is MOLA gridded topography over MOLA hillshade (Smith et al., 2001).

thought to be due to the presence of ice in the subsurface (Mellon, 1997; Mangold et al., 2004b; Mangold, 2005; Levy et al., 2009a, 2010a). This line of evidence indicates that the mantles are probably derived from atmospheric redistribution of the polar ice deposits and that ice represents a significant volume fraction of the LDM, whose subsequent ablation has created the observed erosional landforms (Malin and Edgett, 2001; Mustard et al., 2001; Head et al., 2003; Schon et al., 2009; Mangold, 2011).

Evidence in support of this conclusion comes from the examination of recent impacts, $\sim 10-100$ m in diameter, that have exposed near-surface ice within the deposits, which is observed to disappear over timescales of months (Byrne et al., 2009a) and the ground ice discovered by the Phoenix lander (Smith et al., 2009; Mellon et al., 2009). While the mean annual content of water vapor in the Martian atmosphere is ~ 10 precipitable microns (e.g., Smith, 2004, 2008), water-ice stability models have indicated that ice can persist in the subsurface at latitudes as low as $45°$ if it is buried beneath a dry layer of regolith $\sim 15-50$ cm thick (e.g., Leighton and Murray, 1966; Paige, 1992; Mellon and Jakosky, 1993, 1995; Aharonson and Schorghofer, 2006; Schorghofer and Aharonson, 2005; Dundas and Byrne, 2010). This prediction is supported by Gamma-Ray Spectrometer (GRS) observations indicating the presence of near-surface, ice-rich terrains (with >50 wt% WEH) at mid- to high-latitudes in both hemispheres (Feldman et al., 2011). The GRS results are further discussed in Section 7.3.2.1.

Controversy exists concerning the origin and nature of the ice within the LDM. Two types of ice deposits were observed by Phoenix: a brighter slab-like material composed of nearly pure ice, and darker material composed of a mixture of ice (30 wt%) with soils (Cull et al., 2010). These ground ice deposits could have several possible origins. They may have been part of a more extensive deposit of polar ice that precipitated at high latitudes under different climatic conditions (e.g., Dundas et al. 2014). They could also have formed by the direct condensation of atmospheric water vapor in the regolith pores when diurnal and seasonal temperature variations resulted in local soil temperatures below the H_2O frost point. Finally, they could be due to seasonal melting and refreezing of near-surface ice, perhaps aided by freezing-point depression by salts in the regolith, that migrated from the warmer to the colder regions of the regolith. This increase in water content may have resulted in the formation of ice lenses, needle ice, and associated structures in a process similar to that responsible for frost heave on Earth (Fisher, 2005; Mellon et al., 2008a, 2009; Sizemore et al., 2015).

7.3.1.3 Massive nonpolar ice deposits

As summarized by Squyres and Carr (1986), Viking-era investigations identified a wide range of geomorphic evidence indicative of the presence of ice at mid-latitudes, including VFF. Surface morphologies including parallel arcuate lobate ridges and longitudinal ridges that divert around topography, and deformation and convergence of lobes, have widely been interpreted as evidence that VFFs are of glacial origin (e.g., Head et al., 2010). Mid-latitude VFF can be subdivided into four morphological subtypes: glacier-like forms (GLF) (Arfstrom and Hartmann, 2005; Souness et al., 2012), lobate debris aprons (LDA) (e.g., Squyres, 1979; Lucchitta, 1981, 1984), lineated valley fill (LVF) (e.g., Squyres, 1978, 1979; Kochel and Peake, 1984; Carr, 2001; Head et al., 2006), and concentric crater fill (CCF) (e.g., Squyres, 1979; Head et al., 2005; Dickson et al., 2010, 2012). GLF, LDA, LVF, and CCF commonly form components of integrated glacial landsystems (Fig. 7.8A) on Mars (Head et al., 2010).

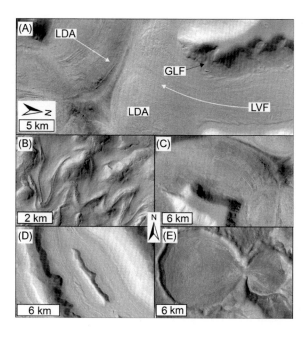

FIGURE 7.8

Context camera image examples of mid-latitude viscous flow features: (A) integrated glacial landsystem comprising GLF, LVF, and LDA. *White arrows* show general flow direction (B05_011513_2241_XI_44N331W), (B) GLFs in Greg crater (G12_022745_1415_XN_38S246W), (C) LDA (G04_019794_2240_XN_44N331W), (D) LVF (D04_028773_2191_XI_39N304W), (E) CCF (B09_013304_1407_XN_39S257W, F07_038490_1424_XI_37S257W). *Image credit NASA/JPL/MSSS.*

GLF (Fig. 7.8B) are morphologically similar to terrestrial valley glaciers (Arfstrom and Hartmann, 2005; Souness et al., 2012). They are typically tongue-shaped, and terminate in distal ridges that resemble terrestrial end-moraines (e.g., Arfstrom and Hartmann, 2005; Hubbard et al., 2011, 2014; Souness et al., 2012), consistent with viscous downslope deformation. GLFs commonly form the lowest-order component of integrated glacial landsystems and flow out of the topographic alcoves in which they originate, into the second-order components of the landsystem: LDA.

LDAs (Fig. 7.8C) are lobate, convex-up features (Squyres, 1978) that extend outward from the footslopes of escarpments. Where LDAs emanate from isolated mesas and massifs (Squyres, 1979), individual flows commonly merge to form radial LDA complexes (Pierce and Crown, 2003). LDAs are typically unconstrained by bounding escarpments and extend tens of kilometers from their source. Their frontal thicknesses can exceed 1 km (Crown et al., 2002, 2003). Where they are topographically confined, LDAs commonly transition into the third-order component of the integrated glacial landsystem, LVF (e.g., Squyres, 1978; Lucchitta, 1984; Kochel and Peake, 1984; Carr, 2001; Head et al., 2010).

LVF (Fig. 7.8D) is characterized by partial infilling of the floors of wide topographic trenches bounded by steep lateral escarpments, such as the fretted terrain. LVF exhibits topographic lineations that parallel the walls of the trenches they occupy, as well as ridges and depressions suggestive of down-valley flow, analogous to similar features on terrestrial debris-covered glaciers (e.g., Squyres, 1978; Lucchitta, 1984; Kochel and Peake, 1984; Head et al., 2006, 2010; Marchant and Head, 2007).

The fourth morphological subtype of VFFs on Mars is CCF (Fig. 7.8E). CCF is a similar type of deformational landform to LVF. CCF comprises viscous flows that emanate from interior crater rims and occupy the base of crater walls or infill crater floors, where it forms a concentric sequence of compressional ridges (e.g., Levy et al., 2010b; Dickson et al., 2010, 2012). In some locations, CCF has become integrated with wider VFF landsystems, either through complete crater infilling and overspill onto exterior crater slopes (e.g., Dickson et al., 2010), or exchange of material through breaches in the crater wall (e.g., Head et al., 2005).

VFFs are distributed throughout the Martian mid-latitudes ($\pm 30°-60°$) (Fig. 7.7), with a cumulative surface area in excess of 0.5% of Mars' total surface area (Levy et al., 2014a). The distribution of CCF is nearly ubiquitous within the mid-latitudes. However, GLF, LDA, and LVF exhibit heterogeneous distributions, being particularly abundant in the Deuteronilus−Protonilus Mensae, Tempe Terra, Phlegra Montes, Nereidum Montes, Charitum Montes, and Promethei Terra (Squyres, 1979; Souness et al., 2012; Levy et al., 2014a). This distribution is in good agreement with regions of preferential mid-latitude snowfall accumulation at intermediate ($\sim 35°$) planetary obliquity predicted by GCM simulations (Madeleine et al., 2009).

Early workers advanced the hypothesis that VFFs formed by ice-assisted mobilization of bulk regolith due to viscous creep of interstitial ice derived from the atmosphere or subsurface into regolith pore spaces in volume fractions $< 30\%$ (rock glacier model; e.g., Sharp, 1973; Squyres, 1979, 1978; Squyres and Carr, 1986; Pierce and Crown, 2003). However, abundant evidence has since accumulated in support of an alternative model, originally proposed by Lucchitta (1984), of flow of bulk ice in volume fractions $> 50\%$ beneath a protective surface lag of debris. Following this hypothesis, the bulk ice is derived from atmospheric precipitation and the relatively thin debris mantle is derived from erosional wasting from adjacent scarps and englacial debris released at the surface due to sublimation of near-surface ice (e.g., Head et al., 2006, 2010; Baker et al., 2010; Dickson et al., 2010; Hubbard et al., 2011). Dielectric values derived from the SHARAD sounding data for LDAs in Promethei Terra (Holt et al., 2008) and Deuteronilus Mensae (Plaut et al., 2009) indicate that they contain a high volumetric content of ice ($\sim 80\%-90\%$) and are mantled by < 10 m of dust and debris. Thus VFFs are widely thought to be analogous to debris-covered glaciers in the Antarctic Dry Valleys on Earth (e.g., Marchant and Head, 2007; Head et al., 2010; Mackay and Marchant, 2017). It is estimated that CCF, LVF, and LDA comprise a globally integrated ice volume of 4.2×10^5 km^3 (Levy et al., 2014a).

Absolute age determinations from crater counting on VFFs consistently find that they formed in an earlier period to that in which the LDM (Section 7.3.1.2) was deposited. Estimates for VFF formation age range over three orders of magnitude, from $\sim 40-500$ My (e.g., Morgan et al., 2009; Baker et al., 2010; Hartmann et al., 2014) to ~ 1 Gy (e.g., Levy et al., 2007; Berman et al., 2015; Baker and Head, 2015). This is consistent with formation prior to a step-change from high ($\sim 35°$) to low ($\sim 25°$) mean planetary spin-axis obliquity $\sim 4-6$ My ago, when prolonged accumulation of ice in the mid-latitudes was possible (see Section 7.2.6; Fastook et al., 2008; Forget et al., 2006; Laskar et al., 2004; Madeleine et al., 2009). However, it must be noted that solutions for planetary orbital and spin-axis parameters > 20 My ago are nonunique such that obliquity variations in the period(s) during which VFF formed are poorly constrained (Laskar et al., 2004). Erosion of VFF during the Amazonian has resulted in a range of degradational and recessional morphologies (Fastook et al., 2014; Levy et al., 2016; Brough et al., 2016).

Volume estimates for LDA, LVF, and CCF imply that these VFF subtypes comprise 0.9−2.6 m GEL of water (Levy et al., 2014a). Reconstruction of a GLF in Greg crater implies that a total ~135 km^3 ice volume may have been lost from GLF alone since glacial maximum (Brough et al., 2016). Evidence for hundreds of meters of surface lowering of other VFF (e.g., Dickson et al., 2008, 2010; Baker and Head, 2015) indicates that far greater volume losses may have occurred from these features.

7.3.1.4 Gullies

Under current climatic conditions, liquid water is unstable at the Martian surface (Ingersoll, 1970; Haberle et al., 2001). However, transient liquid water (i.e., water derived from the melting of surface or subsurface ice, that then undergoes rapid evaporation or is preserved under a thickening cover of ice, Goldspiel and Squyres, 2011) has been proposed as a possible explanation for several landforms. Of particular interest are the geologically recent Martian gullies, discovered by MOC (Malin and Edgett, 2000). These features exhibit characteristic morphologies that begin with an alcove high on a steep slope, associated with a sinuous channel that leads down and terminates in a debris apron (Fig. 7.9). The characteristics led Malin and Edgett (2000) to suggest that the Martian gullies were formed by the seepage of groundwater from shallow underground aquifers.

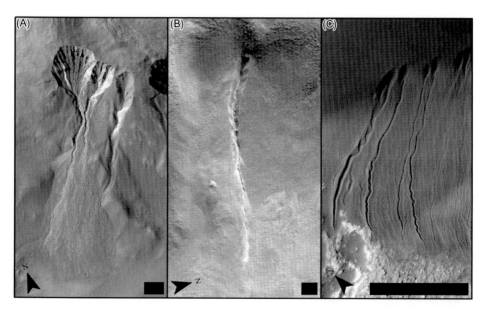

FIGURE 7.9

A variety of scale and morphology of gullies on Mars, each image has been rotated so that downslope is down-image and each scale bar is 200 m. (A) Multiple gully-system on an inner impact crater wall, in the northern part of the Argyre Basin, HiRISE image PSP_006888_1410. (B) Single gully in Nereidum Montes, incised into the Latitude Dependent Mantle, HiRISE image ESP_023700_1415. (C) Sinuous "linear" gullies on Kaiser crater dunefield HiRISE image PSP_006899_1330.

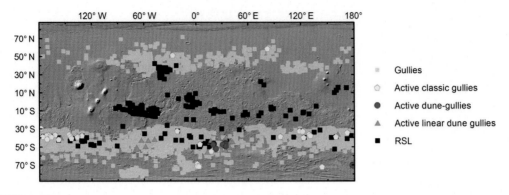

FIGURE 7.10

Global map of active Martian features compiled from the literature, including gullies (Harrison et al., 2015; Pasquon et al., 2016; Dundas et al., 2010, 2012, 2015a) and RSL (Stillman et al., 2017). Background MOLA elevation rendered in shaded relief.

Martian gullies are principally found on steep slopes at mid-latitudes, between 30° and 70° in both hemispheres, as illustrated in Fig. 7.10 (Costard et al., 2002; Heldman and Mellon, 2004; Balme et al., 2006; Dickson et al., 2007; Kneissl et al., 2010; Malin et al., 2010; Harrison et al., 2015). Gullies are preferentially found on poleward-facing slopes, between 30° S and 40° S, and show a clear transition in gully orientation toward equator-facing slopes with increasing latitude (Harrison et al., 2015; Conway et al., 2017). These latitudinal trends in distribution and orientation led to the hypothesis that obliquity-dependent variations in insolation and climate were responsible for their formation (e.g., Costard et al., 2002; Christensen, 2003; Harrison et al., 2015). The observation that 10%−20% of gullies occur on isolated knobs, hills, and central peaks is difficult to reconcile with their formation by the discharge of groundwater, as the aquifer would be isolated and small. The same conclusion is drawn from observations of gullies starting near the top of impact crater walls (Costard et al., 2002; Balme et al., 2006). The two mechanisms of formation that best explain observations to date are overland flow or debris flow produced via ice-melt (Costard et al., 2002; Christensen, 2003; de Haas et al., 2015) and fluidization of the regolith by CO_2 frost sublimation (Cedillo-Flores et al., 2011; Dundas et al., 2015a; Pilorget and Forget, 2016). The latter hypothesis has arisen from the fact that gullies are known to have been active since they were first observed from orbit—first noticed in MOC images by Malin et al. (2006) and later in HiRISE images by Reiss et al. (2010), Diniega et al. (2010, 2013), Dundas et al. (2010, 2015a, 2017), and Raack et al. (2015).

7.3.1.5 Recurring slope lineae

RSL are seasonally active narrow (0.5−5 m) and relatively dark-toned streaks located on steep slopes (ranging from 25° to 40°) that appear and grow during spring and summer and fade during the winter (McEwen et al., 2011). They occur between the latitudes of 30° and 50° in both hemispheres, and a large concentration has also been documented at the equator, particularly in the deep canyons of Valles Marineris, as illustrated in Fig. 7.10 (McEwen et al., 2014; Stillman et al., 2014,

FIGURE 7.11

This image shows RSL extending downhill toward the northwest (north is up) from bedrock cliffs of Hale crater (35.667° S; 323.483° E, ESP_040170_1440). Image is approximately 600 m across. *From HiRISE NASA/JPL/University of Arizona.*

2017; Chojnacki et al., 2016). They appear at approximately the same location every Mars year and favor equatorward slopes, where temperatures may reach up to 250–300 K. Fig. 7.11 illustrates RSL streaks extending downhill from the bedrock cliffs of Hale crater.

One hypothesis is that the high surface temperatures associated with the growth of RSL are sufficient to allow liquid brines to develop near the surface and percolate downhill (Chevrier and Rivera-Valentin, 2012; Martínez and Renno, 2013; Stillman et al., 2017). Supporting this hypothesis, hydrated salt infrared signatures have recently been detected by the CRISM spectrometer onboard MRO. The spectra are consistent with the presence of magnesium perchlorate, magnesium chlorate, and sodium perchlorate (Ojha et al., 2015). Water tracks of the Antarctic environment, where high salinity groundwater circulates above the ice table through the active layer, provide some potential terrestrial analogs for RSL (Levy, 2012; Levy et al., 2014b), but their origin on Mars remains debated. It has been argued that RSL require from 2 to 10 m^3 of water per meter of headwall from their source location, which is typically larger than the expected annual atmospheric cold trapping recharge of subsurface ice (Grimm et al., 2014). The source could then be a briny aquifer for some RSL (Stillman et al., 2016, 2017). However, the fact, that RSL appear to originate simultaneously in large numbers extending from local topographic highs instead of localized source areas, and that they may exploit exactly the same path from year to year but extend to different lengths, argues against the presence of an active aquifer (McEwen et al., 2015; Chojnacki et al., 2016). Analysis of the Melas and Coprates Chasma RSL shows that the atmospheric water volume present over the whole area is an order of magnitude above the volume necessary to generate the RSL, so a mechanism for water vapor concentration could possibly explain the existence of RSL in this region (Chojnacki et al., 2016). Other authors argue that water is unlikely (Edwards and Piqueux, 2016) and possibly not required to form RSL (Schmidt et al., 2017). Schmidt et al. (2017) propose that RSL are granular flows triggered by the thermal gradient between rocky terrain and the regolith via a mechanism called the "Knudsen pump."

Further local and global studies of these active features are paramount to understanding current water reservoir locations and exchanges. Thus, these features are an ongoing focus of orbital investigations.

7.3.1.6 Periglacial landforms

A number of relatively small-scale and therefore young (Amazonian) landforms have been identified on the Martian surface, which have been interpreted as analogous to landforms resulting from thaw on Earth. As a result, these landforms are often used as an indicator of liquid water on Mars. Collectively these landforms are termed "periglacial," a term used in terrestrial geomorphology to indicate landforms and landscapes modified by freeze–thaw processes (e.g., French, 2013). Often the origin of the individual landforms is under debate (as highlighted later), but when many landforms are found together and with similar spatial configurations to those found on Earth (a landscape assemblage), authors have argued that this strengthens the case for thaw through consilience (e.g., Baker, 2001; Soare et al., 2005, 2013, 2016; Balme and Gallagher, 2009, Balme et al., 2013). These landforms are illustrated in Fig. 7.12 and include:

1. Sorted patterned ground, where freeze–thaw cycling sorts the regolith into patterns of clast-poor and clast-rich regions, whose patterns depend on the availability of clasts and the local topography (Kessler and Werner, 2003). Circular or polygonal patterns occur on flat ground and stripes on sloping ground and have been reported in the equatorial (Balme et al., 2009; Balme and Gallagher, 2009), mid- (Mangold, 2005; Hauber et al., 2011) and high-latitudes (Gallagher et al., 2011; Gallagher and Balme, 2011) on Mars. These landforms are meters to tens of meters in scale (although can extend over kilometers) and can only be detected in the highest resolution images. An alternate hypothesis for these features suggests CO_2 slab ice could trap boulders while the underlying ground contracts resulting in boulder-clustering and pattern development (Orloff et al., 2013).
2. So-called scalloped depressions on Mars, which are tens to hundreds of meters wide and meters to tens of meters in depth and can extend across hundreds of kilometers of the surface, are thought to be analogous to thermokarst depressions, or alases (larger depressions with steep sides and flat floor) on Earth. On Earth, these asymmetrically sloped depressions form as a result of runaway destabilization of ground ice in a permafrost zone and are strongly associated with thaw. For the Martian depressions, authors are divided as to whether these depressions form via thaw (Soare et al., 2007, 2008; Ulrich et al., 2010), sublimation (Morgenstern et al., 2007;

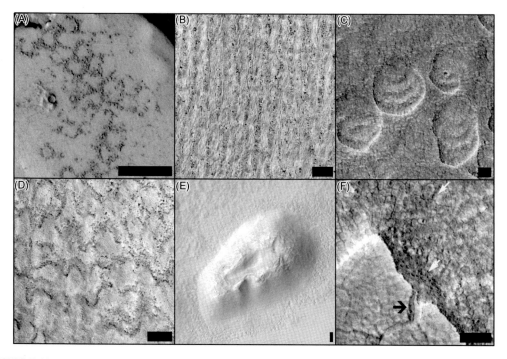

FIGURE 7.12

Images of putative periglacial landforms on Mars; north is up and scale bar is 50 m in all panels. (A) Sorted stone circles identified in Elysium Plantitia by Balme et al. (2009), HiRISE image PSP_004072_1845. (B) Sorted stone stripes on the north-facing wall of Heimdal crater identified by Gallagher et al. (2011), downslope is up-image, HiRISE image PSP_009580_2485. (C) Scalloped terrain in Utopia Planitia in HiRISE image ESP_025277_2275. (D) Lobate forms on pole-facing wall of a crater in the northern plains, HiRISE image ESP_018712_2395. Downslope is up-image. (E) A mound with a summit pit interpreted to be analogous to a pingo in the Argyre Basin (Soare et al., 2014a), HiRISE image ESP_020720_1410. (F) Polygon junction pits (*black arrow*) and low centered polygons (*white arrows*) associated with scalloped depressions in Utopia Planitia, HiRISE image ESP_025277_2275.

Lefort et al., 2009, 2010; Séjourné et al., 2011; Dundas et al., 2015b; Dundas, 2017), or some combination of the two (Soare et al., 2017).

3. Lobate forms analogous in scale (relief of meters, planform of tens to hundreds of meters, stretching over entire hillslopes) and form to solifluction or gelifluction lobes on Earth have been reported in the mid- to high-latitudes of both Martian hemispheres (Gallagher and Balme, 2011; Gallagher et al., 2011; Johnsson et al., 2012; Soare et al., 2016). These slopeforms are uncommon, but remain an important marker of thaw, because no alternate formation hypothesis has been put forward.

4. Small-scale mounds of hundreds of meters in diameter and tens of meters in height with summit pits and/or cracks are hypothesized to be analogous to ice-cored mounds or pingos

formed in periglacial environments on Earth. As summarized in the review by Burr et al. (2009), there are a wealth of alternate hypotheses for these mounds, including volcanic, cryovolcanic, mud volcanoes, erosional remnants, and many others. The strongest contenders for pingos on Mars remain those that occur in association with other putative periglacial landforms (e.g., Soare et al., 2005, 2013, 2014a).

5. Certain morphologies expressed by polygonal patterned ground, including raised margins (Soare et al., 2014b; Levy et al., 2009a), low centered polygons (Soare et al., 2014b), polygon junction pits (Costard et al., 2016), and widened polygon margins (Levy et al., 2009b), which on Earth result from the loss of ice (by thaw), have been interpreted as periglacial on Mars. However, some authors believe that this loss of elevation could be via sublimation rather than thaw (e.g., Marchant et al., 2002; Lefort et al., 2009; Levy et al., 2009c). Again, the periglacial interpretation of these more subtle features relies on the identification of a landscape assemblage (e.g., Soare et al., 2014b; Levy et al., 2009c).

6. Finally, RSL and gullies as discussed earlier often form components of a periglacial landscape assemblage.

Understanding how and why these landforms develop, but also how they come to form landscape assemblages, is key to unlocking the recent evolution of the Martian surface and the role of liquid water therein.

7.3.2 GEOPHYSICAL EVIDENCE FOR DEEPER RESERVOIRS OF ICE AND WATER

7.3.2.1 Gamma ray and neutron spectroscopy

The Gamma-Ray Spectrometer (GRS) instrument suite, which was flown on the 2001 Mars Odyssey spacecraft, consists of three instruments whose primary purpose is to investigate the elemental composition of the Martian surface, with a special emphasis on the content and geographic distribution of hydrogen. Because these measurements are made by instruments on a rapidly moving orbital spacecraft, their spatial resolution is limited to about 500 km. As both gamma rays and thermal neutrons are strongly attenuated by the regolith, these techniques are only useful in assessing the hydrogen mass fraction integrated over the top ~ 0.5 m of the regolith (Boynton et al., 1992).

Although it is the distribution of hydrogen that the GRS instruments actually detect, its most plausible forms in the regolith are as water-ice or hydrated minerals. Ice is typically differentiated from minerals by assessing where it could exist in thermal and diffusive equilibrium with the mean annual water content of the atmosphere (e.g., Mellon and Jakosky, 1995; Schorghofer and Aharonson, 2005). Therefore the GRS results are conventionally expressed as H_2O or water-equivalent hydrogen (WEH).

Fig. 7.13 displays the geographic distribution of WEH measured by GRS. The measurements indicate that the regolith water content varies from 2 wt% H_2O at the equator to nearly pure ice at the poles (Fig. 7.13 adapted from Boynton et al., 2002; Feldman et al., 2004). The high near-surface hydrogen abundances detected near the pole were confirmed by in situ investigations conducted by the Phoenix lander's robotic arm, which revealed the presence of ground ice beneath just a few centimeters of desiccated regolith (Smith et al., 2009). This observation is consistent with the expected distribution of subsurface ice in diffusive equilibrium with the mean annual water content

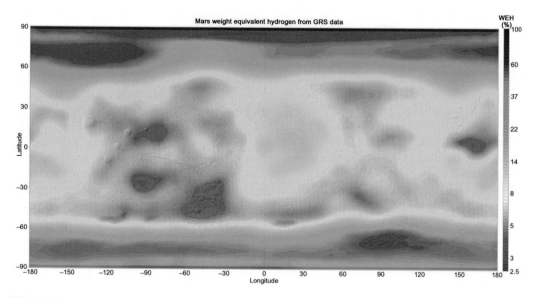

FIGURE 7.13

Superposition map of MOLA shaded Mars altimetric data, Mars Odyssey Neutron Spectrometer (MONS) Hydrogen equivalent water content of the near subsurface. *Adapted from Smith et al. (2001) and Maurice et al. (2011).*

of the atmosphere under present-day climatic conditions (Mellon et al., 2009). See also Section 7.3.1.2 for further details and references.

Near ~55° latitude (both N and S), there is a marked transition between the hydrogen-rich high-latitude regolith and the hydrogen-poor low- to mid-latitudes, at least within the top 0.5 m. This distribution is consistent with the observed distribution of ice-related glacial and periglacial landforms (Mangold et al., 2004b), such as LDA and polygonal terrain (Section 7.3.1). There are two notable exceptions to the generally low abundance of equatorial hydrogen, located in the Sabaea–Arabia and Amazonis–Tharsis regions. These two regions have WEH fractions of up to ~13 wt%, which may be evidence for the presence of large deposits of hydrated minerals (Fialips et al., 2005) or relic ground ice (Jakosky et al., 2005).

Taken altogether, the GRS evidence indicates that the total amount of water present in the near-subsurface (top ~0.5 m) of Mars is equivalent to ~0.11 m GEL (Feldman et al., 2008). Further discussions are available in Chapter 2, Volatiles in Martian Magmas and the Interior: Inputs of Volatiles Into the Crust and Atmosphere; and Chapter 8, Sequestration of Volatiles in the Martian Crust Through Hydrated Minerals: A Significant Planetary Reservoir of Water.

7.3.2.2 Orbital radar sounding

Ground Penetrating Radar is used on Earth to investigate the lithology, structure, and distribution of water in the subsurface, down to depths ranging from ~1 to 10 wavelengths, depending on the electromagnetic properties of the local rock and soil (Olhoeft, 2000). Orbital radar sounders are similar instruments, carried on polar orbiting satellites, to conduct global investigations of the subsurface. Two orbital radar instruments are currently in operation around Mars: the Mars Advanced Radar for Subsurface and Ionospheric Sounding (MARSIS) on Mars Express (Picardi et al., 2005; Jordan et al., 2009) and the SHAllow RADar (SHARAD) onboard NASA's Mars Reconnaissance Orbiter (Seu et al., 2007).

The depths investigated by most remote sensing techniques, such as orbital imaging and spectroscopy, are typically on the order of a few microns. However, at the electromagnetic frequencies of MARSIS and SHARAD, the subsurface can be probed to depths of up to ~ 200 m in lithic environments and $\sim 3-4$ km in nearly pure H_2O ice, like the Martian PLDs (Seu et al., 2007; Jordan et al., 2009). The ability of a sounding radar to investigate the geologic and hydrologic properties of the subsurface is critically dependent on the electromagnetic properties and dielectric contrast between the geologic materials it encounters. In geological settings, one of the largest dielectric contrasts is between liquid water (permittivity of ~ 80) and dry or frozen rock (permittivities of $\sim 6-12$) (Campbell and Ulrichs, 1969; Ulaby et al., 1986).

MARSIS operates at frequencies of $\sim 2-5$ MHz and was designed to investigate the geological and hydrological structures of the deep subsurface, with a particular emphasis on the potential detection of subpermafrost groundwater at depths of up to $\sim 3-5$ km beneath the Martian surface (Picardi et al., 2003, 2004). In practice, MARSIS has achieved this level of sounding performance only in selected environments, which include the ice-rich (and, thus, low dielectric loss) North and South PLD, as well as the Medusae Fossae Formation, whose radar propagation characteristics are consistent with a composition ranging from a dry, high-porosity pyroclastic deposit at one extreme (Kerber et al., 2011) to an ice-rich sedimentary deposit at the other, potentially formed by the redistribution of polar ice at times of high obliquity (Watters et al., 2007a).

Elsewhere, in lithic environments, such as the Martian southern highlands or the volcanic plains surrounding Elysium, there is no unambiguous evidence that MARSIS has detected a subsurface reflection of any sort, at a depth of more than $\sim 200-300$ m below the surface. This result is consistent with several potential explanations (Clifford et al., 2010), the most probable of which is that the dielectric loss and scattering properties of the Martian subsurface are far more attenuating than previously thought (Heggy et al., 2006; Grimm et al., 2006; Boisson et al., 2009; Stillman and Grimm, 2011).

Martian surface permittivity, derived from the strength of the first reflection of selected MARSIS radar pulses, can provide insights into the compositional nature of the subsurface (Mouginot et al., 2010, 2012). Strong reflections result when the dielectric contrast between the Martian atmosphere and surface is high, which is the case when the subsurface consists of low-porosity/high-permittivity rock. Weaker reflections occur when the permittivity of the subsurface is low, due to either its high porosity or the presence of a large volume fraction of water-ice. These derived permittivities represent the volume-averaged dielectric properties of the subsurface over a depth of one radar wavelength, or $\sim 60-80$ m for MARSIS. Interestingly, the permittivity of the Martian northern plains is low, consistent with the occurrence of thick deposits of either highly porous ($\sim 35\%$) or ice-rich (65% by volume) sediment. Support for an ice-related origin is provided by the close agreement between the geographic distribution of low surface permittivities with other potential indicators of subsurface ice, such as the GRS results mentioned earlier, polygonal ground (Carr and Schaber, 1977; Mangold et al., 2004b; Mellon et al., 2008b) and pedestal and layered ejecta craters, whose distinctive lobate morphology is thought to originate from the fluidization of ejecta material during an impact into a water- or ice-rich crust (Carr et al., 1977; Costard, 1989; Barlow et al., 2000; Barlow, 2005).

With an operating frequency of 20 MHz, SHARAD was designed to complement MARSIS by operating at a shorter wavelength to better resolve variations in near-surface composition, stratigraphy, and structure, particularly with regard to the internal layering of the PLD (Seu et al., 2007). The maximum sounding performance of SHARAD is similar to, but slightly less than MARSIS, achieving sounding depths of up to ~2.7 km in the ice-rich PLD, but less than ~100−200 m in lithic environments.

No subsurface reflections were detected with SHARAD, apart from the young (middle to late Amazonian) volcanic or ice-rich units. This could be due to a higher than expected signal scattering by surfaces with meter-scale fractures or absorption by a thin layer of surface clays with adsorbed water (Stillman and Grimm, 2011). Subsurface layers associated with ground ice have been found in mid-latitude debris-covered glaciers (Section 7.3.1.3, Holt et al., 2008; Plaut et al., 2009) and in the northern plains. At the Phoenix landing site, the base ground ice is estimated at 15−66 m depth (Putzig et al., 2014). In Arcadia Planitia, terraces within craters are associated with subsurface reflections and constrain the base of the ice layer to 50 m depth (Bramson et al., 2015). Finally, SHARAD detected an extensive reflector associated with layered mesas 80−170 m thick consistent with >50% water-ice by volume in Utopia Planitia (Stuurman et al., 2016).

7.3.2.3 Methane detections and their implications

The origin of methane detected in the Martian atmosphere has major implications. Should it prove to be of biological origin, our vision of life as a natural process would be revolutionized (e.g., McKay et al., 1996). Alternatively, an abiotic origin can arise from the UV decomposition of organic compounds present in interplanetary dust (Fries et al., 2016) but the MSL rover methane measurements at Gale crater fail to correlate with meteor shower events (Roos-Serote et al., 2016). Finally, methane can be generated locally from the serpentinization of mafic and ultramafic rocks in hydrothermal systems (Oze and Sharma, 2005; Etiope et al., 2013) and can provide a source for abiotic methane gas hydrate reservoirs (e.g., Johnson et al., 2015). The relatively short atmospheric lifetime of the molecule of several centuries (Lefevre and Forget, 2009) implies either active subsurface hydrothermal systems on the planet, or the presence of a cold methane trap possibly in the form of a clathrate cryosphere (e.g., Lasue et al., 2015).

The detection of methane on Mars from orbit and from the ground has proven elusive. During the year 2003, it reached up to 10 ppb (parts per billion) globally with local abundances up to 30 ppb as detected from space with Mars Express (Formisano et al., 2004), as well as from the ground (Krasnopolsky et al., 2004). Mumma et al. (2009) generated maps of the methane abundances on the surface of the planet over 3 Mars years showing spatial and temporal variations of the signal, indicating strongly localized sources with a methane flux estimated to be ≥0.6 kg per second. Such high values were not replicated by later studies and an upper limit of 8 ppb has been obtained from telescopic observation surveys between 2006 and 2010 (Villanueva et al., 2013). The MSL rover at Gale crater initially measured an upper limit of 1.3 ppbv of methane (Webster et al., 2013), but several months later detected an elevated level of methane of 7.2 ± 2.1 ppbv during 2 months, implying that Mars is episodically producing methane from an additional unknown source (Webster et al., 2015).

Such variations could be the result of local sporadic methane release but the exact sources and sinks of methane on Mars on such short timescales remain to be determined. Whether its sources involve biotic or abiotic processes, subsurface sporadic release of methane would involve the presence of water at depth. Current conditions on Mars favor the formation and stability of methane clathrates at depth (Fisk and Giovannoni, 1999; Max and Clifford, 2000), serving as a potential source for the observed releases of atmospheric methane (Chastain and Chevrier, 2007). Large amounts of methane clathrate reservoirs may have formed during the early episodes of alteration of the Martian crust linked to the remnant magnetic field. The total storage capacity of such clathrate reservoirs is estimated to be about $2 \times 10^{19} - 2 \times 10^{20}$ moles of CH_4 (Lasue et al., 2015). Alternatively, current conditions at depth with geothermal fluxes and liquid water in aquifers may be the source of current abiotic methane generation in hydrothermal systems (Oze and Sharma, 2007). Both processes support the presence of subsurface water in the equatorial region of Mars today.

7.4 MARS' EVOLVING WATER BUDGET

The current distribution of water on Mars results from numerous interacting processes. On the one hand, loss mechanisms and sinks include atmospheric erosion by impacts, thermal and nonthermal escape processes (e.g., by photochemical reactions, ion pickup, sputtering, etc., Shizgal and Arkos, 1996), extreme ultraviolet radiation, as well as solar wind forcing (Lundin et al., 2007; Terada et al., 2009; Bougher et al., 2015). Other potential sinks correspond to weathering and hydration processes that can occur at the surface-atmosphere interface (Audouard et al., 2014; Carter et al., 2015). On the other hand, sources of water could be punctually active, for example, volcanic outgassing or delivery of volatiles by impacts (Pham et al., 2009; Lammer et al., 2013; Pham and Karatekin, 2016).

In this section, we review current estimates of how water reservoirs on Mars have evolved with time and assess how each factor may have influenced the evolution of the planet's hydrosphere.

7.4.1 ORIGIN OF MARS' WATER AND POTENTIAL WATER SINKS

It is expected that Mars accreted up to several thousands m GEL of water on its surface during its accretion, either by asteroidal delivery (Lunine et al., 2003) or by accreting water-rich dust from nebular gas adsorption (Drake, 2005). Recent dynamical accretion models estimate that early Mars could have consisted of 0.1% to 0.2% by mass of water (Brasser, 2013).

There are essentially two major stages to the early loss of the Martian atmosphere: atmospheric escape and aqueous alteration of the surface. The primordial epoch lasts about 500 My from the accretion of the planet. It is characterized by the hydrodynamic blow-off of hydrogen and high thermal escape rates due to the strong ultraviolet flux of the young Sun, processes that were aided by Mars' low gravity (Erkaev et al., 2014). After a large fraction of the proto-atmosphere of the planet was lost, a secondary atmosphere formed from the impact of volatile-rich planetesimals and by mantle outgassing (estimated to be a few 10s to 100s of mbar at the end of the Noachian;

Albarede, 2009; Lammer et al., 2013). Since the Noachian, the evolution of the atmosphere has been driven by a complex interplay of thermal escape, sputtering (erosion by the solar wind), and dissociative recombination (Shizgal and Arkos, 1996; Lundin et al., 2007; Terada et al., 2009; Bougher et al., 2015). These effects can be modified by external inputs from impacts, and exchanges with the surface-like volcanic venting and surface loss via alteration or precipitation (Pham et al., 2009; Lammer et al., 2013; Pham and Karatekin, 2016). Overall, it is difficult to assess the global effect on the atmospheric balance. The best estimates of the amount of CO_2 added to the atmosphere are 0.05−0.5 bar by volcanic outgassing and 0−0.3 bar by the impacts of the Late Heavy Bombardment, while the combination of the various loss processes may have removed between 0.004 and 0.3 bar of CO_2. Under those assumptions, if an initial inventory of water several 10−100 m deep GEL was present in the Noachian, it could have been trapped in the crust in the form of water-ice, hydrated minerals, and possibly liquid water under the cryosphere (Lammer et al., 2013). Petrologic studies of Martian meteorites also indicate at least two geochemically distinct crustal reservoirs of water with different water contents and an estimated ∼230 m GEL water content in the Martian crust (McCubbin et al., 2016).

A number of hydrated mineral outcrops have been detected from orbit, indicating crustal alteration due to water activity on early Mars (e.g., Bibring et al., 2006; Ehlmann and Edwards, 2014 and references therein, Section 7.2.5). A global assessment of the Fe/Mg and Al-rich clays stratigraphy relationships on Mars shows evidence of a mineral alteration front indicating an early era (>3.7 Gy) of widespread aqueous alteration of the crust (Carter et al., 2015). However, the amount of water trapped in altered minerals is difficult to assess. Evidence of crustal magnetization and demagnetization is found in association with the planet's Noachian terrain, located in the southern highlands (Acuna et al., 1999; Langlais et al., 2004). In some of the areas with strong magnetic field, surface occurrences of serpentine (Ehlmann et al., 2010) were detected suggesting hydrothermalism during the Noachian that may have aided the preservation of the planet's early magnetic field (Quesnel et al., 2009). The remnant volume of magnetized crust suggests that serpentinization may have acted as a sink to a 500−820 m thick water GEL at the time of the dynamo extinction, which is also consistent with the amount of water lost to explain the present D/H ratio of the atmosphere (Chassefière et al., 2013; Chassefière and Leblanc, 2011).

The duration of the magnetic dynamo is not well constrained. Models of the Martian interior and the lack of significant magnetic field measured over the major impact basins of the planet suggest a relatively short period for the active dynamo, until the mid-Noachian around 4 Gy ago (Lillis et al., 2008; Roberts et al., 2009). However, analysis of gravity and magnetic anomalies around smaller craters and volcanoes shows the presence of sources below younger surfaces, indicating that the dynamo was possibly active during the Late Noachian. Studies on smaller structures seem to indicate that the cessation of the dynamo occurred after the formation of Apollinaris Patera and Antoniadi crater, but before Syrtis Major and Biblis Patera were fully built and Lowell crater was formed, narrowing the timing of the dynamo cessation to between 3.79 and 3.75 Gy (Langlais and Purucker, 2007; Langlais et al., 2012; Milbury et al., 2012). The duration for which the dynamo was active is linked to Mars' climate via the stability of its atmosphere, and hence, is coupled to the extent and duration of surface geologic activity (see Fig. 7.5). If the dynamo remained active until the Late Noachian, then it is very likely that it could have significantly reduced the early atmospheric loss processes.

7.4.2 INFERENCES FROM IN SITU ISOTOPIC MEASUREMENTS

The isotopic ratio of elements, especially the noble gases, is a major marker of the evolution of a planet's atmosphere. Initial measurements of argon ($^{36}Ar/^{38}Ar$) made by the mass spectrometer on the Viking lander gave a poorly constrained range of 4−7 for Mars, while a low limit on krypton seemed to indicate an evolved atmosphere (Biemann et al., 1976). The solar value has been measured by Genesis to be 5.5 ± 0.01 (Pepin et al., 2012), while on average Martian meteorites have a range closer to 4 ± 0.6 (Bogard, 1997). The most recent value of $^{36}Ar/^{38}Ar$ ratio has been measured by the SAM instrument on MSL to a value of 4.2 ± 0.1 (Atreya et al., 2013). This value is significantly lower than the solar primordial one, indicating an argon loss of at least 50% from the atmosphere of Mars (and up to 95% loss depending on the loss rate and the existence of mantle or meteoritic sources) over the past 4 billion years (Atreya et al., 2013 and references therein). Such a conclusion is also reached from the analysis of other atmospheric isotopic ratios of Ar, N, and C (Mahaffy et al., 2013; Mahaffy and Conrad, 2015).

A more direct way to probe the loss of water from the planet is to monitor the deuterium to hydrogen fractionation in its atmosphere. The initial Martian reservoir mantle D/H value from Martian meteorites is 2 times the terrestrial value while the atmospheric value is about 5.2 times the terrestrial value VSMOW (Vienna Standard Mean Ocean Water), suggesting atmospheric escape of volatiles to space over time, and exchange between at least two different reservoirs (Leshin, 2000). Usui et al. (2012) measured a D/H ratio of 1.275 VSMOW in primitive melt inclusions of a Martian shergottite, consistent with a primitive D/H ratio similar to the ones found on Earth and in carbonaceous chondrites. Recent measurements of D/H from the glass melts in shergottite, nakhlite, and chassignite meteorites formed during the ejection of the meteorites exhibit negative linear trends with respect to H_2O indicating mixtures between a surficial water reservoir in contact with the current Mars' atmosphere and a water reservoir in the near-surface that has a D/H intermediate between the mantle and atmospheric values. This trend would be consistent with either (1) a mixture of intermediately deuterated hydrothermally altered products of the crust or (2) a large subsurface ice reservoir equivalent to approximately 130 m GEL (Chapter 4; Kurokawa et al., 2014, Usui et al., 2015, 2017). These large reservoirs of water would be consistent with the detection of low dielectric constant material by radar sounding (Mouginot et al., 2012, Section 7.3.2.2). Similar intermediate values of D/H are measured in situ by MSL on clay-bearing samples at Gale crater (Mahaffy et al., 2015) indicating that water fractionation occurred early, but that the subsurface water reservoir may have preserved a relatively constant D/H ratio relatively constant since the Noachian to Hesperian transition.

Seasonal variations of the D/H content in Mars' atmosphere were detected where during spring time, the northern polar cap sublimates and introduces deuterium into the atmosphere, indicating a deuterium enrichment in the NPLD to about 8 times VSMOW (Villanueva et al., 2015). Since the total North + South PLD is equivalent to 20 m GEL, the early Mars water reservoir must have been at least equivalent to 137 m GEL. This is consistent with the minimum amount of water needed to form a northern ocean on the planet (Carr and Head, 2003). Recent multireservoir model simulations of the D/H ratio exchange between the different water reservoirs of Mars estimate a median 60 m GEL water loss through the Hesperian−Amazonian implying that the larger post-Noachian water inventory is probably trapped below a near-surface cryospheric seal (Grimm et al., 2016).

7.4.3 ATMOSPHERIC ESCAPE

Going a step further, hydrodynamic escape is currently estimated in situ by the Mars Atmosphere and Volatile Evolution (MAVEN) mission orbiting the planet. The average escape rate of the atmosphere has been measured to be around 100 g/s, mostly driven by the loss of the oxygen ions, but this rate can increase by a factor of 10−20 during a solar storm event recorded during an interplanetary coronal mass ejection impact in March 2015. Ion loss early in Mars history, when the young Sun was more active, may have been a major contributor to the long-term evolution of the atmosphere (Jakosky et al., 2015). However, unexpected levels of variability during the measurements complicate the determination of precise rates (Bougher et al., 2015). All these estimates rely on models of H_2O sources and escape rates during the evolution of Mars, which are difficult to constrain. This is especially true of the ancient Martian magnetic field, which probably shielded the primitive atmosphere of Mars from the young Sun. Better estimates from the escape rates by MAVEN in the future will help constrain the history of the planet's atmosphere.

Further measurement analyses from MAVEN data have shown that the upper Mars atmosphere isotopic composition indicates a 66% loss of Ar to space by pickup-ion sputtering. This indicates that a large fraction of the planetary volatile reservoir has been lost to space. It could imply the loss of up to 1 bar of CO_2 or most of the inventory of water depending on which molecular loss is favored, and could explain the change in early Mars climate (Jakosky et al., 2017).

7.5 CURRENT STATE OF THE MARTIAN CRYOSPHERE

The previous sections have highlighted the numerous ancient and current geological features related to water on Mars. In each case, the duration and quantities of water necessary have been discussed. All evidence points to a global inventory of water available for hydrological circulation that decreases with time from the Noachian to the Amazonian. However, without firm direct evidence of the initial or current inventories of water in the subsurface, it could be consistent with either a relatively "dry" early Mars ($\sim$170 m GEL, e.g., Carr and Head, 2015) or a wetter early Mars (>200 m GEL, e.g., Lasue et al., 2013). In the latter case, the available water content would be currently trapped in underground reservoirs that prevent its exchange with the circulating water mobilized on the surface.

Subsurface water on Mars may exist in two thermally distinct reservoirs: as ground ice, within the cryosphere, and as groundwater at greater depth, where the geothermal heat flux elevates the lithospheric temperature above the freezing point (Fanale, 1976; Rossbacher and Judson, 1981; Kuzmin, 1983; Clifford, 1993). The Martian cryosphere encompasses a region of the crust extending from the surface down to a local depth, z, given as a first approximation by the solution to the one-dimensional heat conduction equation:

$$z = k_{(T)} \frac{(T_{mp} - T_{ms})}{Q_g}$$

where $k_{(T)}$ is the temperature-dependent thermal conductivity of the crust, T_{ms} is the mean annual surface temperature (which ranges from about 218 K at the equator to about 154 K at the poles),

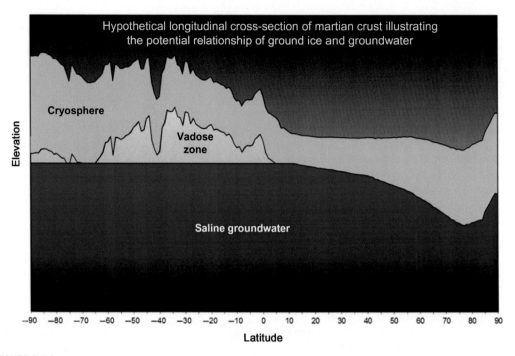

FIGURE 7.14

Hypothetical meridian cross section of the present-day Martian crust (along 203° E, a longitude representative of the global dichotomy between the northern and southern hemispheres and illustrating typical topographic variations), illustrating the potential relationship between surface topography, ground ice, and groundwater. Surface elevations are from the Mars Orbiter Laser Altimeter (MOLA) Mission Experiment Gridded Data Record with a 2° smoothing function (Smith et al., 2003). At those locations where the base of the cryosphere is in contact with the water table, the presence of dissolved salts may reduce the thickness of frozen ground by depressing the freezing point of the groundwater. Where the cryosphere and groundwater are not in direct contact, low-temperature hydrothermal convection within the vadose zone may result in basal melting temperatures closer to 273 K (Clifford, 1993). *From Clifford and Parker (2001).*

T_{mp} is the melting temperature of ice at the base of the cryosphere, and Q_g is the geothermal heat flux (Clifford, 1993; Clifford and Parker, 2001; Clifford et al., 2010). The value for T_{mp} ranges from around 273 K for pure water-ice to lower values depending on the presence of freezing-point depressing salts. Fig. 7.14 illustrates the hypothetical structure of the Martian hydrosphere along a pole-to-pole cross section of the crust at 203° E longitude.

Here we discuss the physical parameters (thermal conductivity, salinity, and geothermal heat flow) that affect the depth, pore volume, and temporal evolution of the cryosphere and examine their implications for the size and state of subsurface water reservoirs over time.

7.5.1 THERMAL CONDUCTIVITY

The thermal conductivity of water-ice increases with decreasing temperature, varying from a low 2.26 W/m/K at 273 K to a high 3.64 W/m/K at 154 K (Hobbs, 1974). The thermal conductivity of basalt exhibits a similar behavior at equivalent temperatures (Clauser and Huenges, 1995; Lee and Deming, 1998) and can be assumed to follow approximately the same expression (Clifford et al., 2010).

However, as discussed in Section 7.4, the detection of atmospheric methane on Mars is consistent with the possible presence and decomposition of gas hydrate in the subsurface. In its pure state, the thermal conductivity of gas hydrate is ~ 0.5 W/m/K (Davidson, 1983; Sloan, 1997), or approximately one-fifth that of water-ice, a value that recent laboratory measurements have demonstrated remains fairly constant over a broad range of subfreezing temperatures (Krivchikov et al., 2005). Thus, where hydrate is present, it may significantly reduce the effective thermal conductivity of the subsurface, especially at shallow depths, where the porosity of the crust is expected to be relatively high ($\sim 20\%-50\%$). This decrease in thermal conductivity would result in a corresponding decrease in the depth of the base of the cryosphere.

Finally, at latitudes below $40°$, where water-ice is diffusively unstable with respect to the relative humidity of the atmosphere, the regolith may be desiccated to depths ranging from several meters to as much as ~ 1 km, depending on the diffusive and thermal properties of the subsurface (Clifford and Hillel, 1983; Fanale et al., 1986; Mellon and Jakosky, 1993). This desiccated region could have an effective thermal conductivity as low as 0.05 W/m/K— or even lower, given a regolith of fine-grained sediments or volcanic ash (Presley and Christensen, 1997), yielding a significant reduction cryosphere thickness at low latitudes. As discussed in Section 7.3, there may still be some water activity on local areas located in this latitudinal range.

7.5.2 EFFECT OF FREEZING POINT DEPRESSING SALTS IN MARTIAN GROUNDWATER

The freezing point of Martian groundwater can be depressed by the presence of dissolved salts. The evolution of Martian groundwater into a highly mineralized brine is an expected consequence of three processes: (1) the leaching of crustal rocks beneath the groundwater table, (2) the increased concentration of dissolved minerals in groundwater as more of the groundwater inventory is cold-trapped by the growth of the cryosphere over geologic time, and (3) the influx of minerals leached from the vadose zone by the low-temperature hydrothermal convection of vapor between the water table and the base of the cryosphere (Clifford, 1991; Clifford et al., 2010).

Plausible saline components of Martian groundwater include NaCl brine that has a freezing point of 252 K at its eutectic and other chloride compounds ($CaCl_2$, $MgCl_2$, etc.) that alone, or in combination, can reduce the freezing point to as little as ~ 210 K (Brass, 1980; Clark and Van Hart, 1981; Clifford, 1993; Burt and Knauth, 2003). The potential for even lower freezing points is supported by the Phoenix lander discovery that the local Martian soil contained up to 1 wt% of magnesium perchlorate ($Mg(ClO_4)_2$), which has a freezing temperature of ~ 203 K at its eutectic (Hecht et al., 2009). On the other hand, the presence of abundant sulfate in the soil may reduce the total freezing point depression of Martian groundwater to just a few degrees (Clark and Van Hart, 1981).

As noted earlier, the presence of a geothermal temperature gradient within the vadose zone will lead to the development of a low-temperature hydrothermal circulation system of ascending water vapor and descending liquid condensate between the local groundwater table and the base of the cryosphere (Clifford, 1993). Given a geothermal gradient of ~15 K/km and reasonable estimates of crustal porosity and pore size, the equivalent of 1 km GEL could be cycled through the vadose zone by this process every 1–10 My. An important mineralogical consequence of this activity would be the depletion of any easily dissolved substance from the vadose zone and their concentration in the underlying groundwater.

Where the base of the cryosphere and saline groundwater are not in direct contact, low-temperature hydrothermal convection is likely to have depleted the intervening crust of any freezing-point depressing salts, resulting in a basal temperature of ~273 K, maximizing the local depth of frozen ground. Conversely, where the cryosphere and saline groundwater are in contact, eutectic concentrations of perchlorate or other dissolved salts may dramatically reduce — and potentially eliminate — the local thickness of the frozen ground (Clifford et al., 2010).

7.5.3 GEOTHERMAL HEAT FLUX

Rheologic estimates of the thickness of the Martian elastic lithosphere suggest that the present day geothermal heat flow ranges from about ~8 to 25 mW/m^2 (Solomon and Head, 1990; McGovern et al., 2004; Phillips et al., 2008), with a probable global mean of ~15 mW/m^2 (Clifford et al., 2010). However, more recent estimates, based on a 3-D spherical model of the thermal evolution of Mars, local crustal thicknesses inferred from gravity and topography data (Neumann et al., 2004) and a geographic distribution of radiogenic heat sources derived from GRS measurements (Taylor et al., 2006), suggest a present-day heat flow range approximately 2× higher (~17.2–49 mW/m^2, depending on location), with an estimated average value of ~23.2 and 27.3 mW/m^2 (Plesa et al., 2016). On Earth, regional geothermal heat flow varies by ±50% about the continental mean (Pollack et al., 1993). Given the evidence for a similar level of crustal diversity on Mars (e.g., Nimmo and Tanaka, 2005; Solomon et al., 2005; Watters et al., 2007b), it is reasonable to expect that the value of the Martian geothermal heat flow will exhibit a comparable range of local variability, which is also obtained on the maps generated by the work of Plesa et al. (2016). At smaller spatial scales, even larger differences in heat flow are possible (e.g., due to local volcanic and magmatic activity).

Clifford et al. (2010) examined the extent and thermal structure of the Martian cryosphere including the temperature-dependent thermal conductivity of ice and rock, lithospheric heat flow, the potential presence of gas hydrate, the exponential decline in crustal porosity with depth, and the long-term (~10^5–10^7 year) astronomically driven variations in insolation. The cryosphere depths were calculated based on average surface temperatures derived from the 20 My nominal insolation history of Laskar et al. (2004), which represents a more relevant (and dynamic) boundary condition than the present-day mean annual surface temperatures. Although these astronomically driven changes in insolation have led to significant variations in mean annual surface temperature over the last 20 My, the mean values over this time period are generally within a few degrees of the current annual mean. The mean annual surface temperature over the last 20 My at different latitudes are 216 K for 0°, 208 K for 30°, 178 K for 60°, and 157 K for 90° given the radiative and thermal properties assumed in the model described in Clifford et al. (2010).

For a cryosphere basal temperature of 273 K, and the present best estimate of lithospheric heat flow (15 mW/m^2), the depth of the cryosphere ranges from ~ 5 km at the equator to ~ 22 km at the poles. Significant reductions in the depth of the cryosphere can occur where its base is in contact with saline groundwater. For example, if the groundwater has a freezing point equal to that of a eutectic brine of NaCl (252 K), the thickness of the equatorial cryosphere could be reduced to as little as 1.5 km. However, if sufficient magnesium perchlorate is present in the crust to lower the freezing point of groundwater to 203 K, then the cryosphere will be absent at all latitudes equatorward of $\sim 35°$ at any elevation where it is in direct contact with groundwater—although it will still exist at higher elevations and higher latitudes (where it would reach a maximum thickness of ~ 9 km at the poles). Barring the saturation in strong freezing point depression salts at the global scale, Fig. 7.14 indicates, like the more elaborate multiphase reservoir exchange simulations from Grimm et al. (2016), that the post-Noachian water inventory could be trapped below a cryospheric seal present on a global scale in the subsurface of the planet.

7.6 CONCLUSIONS AND PERSPECTIVES

In this chapter, we have reviewed the extensive geomorphological evidence for the sustained presence of water on Mars and its evolution with time. Early Mars surfaces have recorded considerable numbers of features related to flowing liquid water, including the valley networks, outflow channels, the paleolake basins and delta fans, and possible shorelines of a former northern ocean. Mineralogical alteration and past glaciation features further attest to surface modification by water in its various states. Estimates of the water necessary to emplace all the sediments related to these features range from 100s to 1000 m GEL. Still, the duration and intensity of the early Mars water cycle remain debated.

As discussed in Section 7.2, the early Mars climate models remain difficult to reconcile with flowing liquid water on the surface of Mars and "cold and icy" early Mars models have been developed by Fastook et al. (2014) and Fastook and Head (2015). While there is no strong evidence for a global highland ice sheet, a limited early global water budget would have also limited the glacial erosion of the highlands. However, as noted by Hynek (2016), it is difficult to reconcile this model with all the geomorphic evidence for flowing water in ancient terrains. As pointed out by Wordsworth (2016), the paradox is all the more striking by the fact that even if early Mars was warm and wet, it must have experienced a transition to the current cold and dry climate state during which wet-based glaciers and ice sheets should have formed in the highlands, but did not leave much visible evidence of glacial erosion. So, the planet probably went through an icy-highland phase in any case, but evidence for it remains scarce.

This paradox could be explained by sporadic and transient warming associated with impact- and volcanism-induced climate change, which led to widespread melting and high erosion rates from liquid water flow (e.g., Irwin et al., 2005a; Halevy and Head, 2014; Batalha et al., 2016). Still, the reasoning and contrast between these lines of evidence will require new measurements provided by the next missions to better constrain the current reservoirs of water on Mars and their implications for the global water inventory of early Mars and its evolution in response to the changes in climate.

Spacecraft investigations have demonstrated that water is present on current Mars in a variety of states. A few tens of precipitable microns are detected in the atmosphere (e.g., Smith, 2004, 2008), however locally higher values could explain seasonally active RSL (e.g., McEwen et al., 2011). The largest visible reservoirs of water on the surface of Mars are the North and South PLDs, which store about 22 m GEL of water as ice (e.g., Lasue et al., 2013; Carr and Head, 2015). The hypothesis that the top ~60−80 m of the Martian subsurface is ice-rich is consistent with the widespread distribution of landforms characteristic of glacial and periglacial environments on Earth (such as LDA and polygonal ground) at mid- to high-latitudes in both hemispheres. Smaller amounts of ice (totaling ~5−10 m GEL) are also present in the LDM and the related shallow subsurface ice deposits (as exposed by recent impacts or detected by radars or GRS measurements).

The amount of water present at greater depths (as ground ice and subpermafrost groundwater) is unknown. As described by the Martian cryosphere models, a significant amount of underground ice and possibly liquid water may still be sealed at depth from the surface (e.g., Lasue et al., 2013; Grimm et al., 2016). Unfortunately, any ice or water present below this depth is beyond the detection capabilities of existing instruments—although transient electromagnetic sounding techniques, conducted from the surface, may enable future assessments of the presence of liquid water down to depths of ~10−20 km (Grimm et al., 2009). The possibility that Martian RSL are tapping a groundwater source is also still an open subject of debate (Stillman et al., 2016, 2017).

The next generation of missions to Mars will better constrain a number of factors important to unraveling the hydrosphere evolution of Mars and to assessing the current presence of water. MAVEN continues to record the escape rates from the Martian atmosphere. ESA's Trace Gas Orbiter mission has been inserted in orbit around Mars and includes a neutron spectrometer that maps the distribution of subsurface hydrogen with a higher spatial resolution, and other instruments that assess the state and variability of minor species in the atmosphere of Mars, especially methane (Zurek et al., 2011). Finally, the alteration state of the subsurface remains unknown. The serpentinization of the crust represents a major potential sink for subsurface water and is likely to be better constrained with improved knowledge of the composition of strata exposed in scarps and outcrops, as observed from orbit and from the next Mars rovers (NASA's Mars 2020 and ESA's ExoMars, equipped with drills and radars).

NASA's Mars 2020 rover's landing site is still under selection, but it should explore an ancient area with a variety of mineralogies related to aqueous alteration and high chances of organic preservation. The goal of the rover is to prepare the first step of a multimission for a sample return from Mars (Witze, 2014, 2017). The ExoMars rover will explore Noachian-aged terrains of Arabia Terra, a region with abundant ancient fluvial landforms (e.g., Davis et al., 2016) and mineralogical evidence for ancient aqueous environments (e.g., Poulet et al., 2008). The rover is equipped with a 2 m long drill that will sample the Martian subsurface to unprecedented depths in search of ancient biomarkers, and provide new insights into the environmental and hydrological conditions on ancient Mars and their implications for habitability (Vago et al., 2015).

REFERENCES

Acuna, M.H., Connerney, J.E.P., Lin, R.P., Mitchell, D., Carlson, C.W., McFadden, J., et al., 1999. Global distribution of crustal magnetization discovered by the Mars Global Surveyor MAG/ER experiment. Science 284 (5415), 790−793.

Adeli, S., Hauber, E., Le Deit, L., Jaumann, R., 2015. Geologic evolution of the eastern Eridania basin: implications for aqueous processes in the southern highlands of Mars. J. Geophys. Res. Planets 120 (11), 1774–1799.

Adeli, S., Hauber, E., Kleinhans, M., Le Deit, L., Platz, T., Fawdon, P., et al., 2016. Amazonian-aged fluvial system and associated ice-related features in Terra Cimmeria, Mars. Icarus 277, 286–299.

Aharonson, O., Schorghofer, N., 2006. Subsurface ice on Mars with rough topography. J. Geophys. Res. Planets 111 (E11). Available from: http://dx.doi.org/10.1029/2005JE002636.

Albarede, F., 2009. Volatile accretion history of the terrestrial planets and dynamic implications. Nature 461 (7268), 1227–1233.

Andrews-Hanna, J.C., Phillips, R.J., 2007. Hydrological modeling of outflow channels and chaos regions on Mars. J. Geophys. Res. Planets 112 (E8). Available from: http://dx.doi.org/10.1029/2006JE002881.

Ansan, V., Mangold, N., 2006. New observations of Warrego Valles, Mars: evidence for precipitation and surface runoff. Planet. Space Sci. 54 (3), 219–242.

Ansan, V., Mangold, N., 2013. 3D morphometry of valley networks on Mars from HRSC/MEX DEMs: implications for climatic evolution through time. J. Geophys. Res. Planets 118 (9), 1873–1894.

Ansan, V., Mangold, N., Masson, P., Gailhardis, E., Neukum, G., 2008. Topography of valley networks on Mars from Mars Express High Resolution Stereo Camera digital elevation models. J. Geophys. Res. Planets 113 (E7). Available from: http://dx.doi.org/10.1029/2007JE002986.

Ansan, V., Loizeau, D., Mangold, N., Le Mouélic, S., Carter, J., Poulet, F., et al., 2011. Stratigraphy, mineralogy, and origin of layered deposits inside Terby Crater, Mars. Icarus 211 (1), 273–304.

Arfstrom, J., Hartmann, W.K., 2005. Martian flow features, moraine-like ridges, and gullies: terrestrial analogs and interrelationships. Icarus 174 (2), 321–335.

Armitage, J.J., Warner, N.H., Goddard, K., Gupta, S., 2011. Timescales of alluvial fan development by precipitation on Mars. Geophys. Res.Lett. 38, L17203. Available from: http://dx.doi.org/10.1029/2011GL048907.

Atreya, S.K., Trainer, M.G., Franz, H.B., Wong, M.H., Manning, H.L., Malespin, C.A., et al., 2013. Primordial argon isotope fractionation in the atmosphere of Mars measured by the SAM instrument on Curiosity and implications for atmospheric loss. Geophys. Res. Lett. 40 (21), 5605–5609.

Audouard, J., Poulet, F., Vincendon, M., Milliken, R.E., Jouglet, D., Bibring, J.P., et al., 2014. Water in the Martian regolith from OMEGA/Mars Express. J. Geophys. Res. Planets 119 (8), 1969–1989.

Baker, D.M., Head, J.W., 2015. Extensive Middle Amazonian mantling of debris aprons and plains in Deuteronilus Mensae, Mars: implications for the record of mid-latitude glaciation. Icarus 260, 269–288.

Baker, D.M., Head, J.W., Marchant, D.R., 2010. Flow patterns of lobate debris aprons and lineated valley fill north of Ismeniae Fossae, Mars: evidence for extensive mid-latitude glaciation in the Late Amazonian. Icarus 207 (1), 186–209.

Baker, V.R., 2001. Water and the Martian landscape. Nature 412 (6843), 228–236.

Baker, V.R., Strom, R.G., Gulick, V.C., Kargel, J.S., Komatsu, G., Kale, V.S., 1991. Ancient oceans, ice sheets and the hydrological cycle on Mars. Nature 352 (6336), 589–594.

Balme, M.R., Gallagher, C., 2009. An equatorial periglacial landscape on Mars. Earth Planet. Sci. Lett. 285 (1), 1–15.

Balme, M.R., Mangold, N., Baratoux, D., Costard, F., Gosselin, M., Masson, P., et al., 2006. Orientation and distribution of recent gullies in the southern hemisphere of Mars: observations from High Resolution Stereo Camera/Mars Express (HRSC/MEX) and Mars Orbiter Camera/Mars Global Surveyor (MOC/MGS) data. J. Geophys. Res. Planets 111 (E5). Available from: http://dx.doi.org/10.1029/2005JE002607.

Balme, M.R., Gallagher, C.J., Page, D.P., Murray, J.B., Muller, J.P., 2009. Sorted stone circles in Elysium Planitia, Mars: implications for recent Martian climate. Icarus 200 (1), 30–38.

Balme, M.R., Gallagher, C.J., Hauber, E., 2013. Morphological evidence for geologically young thaw of ice on Mars: a review of recent studies using high-resolution imaging data. Progress Phy. Geogr. 37 (3), 289–324.

Banfield, D., Donelan, M., Cavaleri, L., 2015. Winds, waves and shorelines from ancient Martian seas. Icarus 250, 368–383.

Banks, M.E., Lang, N.P., Kargel, J.S., McEwen, A.S., Baker, V.R., Grant, J.A., et al., 2009. An analysis of sinuous ridges in the southern Argyre Planitia, Mars using HiRISE and CTX images and MOLA data. J. Geophys. Res. Planets 114 (E9). Available from: http://dx.doi.org/10.1029/2008JE003244.

Banks, M.E., Byrne, S., Galla, K., McEwen, A.S., Bray, V.J., Dundas, C.M., et al., 2010. Crater population and resurfacing of the Martian north polar layered deposits. J. Geophys. Res. Planets 115 (E8). Available from: http://dx.doi.org/10.1029/2009JE003523.

Baratoux, D., Delacourt, C., Allemand, P., 2002. An instability mechanism in the formation of the Martian lobate craters and the implications for the rheology of the ejecta. Geophys. Res. Lett. 29. Available from: http://dx.doi.org/10.1029/2001GL013779.

Barlow, N.G., 1994. Sinuosity of Martian rampart ejecta deposits. J. Geophys. Res. 99, 10927–10935.

Barlow, N.G., 2004. Martian subsurface volatiles concentrations as a function of time: clues from layered ejecta craters. Geophys. Res. Lett. 31. Available from: http://dx.doi.org/10.1029/2003GRL19025.

Barlow, N.G., 2005. A review of Martian impact crater ejecta structures and their implications for target properties. Geol. Soc. Am. Spec. Papers 384, 433–442.

Barlow, N.G., Boyce, J.M., Costard, F.M., Craddock, R.A., Garvin, J.B., Sakimoto, S.E., et al., 2000. Standardizing the nomenclature of Martian impact crater ejecta morphologies. J. Geophys. Res. 105 (E11), 26–733.

Barnouin-Jha, O.S., 1998. Lobateness of impact ejecta deposits from atmospheric interactions. J. Geophys. Res. 103 (E11), 25,739–25, 756.

Barnouin-Jha, O.S., Schultz, P.H., 1996. Ejecta entertainment by impact-generated ring vortices: theory and experiments. J. Geophys. Res. 101 (E9), 21,099–21,115.

Barnouin-Jha, O.S., Baloga, S., Glaze, L., 2005. Comparing landslides to fluidized crater ejecta on Mars. J. Geophys. Res. 110 (E04010). Available from: http://dx.doi.org/10.1029/2003JE002214.

Batalha, N.E., Kopparapu, R.K., Haqq-Misra, J., Kasting, J.F., 2016. Climate cycling on early Mars caused by the carbonate–silicate cycle. Earth Planet. Sci. Lett. 455, 7–13.

Berman, D.C., Crown, D.A., Joseph, E.C., 2015. Formation and mantling ages of lobate debris aprons on Mars: insights from categorized crater counts. Planet. Space Sci. 111, 83–99.

Bernhardt, H., Hiesinger, H., Reiss, D., Ivanov, M., Erkeling, G., 2013. Putative eskers and new insights into glacio-fluvial depositional settings in southern Argyre Planitia, Mars. Planet. Space Sci. 85, 261–278.

Bibring, J.P., Langevin, Y., Gendrin, A., Gondet, B., Poulet, F., Berthé, M., et al., 2005. Mars surface diversity as revealed by the OMEGA/Mars Express observations. Science 307 (5715), 1576–1581.

Bibring, J.P., Langevin, Y., Mustard, J.F., Poulet, F., Arvidson, R., Gendrin, A., et al., 2006. Global mineralogical and aqueous Mars history derived from OMEGA/Mars Express data. Science 312 (5772), 400–404.

Biemann, K., Owen, T., Rushneck, D.R., LaFleur, A.L., Howarth, D.W., 1976. The atmosphere of Mars near the surface: isotope ratios and upper limits on noble gases. Science 194 (4260), 76–78.

Bishop, J.L., et al., 2013. What the ancient phyllosilicates at Mawrth Vallis can tell us about possible habitability on early Mars. Planet. Space Sci. 86, 130–149. Available from: http://dx.doi.org/10.1016/j.pss.2013.05.006.

Black, B.A., Stewart, S.T., 2008. Excess ejecta craters record episodic ice-rich layers at middle latitudes on Mars. J. Geophys. Res. Planets 113 (E2). Available from: http://dx.doi.org/10.1029/2007JE002888.

Bogard, D.D., 1997. A reappraisal of the Martian 36Ar/38Ar ratio. J. Geophys. Res. Planets (1991–2012) 102 (E1), 1653–1661.

Boisson, J., Heggy, E., Clifford, S.M., Frigeri, A., Plaut, J.J., Farrell, W.M., et al., 2009. Sounding the subsurface of Athabasca Valles using MARSIS radar data: exploring the volcanic and fluvial hypotheses for the origin of the rafted plate terrain. J. Geophys. Res. Planets 114 (E8). Available from: http://dx.doi.org/10.1029/2008JE003299.

Bougher, S.W., Cravens, T.E., Grebowsky, J., Luhmann, J., 2015. The aeronomy of Mars: characterization by MAVEN of the upper atmosphere reservoir that regulates volatile escape. Space Sci. Rev. 195 (1-4), 423−456.

Bouley, S., Craddock, R.A., Mangold, N., Ansan, V., 2010. Characterization of fluvial activity in Parana Valles using different age-dating techniques. Icarus 207 (2), 686−698.

Boyce, J.M., Mouginis-Mark, P.J., 2006. Martian craters viewed by the Thermal Emission Imaging System instrument: double-layered ejecta craters. J. Geophys. Res. Planets 111 (E10). Available from: http://dx.doi.org/10.1029/2005JE002638.

Boyce, J., Barlow, N., Mouginis-Mark, P., Stewart, S., 2010. Rampart craters on Ganymede: their implications for fluidized ejecta emplacement. Meteorit. Planet. Sci. 45, 638−661.

Boynton, W.V., Trombka, J.I., Feldman, W.C., Arnold, J.R., Englert, P.A.J., Metzger, A.E., et al., 1992. Science applications of the Mars Observer gamma ray spectrometer. J. Geophys. Res. Planets 97 (E5), 7681−7698.

Boynton, W.V., Feldman, W.C., Squyres, S.W., Prettyman, T.H., Brückner, J., Evans, L.G., et al., 2002. Distribution of hydrogen in the near surface of Mars: evidence for subsurface ice deposits. Science 297 (5578), 81−85.

Bramson, A.M., Byrne, S., Putzig, N.E., Sutton, S., Plaut, J.J., Brothers, T.C., et al., 2015. Widespread excess ice in Arcadia Planitia, Mars. Geophys. Res. Lett. 42 (16), 6566−6574.

Brass, G.W., 1980. Stability of brines on Mars. Icarus 42, 20−28, doi:10.1016/0019-1035(80)90237-7.

Brasser, R., 2013. The formation of Mars: building blocks and accretion time scale. Space Sci. Rev. 174 (1−4), 11−25.

Bristow, T.F., Haberle, R.M., Blake, D.F., Des Marais, D.J., Eigenbrode, J.L., Fairén, A.G., et al., 2017. Low Hesperian PCO2 constrained from in situ mineralogical analysis at Gale Crater, Mars. Proc. Natl Acad. Sci. 2166−2170.

Brough, S., Hubbard, B., Hubbard, A., 2016. Former extent of glacier-like forms on Mars. Icarus 274, 37−49.

Buhler, P.B., Fassett, C.I., Head, J.W., Lamb, M.P., 2013. Evidence for paleolakes in Erythraea Fossa, Mars: implications for an ancient hydrological cycle. Icarus 213, 104−115.

Burr, D.M., Grier, J.A., McEwen, A.S., Keszthelyi, L.P., 2002a. Repeated aqueous flooding from the Cerberus Fossae: evidence for very recently extant, deep groundwater on Mars. Icarus 159 (1), 53−73.

Burr, D.M., McEwen, A.S., Sakimoto, H., Susan, E., 2002b. Recent aqueous floods from the Cerberus Fossae, Mars. Geophys. Res. Lett. 29 (1). Available from: http://dx.doi.org/10.1029/2001GL013345.

Burr, D.M., Tanaka, K.L., Yoshikawa, K., 2009. Pingos on Earth and Mars. Planet. Space Sci. 57 (5), 541−555.

Burr, D.M., Williams, R.M., Wendell, K.D., Chojnacki, M., Emery, J.P., 2010. Inverted fluvial features in the Aeolis/Zephyria Plana region, Mars: formation mechanism and initial paleodischarge estimates. J. Geophys. Res. Planets 115 (E7). Available from: http://dx.doi.org/10.1029/2009JE003496.

Burt, D.M., Knauth, L.P., 2003. Electrically conducting, Ca-rich brines, rather than water, expected in the Martian subsurface. J. Geophys. Res. 108 (E4), 8026. Available from: http://dx.doi.org/10.1029/2002JE001862.

Butcher, F.E.G., Conway, S.J., Arnold, N.S., 2016. Are the Dorsa Argentea on Mars eskers? Icarus 275, 65−84.

Butcher F.E.G., Gallagher, C., Arnold, N.S., Balme, M.R., Conway, S.J., Lewis, S.R., et al., 2017a. Morphometric characterisation of eskers associated with an extant mid-latitude glacier on Mars. In: 48th Lunar and Planetary Science Conference (vol. 48), 2024 March 2017, The Woodlands, TX, Abstract # 1238.

Butcher F.E.G., Gallagher, C., Arnold, N.S., Balme, M.R., Conway, S.J., Hagermann, A., Lewis, S.R., (2017b). Recent basal melting of a mid-latitude glacier on Mars. Journal of Geophysical Research: Planets, 112. https://doi.org/10.1002/2017JE005434.

Byrne, S., 2009b. The polar deposits of Mars. Annu. Rev. Earth Planet. Sci. 37, 535−560.

Byrne, S., Murray, B.C., 2002. North polar stratigraphy and the paleo-erg of Mars. J. Geophys. Res. Planets 107 (E6). Available from: http://dx.doi.org/10.1029/2001JE001615.

Byrne, S., Dundas, C.M., Kennedy, M.R., Mellon, M.T., McEwen, A.S., Cull, S.C., et al., 2009a. Distribution of mid-latitude ground ice on Mars from new impact craters. Science 325 (5948), 1674−1676.

Cabrol, N.A., Grin, E.A., 1999. Distribution, classification, and ages of Martian impact crater lakes. Icarus 142 (1), 160−172.

Cabrol, N.A., Grin, E.A., 2001. The evolution of lacustrine environments on Mars: is Mars only hydrologically dormant? Icarus 149 (2), 291−328.

Cabrol, N.A., Grin, E.A., 2003. Overview on the formation of paleolakes and ponds on Mars. Glob. Planet. Change 35 (3), 199−219.

Cabrol, N.A., Grin, E.A. (Eds.), 2010. Lakes on Mars. Elsevier, Amsterdam.

Cabrol, N.A., Grin, E.A., Pollard, W.H., 2000. Possible frost mounds in an ancient Martian lake bed. Icarus 145 (1), 91−107.

Campbell, M.J., Ulrichs, J., 1969. Electrical properties of rocks and their significance for lunar radar observations. J. Geophys. Res. 74 (25), 5867−5881.

Carr, M.H., 1986. Mars: a water-rich planet? Icarus 68 (2), 187−216.

Carr, M.H., 1996. Water on Mars. Oxford University Press, New York.

Carr, M.H., 2001. Mars Global Surveyor observations of Martian fretted terrain. J. Geophys. Res. Planets 106 (E10), 23571−23593.

Carr, M.H., 2007. The Surface of Mars, vol. 6. Cambridge University Press, Cambridge.

Carr, M.H., Chuang, F.C., 1997. Martian drainage densities. J. Geophys. Res. Planets 102 (E4), 9145−9152.

Carr, M.H., Clow, G.D., 1981. Martian channels and valleys: their characteristics, distribution, and age. Icarus 48 (1), 91−117.

Carr, M.H., Head, J.W., 2003. Oceans on Mars: an assessment of the observational evidence and possible fate. J. Geophys. Res. Planets (1991−2012) 108 (E5). Available from: http://dx.doi.org/10.1029/2002JE001963.

Carr, M.H., Head, J.W., 2010. 2 Acquisition and history of water on Mars. In: Cabrol, Grin (Eds.), Lakes on Mars. Elsevier, Amsterdam, pp. 31−67.

Carr, M.H., Head, J.W., 2015. Martian surface/near-surface water inventory: sources, sinks, and changes with time. Geophys. Res. Lett. 42 (3), 726−732.

Carr, M.H., Schaber, G.G., 1977. Martian permafrost features. J. Geophys. Res. 82 (28), 4039−4054.

Carr, M.H., Crumpler, L., Cutts, J., Greeley, R., Guest, J., Masursky, H., 1977. Martian impact craters and emplacement of ejecta by surface flow. J. Geophys. Res. 82, 4055−4065.

Carter, J., Poulet, F., Bibring, J.P., Murchie, S., 2010. Detection of hydrated silicates in crustal outcrops in the northern plains of Mars. Science 328 (5986), 1682−1686.

Carter, J., Loizeau, D., Mangold, N., Poulet, F., Bibring, J.P., 2015. Widespread surface weathering on early Mars: a case for a warmer and wetter climate. Icarus 248, 373−382.

Casanelli, J.P., Head, J.W., Fastook, J.L., 2015. Sources of water for the outflow channels on Mars: implications of the Late Noachian "icy highlands" model for melting and groundwater recharge on the Tharsis rise. Planet. Space Sci. 108, 54−65.

Cataldo, V., Williams, D.A., Dundas, C.M., Keszthelyi, L.P., 2015. Limited role for thermal erosion by turbulent lava in proximal Athabasca Valles, Mars. J. Geophys. Res. Planets 120 (11), 1800−1819.

Cedillo-Flores, Y., Treiman, A.H., Lasue, J., Clifford, S.M., 2011. CO_2 gas fluidization in the initiation and formation of Martian polar gullies. Geophys. Res. Lett. 38. Available from: http://dx.doi.org/10.1029/2011GL049403.

Chapman, M.G., Neukum, G., Dumke, A., Michael, G., Van Gasselt, S., Kneissl, T., et al., 2010a. Noachian−Hesperian geologic history of the Echus Chasma and Kasei Valles system on Mars: new data and interpretations. Earth Planet. Sci. Lett. 294 (3), 256−271.

Chapman, M.G., Neukum, G., Dumke, A., Michael, G., Van Gasselt, S., Kneissl, T., et al., 2010b. Amazonian geologic history of the Echus Chasma and Kasei Valles system on Mars: new data and interpretations. Earth Planet. Sci. Lett. 294 (3), 238−255.

Chassefière, E., Leblanc, F., 2011. Constraining methane release due to serpentinization by the observed D/H ratio on Mars. Earth Planet. Sci. Lett. 310 (3), 262−271.

Chassefière, E., Langlais, B., Quesnel, Y., Leblanc, F., 2013. The fate of early Mars' lost water: the role of serpentinization. J. Geophys. Res. Planets 118 (5), 1123−1134.

Chastain, B.K., Chevrier, V., 2007. Methane clathrate hydrates as a potential source for Martian atmospheric methane. Planet. Space Sci. 55 (10), 1246−1256.

Chevrier, V.F., Rivera-Valentin, E.G., 2012. Formation of recurring slope lineae by liquid brines on present-day Mars. Geophys. Res. Lett. 39 (21). Available from: http://dx.doi.org/10.1029/2012GL054119.

Chevrier, V.F., Poulet, F., Bibring, J.P., 2007. Early geochemical environment of Mars as determined from thermodynamics of phyllosilicates. Nature 448 (7149), 60−63.

Chojnacki, M., McEwen, A., Dundas, C., Ojha, L., Urso, A., Sutton, S., 2016. Geologic context of recurring slope lineae in Melas and Coprates Chasmata, Mars. J. Geophys. Res. Planets 121 (7), 1204−1231.

Christensen, P.R., 2003. Formation of recent Martian gullies through melting of extensive water-rich snow deposits. Nature 422 (6927), 45−48.

Clark, B.C., Van Hart, D.C., 1981. The salts of Mars. Icarus 45, 370−378, doi:10.1016/0019-1035(81)90041-5.

Clauser, C., Huenges, E., 1995. Thermal conductivity of rocks and minerals. Rock Physics & Phase Relations: A Handbook of Physical Constants. AGU, Washington, DC, pp. 105−126.

Clifford, S.M., 1991. The role of thermal vapor diffusion in the subsurface hydrologic evolution of Mars. Geophys. Res. Lett. 18, 2055−2058. Available from: http://dx.doi.org/10.1029/91GL02469.

Clifford, S.M., 1993. A model for the hydrologic and climatic behavior of water on Mars. J. Geophys. Res. Planets 98 (E6), 10973−11016.

Clifford, S.M., Hillel, D., 1983. The stability of ground ice in the equatorial region of Mars. J. Geophys. Res. 88, 2456−2474. Available from: http://dx.doi.org/10.1029/JB088iB03p02456.

Clifford, S.M., Parker, T.J., 2001. The evolution of the Martian hydrosphere: implications for the fate of a primordial ocean and the current state of the northern plains. Icarus 154 (1), 40−79.

Clifford, S.M., Lasue, J., Heggy, E., Boisson, J., McGovern, P., Max, M.D., 2010. Depth of the Martian cryosphere: revised estimates and implications for the existence and detection of subpermafrost groundwater. J. Geophys. Res. Planets 115 (E7). Available from: http://dx.doi.org/10.1029/2009JE003462.

Committee on Assessing the Solar System Exploration Program, Grading NASA's Solar System Exploration Program: a Midterm Review, (National Academies Press, Washington, 2008). Available from:http://www.nap.edu/catalog.php?record_id=12070 (accessed 19.07.11).

Committee on an Astrobiology Strategy for the Exploration of Mars, An Astrobiology Strategy for the Exploration of Mars (National Academies Press, Washington, 2007). Available from: http://www.nap.edu/catalog.php?record_id = 11937 (accessed 10.09.17).

Committee on the Planetary Science Decadal Survey, Vision and Voyages for Planetary Science in the Decade 2013−2022 (National Academies Press, Washington, 2011). Available from: http://www.nap.edu/catalog.php?record_id = 13117 (accessed 10.09.17).

Conway, S.J., Balme, M.R., 2014. Decameter thick remnant glacial ice deposits on Mars. Geophys. Res. Lett. 41 (15), 5402−5409.

Conway, S.J., Harrison, T.N., Soare, R.J., Britton, A., Steele, L., 2017. A new analysis of global data on orientation and slope of gullies on Mars. Geological Society of London Special Publications (under revision).

Costard, F.M., 1989. The spatial distribution of volatiles in the Martian hydrolithosphere. Earth Moon Planets 45 (3), 265−290.

Costard, F.M., Forget, F., Mangold, N., Peulvast, J.P., 2002. Formation of recent Martian debris flows by melting of near-surface ground ice at high obliquity. Science 295 (5552), 110−113.

Costard, F.M., Sejourne, A., Kargel, J., Godin, E., 2016. Modeling and observational occurrences of near-surface drainage in Utopia Planitia, Mars. Geomorphology 275, 80−89.

Costard, F., Séjourné, A., Kelfoun, K., Clifford, S., Lavigne, F., Di Pietro, I., et al., 2017. Modeling tsunami propagation and the emplacement of thumbprint terrain in an early Mars ocean. J. Geophys. Res.: Planets 122 (3), 633−649.

Craddock, R.A., Howard, A.D., 2002. The case for rainfall on a warm, wet early Mars. J. Geophys. Res. Planets 107 (E11). Available from: http://dx.doi.org/10.1029/2001JE001505.

Craddock, R.A., Maxwell, T.A., 1993. Geomorphic evolution of the Martian highlands through ancient fluvial processes. J. Geophys. Res. Planets 98 (E2), 3453−3468.

Crown, D.A., Pierce, T.L., McElfresh, S.B.Z., Mest, S.C., 2002. Debris aprons in the eastern Hellas region of Mars: implications for styles and rates of highland degradation. In: Lunar and Planetary Science Conference, vol. 33. Houston, TX, USA.

Crown, D.A., McElfresh, S.B.Z., Pierce, T.L., Mest, S.C., 2003. Geomorphology of debris aprons in the Eastern Hellas region of Mars. In: Lunar and Planetary Science Conference, vol. 34. Houston, TX, USA, p. 1126.

Cull, S., Arvidson, R.E., Mellon, M.T., Skemer, P., Shaw, A., Morris, R.V., 2010. Compositions of subsurface ices at the Mars Phoenix landing site. Geophys. Res. Lett. 37 (24). Available from: http://dx.doi.org/10.1029/2010GL045372.

Cutts, J.A., 1973. Nature and origin of layered deposits of the Martian polar regions. J. Geophys. Res. 78 (20), 4231−4249.

Davidson, D., 1983. Gas hydrates as clathrate ices. In: Cox, J.L. (Ed.), Natural Gas Hydrates-Properties, Occurrence and Recovery. Butterworth, Woburn, MA, pp. 1−16.

Davis, J.M., Balme, M., Grindrod, P.M., Williams, R.M.E., Gupta, S., 2016. Extensive Noachian fluvial systems in Arabia Terra: implications for early Martian climate. Geology 44 (10), 847−850.

De Haas, T., Hauber, E., Conway, S.J., Van Steijn, H., Johnsson, A., Kleinhans, M.G., 2015. Earth-like aqueous debris-flow activity on Mars at high orbital obliquity in the last million years. Nat. Commun. 6. Available from: http://dx.doi.org/10.1038/ncomms8543.

De Hon, R.A., 1992. Martian lake basins and lacustrine plains. Earth Moon Planets 56 (2), 95−122.

Dehouck, E., Mangold, N., Le Mouélic, S., Ansan, V., Poulet, F., 2010. Ismenius Cavus, Mars: a deep paleolake with phyllosilicate deposits. Planet. Space Sci. 58 (6), 941−946.

Di Achille, G., Hynek, B.M., 2010a. 8 Deltas and valley networks on Mars: implications for a global hydrosphere. In: Cabrol, Grin (Eds.), Lakes on Mars. Elsevier, Amsterdam, pp. 238−248.

Di Achille, G., Hynek, B.M., 2010b. Ancient ocean on Mars supported by global distribution of deltas and valleys. Nat. Geosci. 3 (7), 459−463.

DiBiase, R.A., Limaye, A.B., Scheingross, J.S., Fischer, W.W., Lamb, M.P., 2013. Deltaic deposits at Aeolis Dorsa: sedimentary evidence for a standing body of water on the northern plains of Mars. J. Geophys. Res. Planets 118 (6), 1285−1302.

Dickson, J.L., Head, J.W., Kreslavsky, M., 2007. Martian gullies in the southern mid-latitudes of Mars: evidence for climate-controlled formation of young fluvial features based upon local and global topography. Icarus 188 (2), 315−323.

Dickson, J.L., Head, J.W., Marchant, D.R., 2008. Late Amazonian glaciation at the dichotomy boundary on Mars: evidence for glacial thickness maxima and multiple glacial phases. Geology 36 (5), 411−414.

Dickson, J.L., Fassett, C.I., Head, J.W., 2009. Amazonian-aged fluvial valley systems in a climatic microenvironment on Mars: melting of ice deposits on the interior of Lyot Crater. Geophys. Res. Lett. 36. Available from: http://dx.doi.org/10.1029/2009GL037472.

Dickson, J.L., Head, J.W., Marchant, D.R., 2010. Kilometer-thick ice accumulation and glaciation in the northern mid-latitudes of Mars: evidence for crater-filling events in the Late Amazonian at the Phlegra Montes. Earth Planet. Sci. Lett. 294 (3), 332–342.

Dickson, J.L., Head, J.W., Fassett, C.I., 2012. Patterns of accumulation and flow of ice in the mid-latitudes of Mars during the Amazonian. Icarus 219 (2), 723–732.

Diniega, S., Byrne, S., Bridges, N.T., Dundas, C.M., McEwen, A.S., 2010. Seasonality of present-day Martian dune-gully activity. Geology 38 (11), 1047–1050.

Diniega, S., Hansen, C.J., McElwaine, J.N., Hugenholtz, C.H., Dundas, C.M., McEwen, A.S., et al., 2013. A new dry hypothesis for the formation of Martian linear gullies. Icarus 225 (1), 526–537.

Drake, M.J., 2005. Origin of water in the terrestrial planets. Meteorit. Planet. Sci. 40 (4), 519–527.

Dundas, C.M., 2017. Effects of varying obliquity on Martian sublimation thermokarst landforms. Icarus 281, 115–120.

Dundas, C.M., Byrne, S., 2010. Modeling sublimation of ice exposed by new impacts in the Martian mid-latitudes. Icarus 206 (2), 716–728.

Dundas, C.M., Keszthelyi, L.P., 2014. Emplacement and erosive effects of lava in south Kasei Valles, Mars. J. Volcanol. Geotherm. Res. 282, 92–102.

Dundas, C.M., McEwen, A.S., Diniega, S., Byrne, S., Martinez-Alonso, S., 2010. New and recent gully activity on Mars as seen by HiRISE. Geophys. Res. Lett. 37 (7). Available from: http://dx.doi.org/10.1029/2009GL041351.

Dundas, C.M., Diniega, S., Hansen, C.J., Byrne, S., McEwen, A.S., 2012. Seasonal activity and morphological changes in Martian gullies. Icarus 220 (1), 124–143.

Dundas, C.M., Byrne, S., McEwen, A.S., Mellon, M.T., Kennedy, M.R., Daubar, I.J., et al., 2014. HiRISE observations of new impact craters exposing Martian ground ice. J. Geophys. Res. Planets 119 (1), 109–127.

Dundas, C.M., Diniega, S., McEwen, A.S., 2015a. Long-term monitoring of Martian gully formation and evolution with MRO/HiRISE. Icarus 251, 244–263.

Dundas, C.M., Byrne, S., McEwen, A.S., 2015b. Modeling the development of Martian sublimation thermokarst landforms. Icarus 262, 154–169.

Dundas, C., McEwen, A., Diniega, S., Hansen, C., Byrne, S., 2017. The formation of gullies on Mars today. Geol. Soc. Lond. (Special Issue on "Martian Gullies and their Earth Analogues"). Geological Society, London, Special Publications, 467, https://doi.org/10.1144/SP467.5.

Edwards, C.S., Piqueux, S., 2016. The water content of recurring slope lineae on Mars. Geophys. Res. Lett. 43 (17), 8912–8919.

Ehlmann, B.L., Edwards, C.S., 2014. Mineralogy of the Martian surface. Annu. Rev. Earth Planet. Sci. 42, 291–315.

Ehlmann, B.L., Mustard, J.F., Murchie, S.L., Poulet, F., Bishop, J.L., Brown, A.J., et al., 2008. Orbital identification of carbonate-bearing rocks on Mars. Science 322 (5909), 1828–1832.

Ehlmann, B.L., Mustard, J.F., Murchie, S.L., 2010. Geologic setting of serpentine deposits on Mars. Geophys. Res. Lett. 37 (6). Available from: http://dx.doi.org/10.1029/2010GL042596.

Ehlmann, B.L., Mustard, J.F., Murchie, S.L., Bibring, J.P., Meunier, A., Fraeman, A.A., et al., 2011. Subsurface water and clay mineral formation during the early history of Mars. Nature 479 (7371), 53–60.

Erkaev, N.V., Lammer, H., Elkins-Tanton, L.T., Stökl, A., Odert, P., Marcq, E., et al., 2014. Escape of the Martian protoatmosphere and initial water inventory. Planet. Space Sci. 98, 106–119.

Etiope, G., Ehlmann, B.L., Schoell, M., 2013. Low temperature production and exhalation of methane from serpentinized rocks on Earth: a potential analog for methane production on Mars. Icarus 224 (2), 276–285.

Fairén, A.G., Dohm, J.M., Baker, V.R., de Pablo, M.A., Ruiz, J., Ferris, J.C., et al., 2003. Episodic flood inundations of the northern plains of Mars. Icarus 165 (1), 53–67.

Fairén, A.G., Davila, A.F., Gago-Duport, L., Haqq-Misra, J.D., Gil, C., McKay, C.P., et al., 2011. Cold glacial oceans would have inhibited phyllosilicate sedimentation on early Mars. Nat. Geosci. 4 (10), 667—670.

Fanale, F.P., 1976. Martian volatiles: their degassing history and geochemical fate. Icarus 28 (2), 179—202.

Fanale, F.P., Salvail, J.R., Zent, A.P., Postawko, S.E., 1986. Global distribution and migration of subsurface ice on Mars. Icarus 67, 1—18, doi:10.1016/0019-1035(86)90170-3.

Fassett, C.I., Head, J.W., 2005. Fluvial sedimentary deposits on Mars: ancient deltas in a crater lake in the Nili Fossae region. Geophys. Res. Lett. 32 (14). Available from: http://dx.doi.org/10.1029/2005GL023456.

Fassett, C.I., Head, J.W., 2008a. The timing of Martian valley network activity: constraints from buffered crater counting. Icarus 195 (1), 61—89.

Fassett, C.I., Head, J.W., 2008b. Valley network-fed, open-basin lakes on Mars: distribution and implications for Noachian surface and subsurface hydrology. Icarus 198 (1), 37—56.

Fassett, C.I., Dickson, J.L., Head, J.W., Levy, J.S., Marchant, D.R., 2010. Supraglacial and proglacial valleys on Amazonian Mars. Icarus 208, 86—100. Available from: http://dx.doi.org/10.1016/j.icarus.2010.02.021.

Fastook, J.L., Head, J.W., 2015. Glaciation in the Late Noachian Icy Highlands: ice accumulation, distribution, flow rates, basal melting, and top-down melting rates and patterns. Planet. Space Sci. 106, 82—98.

Fastook, J.L., Head, J.W., Marchant, D.R., Forget, F., 2008. Tropical mountain glaciers on Mars: altitude-dependence of ice accumulation, accumulation conditions, formation times, glacier dynamics, and implications for planetary spin-axis/orbital history. Icarus 198 (2), 305—317.

Fastook, J.L., Head, J.W., Marchant, D.R., Forget, F., Madeleine, J.B., 2012. Early Mars climate near the Noachian—Hesperian boundary: independent evidence for cold conditions from basal melting of the south polar ice sheet (Dorsa Argentea Formation) and implications for valley network formation. Icarus 219 (1), 25—40.

Fastook, J.L., Head, J.W., Marchant, D.R., 2014. Formation of lobate debris aprons on Mars: assessment of regional ice sheet collapse and debris-cover armoring. Icarus 228, 54—63.

Feldman, W.C., Prettyman, T.H., Maurice, S., Plaut, J.J., Bish, D.L., Vaniman, D.T., et al., 2004. Global distribution of near-surface hydrogen on Mars. J. Geophys. Res. Planets 109 (E9). Available from: http://dx.doi.org/10.1029/2003JE002160.

Feldman, W.C., Mellon, M.T., Gasnault, O., Maurice, S., Prettyman, T.H., 2008. Volatiles on Mars: scientific results from the Mars Odyssey neutron spectrometer, The Martian Surface-Composition, Mineralogy, and Physical Properties, 1. Cambridge University Press, Cambridge, pp. 125—148.

Feldman, W.C., Pathare, A., Maurice, S., Prettyman, T.H., Lawrence, D.J., Milliken, R.E., et al., 2011. Mars Odyssey neutron data: 2. Search for buried excess water ice deposits at nonpolar latitudes on Mars. J. Geophys. Res. Planets 116 (E11). Available from: http://dx.doi.org/10.1029/2011JE003806.

Fialips, C.I., Carey, J.W., Vaniman, D.T., Bish, D.L., Feldman, W.C., Mellon, M.T., 2005. Hydration state of zeolites, clays, and hydrated salts under present-day Martian surface conditions: can hydrous minerals account for Mars Odyssey observations of near-equatorial water-equivalent hydrogen? Icarus 178 (1), 74—83.

Fishbaugh, K.E., Head, J.W., 2005. Origin and characteristics of the Mars north polar basal unit and implications for polar geologic history. Icarus 174 (2), 444—474.

Fisher, D.A., 2005. A process to make massive ice in the Martian regolith using long-term diffusion and thermal cracking. Icarus 179 (2), 387—397.

Fisk, M.R., Giovannoni, S.J., 1999. Sources of nutrients and energy for a deep biosphere on Mars. J. Geophys. Res. Planets 104 (E5), 11805—11815.

Forget, F., Haberle, R.M., Montmessin, F., Levrard, B., Head, J.W., 2006. Formation of glaciers on Mars by atmospheric precipitation at high obliquity. Science 311 (5759), 368—371.

Formisano, V., Atreya, S., Encrenaz, T., Ignatiev, N., Giuranna, M., 2004. Detection of methane in the atmosphere of Mars. Science 306 (5702), 1758—1761.

French, H.M., 2013. The Periglacial Environment. John Wiley & Sons, Chichester.

Fries, M., Christou, A., Archer, D., Conrad, P., Cooke, W., Eigenbrode, J., et al., 2016. A cometary origin for Martian atmospheric methane. Geochem. Perspect. Lett. 2, 10−23.

Gallagher, C.J., Balme, M., 2015. Eskers in a complete, wet-based glacial system in the Phlegra Montes region, Mars. Earth Planet. Sci. Lett. 431, 96−109.

Gallagher, C.J., Balme, M.R., 2011. Landforms indicative of ground-ice thaw in the northern high latitudes of Mars. Geol. Soc. Lond. Spec. Publ. 356 (1), 87−110.

Gallagher, C.J., Balme, M.R., Conway, S.J., Grindrod, P.M., 2011. Sorted clastic stripes, lobes and associated gullies in high-latitude craters on Mars: landforms indicative of very recent, polycyclic ground-ice thaw and liquid flows. Icarus 211 (1), 458−471.

Ghatan, G.J., Head, J.W., 2004. Regional drainage of meltwater beneath a Hesperian-aged south circumpolar ice sheet on Mars. J. Geophys. Res. Planets 109 (E7). Available from: http://dx.doi.org/10.1029/2003JE002196.

Ghatan, G.J., Zimbelman, J.R., 2006. Paucity of candidate coastal constructional landforms along proposed shorelines on Mars: implications for a northern lowlands-filling ocean. Icarus 185 (1), 171−196.

Goldspiel, J.M., Squyres, S.W., 2011. Groundwater discharge and gully formation on Martian slopes. Icarus 211 (1), 238−258.

Goudge, T.A., Head, J.W., Mustard, J.F., Fassett, C.I., 2012. An analysis of open-basin lake deposits on Mars: evidence for the nature of associated lacustrine deposits and post-lacustrine modification processes. Icarus 219 (1), 211−229.

Goudge, T.A., Aureli, K.L., Head, J.W., Fassett, C.I., Mustard, J.F., 2015. Classification and analysis of candidate impact crater-hosted closed-basin lakes on Mars. Icarus 260, 346−367.

Goudge, T.A., Fassett, C.I., Head, J.W., Mustard, J.F., Aureli, K.L., 2016. Insights into surface runoff on early Mars from paleolake basin morphology and stratigraphy. Geology 44 (6), 419−422.

Goudge, T.A., Milliken, R.E., Head, J.W., Mustard, J.F., Fassett, C.I., 2017. Sedimentological evidence for a deltaic origin of the western fan deposit in Jezero Crater, Mars and implications for future exploration. Earth Planet. Sci. Lett. 458, 357−365.

Grant, J.A., Wilson, S.A., Mangold, N., Calef III, F., Grotzinger, J.P., 2014. The timing of alluvial activity in Gale Crater, Mars. Geophys. Res. Lett. 41, 1142−1148. Available from: http://dx.doi.org/10.1002/2013GL058909.

Grima, C., Kofman, W., Mouginot, J., Phillips, R.J., Hérique, A., Biccari, D., et al., 2009. North polar deposits of Mars: extreme purity of the water ice. Geophys. Res. Lett. 36, 3. Available from: http://dx.doi.org/10.1029/2008GL036326.

Grimm, R.E., Heggy, E., Clifford, S., Dinwiddie, C., McGinnis, R., Farrell, D., 2006. Absorption and scattering in ground-penetrating radar: analysis of the Bishop Tuff. J. Geophys. Res. Planets 111 (E6). Available from: http://dx.doi.org/10.1029/2005JE002619.

Grimm, R.E., Berdanier, B., Warden, R., Harrer, J., Demara, R., Pfeiffer, J., et al., 2009. A time-domain electromagnetic sounder for detection and characterization of groundwater on Mars. Planet. Space Sci. 57 (11), 1268−1281.

Grimm, R.E., Harrison, K.P., Stillman, D.E., 2014. Water budgets of Martian recurring slope lineae. Icarus 233, 316−327.

Grimm, R.E., Harrison, K.P., Stillman, D.E., Kirchoff, M.R., 2016. On the secular retention of ground water and ice on Mars. J. Geophys. Res. Planets 122 (1), 94−109. Available from: http://dx.doi.org/10.1002/2016JE005132.

Grotzinger, J.P., Sumner, D.Y., Kah, L.C., Stack, K., Gupta, S., Edgar, L., et al., 2014. A habitable fluvio-lacustrine environment at Yellowknife Bay, Gale Crater, Mars. Science 343 (6169). Available from: http://dx.doi.org/10.1126/science.1242777.

Grotzinger, J.P., Gupta, S., Malin, M.C., Rubin, D.M., Schieber, J., Siebach, K., et al., 2015. Deposition, exhumation, and paleoclimate of an ancient lake deposit, Gale Crater, Mars. Science 350 (6257). Available from: http://dx.doi.org/10.1126/science.aac7575.

Gourronc, M., Bourgeois, O., Mège, D., Pochat, S., Bultel, B., Massé, M., et al., 2014. One million cubic kilometers of fossil ice in Valles Marineris: relics of a 3.5 Gy old glacial landsystem along the Martian equator. Geomorphology 204, 235−255.

Gulick, V.C., Baker, V.R., 1990. Origin and evolution of valleys on Martian volcanoes. J. Geophys. Res. 95, 14325−14344.

Haberle, R.M., McKay, C.P., Schaeffer, J., Cabrol, N.A., Grin, E.A., Zent, A.P., et al., 2001. On the possibility of liquid water on present-day Mars. J. Geophys. Res. Planets 106 (E10), 23317−23326.

Halevy, I., Head III, J.W., 2014. Episodic warming of early Mars by punctuated volcanism. Nat. Geosci. 7 (12), 865−868.

Hare, T.M., G. Cushing, J. Shinamen, B. Day, E. Law 2016. Context Camera (CTX) Image Mosaics for Mars Human Exploration Zones, Astrogeology, U.S. Geological Survey. Available from: http://bit.ly/CTX_EZs.

Harrison, T.N., Osinski, G.R., Tornabene, L.L., Jones, E., 2015. Global documentation of gullies with the Mars Reconnaissance Orbiter Context Camera and implications for their formation. Icarus 252, 236−254.

Hartmann, W.K., Neukum, G., 2001. Cratering chronology and the evolution of Mars. Space Sci. Rev. 96 (1-4), 165−194.

Hartmann, W.K., Ansan, V., Berman, D.C., Mangold, N., Forget, F., 2014. Comprehensive analysis of glaciated Martian crater Greg. Icarus 228, 96−120.

Hauber, E., van Gasselt, S., Ivanov, B., Werner, S., Head, J.W., Neukum, G., 2005. Discovery of a flank caldera and very young glacial activity at Hecates Tholus, Mars, HRSC Co-Investigator Team. Nature 434 (7031), 356−361.

Hauber, E., Gwinner, K., Kleinhans, M., Reiss, D., Di Achille, G., Ori, G.G., et al., 2009. Sedimentary deposits in Xanthe Terra: implications for the ancient climate on Mars. Planet. Space Sci. 57 (8), 944−957.

Hauber, E., Reiss, D., Ulrich, M., Preusker, F., Trauthan, F., Zanetti, M., et al., 2011. Landscape evolution in Martian mid-latitude regions: insights from analogous periglacial landforms in Svalbard. Geol. Soc. Lond. Spec. Publ. 356 (1), 111−131.

Head, J.W., Marchant, D.R., 2003. Cold-based mountain glaciers on Mars: western Arsia Mons. Geology 31 (7), 641−644.

Head, J.W., Pratt, S., 2001. Extensive Hesperian-aged south polar ice sheet on Mars: evidence for massive melting and retreat, and lateral flow and ponding of meltwater. J. Geophys. Res. Planets 106 (E6), 12275−12299.

Head, J.W., Hiesinger, H., Ivanov, M.A., Kreslavsky, M.A., Pratt, S., Thomson, B.J., 1999. Possible ancient oceans on Mars: evidence from Mars Orbiter Laser Altimeter data. Science 286 (5447), 2134−2137.

Head, J.W., Mustard, J.F., Kreslavsky, M.A., Milliken, R.E., Marchant, D.R., 2003. Recent ice ages on Mars. Nature 426 (6968), 797−802.

Head, J.W., Neukum, G., Jaumann, R., Hiesinger, H., Hauber, E., Carr, M., et al., 2005. Tropical to mid-latitude snow and ice accumulation, flow and glaciation on Mars. Nature 434 (7031), 346−351.

Head, J.W., Marchant, D.R., Agnew, M.C., Fassett, C.I., Kreslavsky, M.A., 2006. Extensive valley glacier deposits in the northern mid-latitudes of Mars: evidence for Late Amazonian obliquity-driven climate change. Earth Planet. Sci. Lett. 241 (3), 663−671.

Head, J.W., Marchant, D.R., Dickson, J.L., Kress, A.M., Baker, D.M., 2010. Northern mid-latitude glaciation in the Late Amazonian period of Mars: criteria for the recognition of debris-covered glacier and valley glacier landsystem deposits. Earth Planet. Sci. Lett. 294 (3), 306−320.

Hecht, M.H., et al., 2009. Detection of perchlorate and the soluble chemistry of Martian soil at the Phoenix Lander site. Science 325, 64−67. Available from: http://dx.doi.org/10.1126/science.1172466.

Heggy, E., Clifford, S.M., Grimm, R.E., Dinwiddie, C.L., Wyrick, D.Y., Hill, B.E., 2006. Ground-penetrating radar sounding in mafic lava flows: assessing attenuation and scattering losses in Mars-analog volcanic terrains. J. Geophys. Res. Planets 111 (E6). Available from: http://dx.doi.org/10.1029/2005JE002589.

Heldmann, J.L., Mellon, M.T., 2004. Observations of Martian gullies and constraints on potential formation mechanisms. Icarus 168 (2), 285−304.

Herkenhoff, K.E., Plaut, J.J., 2000. Surface ages and resurfacing rates of the polar layered deposits on Mars. Icarus 144 (2), 243−253.

Hobbs, P.V., 1974. Ice Physics. Clarendon, Oxford.

Hoke, M.R., Hynek, B.M., 2009. Roaming zones of precipitation on ancient Mars as recorded in valley networks. J. Geophys. Res. Planets 114 (E8). Available from: http://dx.doi.org/10.1029/2008JE003247.

Hoke, M.R., Hynek, B.M., Tucker, G.E., 2011. Formation timescales of large Martian valley networks. Earth Planet. Sci. Lett. 312 (1), 1−12.

Holt, J.W., Safaeinili, A., Plaut, J.J., Head, J.W., Phillips, R.J., Seu, R., et al., 2008. Radar sounding evidence for buried glaciers in the southern mid-latitudes of Mars. Science 322 (5905), 1235−1238.

Howard, A.D., 1981. Etched plains and braided ridges of the south polar region of Mars: features produced by basal melting of ground ice? Reports of Planetary Geology Program, pp. 286−288.

Howard, A.D., Moore, J.M., Irwin, R.P., 2005. An intense terminal epoch of widespread fluvial activity on early Mars: 1. Valley network incision and associated deposits. J. Geophys. Res. Planets 110 (E12). Available from: http://dx.doi.org/10.1029/2005JE002459.

Hubbard, B., Milliken, R.E., Kargel, J.S., Limaye, A., Souness, C., 2011. Geomorphological characterisation and interpretation of a mid-latitude glacier-like form: Hellas Planitia, Mars. Icarus 211 (1), 330−346.

Hubbard, B., Souness, C., Brough, S., 2014. Glacier-like forms on Mars. The Cryosphere 8 (6), 2047.

Hynek, B., 2016. RESEARCH FOCUS: the great climate paradox of ancient Mars. Geology 44 (10), 879−880.

Hynek, B.M., Beach, M., Hoke, M.R., 2010. Updated global map of Martian valley networks and implications for climate and hydrologic processes. J. Geophys. Res. Planets 115 (E9). Available from: http://dx.doi.org/10.1029/2009JE003548.

Ingersoll, A.P., 1970. Mars: occurrence of liquid water. Science 168 (3934), 972−973.

Irwin, R.P., Maxwell, T.A., Howard, A.D., Craddock, R.A., Leverington, D.W., 2002. A large paleolake basin at the head of Ma'adim Vallis, Mars. Science 296 (5576), 2209−2212.

Irwin, R.P., Howard, A.D., Craddock, R.A., Moore, J.M., 2005a. An intense terminal epoch of widespread fluvial activity on early Mars: 2. Increased runoff and paleolake development. J. Geophys. Res. Planets 110 (E12). Available from: http://dx.doi.org/10.1029/2005JE002460.

Irwin, R.P., Craddock, R.A., Howard, A.D., 2005b. Interior channels in Martian valley networks: discharge and runoff production. Geology 33 (6), 489−492.

Irwin, R.P., Lewis, K., Howard, A.H., Grant, J., et al., 2011. Timing, duration, and hydrology of the Eberswalde Crater paleolake, Mars. Geomorphology 240 (2015), 83−101.

Irwin, R.P., Lewis, K.W., Howard, A.D., Grant, J.A., 2015. Paleohydrology of Eberswalde Crater, Mars. Geomorphology 240, 83−101.

Jakosky, B.M., Mellon, M.T., Varnes, E.S., Feldman, W.C., Boynton, W.V., Haberle, R.M., 2005. Mars low-latitude neutron distribution: possible remnant near-surface water ice and a mechanism for its recent emplacement. Icarus 175 (1), 58−67.

Jakosky, B.M., Grebowsky, J.M., Luhmann, J.G., Connerney, J., Eparvier, F., Ergun, R., et al., 2015. MAVEN observations of the response of Mars to an interplanetary coronal mass ejection. Science 350 (6261). Available from: http://dx.doi.org/10.1126/science.aad0210.

Jakosky, B.M., Slipski, M., Benna, M., Mahaffy, P., Elrod, M., Yelle, R., et al., 2017. Mars' atmospheric history derived from upper-atmosphere measurements of 38Ar/36Ar. Science 355 (6332), 1408−1410.

Jaumann, R., Nass, A., Tirsch, D., Reiss, D., Neukum, G., 2010. The Western Libya Montes Valley System on Mars: evidence for episodic and multi-genetic erosion events during the Martian history. Earth Planet. Sci. Lett. 294 (3), 272–290.

Johnsson, A., Reiss, D., Hauber, E., Zanetti, M., Hiesinger, H., Johansson, L., et al., 2012. Periglacial mass-wasting landforms on Mars suggestive of transient liquid water in the recent past: insights from solifluction lobes on Svalbard. Icarus 218 (1), 489–505.

Johnson, J.E., Mienert, J., Plaza-Faverola, A., Vadakkepuliyambatta, S., Knies, J., Bünz, S., et al., 2015. Abiotic methane from ultraslow-spreading ridges can charge Arctic gas hydrates. Geology 43 (5), 371–374.

Jones, A.P., McEwen, A.S., Tornabene, L.L., Baker, V.R., Melosh, H.J., Berman, D.C., 2011. A geomorphic analysis of Hale Crater, Mars: the effects of impact into ice-rich crust. Icarus 211, 259–272.

Jones, E., Osinski, G.R., 2015. Using Martian single and double layered ejecta craters to probe subsurface stratigraphy. Icarus 247, 260–278.

Jordan, R., Picardi, G., Plaut, J., Wheeler, K., Kirchner, D., Safaeinili, A., et al., 2009. The Mars express MARSIS sounder instrument. Planet. Space Sci. 57 (14), 1975–1986.

Kadish, S.J., Barlow, N.G., Head, J.W., 2009. Latitude dependence of Martian pedestal craters: evidence for a sublimation-driven formation mechanism. J. Geophys. Res. Planets 114 (E10). Available from: http://dx.doi.org/10.1029/2008JE003318.

Kargel, J.S., Strom, R.G., 1992. Ancient glaciation on Mars. Geology 20 (1), 3–7.

Kerber, L., Head, J.W., Madeleine, J.B., Forget, F., Wilson, L., 2011. The dispersal of pyroclasts from Apollinaris Patera, Mars: implications for the origin of the Medusae Fossae Formation. Icarus 216 (1), 212–220.

Kessler, M.A., Werner, B.T., 2003. Self-organization of sorted patterned ground. Science 299 (5605), 380–383.

Kneissl, T., Reiss, D., Van Gasselt, S., Neukum, G., 2010. Distribution and orientation of northern-hemisphere gullies on Mars from the evaluation of HRSC and MOC-NA data. Earth Planet. Sci. Lett. 294 (3), 357–367.

Kochel, R.C., Peake, R.T., 1984. Quantification of waste morphology in Martian fretted terrain. J. Geophys. Res. Solid Earth 89 (S01). Available from: http://dx.doi.org/10.1029/JB089iS01p0C336.

Kolb, E.J., Tanaka, K.L., 2006. Accumulation and erosion of south polar layered deposits in the Promethei Lingula region, Planum Australe, Mars. Int. J. Mars Sci. Explor. 2, 1–9.

Komatsu, G., Ori, G.G., Di Lorenzo, S., Rossi, A.P., Neukum, G., 2007. Combinations of processes responsible for Martian impact crater "layered ejecta structures" emplacement. J. Geophys. Res. Planets 112 (E6). Available from: http://dx.doi.org/10.1029/2006JE002787.

Kostama, V.P., Kreslavsky, M.A., Head, J.W., 2006. Recent high-latitude icy mantle in the northern plains of Mars: characteristics and ages of emplacement. Geophys. Res. Lett. 33 (11). Available from: http://dx.doi.org/10.1029/2006GL025946.

Koutnik, M., Byrne, S., Murray, B., 2002. South polar layered deposits of Mars: the cratering record. J. Geophys. Res. Planets 107 (E11). Available from: http://dx.doi.org/10.1029/2001JE001805.

Kraal, E.R., Asphaug, E., Moore, J.M., Lorenz, R.D., 2006. Quantitative geomorphic modeling of Martian bedrock shorelines. J. Geophys. Res. Planets 111 (E3). Available from: http://dx.doi.org/10.1029/2005JE002567.

Kraal, E.R., Asphaug, E., Moore, J.M., Howard, A., Bredt, A., 2008a. Catalogue of large alluvial fans inMartian impact craters. Icarus 194, 101–110.

Kraal, E.R., Van Dijk, M., Postma, G., Kleinhans, M.G., 2008b. Martian stepped-delta formation by rapid water release. Nature 451 (7181), 973–976.

Krasnopolsky, V.A., Maillard, J.P., Owen, T.C., 2004. Detection of methane in the Martian atmosphere: evidence for life? Icarus 172 (2), 537–547.

Kreslavsky, M.A., Head, J.W., 1999. Kilometer-scale slopes on Mars and their correlation with geologic units: initial results from Mars Orbiter Laser Altimeter (MOLA) data. J. Geophys. Res. Planets 104 (E9), 21911–21924.

Kreslavsky, M.A., Head, J.W., 2000. Kilometer-scale roughness of Mars—results from MOLA data analysis. J. Geophys. Res. 105 (E11), 26695–26711.

Kreslavsky, M.A., Head, J.W., 2002. Fate of outflow channel effluents in the northern lowlands of Mars: the Vastitas Borealis Formation as a sublimation residue from frozen ponded bodies of water. J. Geophys. Res. Planets 107 (E12). Available from: http://dx.doi.org/10.1029/2001JE001831.

Kress, A.M., Head, J.W., 2015. Late Noachian and Early Hesperian ridge systems in the south circumpolar Dorsa Argentea Formation, Mars: evidence for two stages of melting of an extensive Late Noachian ice sheet. Planet. Space Sci. 109. pp. 1–20.

Krivchikov, A.I., Gorodilov, B.Y., Korolyuk, O.A., Manzhelii, V.G., Conrad, H., Press, W., 2005. The thermal conductivity of methane hydrate. J. Low Temp. Phys. 139, 693–702. Available from: http://dx.doi.org/10.1007/s10909-005-5481-z.

Kurokawa, H., Sato, M., Ushioda, M., Matsuyama, T., Moriwaki, R., Dohm, J.M., et al., 2014. Evolution of water reservoirs on Mars: constraints from hydrogen isotopes in Martian meteorites. Earth Planet. Sci. Lett. 394, 179–185.

Kuzmin, R.O., 1983. Cryolithosphere of Mars. Nauka, Moscow.

Lammer, H., Chassefière, E., Karatekin, Í., Morschhauser, A., Niles, P.B., Mousis, O., et al., 2013. Outgassing history and escape of the Martian atmosphere and water inventory. Space Sci. Rev. 174 (1-4), 113–154.

Landis, M.E., Byrne, S., Daubar, I.J., Herkenhoff, K.E., Dundas, C.M., 2016. A revised surface age for the North polar layered deposits of Mars. Geophys. Res. Lett. 43 (7), 3060–3068.

Langlais, B., Purucker, M., 2007. A polar magnetic paleopole associated with Apollinaris Patera, Mars. Planet. Space Sci. 55 (3), 270–279.

Langlais, B., Purucker, M.E., Mandea, M., 2004. Crustal magnetic field of Mars. J. Geophys. Res. Planets (1991–2012) 109 (E2). Available from: http://dx.doi.org/10.1029/2003JE002048.

Langlais, B., Thébault, E., Ostanciaux, E., Mangold, N., 2012. A late Martian dynamo cessation time 3.77 Gy ago. LPI Contrib. 1680, 7067.

Laskar, J., Robutel, P., 1993. The chaotic obliquity of the planets. Nature 361 (6413), 608–612.

Laskar, J., Correia, A.C.M., Gastineau, M., Joutel, F., Levrard, B., Robutel, P., 2004. Long term evolution and chaotic diffusion of the insolation quantities of Mars. Icarus 170, 343–364. Available from: http://dx.doi.org/10.1016/j.icarus.2004.04.005.

Lasue, J., Mangold, N., Hauber, E., Clifford, S., Feldman, W., Gasnault, O., et al., 2013. Quantitative assessments of the Martian hydrosphere. Space Sci. Rev. 174 (1-4), 155–212.

Lasue, J., Quesnel, Y., Langlais, B., Chassefière, E., 2015. Methane storage capacity of the early Martian cryosphere. Icarus 260, 205–214.

Lee, Y., Deming, D., 1998. Evaluation of thermal conductivity temperature corrections applied in terrestrial heat flow studies. J. Geophys. Res. 103, 2447–2454.

Lefevre, F., Forget, F., 2009. Observed variations of methane on Mars unexplained by known atmospheric chemistry and physics. Nature 460 (7256), 720–723.

Lefort, A., Russell, P.S., Thomas, N., McEwen, A.S., Dundas, C.M., Kirk, R.L., 2009. Observations of periglacial landforms in Utopia Planitia with the high resolution imaging science experiment (HiRISE). J. Geophys. Res. Planets 114 (E4). Available from: http://dx.doi.org/10.1029/2008JE003264.

Lefort, A., Russell, P.S., Thomas, N., 2010. Scalloped terrains in the Peneus and Amphitrites Paterae region of Mars as observed by HiRISE. Icarus 205 (1), 259–268.

Leighton, R.B., Murray, B.C., 1966. Behavior of carbon dioxide and other volatiles on Mars. Science 153 (3732), 136–144.

Leshin, L.A., 2000. Insights into Martian water reservoirs from analyses of Martian meteorite QUE94201. Geophys. Res. Lett 27 (14), 2017–2020.

Leverington, D.W., 2004. Volcanic rilles, streamlined islands, and the origin of outflow channels on Mars. J. Geophys. Res. Planets 109 (E10). Available from: http://dx.doi.org/10.1029/2004JE002311.

Leverington, D.W., 2009. Reconciling channel formation processes with the nature of elevated outflow systems at Ophir and Aurorae Plana, Mars. J. Geophys. Res. Planets 114 (E10). Available from: http://dx.doi.org/10.1029/2009JE003398.

Leverington, D.W., 2011. A volcanic origin for the outflow channels of Mars: key evidence and major implications. Geomorphology 132 (3), 51–75.

Levrard, B., Forget, F., Montmessin, F., Laskar, J., 2004. Recent ice-rich deposits formed at high latitudes on Mars by sublimation of unstable equatorial ice during low obliquity. Nature 431 (7012), 1072–1075.

Levrard, B., Forget, F., Montmessin, F., Laskar, J., 2007. Recent formation and evolution of northern Martian polar layered deposits as inferred from a Global Climate Model. J. Geophys. Res. Planets 112 (E6). Available from: http://dx.doi.org/10.1029/2006JE002772.

Levy, J.S., 2012. Hydrological characteristics of recurrent slope lineae on Mars: evidence for liquid flow through regolith and comparisons with Antarctic terrestrial analogs. Icarus 219 (1), 1–4.

Levy, J.S., Head, J.W., Marchant, D.R., 2007. Lineated valley fill and lobate debris apron stratigraphy in Nilosyrtis Mensae, Mars: evidence for phases of glacial modification of the dichotomy boundary. J. Geophys. Res. Planets 112 (E8). Available from: http://dx.doi.org/10.1029/2006JE002852.

Levy, J.S., Head, J., Marchant, D., 2009a. Thermal contraction crack polygons on Mars: classification, distribution, and climate implications from HiRISE observations. J. Geophys. Res. Planets 114 (E1). Available from: http://dx.doi.org/10.1029/2008JE003273.

Levy, J.S., Head, J.W., Marchant, D.R., Dickson, J.L., Morgan, G.A., 2009b. Geologically recent gully–polygon relationships on Mars: insights from the Antarctic Dry Valleys on the roles of permafrost, microclimates, and water sources for surface flow. Icarus 201 (1), 113–126.

Levy, J.S., Head, J.W., Marchant, D.R., 2009c. Concentric crater fill in Utopia Planitia: history and interaction between glacial "brain terrain" and periglacial mantle processes. Icarus 202 (2), 462–476.

Levy, J.S., Marchant, D.R., Head, J.W., 2010a. Thermal contraction crack polygons on Mars: a synthesis from HiRISE, Phoenix, and terrestrial analog studies. Icarus 206 (1), 229–252.

Levy, J.S., Head, J.W., Marchant, D.R., 2010b. Concentric crater fill in the northern mid-latitudes of Mars: formation processes and relationships to similar landforms of glacial origin. Icarus 209 (2), 390–404.

Levy, J.S., Fassett, C.I., Head, J.W., Schwartz, C., Watters, J.L., 2014a. Sequestered glacial ice contribution to the global Martian water budget: geometric constraints on the volume of remnant, midlatitude debris-covered glaciers. J. Geophys. Res. Planets 119 (10), 2188–2196.

Levy, J.S., Fountain, A.G., Gooseff, M.N., Barrett, J.E., Vantreese, R., Welch, K.A., et al., 2014b. Water track modification of soil ecosystems in the Lake Hoare basin, Taylor Valley, Antarctica. Antarctic Sci. 26 (02), 153–162.

Levy, J.S., Fassett, C.I., Head, J.W., 2016. Enhanced erosion rates on Mars during Amazonian glaciation. Icarus 264, 213–219.

Lillis, R.J., Frey, H.V., Manga, M., 2008. Rapid decrease in Martian crustal magnetization in the Noachian era: implications for the dynamo and climate of early Mars. Geophys. Res. Lett. 35 (14). Available from: http://dx.doi.org/10.1029/2008GL034338.

Loizeau, D., Mangold, N., Poulet, F., Bibring, J.P., Gendrin, A., Ansan, V., et al., 2007. Phyllosilicates in the Mawrth Vallis region of Mars. J. Geophys. Res. Planets 112 (E8). Available from: http://dx.doi.org/10.1029/2006JE002877.

Lucchitta, B.K., 1981. Mars and Earth: comparison of cold-climate features. Icarus 45 (2), 264–303.

Lucchitta, B.K., 1984. Ice and debris in the fretted terrain, Mars. J. Geophys. Res. Solid Earth 89 (S02). Available from: http://dx.doi.org/10.1029/JB089iS02p0B409.

Lundin, R., Lammer, H., Ribas, I., 2007. Planetary magnetic fields and solar forcing: implications for atmospheric evolution. Space Sci. Rev. 129 (1-3), 245−278.

Lunine, J.I., Chambers, J., Morbidelli, A., Leshin, L.A., 2003. The origin of water on Mars. Icarus 165 (1), 1−8.

Luo, W., Stepinski, T.F., 2009. Computer-generated global map of valley networks on Mars. J. Geophys. Res. Planets 114 (E11). Available from: http://dx.doi.org/10.1029/2009JE003357.

Mackay, S.L., Marchant, D.R., 2017. Obliquity-paced climate change recorded in Antarctic debris-covered glaciers. Nat. Commun. 8. Available from: http://dx.doi.org/10.1038/ncomms14194.

Madeleine, J.B., Forget, F., Head, J.W., Levrard, B., Montmessin, F., Millour, E., 2009. Amazonian northern mid-latitude glaciation on Mars: a proposed climate scenario. Icarus 203 (2), 390−405.

Mahaffy, P.R., Conrad, P.G., 2015. Volatile and isotopic imprints of ancient Mars. Elements 11 (1), 51−56.

Mahaffy, P.R., Webster, C.R., Stern, J.C., Brunner, A.E., Atreya, S.K., Conrad, P.G., et al., 2015. The imprint of atmospheric evolution in the D/H of Hesperian clay minerals on Mars. Science 347 (6220), 412−414.

Mahaffy, P.R., Webster, C.R., Atreya, S.K., Franz, H., Wong, M., Conrad, P.G., et al., 2013. Abundance and isotopic composition of gases in the Martian atmosphere from the Curiosity rover. Science 341 (6143), 263−266.

Malin, M.C., Edgett, K.S., 2000. Evidence for recent groundwater seepage and surface runoff on Mars. Science 288 (5475), 2330−2335.

Malin, M.C., Edgett, K.S., 2001. Mars global surveyor Mars orbiter camera: interplanetary cruise through primary mission. J. Geophys. Res. Planets 106 (E10), 23429−23570.

Malin, M.C., Edgett, K.S., 2003. Evidence for persistent flow and aqueous sedimentation on early Mars. Science 302, 1931−1935.

Malin, M.C., Edgett, K.S., Posiolova, L.V., McColley, S.M., Dobrea, E.Z.N., 2006. Present-day impact cratering rate and contemporary gully activity on Mars. Science 314 (5805), 1573−1577.

Malin, M.C., Edgett, K.S., Cantor, B.A., Caplinger, M.A., Danielson, G.E., Jensen, E.H., et al., 2010. An overview of the 1985−2006 Mars Orbiter Camera science investigation. Mars 5, 1−60.

Manga, M., 2004. Martian floods at Cerberus Fossae can be produced by groundwater discharge. Geophys. Res. Lett. 31 (2). Available from: http://dx.doi.org/10.1029/2003GL018958.

Mangold, N., 2005. High latitude patterned grounds on Mars: classification, distribution and climatic control. Icarus 174 (2), 336−359.

Mangold, N., 2011. Ice sublimation as a geomorphic process: a planetary perspective. Geomorphology 126 (1), 1−17.

Mangold, N., 2012. Fluvial landforms on fresh ejecta craters. Planet. Space Sci. 62, 69−85.

Mangold, N., Ansan, V., 2006. Detailed study of an hydrological system of valleys, a delta and lakes in the Southwest Thaumasia region, Mars. Icarus 180 (1), 75−87.

Mangold, N., Howard, A.D., 2013. Outflow channels with deltaic deposits in Ismenius Lacus, Mars. Icarus. Available from: http://dx.doi.org/10.1016/j.icarus.2013.05.040.

Mangold, N., Quantin, C., Ansan, V., Delacourt, C., Allemand, P., 2004a. Evidence for precipitation on Mars from dendritic valleys in the Valles Marineris area. Science 305 (5680), 78−81.

Mangold, N., Maurice, S., Feldman, W.C., Costard, F., Forget, F., 2004b. Spatial relationships between patterned ground and ground ice detected by the Neutron Spectrometer on Mars. J. Geophys. Res. Planets 109 (E8). Available from: http://dx.doi.org/10.1029/2004JE002235.

Mangold, N., Poulet, F., Mustard, J.F., Bibring, J.P., Gondet, B., Langevin, Y., et al., 2007. Mineralogy of the Nili Fossae region with OMEGA/Mars Express data: 2. Aqueous alteration of the crust. J. Geophys. Res. Planets 112 (E8). Available from: http://dx.doi.org/10.1029/2006JE002835.

Mangold, N., Kite, E.S., Kleinhans, M.G., Newsom, H., Ansan, V., Hauber, E., et al., 2012. The origin and timing of fluvial activity at Eberswalde Crater, Mars. Icarus 220 (2), 530−551.

Marchant, D.R., Head, J.W., 2007. Antarctic Dry Valleys: microclimate zonation, variable geomorphic processes, and implications for assessing climate change on Mars. Icarus 192 (1), 187−222.

Marchant, D.R., Lewis, A.R., Phillips, W.M., Moore, E.J., Souchez, R.A., Denton, G.H., et al., 2002. Formation of patterned ground and sublimation till over Miocene glacier ice in Beacon Valley, southern Victoria Land, Antarctica. Geological Society of America Bulletin 114 (6), 718–730.

Marra, W.A., Braat, L., Baar, A.W., Kleinhans, M.G., 2014. Valley formation by groundwater seepage, pressurized groundwater outbursts and crater-lake overflow in flume experiments with implications for Mars. Icarus 232, 97–117.

Martínez, G.M., Renno, N.O., 2013. Water and brines on Mars: current evidence and implications for MSL. Space Sci. Rev. 175 (1-4), 29–51.

Maurice, S., Feldman, W., Diez, B., Gasnault, O., Lawrence, D.J., Pathare, A., et al., 2011. Mars Odyssey neutron data: 1. Data processing and models of water-equivalent-hydrogen distribution. J. Geophys. Res. Planets 116 (E11). Available from: http://dx.doi.org/10.1029/2011JE003810.

Max, M.D., Clifford, S.M., 2000. The state, potential distribution, and biological implications of methane in the Martian crust. J. Geophys. Res. Planets 105 (E2), 4165–4171.

McCubbin, F.M., Boyce, J.W., Srinivasan, P., Santos, A.R., Elardo, S.M., Filiberto, J., et al., 2016. Heterogeneous distribution of H_2O in the Martian interior: implications for the abundance of H_2O in depleted and enriched mantle sources. Meteorit. Planet. Sci. 51 (11), 2036–2060.

McEwen, A.S., Ojha, L., Dundas, C.M., Mattson, S.S., Byrne, S., Wray, J.J., et al., 2011. Seasonal flows on warm Martian slopes. Science 333 (6043), 740–743.

McEwen, A.S., Dundas, C.M., Mattson, S.S., Toigo, A.D., Ojha, L., Wray, J.J., et al., 2014. Recurring slope lineae in equatorial regions of Mars. Nat. Geosci. 7 (1), 53–58.

McEwen, A., Chojnacki, M., Dundas, C., Ojha, L., Masse, M., Schaefer, E., et al., 2015. Recurring slope lineae on Mars: atmospheric origin. EPSC Abstracts 10.

McGovern, P.J., Solomon, S.C., Smith, D.E., Zuber, M.T., Simons, M., Wieczorek, M.A., et al., 2004. Correction to "Localized gravity/topography admittance and correlation spectra on Mars: implications for regional and global evolution. J. Geophys. Res. 109, E07007. Available from: http://dx.doi.org/10.1029/2004JE002286.

McKay, D.S., Gibson Jr, E.K., Thomas-Keprta, K.L., Vali, H., 1996. Search for past life on Mars: possible relic biogenic activity in Martian meteorite ALH84001. Science 273 (5277), 924.

McKeown, N.K., et al., 2009. Characterization of phyllosilicates observed in the central Mawrth Vallis region, Mars, their potential formational processes, and implications for past climate. J. Geophys. Res. 114, E00D10. Available from: http://dx.doi.org/10.1029/2008JE003301.

Mellon, M.T., 1997. Small-scale polygonal features on Mars: seasonal thermal contraction cracks in permafrost. J. Geophys. Res. Planets 102 (E11), 25617–25628.

Mellon, M.T., Jakosky, B.M., 1993. Geographic variations in the thermal and diffusive stability of ground ice on Mars. J. Geophys. Res. 98, 3345–3364. Available from: http://dx.doi.org/10.1029/92JE02355.

Mellon, M.T., Jakosky, B.M., 1995. The distribution and behavior of Martian ground ice during past and present epochs. J. Geophys. Res. Planets 100 (E6), 11781–11799.

Mellon, M.T., Boynton, W.V., Feldman, W.C., Arvidson, R.E., Titus, T.N., Bandfield, J.L., et al., 2008a. A prelanding assessment of the ice table depth and ground ice characteristics in Martian permafrost at the Phoenix landing site. J. Geophys. Res. Planets 113 (E3). Available from: http://dx.doi.org/10.1029/2007JE003067.

Mellon, M.T., Arvidson, R.E., Marlow, J.J., Phillips, R.J., Asphaug, E., 2008b. Periglacial landforms at the Phoenix landing site and the northern plains of Mars. J. Geophys. Res. Planets 113 (E3). Available from: http://dx.doi.org/10.1029/2007JE003039.

Mellon, M.T., Arvidson, R.E., Sizemore, H.G., Searls, M.L., Blaney, D.L., Cull, S., et al., 2009. Ground ice at the Phoenix landing site: stability state and origin. J. Geophys. Res. Planets 114 (E1). Available from: http://dx.doi.org/10.1029/2009JE003417.

Meresse, S., Costard, F., Mangold, N., Baratoux, D., Boyce, J.M., 2006. Martian perched craters and large ejecta volume: evidence of episodes of deflation of the northern lowlands. Meteorit. Planet. Sci. 41, 147–1658.

Metz, J.M., Grotzinger, J.P., Mohrig, D., Milliken, R., Prather, B., Pirmez, C., et al., 2009. Sublacustrine depositional fans in southwest Melas Chasma. J. Geophys. Res. E 114, E10002, ISSN 0148-0227.

Milbury, C., Schubert, G., Raymond, C.A., Smrekar, S.E., Langlais, B., 2012. The history of Mars' dynamo as revealed by modeling magnetic anomalies near Tyrrhenus Mons and Syrtis Major. J. Geophys. Res. Planets (1991–2012) 117 (E10). Available from: http://dx.doi.org/10.1029/2012JE004099.

Milliken, R.E., Bish, D.L., 2010. Sources and sinks of clay minerals on Mars. Philos. Mag. 90 (17-18), 2293–2308.

Moore, J.M., Edgett, K.S., 1993. Hellas Planitia, Mars- Site of net dust erosion and implications for the nature of basin floor deposits. Geophys. Res. Lett. 20 (15), 1599–1602.

Moore, J.M., Howard, A.D., 2005. Large alluvial fans on Mars. J. Geophys. Res. 110, E04005. Available from: http://dx.doi.org/10.1029/2004JE002352.

Moore, J.M., Howard, A.D., Dietrich, W.E., Schenk, P.M., 2003. Martian layered fluvial deposits: implications for Noachian climate scenarios. Geophys. Res. Lett. 30 (24), 2292.

Morgan, A.M., Howard, A.D., Hobley, D.E.J., Moore, J.M., Dietrich, W.E., Williams, R.M.E., et al., 2013. Sedimentology and climatic environment of alluvial fans in the Martian Saheki Crater and a comparison with terrestrial fans in the Atacama Desert. Icarus 229, 131–156.

Morgan, G.A., Head, J.W., 2009. Sinton Crater, Mars: evidence for impact into a plateau icefield and melting to produce valley networks at the Hesperian–Amazonian boundary. Icarus 202 (1), 39–59.

Morgan, G.A., Head, J.W., Marchant, D.R., 2009. Lineated valley fill (LVF) and lobate debris aprons (LDA) in the Deuteronilus Mensae northern dichotomy boundary region, Mars: constraints on the extent, age and episodicity of Amazonian glacial events. Icarus 202 (1), 22–38.

Morgenstern, A., Hauber, E., Reiss, D., van Gasselt, S., Grosse, G., Schirrmeister, L., 2007. Deposition and degradation of a volatile-rich layer in Utopia Planitia and implications for climate history on Mars. J. Geophys. Res. Planets 112 (E6). Available from: http://dx.doi.org/10.1029/2006JE002869.

Morris, R.V., Ruff, S.W., Gellert, R., Ming, D.W., Arvidson, R.E., Clark, B.C., et al., 2010. Identification of carbonate-rich outcrops on Mars by the Spirit rover. Science 329 (5990), 421–424.

Mouginis-Mark, P.J., 1979. Martian fluidized crater morphology: variations with crater size, latitude, altitude and target material. J. Geophys. Res. 84, 8011–8022.

Mouginis-Mark, P.J., 1987. Water or ice in the Martian regolith? Clues from rampart craters seen at very high resolution. Icarus 71 (2), 268–286.

Mouginot, J., Pommerol, A., Kofman, W., Beck, P., Schmitt, B., Herique, A., et al., 2010. The 3–5 MHz global reflectivity map of Mars by MARSIS/Mars Express: implications for the current inventory of subsurface H_2O. Icarus 210 (2), 612–625.

Mouginot, J., Pommerol, A., Beck, P., Kofman, W., Clifford, S.M., 2012. Dielectric map of the Martian northern hemisphere and the nature of plain filling materials. Geophys. Res. Lett. 39 (2). Available from: http://dx.doi.org/10.1029/2011GL050286.

Mumma, M.J., Villanueva, G.L., Novak, R.E., Hewagama, T., Bonev, B.P., DiSanti, M.A., et al., 2009. Strong release of methane on Mars in northern summer 2003. Science 323 (5917), 1041–1045.

Murchie, S.L., Mustard, J.F., Ehlmann, B.L., Milliken, R.E., Bishop, J.L., McKeown, N.K., et al., 2009. A synthesis of Martian aqueous mineralogy after 1 Mars year of observations from the Mars Reconnaissance Orbiter. J. Geophys. Res. Planets 114 (E2). Available from: http://dx.doi.org/10.1029/2009JE003342.

Mustard, J.F., Cooper, C.D., Rifkin, M.K., 2001. Evidence for recent climate change on Mars from the identification of youthful near-surface ground ice. Nature 412 (6845), 411–414.

Mustard, J.F., Murchie, S.L., Pelkey, S.M., Ehlmann, B.L., Milliken, R.E., Grant, J.A., et al., 2008. Hydrated silicate minerals on Mars observed by the Mars Reconnaissance Orbiter CRISM instrument. Nature 454 (7202), 305–309.

Neumann, G.A., Abshire, J.B., Aharonson, O., Garvin, J.B., Sun, X., Zuber, M.T., 2003. Mars Orbiter Laser Altimeter pulse width measurements and footprint-scale roughness. Geophys. Res. Lett. 30 (11). Available from: http://dx.doi.org/10.1029/2003GL017048.

Neumann, G.A., Zuber, M.T., Wieczorek, M.A., McGovern, P.J., Lemoine, F.G., Smith, D.E., 2004. Crustal structure of Mars from gravity and topography. J. Geophys. Res. Planets 109 (E8). Available from: http://dx.doi.org/10.1029/2004JE002262.

Newsom, H.E., Brittelle, G.E., Hibbitts, C.A., Crossey, L.J., Kudo, A.M., 1996. Impact crater lakes on Mars. J. Geophys. Res. Planets 101 (E6), 14951–14955.

Nimmo, F., Tanaka, K., 2005. Early crustal evolution of Mars. Annu. Rev. Earth Planet. Sci. 33, 133–161. Available from: http://dx.doi.org/10.1146/annurev.earth.33.092203.122637.

Okubo, C.H., McEwen, A.S., 2007. Fracture-controlled paleo-fluid flow in Candor Chasma, Mars. Science 315 (5814), 983–985.

Ojha, L., Wilhelm, M.B., Murchie, S.L., McEwen, A.S., Wray, J.J., Hanley, J., et al., 2015. Spectral evidence for hydrated salts in recurring slope lineae on Mars. Nat. Geosci. 8 (11), 829–832.

Olhoeft, G.R., 2000. Maximizing the information return from ground penetrating radar. J. Appl. Geophys. 43 (2), 175–187.

Ori, G.G., Marinangeli, L., Baliva, A., 2000. Terraces and Gilbert-type deltas in crater lakes in Ismenius Lacus and Memnonia (Mars). J. Geophys. Res. 105 (E7), 17629–17641.

Orloff, T.C., Kreslavsky, M.A., Asphaug, E.I., 2013. Possible mechanism of boulder clustering on Mars. Icarus 225 (2), 992–999.

Oze, C., Sharma, M., 2005. Have olivine, will gas: serpentinization and the abiogenic production of methane on Mars. Geophys. Res. Lett. 32 (10). Available from: http://dx.doi.org/10.1029/2005GL022691.

Oze, C., Sharma, M., 2007. Serpentinization and the inorganic synthesis of H 2 in planetary surfaces. Icarus 186 (2), 557–561.

Paige, D.A., 1992. The thermal stability of near-surface ground ice on Mars. Nature 356 (6364), 43–45.

Palucis, M.C., Dietrich, W.E., Hayes, A.G., Williams, R.M.E., Gupta, S., Mangold, N., et al., 2014. The origin and evolution of the Peace Vallis fan system that drains to the Curiosity landing area, Gale Crater, Mars. J. Geophys. Res. Planets 119, 705–728. Available from: http://dx.doi.org/10.1002/2013JE004583.

Parker, T.J., Calef, F.J., 2012. Digital global map of potential ocean paleoshorelines on Mars. LPI Contrib. 1680, 7085.

Parker, T.J., Saunders, R.S., Schneeberger, D.M., 1989. Transitional morphology in west Deuteronilus Mensae, Mars: implications for modification of the lowland/upland boundary. Icarus 82 (1), 111–145.

Parker, T.J., Gorsline, D.S., Saunders, R.S., Pieri, D.C., Schneeberger, D.M., 1993. Coastal geomorphology of the Martian northern plains. J. Geophys. Res. Planets 98 (E6), 11061–11078.

Parker, T.J., Grant, J.A., Brenda, J.F., 2010. The northern plains: a Martian oceanic basin? In: Cabrol, Grin (Eds.), Lakes on Mars. Elsevier, Amsterdam, pp. 249–273.

Parsons, R.A., Moore, J.M., Howard, A.D., 2013. Evidence for a short period of hydrologic activity in Newton Crater, Mars, near the Hesperian-Amazonian transition. J. Geophys. Res. Planets 118 (5), 1082–1093.

Pasquon, K., Gargani, J., Massé, M., Conway, S.J., 2016. Present-day formation and seasonal evolution of linear dune gullies on Mars. Icarus 274, 195–210.

Penido, J.C., Fassett, C.I., Som, S.M., 2013. Scaling relationships and concavity of small valley networks on Mars. Planet. Space Sci. 75, 105–116.

Pepin, R.O., Schlutter, D.J., Becker, R.H., Reisenfeld, D.B., 2012. Helium, neon, and argon composition of the solar wind as recorded in gold and other Genesis collector materials. Geochim. Cosmochim. Acta 89, 62–80.

Perron, J.T., Mitrovica, J.X., Manga, M., Matsuyama, I., Richards, M.A., 2007. Evidence for an ancient Martian ocean in the topography of deformed shorelines. Nature 447 (7146), 840–843.

Pham, L.B.S., Karatekin, Ö., 2016. Scenarios of atmospheric mass evolution on Mars influenced by asteroid and comet impacts since the Late Noachian. Planet. Space Sci. 125, 1–11.

Pham, L.B.S., Karatekin, Ö., Dehant, V., 2009. Effects of meteorite impacts on the atmospheric evolution of Mars. Astrobiology 9 (1), 45–54.

Phillips, R.J., Zuber, M.T., Smrekar, S.E., Mellon, M.T., Head, J.W., Tanaka, K.L., et al., 2008. Mars north polar deposits: stratigraphy, age, and geodynamical response. Science 320 (5880), 1182−1185.

Picardi, G., Biccari, D., Cicchetti, A., Seu, R., Plaut, J., Johnson, W.T.K., et al., 2003. Mars Advanced Radar for subsurface and ionosphere sounding (MARSIS). In EGS-AGU-EUG Joint Assembly, 9597.

Picardi, G., Biccari, D., Seu, R., Marinangeli, L., Johnson, W.T.K., Jordan, R.L., et al., 2004. Performance and surface scattering models for the Mars Advanced Radar for Subsurface and Ionosphere Sounding (MARSIS). Planet. Space Sci. 52 (1), 149−156.

Picardi, G., Plaut, J.J., Biccari, D., Bombaci, O., Calabrese, D., Cartacci, M., et al., 2005. Radar soundings of the subsurface of Mars. Science 310 (5756), 1925−1928.

Pierce, T.L., Crown, D.A., 2003. Morphologic and topographic analyses of debris aprons in the eastern Hellas region, Mars. Icarus 163 (1), 46−65.

Pilorget, C., Forget, F., 2016. Formation of gullies on Mars by debris flows triggered by CO_2 sublimation. Nat. Geosci. 9 (1), 65−69.

Plaut, J.J., Picardi, G., Safaeinili, A., Ivanov, A.B., Milkovich, S.M., Cicchetti, A., et al., 2007. Subsurface radar sounding of the south polar layered deposits of Mars. Science 316 (5821), 92−95.

Plaut, J.J., Safaeinili, A., Holt, J.W., Phillips, R.J., Head, J.W., Seu, R., et al., 2009. Radar evidence for ice in lobate debris aprons in the mid-northern latitudes of Mars. Geophys. Res. Lett. 36, 2. Available from: http://dx.doi.org/10.1029/2008GL036379.

Plesa, A.C., Grott, M., Tosi, N., Breuer, D., Spohn, T., Wieczorek, M.A., 2016. How large are present-day heat flux variations across the surface of Mars? J. Geophys. Res. Planets 121 (12), 2386−2403. Available from: http://dx.doi.org/10.1002/2016JE005126.

Pollack, H.N., Hurter, S.J., Johnson, J.R., 1993. Heat flow from the Earth's interior—analysis of the global data set. Rev. Geophys. 31, 267−280. Available from: http://dx.doi.org/10.1029/93RG01249.

Pollack, J.B., Colburn, D.S., Flasar, F.M., Kahn, R., Carlston, C.E., Pidek, D., 1979. Properties and effects of dust particles suspended in the Martian atmosphere. J. Geophys. Res. 84 (B6), 2929−2945.

Pondrelli, M., Rossi, A.P., Marinangeli, L., Hauber, E., Gwinner, K., Baliva, A., et al., 2008. Evolution and depositional environments of the Eberswalde fan delta, Mars. Icarus 197 (2), 429−451.

Poulet, F., Bibring, J.P., Mustard, J.F., Gendrin, A., Mangold, N., Langevin, Y., et al., 2005. Phyllosilicates on Mars and implications for early Martian climate. Nature 438 (7068), 623−627.

Poulet, F., Arvidson, R.E., Gomez, C., Morris, R.V., Bibring, J.-P., Langevin, Y., et al., 2008. Mineralogy of Terra Meridiani and western Arabia Terra from OMEGA.Mex and implications for their formation. Icarus 195, 106−130.

Presley, M.A., Christensen, P.R., 1997. The effect of bulk density and particle size sorting on the thermal conductivity of particulate materials under Martian atmospheric pressures. J. Geophys. Res. 102, 9221−9229. Available from: http://dx.doi.org/10.1029/97JE00271.

Putzig, N.E., Phillips, R.J., Campbell, B.A., Holt, J.W., Plaut, J.J., Carter, L.M., et al., 2009. Subsurface structure of Planum Boreum from Mars Reconnaissance Orbiter Shallow Radar soundings. Icarus 204 (2), 443−457.

Putzig, N.E., Phillips, R.J., Campbell, B.A., Mellon, M.T., Holt, J.W., Brothers, T.C., 2014. SHARAD soundings and surface roughness at past, present, and proposed landing sites on Mars: reflections at Phoenix may be attributable to deep ground ice. J. Geophys. Res. Planets 119 (8), 1936−1949.

Quantin, C., Allemand, P., Mangold, N., Dromart, G., Delacourt, C., 2005. Fluvial and lacustrine activity on layered deposits in Melas Chasma, Valles Marineris, Mars. J. Geophys. Res. Vol. 110 (E12), E12S19, 2005. Available from: http://dx.doi.org/10.1029/2005JE002440.

Quesnel, Y., Sotin, C., Langlais, B., Costin, S., Mandea, M., Gottschalk, M., et al., 2009. Serpentinization of the Martian crust during Noachian. Earth Planet. Sci. Lett. 277 (1), 184−193.

Raack, J., Reiss, D., Appéré, T., Vincendon, M., Ruesch, O., Hiesinger, H., 2015. Present-day seasonal gully activity in a south polar pit (Sisyphi Cavi) on Mars. Icarus 251, 226−243.

Ramirez, R.M., Kopparapu, R., Zugger, M.E., Robinson, T.D., Freedman, R., Kasting, J.F., 2014. Warming early Mars with CO_2 and H_2. Nat. Geosci. 7 (1), 59−63.

Reiss, D., Erkeling, G., Bauch, K.E., Hiesinger, H., 2010. Evidence for present day gully activity on the Russell Crater dune field, Mars. Geophys. Res. Lett. 37 (6). Available from: http://dx.doi.org/10.1029/2009GL042192.

Roberts, J.H., Lillis, R.J., Manga, M., 2009. Giant impacts on early Mars and the cessation of the Martian dynamo. J. Geophys. Res. Planets (1991−2012) 114 (E4). Available from: http://dx.doi.org/10.1029/2008JE003287.

Rodriguez, J.A.P., Gulick, V.C., Baker, V.R., Platz, T., Fairén, A.G., Miyamoto, H., et al., 2014. Evidence for Middle Amazonian catastrophic flooding and glaciation on Mars. Icarus 242, 202−210.

Rodriguez, J.A.P., Platz, T., Gulick, V., Baker, V.R., Fairén, A.G., Kargel, J., et al., 2015. Did the Martian outflow channels mostly form during the Amazonian Period? Icarus 257, 387−395.

Rodriguez, J.A.P., Fairén, A.G., Tanaka, K.L., Zarroca, M., Linares, R., Platz, T., et al., 2016. Tsunami waves extensively resurfaced the shorelines of an early Martian ocean. Scientific Reports 6. Available from: http://dx.doi.org/10.1038/srep25106.

Roos-Serote, M., Atreya, S.K., Webster, C.R., Mahaffy, P.R., 2016. Cometary origin of atmospheric methane variations on Mars unlikely. J. Geophys. Res. Planets 121 (10), 2108−2119.

Rossbacher, L.A., Judson, S., 1981. Ground ice on Mars: inventory, distribution, and resulting landforms. Icarus 45 (1), 39−59.

Salese, F., Ansan, V., Mangold, N., Carter, J., Ody, A., Poulet, F., et al., 2016. A sedimentary origin for inter-crater plains north of the Hellas basin: implications for climate conditions and erosion rates on early Mars. J. Geophys. Res. Planets 121 (11), 2239−2267.

Scanlon, K.E., Head, J.W., 2015. The recession of the Dorsa Argentea Formation ice sheet: geologic evidence and climate simulations. In: Lunar and Planetary Science Conference, Vol. 46. Houston, TX, USA, p. 2247.

Schenk, P.M., Moore, J.M., 2000. Stereo topography of the south polar region of Mars: volatile inventory and Mars Polar Lander landing site. J. Geophys. Res. Planets 105 (E10), 24529−24546.

Schmidt, F., Andrieu, F., Costard, F., Kocifaj, M., Meresescu, A.G., 2017. Formation of recurring slope lineae on Mars by rarefied gas-triggered granular flows. Nat. Geosci. 10 (4), 270−273.

Schon, S.C., Head, J.W., Milliken, R.E., 2009. A recent ice age on Mars: evidence for climate oscillations from regional layering in mid-latitude mantling deposits. Geophys. Res. Lett. 36 (15). Available from: http://dx.doi.org/10.1029/2009GL038554.

Schon, S.C., Head, J.W., Fassett, C.I., 2012. An overfilled lacustrine system and progradational delta in Jezero Crater, Mars: implications for Noachian climate. Planet. Space Sci. 67 (1), 28−45.

Schorghofer, N., Aharonson, O., 2005. Stability and exchange of subsurface ice on Mars. J. Geophys. Res. Planets 110 (E5). Available from: http://dx.doi.org/10.1029/2004JE002350.

Schultz, P.H., Gault, D.E., 1979. Atmospheric effects on Martian ejecta emplacement. J. Geophys. Res. 84, 7669−7687.

Séjourné, A., Costard, F., Gargani, J., Soare, R.J., Fedorov, A., Marmo, C., 2011. Scalloped depressions and small-sized polygons in western Utopia Planitia, Mars: a new formation hypothesis. Planet. Space Sci. 59 (5), 412−422.

Segura, T.L., Toon, O.B., Colaprete, A., Zahnle, K., 2002. Environmental effects of large impacts on Mars. Science 298 (5600), 1977−1980.

Selvans, M.M., Plaut, J.J., Aharonson, O., Safaeinili, A., 2010. Internal structure of Planum Boreum, from Mars Advanced Radar for Subsurface and Ionospheric Sounding data. J. Geophys. Res. Planets 115 (E9). Available from: http://dx.doi.org/10.1029/2009JE003537.

Senft, L.E., Stewart, S.T., 2008. Impact crater formation in icy layered terrains on Mars. Meteorit. Planet. Sci. 43 (12), 1993−2013.

Seu, R., Phillips, R.J., Biccari, D., Orosei, R., Masdea, A., Picardi, G., et al., 2007. SHARAD sounding radar on the Mars Reconnaissance Orbiter. J. Geophys. Res. Planets 112 (E5). Available from: http://dx.doi.org/10.1029/2006JE002745.

Sharp, R.P., 1973. Mars: fretted and chaotic terrains. J. Geophys. Res. 78 (20), 4073–4083.

Sharp, R.P., Malin, M.C., 1975. Channels on Mars. Geolog. Soc. Am. Bull. 86 (5), 593–609.

Shaw, G.H., 2016. The faint young sun problem. Earth's Early Atmosphere and Oceans, and The Origin of Life. Springer International Publishing, Switzerland, pp. 69–74.

Shean, D.E., 2010. Candidate ice-rich material within equatorial craters on Mars. Geophys. Res. Lett. 37 (24). Available from: http://dx.doi.org/10.1029/2010GL045181.

Shean, D.E., Head, J.W., Marchant, D.R., 2005. Origin and evolution of a cold-based tropical mountain glacier on Mars: the Pavonis Mons fan-shaped deposit. J. Geophys. Res. Planets 110 (E5). Available from: http://dx.doi.org/10.1029/2004JE002360.

Shizgal, B.D., Arkos, G.G., 1996. Nonthermal escape of the atmospheres of Venus, Earth, and Mars. Rev. Geophys. 34 (4), 483–505.

Sizemore, H.G., Zent, A.P., Rempel, A.W., 2015. Initiation and growth of Martian ice lenses. Icarus 251, 191–210.

Sloan Jr., E.D., 1997. Clathrate Hydrates of Natural Gases. Marcel Dekker, New York.

Smith, D.E., Zuber, M.T., Solomon, S.C., Phillips, R.J., Head, J.W., Garvin, J.B., et al., 1999. The global topography of Mars and implications for surface evolution. Science 284 (5419), 1495–1503.

Smith, D.E., Zuber, M.T., Frey, H.V., Garvin, J.B., Head, J.W., Muhleman, D.O., et al., 2001. Mars Orbiter Laser Altimeter: experiment summary after the first year of global mapping of Mars. J. Geophys. Res. Planets 106 (E10), 23689–23722.

Smith, D.E., G. Neumann, R.E. Arvidson, E.A. Guinness, and S. Slavney, 2003. Mars Global Surveyor Laser Altimeter Mission Experiment Gridded Data Record. Available from: http://starbrite.jpl.nasa.gov/pds/viewDataset.jsp? dsid=MGS-M-MOLA-5-MEGDR-L3-V1.0, NASA Planet. Data Syst., Greenbelt, Md.

Smith, I.B., Putzig, N.E., Holt, J.W., Phillips, R.J., 2016. An ice age recorded in the polar deposits of Mars. Science 352 (6289), 1075–1078.

Smith, M.D., 2004. Interannual variability in TES atmospheric observations of Mars during 1999–2003. Icarus 167 (January), 148–165.

Smith, M.D., 2008. Spacecraft observations of the Martian atmosphere. Annu. Rev. Earth Planet. Sci. 36, 191–219.

Smith, P.H., Tamppari, L.K., Arvidson, R.E., Bass, D., Blaney, D., Boynton, W.V., et al., 2009. H$_2$O at the Phoenix landing site. Science 325 (5936), 58–61.

Soare, R.J., Burr, D.M., Tseung, J.M.W.B., 2005. Possible pingos and a periglacial landscape in northwest Utopia Planitia. Icarus 174 (2), 373–382.

Soare, R.J., Kargel, J.S., Osinski, G.R., Costard, F., 2007. Thermokarst processes and the origin of crater-rim gullies in Utopia and western Elysium Planitia. Icarus 191 (1), 95–112.

Soare, R.J., Osinski, G.R., Roehm, C.L., 2008. Thermokarst lakes and ponds on Mars in the very recent (Late Amazonian) past. Earth Planet. Sci. Lett. 272 (1), 382–393.

Soare, R.J., Conway, S.J., Pearce, G.D., Dohm, J.M., Grindrod, P.M., 2013. Possible crater-based pingos, paleolakes and periglacial landscapes at the high latitudes of Utopia Planitia, Mars. Icarus 225 (2), 971–981.

Soare, R.J., Conway, S.J., Dohm, J.M., El-Maarry, M.R., 2014a. Possible open-system (hydraulic) pingos in and around the Argyre impact region of Mars. Earth Planet. Sci. Lett. 398, 25–36.

Soare, R.J., Conway, S.J., Dohm, J.M., 2014b. Possible ice-wedge polygons and recent landscape modification by "wet" periglacial processes in and around the Argyre impact basin, Mars. Icarus 233, 214–228.

Soare, R.J., Conway, S.J., Gallagher, C., Dohm, J.M., 2016. Sorted (clastic) polygons in the Argyre region, Mars, and possible evidence of pre-and post-glacial periglaciation in the Late Amazonian Epoch. Icarus 264, 184–197.

Soare, R.J., Conway, S.J., Gallagher, C., Dohm, J.M., 2017. Ice-rich (periglacial) vs icy (glacial) depressions in the Argyre region, Mars: a proposed cold-climate dichotomy of landforms. Icarus 282, 70−83.

Soderblom, L.A., Malin, M.C., Cutts, J.A., Murray, B.C., 1973. Mariner 9 observations of the surface of Mars in the north polar region. J. Geophys. Res. 78 (20), 4197−4210.

Solar System Exploration Survey, New Frontiers in Solar System Exploration, National Academies Press, Washington, 2003. Available from: http://www.nap.edu/catalog.php?record_id = 10898 (accessed 10.09.17).

Solomon, S.C., Head, J.W., 1990. Heterogeneities in the thickness of the elastic lithosphere of Mars: constraints on heat flow and internal dynamics. J. Geophys. Res. 95, 11,073−11,083. Available from: http://dx.doi.org/10.1029/JB095iB07p11073.

Solomon, S.C., et al., 2005. New perspectives on ancient Mars. Science 307, 1214−1220. Available from: http://dx.doi.org/10.1126/science.1101812.

Som, S.M., Montgomery, D.R., Greenberg, H.M., 2009. Scaling relations for large Martian valleys. J. Geophys. Res. Planets 114 (E2). Available from: http://dx.doi.org/10.1029/2008JE003132.

Souness, C., Hubbard, B., Milliken, R.E., Quincey, D., 2012. An inventory and population-scale analysis of Martian glacier-like forms. Icarus 217 (1), 243−255.

Squyres, S.W., 1978. Martian fretted terrain: flow of erosional debris. Icarus 34 (3), 600−613.

Squyres, S.W., 1979. The distribution of lobate debris aprons and similar flows on Mars. J. Geophys. Res. Solid Earth 84 (B14), 8087−8096.

Squyres, S.W., Carr, M.H., 1986. Geomorphic evidence for the distribution of ground ice on Mars. Science 231 (4735), 249−252.

Stepinski, T.F., Stepinski, A.P., 2005. Morphology of drainage basins as an indicator of climate on early Mars. J. Geophys. Res. Planets 110 (E12). Available from: http://dx.doi.org/10.1029/2005JE002448.

Stillman, D.E., Grimm, R.E., 2011. Radar penetrates only the youngest geological units on Mars. J. Geophys. Res. Planets 116 (E3). Available from: http://dx.doi.org/10.1029/2010JE003661.

Stillman, D.E., Michaels, T.I., Grimm, R.E., Harrison, K.P., 2014. New observations of Martian southern mid-latitude recurring slope lineae (RSL) imply formation by freshwater subsurface flows. Icarus 233, 328−341.

Stillman, D.E., Michaels, T.I., Grimm, R.E., Hanley, J., 2016. Observations and modeling of northern mid-latitude recurring slope lineae (RSL) suggest recharge by a present-day Martian briny aquifer. Icarus 265, 125−138.

Stillman, D.E., Michaels, T.I., Grimm, R.E., 2017. Characteristics of the numerous and widespread recurring slope lineae (RSL) in Valles Marineris, Mars. Icarus 285, 195−210.

Stuurman, C.M., Osinski, G.R., Holt, J.W., Levy, J.S., Brothers, T.C., Kerrigan, M., et al., 2016. SHARAD detection and characterization of subsurface water ice deposits in Utopia Planitia, Mars. Geophys. Res. Lett. 43 (18), 9484−9491.

Tanaka, K.L., 1986. The straitigraphy of Mars. J. Geophys. Res. 91 (B13), E139−E158.

Tanaka, K.L., 2000. Dust and ice deposition in the Martian geologic record. Icarus 144 (2), 254−266.

Tanaka, K.L., Leonard, G.J., 1995. Geology and landscape evolution of the Hellas region of Mars. J. Geophys. Res. Planets 100 (E3), 5407−5432.

Tanaka, K.L., Kargel, J.S., MacKinnon, D.J., Hare, T.M., Hoffman, N., 2002. Catastrophic erosion of Hellas basin rim on Mars induced by magmatic intrusion into volatile-rich rocks. Geophys. Res. Lett. 29 (8). Available from: http://dx.doi.org/10.1029/2001GL013885.

Tanaka, K.L., Rodriguez, J.A.P., Skinner, J.A., Bourke, M.C., Fortezzo, C.M., Herkenhoff, K.E., et al., 2008. North polar region of Mars: advances in stratigraphy, structure, and erosional modification. Icarus 196 (2), 318−358.

Tanaka, K.L., Skinner, J.A.J., Dohm, J.M., Irwin III, R.P., Kolb, E.J., Fortezzo, C.M., et al., 2014. Geologic map of Mars: US Geological Survey Scientific Investigations Map 3292, scale 1, 20,000,000.

Taylor, G.J., Stopar, J.D., Boynton, W.V., Karunatillake, S., Keller, J.M., Brückner, J., et al., 2006. Variations in K/Th on Mars. J. Geophys. Res. Planets 111 (E3). Available from: http://dx.doi.org/10.1029/2006JE002676.

Terada, N., Kulikov, Y.N., Lammer, H., Lichtenegger, H.I., Tanaka, T., Shinagawa, H., et al., 2009. Atmosphere and water loss from early Mars under extreme solar wind and extreme ultraviolet conditions. Astrobiology 9 (1), 55–70.

Touma, J.R., Wisdom, J., 1993. The chaotic obliquity of mars. (doctoral dissertation), Massachusetts Institute of Technology, Department of Mathematics.

Ulaby, F.T., Moore, R.K., Fung, A.K., 1986. Microwave Remote Sensing: Active and Passive, Vol. III: Volume Scattering and Emission Theory, Advanced Systems and Applications. Artech House, Dedham.

Ulrich, M., Morgenstern, A., Günther, F., Reiss, D., Bauch, K.E., Hauber, E., et al., 2010. Thermokarst in Siberian ice-rich permafrost: comparison to asymmetric scalloped depressions on Mars. J. Geophys. Res. Planets 115 (E10). Available from: http://dx.doi.org/10.1029/2010JE003640.

Usui, T., Alexander, C.M.D., Wang, J., Simon, J.I., Jones, J.H., 2012. Origin of water and mantle–crust interactions on Mars inferred from hydrogen isotopes and volatile element abundances of olivine-hosted melt inclusions of primitive shergottites. Earth Planet. Sci. Lett. 357, 119–129.

Usui, T., Alexander, C.M.D., Wang, J., Simon, J.I., Jones, J.H., 2015. Meteoritic evidence for a previously unrecognized hydrogen reservoir on Mars. Earth Planet. Sci. Lett. 410, 140–151.

Usui, T., Kurokawa, H., Wang, J., Alexander, C.M., Simon, J.I., & Jones, J.H., 2017. Hydrogen isotopic constraints on the evolution of surface and subsurface water on Mars. In: 48th Lunar and Planetary Science Conference, Houston, TX, USA, #1278.

Vago, J., Witasse, O., Svedhem, H., Baglioni, P., Haldemann, A., Gianfiglio, G., et al., 2015. ESA ExoMars program: the next step in exploring Mars. Sol. Syst. Res. 49, 518–528. Available from: http://dx.doi.org/10.1134/S0038094615070199.

Vaniman, D.T., Bish, D.L., Ming, D.W., Bristow, T.F., Morris, R.V., Blake, D.F., et al., 2014. Mineralogy of a mudstone at Yellowknife Bay, Gale Crater, Mars. Science 343 (6169). Available from: http://dx.doi.org/10.1126/science.1243480.

Villanueva, G.L., Mumma, M.J., Novak, R.E., Radeva, Y.L., Käufl, H.U., Smette, A., et al., 2013. A sensitive search for organics (CH_4, CH_3OH, H_2CO, C_2H_6, C_2H_2, C_2H_4), hydroperoxyl (HO_2), nitrogen compounds (N_2O, NH_3, HCN) and chlorine species (HCl, CH_3Cl) on Mars using ground-based high-resolution infrared spectroscopy. Icarus 223 (1), 11–27.

Villanueva, G.L., Mumma, M.J., Novak, R.E., Käufl, H.U., Hartogh, P., Encrenaz, T., et al., 2015. Strong water isotopic anomalies in the Martian atmosphere: probing current and ancient reservoirs. Science 348 (6231), 218–221.

Ward, W.R., 1992. Long-term orbital and spin dynamics of Mars. Mars 1, 298–320.

Watters, T.R., Campbell, B., Carter, L., Leuschen, C.J., Plaut, J.J., Picardi, G., et al., 2007a. Radar sounding of the Medusae Fossae Formation Mars: equatorial ice or dry, low-density deposits? Science 318 (5853), 1125–1128.

Watters, T.R., McGovern, P.J., Irwin III, R.P., 2007b. Hemispheres apart: the crustal dichotomy on Mars. Annu. Rev. Earth Planet. Sci. 35, 621–652. Available from: http://dx.doi.org/10.1146/annurev.earth.35.031306.140220.

Webster, C.R., Mahaffy, P.R., Atreya, S.K., Flesch, G.J., Farley, K.A., Kemppinen, O., et al., 2013. Low upper limit to methane abundance on Mars. Science 342 (6156), 355–357.

Webster, C.R., Mahaffy, P.R., Atreya, S.K., Flesch, G.J., Mischna, M.A., Meslin, P.Y., et al., 2015. Mars methane detection and variability at Gale Crater. Science 347 (6220), 415–417.

Weiss, D.K., Head, J.W., 2013. Formation of double-layered ejecta craters on Mars: a glacial substrate model. Geophys. Res. Lett. 40 (15), 3819–3824.

Weiss, D.K., Head, J.W., 2017. Testing landslide and atmospheric-effects models for the formation of double-layered ejecta craters on Mars. Meteorit Planet Sci. Available from: https://doi.org/10.1111/maps.12859.

Werner, S.C., Ody, A., Poulet, F., 2014. The source crater of Martian shergottite meteorites. Science 343 (6177), 1343–1346.

Williams, R.M., Phillips, R.J., 2001. Morphometric measurements of Martian valley networks from Mars Orbiter Laser Altimeter (MOLA) data. J. Geophys. Res. Planets 106 (E10), 23737–23751.

Williams, R.M., Malin, M.C., 2004. Evidence for late stage fluvial activity in Kasei Valles, Mars. J. Geophys. Res. Planets 109 (E6). Available from: http://dx.doi.org/10.1029/2003JE002178.

Williams, R.M., Malin, M.C., 2008. Sub-kilometer fans at Mojave Crater. Icarus 198, 365–383.

Williams, R.M.E., et al., 2013. Martian fluvial conglomerates at Gale Crater. Science 340, 1068–1072. Available from: http://dx.doi.org/10.1126/science.1237317.

Willmes, M., Reiss, D., Hiesinger, H., Zanetti, M., 2012. Surface age of the ice–dust mantle deposit in Malea Planum, Mars. Planet. Space Sci. 60 (1), 199–206.

Wilson, S.A., Howard, A.D., Moore, J.M., Grant, J.A., 2007. Geomorphic and stratigraphic analysis of Crater Terby and layered deposits north of Hellas basin, Mars. J. Geophys. Res. Planets 112 (E8). Available from: http://dx.doi.org/10.1029/2006JE002830.

Witze, A., 2014. NASA plans Mars sample-return rover: agency to narrow down list of landing sites for 2020 mission. Nature 509 (7500), 272–273.

Witze, A., 2017. NEXT STOP, MARS. Nature 541, 274–278.

Wordsworth, R.D., 2016. The climate of early Mars. Annu. Rev. Earth Planet. Sci. 44, 381–408.

Wordsworth, R.D., Forget, F., Millour, E., Head, J.W., Madeleine, J.B., Charnay, B., 2013. Global modelling of the early Martian climate under a denser CO_2 atmosphere: water cycle and ice evolution. Icarus 222 (1), 1–19.

Wordsworth, R.D., Kerber, L., Pierrehumbert, R.T., Forget, F., Head, J.W., 2015. Comparison of "warm and wet" and "cold and icy" scenarios for early Mars in a 3-D climate model. J. Geophys. Res. Planets 120 (6), 1201–1219.

Wordsworth, R.D., Kalugina, Y., Lokshtanov, S., Vigasin, A., Ehlmann, B., Head, J., et al., 2017. Transient reducing greenhouse warming on early Mars. Geophys. Res. Lett. Available from: http://dx.doi.org/10.1002/2016GL071766.

Zanetti, M., Hiesinger, H., Reiss, D., Hauber, E., Neukum, G., 2010. Distribution and evolution of scalloped terrain in the southern hemisphere, Mars. Icarus 206 (2), 691–706.

Zent, A.P., Fanale, F.P., Salvail, J.R., Postawko, S.E., 1986. Distribution and state of H_2O in the high-latitude shallow subsurface of Mars. Icarus 67 (1), 19–36.

Zuber, M.T., Smith, D.E., Solomon, S.C., Abshire, J.B., Afzal, R.S., Aharonson, O., et al., 1998. Observations of the north polar region of Mars from the Mars Orbiter Laser Altimeter. Science 282 (5396), 2053–2060.

Zurek, R.W., Chicarro, A., Allen, M.A., Bertaux, J.L., Clancy, R.T., Daerden, F., et al., 2011. Assessment of a 2016 mission concept: The search for trace gases in the atmosphere of Mars. Planet. Space Sci. 59 (2), 284–291.

SEQUESTRATION OF VOLATILES IN THE MARTIAN CRUST THROUGH HYDRATED MINERALS: A SIGNIFICANT PLANETARY RESERVOIR OF WATER

8

John F. Mustard

Brown University, Providence, RI, United States

Where and in what form water currently resides in reservoirs on Mars has important implications for understanding modern processes and the evolution of water on the planet. Estimates of water stored in reservoirs on Mars are typically expressed as the equivalent thickness of water spread over the planet in meters, or a global equivalent layer (GEL). Geological and geomorphological estimates for the total amount of water that may have been present on or passed over the Martian surface or that resides in the crust range from a minimum of 600 m to upward of 3000 m GEL (Carr, 1996; Baker, 2001), while accretion models suggest a total water inventory of 600–2600 m GEL (Lunine et al., 2003; Lammer et al., 2013). While these estimates of the GEL of water throughout Martian history are intriguing and result in a rich set of possibilities of a water-rich planet, in many ways, quantitative observations that provide constraints on these estimates are lacking. A recent synthesis of current reservoirs and an analysis of how those reservoirs may have changed going back in Mars geologic time provided an excellent assessment of water in the form of liquid, ice, and estimates of additions through volcanic processes and losses through escape and other mechanisms (Carr and Head, 2015). Lasue et al. (2012) also provide an updated assessment of GEL of water based on geomorphic and radar sounding data. The starting point for Carr and Head's (2015) analysis was based on an inventory of present water that existed in six reservoirs: (1) surface ice such as the exposed polar deposits, (2) shallow sequestered ice buried beneath regolith cover, (3) the global permafrost layer (Feldman et al., 2004; Mellon et al., 2004), (4) hypothesized groundwater beneath the base of the cryosphere (Clifford et al., 2010), (5) water vapor, and (6) water sequestered in hydrated mineral phases or consumed through alteration reactions (Mustard et al., 2012; Ehlmann et al., 2011).

It is reasonable to estimate the GEL of water for reservoirs that can be measured and quantified with existing spacecraft data. Using this approach, Carr and Head (2015) quantified the current reservoirs as approximately 22 m GEL for the layered terrains of the polar caps, which are very well constrained by Mars Orbiter Laser Altimeter (MOLA) (Smith et al., 1999) and Mars Advanced Radar for Subsurface and Ionosphere Sounding (MARSIS) (Plaut et al., 2007), a 7 m GEL for shallowly

Volatiles in the Martian Crust. DOI: http://dx.doi.org/10.1016/B978-0-12-804191-8.00008-8

buried ground ice as detected by GRS on Odyssey (Boynton et al., 2002; Feldman et al., 2004), MARSIS on Mars Express (Mouginot et al., 2010), and confirmed by the Phoenix lander (Smith et al., 2009) and observations of fresh impact craters (Byrne et al., 2009), and water sequestered among the rich collection of landforms between 30 and 60° N and S interpreted to host water-ice deposits such as the latitude-dependent mantle (Mustard et al., 2001), lobate debris aprons and possible glaciers (Head et al., 2003, 2005, 2010; Hauber et al., 2008) estimated in total to contain approximately 2.6 GEL of water (Levy et al., 2014).

Significant water may be stored in hypothesized deep aquifers (Clifford, 1993; Clifford et al., 2010) or as buried ice (Clifford and McCubbin, 2016), but direct observational evidence of these hypothesized water bodies from gravity or radar sounding is lacking and these remain purely hypothetical. See Chapter 7, The hydrology of Mars including a Potential Cryosphere for a full description.

Currently defined water reservoirs are insufficient to account for the estimated water inventory for the planet based on accretion and planetary evolution (Lunine et al., 2003; Lammer et al., 2013). Water has likely been lost through processes, such as impact erosion or solar wind sputtering, and may be as much as 99% of the original inventory (Brain and Jakosky, 1998). The MAVEN mission is in the midst of quantifying the modern hydrodynamic escape of water with important results published in 2017 showing that a large fraction, if not most of its atmosphere has been lost by interaction with the solar wind (Jakosky et al., 2017). A major reservoir of water that has been left unexamined is associated with hydrated minerals. Observations of hydrated minerals, their geologic contexts, and distribution (e.g., clays and sulfates) (Poulet et al., 2005; Gendrin et al., 2005; Mustard et al., 2008; Murchie et al., 2009; Ehlmann et al., 2009; Ehlmann and Edwards 2014; Carter et al., 2013) make it possible to derive a first order estimate of the amount of water that may be stored in the Martian crust in the form of hydrous minerals. Here, I examine data from the recent Mars missions to accomplish this goal.

The evidence for hydrated minerals from meteorites, landers, and orbiters will be briefly reviewed, and I will also consider the evidence for the hydration state of the Martian regolith. I then consider the global distribution of hydrated minerals across Mars and their distribution with depth, assess estimates of the abundance of hydrated minerals, and their capacity to store water. I then derive first order estimates of the water that is likely to be sequestered in the form of hydrated minerals on Mars.

8.1 HYDRATED MINERALS IN METEORITES

Prior to the first results from the Mars exploration rover Opportunity (Squyres et al., 2004), showing the presence of hydrated mineral phases in bedrock outcrops, the mineralogy of Mars was considered dominated by crystalline igneous minerals typical of basalt. While many hydrated mineral phases have been detected in Martian meteorites (e.g., McSween, 1994; Leshin and Vicenzi, 2006) how those observations translated to the planet, and thus the extent of alteration on Mars, was largely unknown or considered to be minimal. The meteorite record does show that Martian meteorites across the full range of ages show aqueous alteration, at some level, which occurred while the rocks were on Mars (Leshin and Vicenzi, 2006). However the degree of alteration is quite small, and that is to be expected given the mechanism of excavation, launch, and delivery of rocks from Mars to the

Earth. If a rock was extensively altered it is unlikely to have sufficient internal cohesion and strength to survive launch from Mars and delivery to Earth intact. Some rocks such as ALH 84001 show evidence for a complex period of aqueous alteration in a near-surface setting involving fluids in contact with the atmosphere (Niles et al., 2005) recorded in zoned carbonate nodules indicating that for some rocks a complex record of aqueous alteration is preserved. See Chapter 5, Carbonates on Mars for a full account of the carbonates in Martian meteorites.

A unique Martian meteorite, NWA 7034, and related meteorites provide novel insight into early Mars and regolith processes. Its bulk composition is comparable to estimates of average Martian surface composition by the gamma ray spectrometer (GRS) on the Mars Odyssey spacecraft and crustal rocks and soils measured by the MER and Curiosity rovers. These regolith breccia's contain a diverse suite of mineral and rock fragments and the highest water content of any Martian meteorite, 6000 ppm (Agee et al., 2013; Humayun et al., 2013). Thus, these meteorites are considered to be the first direct measurements of aqueous processes representative of the Martian regolith (Muttik et al., 2014). A recent thorough investigation of the distribution of water in NWA 7034 using mass balance calculations and modal abundances of minerals constrained by powder X-ray diffraction and petrography showed that the H_2O measured in the bulk of NWA 7034 is evenly distributed between hydrous Fe-rich oxides and phyllosilicate phases (Muttik et al., 2014). An important implication of this result is that Fe oxides could be as important as phyllosilicates and sulfates for H_2O storage in Martian surface material.

8.2 HYDRATED STATE OF MARTIAN REGOLITH/SOILS

Every near infrared spectrum of Mars acquired since the first telescopic IR observations (Sinton, 1967) shows clear evidence for the presence of water. This is indicated by a strong absorption near 3 μm due to fundamental stretching vibrations of OH either as −OH bond in mineral structures or in H_2O (Bishop et al., 1994). Similarly mid-infrared spectra show an emissivity maximum near 6 μm due to the fundamental H_2O band vibration. In a thorough suite of laboratory studies, Milliken and Mustard (2005) showed that the absolute water content of many hydrous phases (e.g., clay minerals, sulfate, and palagonite) can be estimated within ± 1 wt% using the 3-μm absorption analyzed with the radiative transfer formulation of the ESPAT function (Hapke, 1993). Leveraging the quantitative laboratory characterization of the hydration absorption, the abundance of water in surface soils was mapped across Mars with OMEGA data by Milliken et al. (2007) that compared favorably with an empirical formulation of water abundance by Jouglet et al. (2007). These studies estimated the surfaces to have 2–4 wt% H_2O at equatorial and mid-latitudes (Fig. 8.1). At high latitudes, typically above 40° N and S, there was an increase in water related to the seasonal polar hood and other processes and thus not necessarily indicative of water hosted in the soil mineralogy. In these polar regions, the presence of near surface to surface water-ice was confirmed though the Phoenix mission (Smith et al., 2009) and by exposure water-ice at latitudes as low as 50° by recent impact craters (Byrne et al., 2009).

The water content mapping by OMEGA showed an overall low abundance of water in the ≈4% range for most soils outside the polar region. There are some notable regions with large exposures of hydrous minerals detected from orbit in Mawrth Vallis (Loizeau et al., 2012) and Nili Fossae (Mangold et al., 2007; Ehlmann et al., 2009). Distinct from other typical equatorial and

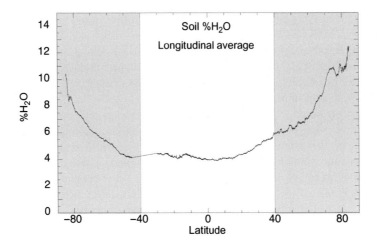

FIGURE 8.1

Longitudinally averaged estimate of rock, soil, and mineral hosted water abundance in the upper meter of surface soils determined using OMEGA 3-μm data (Milliken et al., 2007). The uncertainties in the OMEGA data are ± 1% and the precise water abundance for latitudes poleward of 40° (*shaded regions* of the graph) are likely affected by seasonal water-ice on the surface or in pores (Feldman et al., 2011). Water content calculated from the Mars Observer Neutron Spectrometer (MONS) averages ≈ 2% in similar longitudinally averaged syntheses though the MONS footprint is ≈ 900-km diameter spots vs the 1-km average surface resolution of OMEGA data.

mid-latitude regions these regions of high phyllosilicate abundance show enrichment in water by 2−3 × the commonly estimate amount of 2%−4%. High-latitude surfaces (polarwards 55°) exhibited water contents up to ∼ 15 wt%. Water abundance determined independently with the epithermal neutrons by the GRS showed a very similar abundance of water equivalent hydrogen (WEH) in the equatorial to mid-latitudes (Feldman et al., 2011). Direct comparison of longitudinally averaged profiles of water content from pole to pole (Fig. 8.1) show this quite well where estimates from the Mars Odyssey Neutron Spectrometer (MONS) show a concentration of 1%−3% WEH in the equatorial to mid-latitudes (Feldman et al., 2011). Thus, these global estimates from reflectance and gamma ray spectroscopy are very consistent with the determination of water abundance in the upper 1 m of the Martian regolith on the order of 2%−4%.

The detailed analyses of surface soil and rock chemistry by the Mars exploration rovers (Squyres et al., 2004) and the Curiosity rover (Ming et al., 2014) show derived water contents that are consistent with the remotely sensed water abundances. Glotch et al. (2006) used Mini-TES mineralogical model deconvolutions to estimate the sulfate-rich Burns formation outcrop water abundance at ∼ 5.5%. Not representative of the bulk regolith, but in line with higher abundances reported for sulfate-rich regions in Visible and Near Infrared (VNIR) remote sensing data. The most detailed analyses are for the John Klein and Cumberland drill core samples using the SAM evolved gas abundances measured during gas release of pyrolysis runs of samples. These show H_2O abundances ranging from 1.7% to 2.6% with uncertainties of ± 1.4% are within error of the Rocknest regolith samples (Ming et al., 2014). The Dynamic Albedo of Neutrons (DAN) experiment on the Curiosity rover has estimated WEH along its traverse during the first 200 Sols of the mission (Tate et al., 2015). WEH content was determined in both a fixed mode, typically measured overnight, and while in transit between fixed locations. The DAN results have a high degree of variability along the traverse. The fixed location estimates range from 0.5 ± 0.1 to 3.9 ± 0.2 wt% WEH and 0.6 ± 0.2 to 7.6 ± 1.3 wt% for areas measured while the rover was in transit measuring neutrons continuously.

8.3 GLOBAL DISTRIBUTION OF HYDRATED MINERALS

In addition to hydrous phases observed in Martian meteorites (Leshin and Vicenzi, 2006) and landers/rovers (Squyres et al., 2004; Ming et al., 2014), the diversity of hydrous minerals or minerals formed in the presence of water that have been positively identified on Mars from orbit encompass phyllosilicates (Poulet et al., 2005; Mustard et al., 2008), sulfates (Gendrin et al., 2005; Squyres et al., 2004), hydrated silica (Milliken et al., 2008), halides (Osterloo et al., 2008), and iron oxy-hydroxides such as goethite (Squyres et al., 2004; Christensen et al., 2000; Bibring et al., 2007). A list of these minerals detected is shown in Table 8.1. These minerals are found across Mars (Fig. 8.2) in a variety of depositional environments and host formations (Murchie et al., 2009; Ehlmann and Edwards, 2014; Carter et al., 2013).

An important characteristic of the orbital detections, noted in the earliest synthesis of the OMEGA results (Bibring et al., 2006), is that they are most common in outcrops throughout Noachian-aged terrain (Ehlmann et al., 2011; Carter et al., 2013). They are also present in crustal materials excavated from beneath Hesperian ridges plains in the northern lowlands (Carter et al., 2010). Hydrous minerals exist in Hesperian and younger terrains (Mangold et al., 2010; Skok et al., 2010; Milliken et al., 2008;

Table 8.1 Secondary Hydrated Minerals Detected on Mars from Orbital and Landed Instruments from the Recent Review by Ehlmann and Edwards (2014)

Mineral/Phase	Formula
Goethite	$FeO(OH)$
Akaganeite	$Fe(O,OH,C)$
Fe/Mg smectites (e.g., nontronite, saponite)	$(Ca,Na)_{0.3-0.5}(Fe,Mg,Al)_{2-3}(Al,Si)_4O_{10}(OH)_2 \cdot nH_2O$
Al smectites (e.g., montmorillonite, beidellite)	$(Na,Ca)_{0.3-0.5}(Al,Mg)_2(Al,Si)_4O_{10}(OH)_2 \cdot nH_2O$
Kaolin group minerals	$Al_2Si_2O_5(OH)_4$
Chlorite	$(Mg,Fe)_5Al(Si_3Al)O_{10}(OH)_8$
Serpentine	$(Mg, Fe)_3Si_2O_5(OH)_4$
Prehnite	$Ca_2Al(AlSi_3O_{10})(OH)_2$
Analcime	$NaAlSi_2O_6 \cdot H_2O$
Opaline silica	$SiO_2 \cdot nH_2O$
Kieserite, szomolnokite, polyhydrated sulfates	$(Mg,Fe)SO_4 \cdot nH_2O$
Gypsum	$Ca_2SO_4 \cdot 2H_2O$
Alunite	$KAl_3(SO_4)_2(OH)_6$
Jarosite	$KFe^{3+}_3(OH)_6(SO_4)_2$

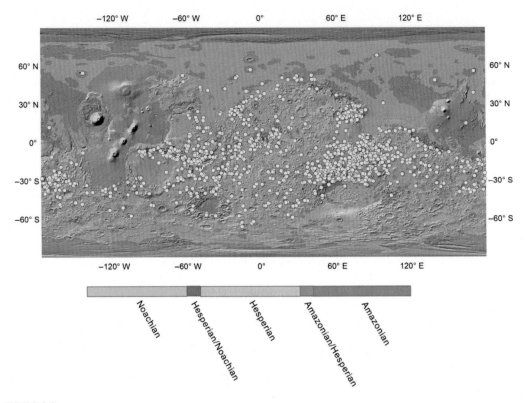

FIGURE 8.2

Global distribution of hydrated/aqueous minerals from the synthesis of Ehlmann and Edwards (2014) mapped on the MOLA shaded relief map with the distribution of geologic units colored by age. Detections are indicated by *dots* and include phyllosilicates and sulfates but do not discriminate among these hydrated mineral types. It also does not include minerals for which aqueous alteration was the likely process of formation (e.g., carbonates). The paucity of detections poleward of 40° N and S is due to the presence of a latitude-dependent layer (Mustard et al., 2001) that obscures crustal outcrops.

Sun and Milliken, 2015) but are less common, the outcrops more restricted in size and the deposits are in very specific geologic settings. In general, the most commonly detected hydrous mineral group is Fe/Mg phyllosilicates that include smectite clays and chlorite minerals, shown to encompass 70% of hydrous mineral detections (Carter et al., 2013). As shown in Fig. 8.2, the distribution of outcrops containing hydrous minerals is global and where exposure of Noachian crust is not obscured by alluvium, dust, or volcanic flows. Notable concentrations of aqueous minerals are observed in the very-well-exposed terrains in Nili Fossae (Mangold et al., 2007; Ehlmann et al., 2009; Mustard et al., 2009), Mawrth Valles (Bishop et al., 2008; Loizeau et al., 2012), Meridiani Planum (Gendrin et al., 2005), and Valles Marineris and circum-Chryse chaos terrains (Gendrin et al., 2005; Roach et al., 2010; Murchie et al., 2009). Because the detection of hydrous minerals in outcrop is biased by exposure (dust and other surficial deposits can obscure the spectral signatures of the bedrock), the mapped occurrences of hydrous mineral deposits likely represent a lower limit for the global distribution. This finding is supported by the detection of hydrated minerals in rocks excavated by impact craters in the northern plains that are otherwise covered by anhydrous volcanic plains of Hesperian age (Carter et al., 2010).

8.3.1 TO WHAT DEPTH IN THE MARTIAN CRUST ARE HYDRATED MINERALS OBSERVED?

Some of the thickest sections of crust exposed on Mars are along the walls of Valles Marineris. Here phyllosilicates have been observed in exposures of bedrock at the base of the walls of the canyon (Mustard et al., 2008; Flahaut et al., 2012). These outcrops are up to 8 km below the plateaus. Impact craters also provide a means to explore the deeper crust of Mars, and a survey of phyllosilicates exposed by impact craters in Terra Tyrrhena showed that craters up to 100 km in diameter contain phyllosilicates in their ejecta or excavated deposits (Fraeman et al., 2009) indicating an approximate depth of excavation of 10 km for phyllosilciate-bearing crust. Sun and Milliken (2015) conducted a thorough mineralogic and geologic analysis of 633 central peaks in Martian craters and using crater scaling algorithms estimated that the Martain crust hosts clays to depths of at least 7 km, in broad agreement with the observations from the unique outcrops in Valles Marineris. The data of Sun and Milliken (2015) indicate depths of excavation as large as 17 km for one impact. Together, these observations suggest that at least the upper 10 km of crust contain hydrated minerals.

8.3.1.1 Abundance of Aqueous Minerals

Phyllosilicate abundances for a range of aqueous mineral-bearing deposits across Mars have been estimated using radiative transfer modeling of CRISM and OMEGA spectra (e.g., Poulet et al., 2008, 2014; Goudge et al., 2015). Radiative transfer models for VNIR spectroscopy are largely based on the models of Hapke (1993) or Shkruatov et al., (1999). These models have a typical accuracy of 5%−10% under ideal laboratory conditions but with unknown accuracy applied to remotely acquired data. A significant challenge in deriving quantitative mineral abundance estimates is the degree of exposure. In many regions on Mars, thin to thick coatings of dust obscure the spectral signature of the underlying bedrock, impairing or precluding abundance determinations. Nevertheless, there are some excellent exposures for which abundance estimates have been derived. The most phyllosilicate-rich region of the planet identified with orbital data is Mawrth Valles, where more than 50% phyllosilicate has been calculated for some outcrops with excellent exposure (Poulet et al., 2008). Mawrth Valles shows a wide range of phyllosilicates from nontronite and saponite to kaolinite, montmorillonite, and opal, as well as sulfates (Farrand et al., 2009, 2014) and hydrated ferric oxides like ferrihydrite (Poulet et al., 2014).

The modal mineralogy abundances calculated by Poulet et al. (2014) for the four priority landing sites considered for the Mars Science Laboratory ranged from 50% in Mawrth Valles to 10% or less at some of the other sites like Eberswalde and Gale crater. Gale crater, however, was suspected to have an obscuring dust cover, confirmed by the landed science analyses (e.g., Wellington et al., 2017). Using an innovative unmixing approach, combining spectra from laboratory minerals and remotely sensed components (Goudge et al., 2015) showed the deposits in Kashira crater Mars, rich in kaolinite group minerals, had ≈ 30% halloysite. The results were consistent for both visible-near infrared reflectance data and mid-infrared emission spectra. The ChemMin X-ray diffraction instrument on the Curiosity rover has quantitatively determined the mineralogy of carefully selected mudstones on Mars (Vaniman et al., 2014). These results showed approximately 20% phyllosilicates and a further 20% amorphous components that were of uncertain hydration. These amorphous phases could be the initial breakdown products of weathering or immature phyllosilicates and hydrated ferric oxides.

The vast majority of rock exposures on Mars, however, do not show evidence of hydrated mineral phases. Instead they are dominated by the anhydrous rock-forming minerals pyroxene, plagioclase, and olivine (e.g., Ody et al., 2013; Skok et al., 2012; Mustard et al., 2005; Rogers and Christensen, 2007), as well as abundant glassy or amorphous phases (Schultz and Mustard, 2004; Cannon et al., 2016). Quantitative mineral abundance determinations for these regions do not require hydrated mineral phases (e.g., Poulet et al., 2009; Rogers and Christensen, 2007). While the global distribution maps (Fig. 8.2) of hydrated mineral occurrences show the minerals are quite common and globally distributed (Ehlmann and Edwards, 2014; Carter et al., 2013), hydrated minerals nevertheless make up only a small fraction of the bulk crust, but in exceptionally well-exposed locations like Mawrth Valles and Nili Fossae they make up the bulk of the rock. This makes precise determination of the bulk abundance of hydrated minerals in typical Martian Noachian crust somewhat problematic, but the global occurrence, high concentration in some regions, and consistent characterization across landed in situ and orbital science provide a compelling finding pointing to global distribution and populating the top 10 km of crust. Using a thorough compilation of all occurrences of hydrated minerals, Carter et al. (2013) performed a statistical analysis of the detection rate and areal coverage of hydrated minerals. This detailed analysis resulted in a result that hydrous silicates in the Noachian-aged southern highlands cover roughly 3% of the surface.

Earth's oceanic crust provides an interesting analog with which to compare the abundance of hydrous minerals in the Martian crust. Analysis of samples from deep drilling, dredging, and sampling of the oceanic crust for the abundance of alteration minerals show typical abundances of 10% hydrous minerals (e.g., Alt et al., 1986). In the vertical sections of ocean crust reconstructed from sample and seismic analyses, the type of phyllosilicate changes from smectite clays and chlorite in the upper sections of the crust to amphiboles in the lower sections due to increasing temperature and pressure. It is notable that except for rare outcrops containing prehnite (a phyllosilicate formed between 250 °C and 350 °C), there have been no high temperature and/or high pressure phyllosilicates identified on Mars from orbital data, suggesting no phyllosilicate-bearing rocks from deep terrains have been excavated or that high temperature and pressure phases are not commonly formed on Mars.

From the analyses of oceanic basalts, it has been estimated that the top 5 km of oceanic crust is altered and typically contains ≈ 10% hydrous minerals. Converting this phyllosilicate abundance to bulk water fixed in the rocks results in an average of 1%−3% water depending on the exact abundance and the most common hydrous mineral phases (Carlson, 2003). The persistent presence of alteration in the Earth's oceanic crust is a product of hydrothermal alteration at mid-ocean ridges and as newly formed crust is carried off the ridge axis by seafloor spreading (Alt et al., 1986). While the process of hydrothermal alteration that occurs on the ocean floor is not analogous to Mars, because of the lack of plate tectonics and an ocean, phyllosilicate-bearing rocks are common in the bulk of Noachian-aged rocks exposed by erosion, faulting, and impact cratering either in the crater walls or in ejecta.

It has been well established that Noachian-aged crust on Mars commonly exhibits the presence of phyllosilicates (e.g., Bibring et al., 2006; Ehlmann and Edwards, 2014; Carter et al., 2013). Using spatial statistics, Carter et al. (2013) evaluated the areal coverage of Noachian outcrops showing the presence of hydrated minerals and estimated that 3% of the exposed surface that is not covered by dust exhibited spectral evidence for the presence of aqueous minerals. The processes of formation for the phyllosilicates are widely debated and include low-grade hydrothermal alteration in the crust, alteration by hydrothermal cells stimulated by impact processes, weathering at or near the surface (Ehlmann et al., 2011; Carter et al., 2013), and alteration by supercritical fluids during condensation of the original Martian atmosphere during accretion (Cannon et al., 2016). Estimates of phyllosilcate mineral abundance for these typical terrains are on the order of 5%−15% with locally very high concentrations (Poulet et al., 2009, 2014). Combining estimates of H_2O abundance from the Odyssey Neutron Spectrometer, phyllosilicate mineral abundance with maps of aqueous mineral distributions, we conservatively estimate the abundance of hydrated minerals in Noachian crust to range from 1% to 10% when averaged over all Noachian terrains. This value is comparable to, though likely on the lower end of, the average concentration of hydrated minerals in average oceanic crust.

8.3.2 TRANSLATING AQUEOUS MINERAL ABUNDANCE TO WATER CONTENT

The water content of phyllosilicate minerals varies widely. For minerals such as kaolinite, that actually do not have water in their structure but instead hydroxyl groups, the equivalent water content is 13%. This represents the amount of water that would be sequestered in kaolin minerals due to their formation from primary igneous minerals through alteration by water and subsequent burial. Smectite clays such as montmorillonite, saponite, and nontronite have a capacity to exchange water in their interlayer regions and can thus have quite variable water content. For example, montmorillonite has 2 hydroxyl molecules per formula unit that are structural (not exchangeable) and up to 10 additional water molecules in the interlayer region. In the maximum case, montmorillonite is 36% water. However, estimates of the water content of phyllosilicate-bearing rock based on the identification of the primary aqueous mineral phase and the abundance of that phase is predicated on a number of assumptions such as (1) the mineral phases detected from remotely sensed data are the most common alteration phases present, (2) they follow the textbook mineral formulation, and (3) the surface areal abundance measured by remote sensing is representative of the volume abundance to some reasonable depth in the crust. I can nevertheless conduct a first-order validation of this approach to the translation of mineral abundance to water content using spacecraft observations. I start with the abundance determination of 50% montmorillonite for the Mawrth Valles region (Poulet et al., 2014). Using the range in montmorillonite formula abundance of water, this results in a prediction of the water content of the outcrops in Mawrth Valles to be 7.8%−9.1%. The H_2O content of Mawrth Valles region with the strongest montmorillonite absorptions bands (and thus the highest concentration of montmorillonite exposed on the surface) was estimated by Milliken et al. (2007) using the strength of the fundamental OH absorption near 3.0 μm in thermally corrected OMEGA data to be 6%−8%. This is remarkably consistent with the abundance of water derived from clay abundance estimates for the same region of 7.8%−9.1%. This mineral abundance estimate from Poulet et al. (2014) does not use the same spectral features or wavelength region as the spectral region used to estimate the water content (Milliken et al., 2007). The good correspondence between the remotely detected H_2O abundance and that estimated by the abundance of hydrous minerals provides confidence in using mineral abundance to estimate water content of the Martian crustal materials. The use of VNIR spectroscopy to estimate water content is further supported by the excellent correspondence between WEH from remotely sensed neutrons and water content estimated from VNIR spectroscopy (Feldman et al., 2011).

8.3.3 SIZE OF THE HYDROUS MINERAL CRUST RESERVOIR

Synthesis of orbital, landed, and meteoritical observations and measurements has established that hydrated minerals are commonly present in Noachian-aged crust, but that the abundance of the minerals ranges widely from undetected to >50% by weight. More commonly, the abundance of phyllosilicates is in the range of 1%–10%, while the surface soils, the product of physical and chemical weathering, consistently show a water content of 2%–4%. Hydrated minerals are distributed in the bulk of the crust as shown by excavation by meteorite impact to at least 10 km. Based on the presence of hydrated minerals throughout Noachian-aged terrains and the depth to which they are observed, I can then estimate the size of the crustal reservoir for water sequestered in hydrous minerals. Using the density of typical Martian crust of $3100\ kg/m^3$ (Baratoux et al., 2014), estimates of water content by percentage are then used to integrate the mass of water present in a column of Mars rock 1, 5, and 10 km thick (Table 8.2). I develop these estimates for three concentrations of water: 0.5, 1, and 3 wt% water. The lower range acknowledges that the most common spectroscopic observation of Mars lacks a combination tone absorption band at 1.9 μm that is the definitive indicator for the presence of water-bearing minerals. However, thin coatings of airfall dust readily prevent the detection of this and other diagnostic absorptions and every VNIR spectrum of Mars does show a 3.0-μm absorption showing water is present at a low abundance, perhaps as thin films of adsorbed water or within and associated with the ubiquitous amorphous component of the soils. The value of 3% is an upper limit that is supported by the statistical analyses of outcrops bearing hydrated minerals (Carter et al., 2013).

The lower range represents the volume of water that Andrews-Hanna and Lewis (2011) estimated would need to have been removed from the active hydrological system to have changed the

Table 8.2 Variance in the Estimated Global Equivalent Layer of Water Sequestered in Hydrated Minerals for Different Depths of Crust Altered and Intensity of Alteration

Wt% H_2O	Thickness of Altered Crust (km)	GEL of Water (m)
0.5	1	15.5
	5	77.5
	10	155
1	1	31
	5	155
	10	310
3	1	93
	5	465
	10	930

hydrology of early Mars from one with abundant surface activity (e.g., valley networks) to one dominated by groundwater processes. The upper range would accommodate the total water budget for the planet estimated by some recent models (Lunine et al., 2003). Regardless, a significant fraction of this "water" may represent a reservoir that is largely sequestered from participating in hydrological processes unless liberated by impact or heating from volcanism (e.g., structural OH in phyllosilicates). The calculations assume an average water content that is uniform with depth. If the water content instead varies with depth, as is likely, the reservoir size could be enhanced or diminished. Did the alteration of the crust happen slowly, leading to a declining abundance of free water? Was this related to the Noachian–Hesperian transition? How did this affect the atmosphere? I will be refining my analyses factoring in new calculations and evaluating the implications for planetary evolution.

8.4 POSSIBLE IMPLICATIONS

With the discovery of hydrated minerals on Mars and the decade of detailed analyses of orbital and landed mission data, we now recognize that a reservoir of water exists sequestered in hydrated mineral phases. The existence of the reservoir is supported by analyses of meteorites, global geochemical measurements by gamma ray and neutron spectroscopy, visible-near infrared and thermal emission spectroscopy, and landed science packages. The size of the reservoir is a function of the abundance of hydrated mineral phases and their distribution in the crust of Mars. Because hydrated mineral phases are a ubiquitous presence in Noachian-aged crust distributed in the top 10 km, it is then possible to estimate the reservoir size. The reservoir is estimated here (Table 8.2) to range in size from a low of 10s of meters GEL of water if only the top 1 km of crust is altered with an abundance of water hosted in hydrated minerals phases of 0.5% to a high of nearly 1 km GEL if the abundance of water hosted in hydrated mineral phases is 3% distributed over the top 10 km of Mars' crust.

While the existence of this reservoir has been widely acknowledged (e.g., Carr and Head, 2015), this is the first attempt to establish the size of the reservoir. The smallest volume derived here, 15 m GEL, is comparable to the volume of water currently estimated to be stored in the polar caps, while the larger volume is well within the range of volumes proposed for the planet at the time for formation and early evolution (e.g., Lunine et al., 2003; Kurokawa et al., 2014). The difference is that the current volume of water in the polar caps is well defined, while the volumes estimated in planetary evolution models are not specific of the fate of that water, where it is today, and in what volumes. The 15 m GEL estimate is too small; it seems to represent the total volume of water that might be stored in the crust. But for the larger volume estimate, there are significant questions about its current state and the processes and timeline that led to the formation of the alteration minerals in the crust.

Is it possible that the water now sequestered in the crust could be liberated, now or in the past, to become reunited with the surface and near-surface reservoirs of water? For example, impact heating could elevate the temperate of aqueous mineral phases to the point that water is exsolved. In the case of smectite clays, this is on the order of 100 °C–200 °C. Or heating from ascent and eruption of magma in a volcanic province emplaced on Noachian-aged terrains, like the

Hesperian-aged Syrtis Major, could significantly increase heat flow and elevate the temperature of the crust sufficiently to release water from weakly held water in sulfate and smectite, or similar phyllosilicate, minerals. While the source of the Hesperian-aged outflow channels are widely believed to be from overpressurization of buried and frozen aquifers (Clifford et al., 2010), the Amazonian-aged outflow channels associated with Cerberus Fossae are hypothesized to be from buried aquifers mobilized by volcanic eruptions (Burr et al., 2002). However, the source of the water for the Cerberus Fossae outflows is largely unconstrained and sourcing the water from hydrated minerals is a process that has been unexamined.

The volume of water involved could have large implications for magmatic process and the evolution of the Mars interior (Grott et al., 2013). The apparent differences in water in ancient Noachian crust versus the more anhydrous Hesperian crust has implications for the evolution of magma source regions and how the mantle may have evolved with time. This will clearly have an effect on our understanding of the new igneous lithologies found by the Curiosity rover (Stolper et al., 2013; Schmidt et al., 2014).

When and by what processes the aqueous alteration occurred are significant unknowns about this hydrated mineral water reservoir. Three hypotheses have been proposed: low grade metamorphism in the shallow crust (Ehlmann et al., 2011), surface weathering and alteration mixed into the crust through impact processes (Carter et al., 2013) and alteration by supercritical fluids during accretion, and formation of the first atmosphere (Elkins-Tanton, 2008; Cannon et al., 2016). The processes of alteration likely encompassed a large span of time and involved a number of different environments and geologic settings (Murchie et al., 2009). This is still an evolving area of research but when and how alteration occurred would have had some effect on the evolution of aqueous activity on the surface and in the shallow subsurface. For example, the model proposed by Andrews-Hanna and Lewis (2011) to explain the transition from a hypothesized warmer and wetter early Mars to a cold and arid Hesperian Mars required loss of water from the surface and near surface reservoirs of on the order of 60 m GEL. The process of water loss is not important in their model, and the water could be lost due to atmospheric stripping or storage into long-term reservoirs. The altered mineral reservoir is not examined in detail in their model but could readily accommodate the volume of water needed for the transition if the sequestration occurred at the same time as the environmental change.

Further refinement of the types and abundance of alteration minerals present in the crust of Mars, and distribution in different geologic settings and the crust by existing and future missions will greatly improve the estimate of crustal reservoir of water by hydrated minerals. As the prospects of sample return and future exploration of Mars by humans become more well defined and closer to being possible, these crustal reservoirs of water may be significant resources for robotic and human missions.

REFERENCES

Agee, C.B., Wilson, N.V., McCubbin, F.M., Ziegler, K., Polyak, V.J., Sharp, Z.D., et al., 2013. Unique meteorite from early Amazonian Mars: water-rich basaltic breccia Northwest Africa 7034. Science 339, 780–785.

Alt, J.C., Honnorez, J., Laverne, C., Emmermann, R., 1986. Hydrothermal alteration of a 1 km section through the upper oceanic crust, Deep Sea Drilling Project Hole 504B: mineralogy, chemistry and evolution of seawater-basalt interactions. J. Geophys. Res.: Solid Earth 91, 10309–10335.

Andrews-Hanna, J.C., Lewis, K.W., 2011. Early Mars hydrology: 2. Hydrological evolution in the Noachian and Hesperian epochs. J. Geophys. Res. 116. Available from: http://dx.doi.org/doi:10.1029/2010JE003709.

Baker, V.R., 2001. Water and the Martian landscape. Nature. 412, 228–236.

Baratoux, D., Samuel, H., Michaut, C., Toplis, M.J., Monnereau, M., Wieczorek, M., et al., 2014. Petrological constraints on the density of the Martian crust. J. Geophys. Res.: Planets 119, 1707–1727.

Bibring, J.P., Langevin, Y., Mustard, J.F., Poulet, F., Arvidson, R., Gendrin, A., et al., 2006. Global mineralogical and aqueous Mars history derived from OMEGA/Mars Express data. Science 312, 400–404.

Bibring, J.P., Arvidson, R.E., Gendrin, A., Gondet, B., Langevin, Y., Le Mouelic, S., et al., 2007. Coupled ferric oxides and sulfates on the Martian surface. Science 317, 1206–1210.

Bishop, J.L., Pieters, C.M., Edwards, J.O., 1994. Infrared spectroscopic analyses on the nature of water in montmorillonite. Clay Clay Miner. 42, 702–716.

Bishop, J.L., Dobrea, E.Z.N., McKeown, N.K., Parente, M., Ehlmann, B.L., Michalski, J.R., et al., 2008. Phyllosilicate diversity and past aqueous activity revealed at Mawrth Vallis, Mars. Science 321, 830–833.

Boynton, W.V., Feldman, W., Squyres, S., Prettyman, T., Brückner, J., Evans, L., et al., 2002. Distribution of hydrogen in the near surface of Mars: evidence for subsurface ice deposits. Science 297, 81–85.

Brain, D.A., Jakosky, B.M., 1998. Atmospheric loss since the onset of the Martian geologic record: combined role of impact erosion and sputtering. J. Geophys. Res.: Planets 103, 22689–22694.

Burr, D.M., Grier, J.A., McEwen, A.S., Keszthelyi, L.P., 2002. Repeated aqueous flooding from the Cerberus Fossae: evidence for very recently extant, deep groundwater on Mars. Icarus. 159, 53–73.

Byrne, S., Dundas, C.M., Kennedy, M.R., Mellon, M.T., McEwen, A.S., Cull, S.C., et al., 2009. Distribution of mid-latitude ground ice on Mars from new impact craters. Science 325, 1674–1676.

Cannon, K.M., Parman, S.W., Mustard, J.F., 2016. Hot and steamy: alteration of the primordial Martian crust by supercritical fluids during magma ocean cooling. In: Lunar and Planetary Science Conference. Abstract #1265.

Carlson, R.L., 2003. Bound water content of the lower oceanic crust estimated from modal analyses and seismic velocities of oceanic diabase and gabbro. Geophys. Res. Lett. 30.

Carr, M., 1996. Water on Mars. Oxford University Press, New York, 229 pp.

Carr, M., Head, J., 2015. Martian surface/near-surface water inventory: sources, sinks, and changes with time. Geophys. Res. Lett. 42, 726–732.

Carter, J., Poulet, F., Bibring, J.-P., Murchie, S., 2010. Detection of hydrated silicates in crustal outcrops in the northern plains of Mars. Science 328, 1682–1686.

Carter, J., Poulet, F., Bibring, J.P., Mangold, N., Murchie, S., 2013. Hydrous minerals on Mars as seen by the CRISM and OMEGA imaging spectrometers: updated global view. J. Geophys. Res.: Planets 118, 831–858.

Christensen, P., Bandfield, J., Clark, R., Edgett, K., Hamilton, V., Hoefen, T., et al., 2000. Detection of crystalline hematite mineralization on Mars by the thermal emission spectrometer: evidence for near-surface water. J. Geophys. Res.: Planets 105, 9623–9642.

Clifford, S.M., 1993. A model for the hydrologic and climatic behavior of water on Mars. J. Geophys. Res.: Planets 98, 10973–11016.

Clifford, S.M., McCubbin, F.M., 2016. How well does the present surface inventory of water on Mars constrain the past? In: 47th Lunar and Planetary Science Conference, Houston, TX. Abstract #2388.

Clifford, S.M., Lasue, J., Heggy, E., Boisson, J., McGovern, P., Max, M.D., 2010. Depth of the Martian cryosphere: revised estimates and implications for the existence and detection of subpermafrost groundwater. J. Geophys. Res. 115. Available from: http://dx.doi.org/doi:10.1029/2009JE003462.

Ehlmann, B.L., Edwards, C.S., 2014. Mineralogy of the Martian surface. Annu. Rev. Earth. Planet. Sci. 42, 291–315.

Ehlmann, B.L., Mustard, J.F., Swayze, G.A., Clark, R.N., Bishop, J.L., Poulet, F., et al., 2009. Identification of hydrated silicate minerals on Mars using MRO-CRISM: geologic context near Nili Fossae and implications for aqueous alteration. J. Geophys. Res.-Planets 114. Available from: http://dx.doi.org/doi:10.1029/2009JE003339.

Ehlmann, B.L., Mustard, J.F., Clark, R.N., Swayze, G.A., Murchie, S.L., 2011. Evidence for low-grade metamorphism, hydrothermal alteration, and diagenesis on Mars from phyllosilicate mineral assemblages. Clay Clay Miner. 59, 359–377.

Elkins-Tanton, L.T., 2008. Linked magma ocean solidification and atmospheric growth for Earth and Mars. Earth Planet. Sci. Lett. 271, 181–191.

Farrand, W.H., Glotch, T.D., Rice, J.W., Hurowitz, J.A., Swayze, G.A., 2009. Discovery of jarosite within the Mawrth Vallis region of Mars: implications for the geologic history of the region. Icarus 204 (2), 478–488.

Farrand, W.H., Glotch, T.D., Horgan, B., 2014. Detection of copiapite in the northern Mawrth Vallis region of Mars: evidence of acid sulfate alteration. Icarus 241, 346–357.

Feldman, W.C., Prettyman, T.H., Maurice, S., Plaut, J., Bish, D., Vaniman, D., et al., 2004. Global distribution of near-surface hydrogen on Mars. J. Geophys. Res.: Planets 109. Available from: http://dx.doi.org/doi:10.1029/2003JE002160.

Feldman, W.C., Pathare, A., Maurice, S., Prettyman, T.H., Lawrence, D.J., Milliken, R.E., et al., 2011. Mars Odyssey neutron data: 2. Search for buried excess water ice deposits at nonpolar latitudes on Mars. J. Geophys. Res. 116, E11009. Available from: http://dx.doi.org/doi:10.1029/2011JE003806.

Flahaut, J., Quantin, C., Clenet, H., Allemand, P., Mustard, J.F., Thomas, P., 2012. Pristine Noachian crust and key geologic transitions in the lower walls of Valles Marineris: insights into early igneous processes on Mars. Icarus 221, 420–435.

Fraeman, A.A., Mustard, J.F., Ehlmann, B.L., Roach, L.H., Milliken, R.E., Murchie, S.L., 2009. Evaluating models of crustal cooling using CRISM observations of impact craters in Terra Tyrrhena and Noachis Terra In: 40th Lunar and Planetary Science Conference. Lunar and Planetary Institute, Houston, TX. Abstract #2320..

Gendrin, A., Mangold, N., Bibring, J.P., Langevin, Y., Gondet, B., Poulet, F., et al., 2005. Suffates in Martian layered terrains: the OMEGA/Mars Express view. Science 307, 1587–1591.

Glotch, T.D., Bandfield, J.L., Christensen, P.R., Calvin, W.M., McLennan, S.M., Clark, B.C., et al., 2006. Mineralogy of the light-toned outcrop at Meridiani Planum as seen by the Miniature Thermal Emission Spectrometer and implications for its formation. J. Geophys. Res.: Planets 111 (E12), E12S03. Available from: http://dx.doi.org/doi:10.1029/2005JE002672.

Goudge, T.A., Mustard, J.F., Head, J.W., Salvatore, M.R., Wiseman, S.M., 2015. Integrating CRISM and TES hyperspectral data to characterize a halloysite-bearing deposit in Kashira Crater, Mars. Icarus 250, 165–187.

Grott, M., Baratoux, D., Hauber, E., Sautter, V., Mustard, J., Gasnault, O., et al., 2013. Long-term evolution of the Martian crust-mantle system. Space. Sci. Rev. 174 (1–4), 49–111.

Hapke, B., 1993. Theory of Emittance and Reflectance Spectroscopy. Cambridge University Press, New York.

Hauber, E., van Gasselt, S., Chapman, M., Neukum, G., 2008. Geomorphic evidence for former lobate debris aprons at low latitudes on Mars: indicators of the Martian paleoclimate. J. Geophys. Res.: Planets 113. Available from: http://dx.doi.org/doi:10.1029/2007JE002897.

Head, J.W., Mustard, J.F., Kreslavsky, M.A., Milliken, R.E., Marchant, D.R., 2003. Recent ice ages on Mars. Nature 426, 797–802.

Head, J.W., Neukum, G., Jaumann, R., Hiesinger, H., Hauber, E., Carr, M., et al., 2005. Tropical to mid-latitude snow and ice accumulation, flow and glaciation on Mars. Nature 434, 346–351.

Humayun, M., Nemchin, A., Zanda, B., Hewins, R., Grange, M., Kennedy, A., et al., 2013. Origin and age of the earliest Martian crust from meteorite NWA [thinsp] 7533. Nature 503, 513−516.

Jakosky, B.M., Slipski, M., Benna, M., Mahaffy, P., Elrod, M., Yelle, R., et al., 2017. Mars' atmospheric history derived from upper-atmosphere measurements of 38Ar/36Ar. Science 355 (6332), 1408−1410.

Jouglet, D., Poulet, F., Milliken, R.E., Mustard, J.F., Bibring, J.P., Langevin, Y., et al., 2007. Hydration state of the Martian surface as seen by Mars Express OMEGA: 1. Analysis of the $3\,\mu m$ hydration feature. J. Geophys. Res.-Planets 112. Available from: http://dx.doi.org/doi:10.1029/2006JE002846.

Kurokawa, H., Sato, M., Ushioda, M., Matsuyama, T., Moriwaki, R., Dohm, J.M., et al., 2014. Evolution of water reservoirs on Mars: constraints from hydrogen isotopes in Martian meteorites. Earth Planet. Sci. Lett. 394, 179−185.

Lammer, H., Chassefière, E., Karatekin, Ö., Morschhauser, A., Niles, P.B., Mousis, O., et al., 2013. Outgassing history and escape of the Martian atmosphere and water inventory. Space. Sci. Rev. 174, 113−154.

Lasue, J., Mangold, N., Hauber, E., Clifford, S., Feldman, W., Gasnault, O., et al., 2012. Quantitative assessments of the Martian hydrosphere. Space. Sci. Rev. 174, 155−212.

Leshin, L.A., Vicenzi, E., 2006. Aqueous processes recorded by Martian meteorites: analyzing Martian water on Earth. Elements 2, 157−162.

Levy, J.S., Fassett, C.I., Head, J.W., Schwartz, C., Watters, J.L., 2014. Sequestered glacial ice contribution to the global Martian water budget: geometric constraints on the volume of remnant, midlatitude debris-covered glaciers. J. Geophys. Res.-Planets 119, 2188−2196.

Loizeau, D., Werner, S.C., Mangold, N., Bibring, J.P., Vago, J.L., 2012. Chronology of deposition and alteration in the Mawrth Vallis region, Mars. Planet. Space. Sci. 72, 31−43.

Lunine, J.I., Chambers, J., Morbidelli, A., Leshin, L.A., 2003. The origin of water on Mars. Icarus 165, 1−8.

Mangold, N., Poulet, F., Mustard, J.F., Bibring, J.P., Gondet, B., Langevin, Y., et al., 2007. Mineralogy of the Nili Fossae region with OMEGA/Mars Express data: 2. Aqueous alteration of the crust. J. Geophys. Res.-Planets 112, E08S04. Available from: http://dx.doi.org/doi:10.1029/2006JE002835.

Mangold, N., Roach, L., Milliken, R., Le Mouelic, S., Ansan, V., Bibring, J.P., et al., 2010. A Late Amazonian alteration layer related to local volcanism on Mars. Icarus 207, 265−276.

McSween, H.Y., 1994. What we have learned about Mars from SNC meteorites. Meteoritics 29, 757−779.

Mellon, M.T., Feldman, W.C., Prettyman, T.H., 2004. The presence and stability of ground ice in the southern hemisphere of Mars. Icarus 169, 324−340.

Milliken, R.E., Mustard, J.F., 2005. Quantifying absolute water content of minerals using near-infrared reflectance spectroscopy. J. Geophys. Res.-Planets 110. Available from: http://dx.doi.org/doi:10.1029/2005je002534.

Milliken, R.E., Mustard, J.F., 2007. Estimating the water content of hydrated minerals using reflectance spectroscopy. II. Effects of particle size. Icarus 189, 574−588.

Milliken, R.E., Mustard, J.F., Poulet, F., Jouglet, D., Bibring, J.-P., Gondet, B., et al., 2007. Hydration state of the Martian surface as seen by Mars Express OMEGA: 2. H_2O content of the surface. J. Geophys. Res.-Planets 112. Available from: http://dx.doi.org/doi:10.1029/2006je002853.

Milliken, R.E., Swayze, G.A., Arvidson, R.E., Bishop, J.L., Clark, R.N., Ehlmann, B.L., et al., 2008. Opaline silica in young deposits on Mars. Geology 36, 847−850.

Ming, D.W., Archer, P.D., Glavin, D.P., et al., 2014. Volatile and organic compositions of sedimentary rocks in Yellowknife Bay, Gale Crater, Mars. Science 343. Available from: http://dx.doi.org/doi:10.1126/science.1245267.

Mouginot, J., Pommerol, A., Kofman, W., Beck, P., Schmitt, B., Herique, A., et al., 2010. The 3−5 MHz global reflectivity map of Mars by MARSIS/Mars Express: implications for the current inventory of sub-surface H_2O. Icarus 210, 612−625.

Murchie, S.L., Mustard, J.F., Ehlmann, B.L., Milliken, R.E., Bishop, J.L., McKeown, N.K., et al., 2009. A synthesis of Martian aqueous mineralogy after 1 Mars year of observations from the Mars Reconnaissance Orbiter. J. Geophys. Res.-Planets 114. Available from: http://dx.doi.org/doi:10.1029/2009je003342.

Mustard, J.F., Cooper, C.D., Rifkin, M.K., 2001. Evidence for recent climate change on Mars from the identification of youthful near-surface ground ice. Nature 412, 411−414.

Mustard, J.F., Poulet, F., Gendrin, A., Bibring, J.P., Langevin, Y., Gondet, B., et al., 2005. Olivine and pyroxene, diversity in the crust of Mars. Science 307, 1594−1597.

Mustard, J.F., Murchie, S.L., Pelkey, S.M., Ehlmann, B.L., Milliken, R.E., Grant, J.A., et al., 2008. Hydrated silicate minerals on Mars observed by the Mars Reconnaissance Orbiter CRISM instrument. Nature 454, 305−309.

Mustard, J.F., Ehlmann, B.L., Murchie, S.L., Poulet, F., Mangold, N., Head, J.W., et al., 2009. Composition, morphology, and stratigraphy of Noachian crust around the Isidis basin. J. Geophys. Res.-Planets 114. Available from: http://dx.doi.org/doi:10.1029/2009je003349.

Mustard, J.F., Poulet, F., Ehlman, B.E., Milliken, R.E., Fraeman, A., 2012. Sequestration of volatiles in the Martian crust through hydrated minerals: a significant planetary reservoir of water. In: 43rd Lunar and Planetary Science Conference. Abstract #1539.

Muttik, N., McCubbin, F.M., Keller, L.P., Santos, A.R., McCutcheon, W.A., Provencio, P.P., et al., 2014. Inventory of H2O in the ancient Martian regolith from Northwest Africa 7034: the important role of Fe oxides. Geophys. Res. Lett. 41, 8235−8244.

Niles, P., Leshin, L., Guan, Y., 2005. Microscale carbon isotope variability in ALH84001 carbonates and a discussion of possible formation environments. Geochim. Cosmochim. Acta 69, 2931−2944.

Ody, A., Poulet, F., Bibring, J.P., Loizeau, D., Carter, J., Gondet, B., et al., 2013. Global investigation of olivine on Mars: insights into crust and mantle compositions. J. Geophys. Res.: Planets 118, 234−262.

Osterloo, M., Hamilton, V., Bandfield, J., Glotch, T., Baldridge, A., Christensen, P., et al., 2008. Chloride-bearing materials in the southern highlands of Mars. Science 319, 1651−1654.

Plaut, J.J., Picardi, G., Safaeinili, A., Ivanov, A.B., Milkovich, S.M., Cicchetti, A., et al., 2007. Subsurface radar sounding of the south polar layered deposits of Mars. Science 316, 92−95.

Poulet, F., Bibring, J.P., Mustard, J.F., Gendrin, A., Mangold, N., Langevin, Y., et al., 2005. Phyllosilicates on Mars and implications for early Martian climate. Nature 438, 623−627.

Poulet, F., Mangold, N., Loizeau, D., Bibring, J.-P., Langevin, Y., Michalski, J., et al., 2008. Abundance of minerals in the phyllosilicate-rich units on Mars. Astron. Astrophys. 487, L41−L44.

Poulet, F., Mangold, N., Platevoet, B., Bardintzeff, J.M., Sautter, V., Mustard, J.F., et al., 2009. Quantitative compositional analysis of Martian mafic regions using the MEx/OMEGA reflectance data 2. Petrological implications. Icarus 201, 84−101.

Poulet, F., Carter, J., Bishop, J.L., Loizeau, D., Murchie, S.M., 2014. Mineral abundances at the final four curiosity study sites and implications for their formation. Icarus 231, 65−76.

Roach, L.H., Mustard, J.F., Swayze, G., Milliken, R.E., Bishop, J.L., Murchie, S.L., et al., 2010. Hydrated mineral stratigraphy of Ius Chasma, Valles Marineris. Icarus 206, 253−268.

Rogers, A.D., Christensen, P.R., 2007. Surface mineralogy of Martian low-albedo regions from MGS-TES data: implications for upper crustal evolution and surface alteration. J. Geophys. Res.: Planets 112. Available from: http://dx.doi.org/doi:10.1029/2006JE002727.

Schmidt, M.E., Campbell, J.L., Gellert, R., Perrett, G.M., Treiman, A.H., Blaney, D.L., et al., 2014. Geochemical diversity in first rocks examined by the Curiosity rover in Gale Crater: evidence for and significance of an alkali and volatile-rich igneous source. J. Geophys. Res.: Planets 119 (1), 64−81.

Schultz, P.H., Mustard, J.F., 2004. Impact melts and glasses on Mars. J. Geophys. Res.: Planets 109 (E1). Available from: http://dx.doi.org/doi:10.1029/2002je002025.

Shkuratov, Y., Starukhina, L., Hoffmann, L., Arnold, G., 1999. A model of spectral albedo of particulate surfaces: implications for optical properties of the moon. Icarus 137, 235−246.

Sinton, W., 1967. On the composition of Martian surface materials. Icarus 6, 222−228.

Skok, J.R., Mustard, J.F., Ehlmann, B.L., Milliken, R.E., Murchie, S.L., 2010. Silica deposits in the Nili Patera caldera on the Syrtis Major volcanic complex on Mars. Nat. Geosci. 3, 838−841.

Skok, J., Mustard, J., Tornabene, L., Pan, C., Rogers, D., Murchie, S., 2012. A spectroscopic analysis of Martian Crater central peaks: formation of the ancient crust. J. Geophys. Res.: Planets 117. Available from: http://dx.doi.org/doi:10.1029/2012JE004148.

Smith, D.E., Zuber, M.T., Solomon, S.C., Phillips, R.J., Head, J.W., Garvin, J.B., et al., 1999. The global topography of Mars and implications for surface evolution. Science 284, 1495−1503.

Smith, P., Tamppari, L., Arvidson, R., Bass, D., Blaney, D., Boynton, W.V., et al., 2009. H_2O at the Phoenix landing site. Science 325, 58−61.

Squyres, S.W., Grotzinger, J.P., Arvidson, R.E., Bell, J.F., Calvin, W., Christensen, P.R., et al., 2004. In situ evidence for an ancient aqueous environment at Meridiani Planum, Mars. Science 306, 1709−1714.

Stolper, E.M., Baker, M.B., Newcombe, M.E., Schmidt, M.E., Treiman, A.H., Cousin, A., et al., 2013. The petrochemistry of Jake_M: a Martian mugearite. Science 341 (6153). Available from: http://dx.doi.org/doi:10.1126/science.1239463.

Sun, V.Z., Milliken, R.E., 2015. Ancient and recent clay formation on Mars as revealed from a global survey of hydrous minerals in crater central peaks. J. Geophys. Res.: Planets 120, 2293−2332.

Tate, C.G., Moersch, J., Jun, I., Ming, D.W., Mitrofanov, I., Litvak, M., et al., 2015. Water equivalent hydrogen estimates from the first 200 sols of Curiosity's traverse (Bradbury Landing to Yellowknife Bay): results from the Dynamic Albedo of Neutrons (DAN) passive mode experiment. Icarus 262, 102−123.

Vaniman, D.T., Bish, D.L., Ming, D.W., Bristow, T.F., Morris, R.V., Blake, D.F., et al., 2014. Mineralogy of a Mudstone at Yellowknife Bay, Gale Crater, Mars. Science 343. Available from: http://dx.doi.org/doi:10.1126/science.1243480.

Wellington, D.F., Bell, J.F., Johnson, J.R., Kinch, K.M., Rice, M.S., Godber, A., et al., 2017. Visible to near-infrared MSL/Mastcam multispectral imaging: Initial results from select high-interest science targets within Gale Crater, Mars. Am. Mineral. 102 (6), 1202−1217.

VOLATILES MEASURED BY THE PHOENIX LANDER AT THE NORTHERN PLAINS OF MARS

9

Samuel P. Kounaves and Elizabeth A. Oberlin

Tufts University, Medford, MA, United States

9.1 INTRODUCTION

The Phoenix Mars Scout lander touched down in the Vastitas Borealis (VB) plains on the eroded ejecta of Heimdal crater on May 25, 2008 and was operational starting in the late Martian spring for 152 sols ($L_s78° − L_s148°$). Located at 68.22° N latitude, $\sim 20°$ farther north than the Viking II lander, it was the first mission to study the Martian high latitude region (Smith et al., 2009). The Phoenix science goals, partially motivated by the discovery of possible near-surface ice deposits surrounding the northern polar cap as far south as 60° N (Mellon et al., 2008; Boynton et al., 2002), were to verify this buried ice reservoir, characterize the shallow icy soil and the overlying soil deposits, document the high latitude surface and atmospheric environments (Smith et al., 2008), and evaluate the ability of the region to support life. These goals guided the development and selection of Phoenix's instrument payload and provided the capability for measuring a variety of volatiles at the landing site, including sulfur, carbon, chlorine, and H_2O. The following is a discussion of the measurements taken during the Phoenix mission that put these volatiles in context of their physical states and mineral phases, allowing for insights into the geochemical mechanisms that produced them and the implications that these mechanisms have for the aqueous history of the region. The primary instruments for measurement of volatiles were the Thermal and Evolved Gas Analyzer (TEGA), the Wet Chemistry Laboratory (WCL), and the Thermal and Electrical Conductivity Probe (TECP). Their operations have been previously described in detail in the literature and are only briefly reviewed below.

The TEGA, shown in Fig. 9.1, was designed not only to identify and quantify organics, but also inorganic volatiles, such as H_2O, O_2, Cl_2, SO_2, and CO_2, present in the atmosphere and released from thermal decomposition of minerals. TEGA consisted of eight differential scanning calorimetry cells coupled to a magnetic-sector mass spectrometer (MS) acting as the evolved gas analyzer (Hoffman et al., 2008; Boynton et al., 2001). Thermal decomposition analyses were conducted on individual samples by heating them from ambient temperature to 1000 °C while monitoring changes in power output. Both endothermic and exothermic phase transitions were used to identify the phases present. By correlating gas releases with calorimetry, the amount of each volatile compound associated with a specific mineral phase could be determined. The volatiles released during heating

Volatiles in the Martian Crust. DOI: http://dx.doi.org/10.1016/B978-0-12-804191-8.00009-X

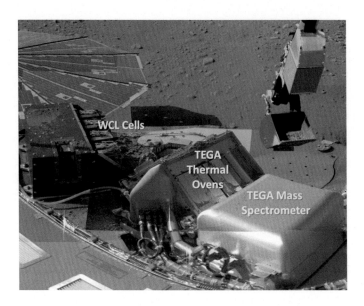

FIGURE 9.1

Deck of the Phoenix lander on Mars showing the four WCL units for measuring the soluble ions and other parameters in the soil/water mixture, four of the eight Thermal Evolved Gas Analyzer (TEGA) inlets to the thermal ovens, and the TEGA evolved gas MS for identifying and quantifying organic molecules and inorganic volatiles. *Image Credit: NASA/JPL.*

were carried by N_2 gas maintained at 12-mbar pressure to the MS for molecular identification and correlation to their release temperature.

The four WCL units, shown in Fig. 9.1, each consisted of an array of sensors embedded in the wall of a cell (beaker) capped by an actuator unit that provided for stirring and the addition of reagents, leaching solution, and a soil sample. The sensor array included ion selective electrodes (ISE) for measuring the concentration in the soil/solution mixture for K^+, Na^+, Mg^{2+}, Ca^{2+}, NH_4^+, Ba^{2+} (for SO_4^{2-}), Cl^-, Br^-, I^-, NO_3^-/ClO_4^-, and H^+(pH), electrodes for conductivity, redox potential (Eh), for performing cyclic voltammetry (CV), and chronopotentiometry (CP) (Kounaves et al., 2009). The goal of the WCL was to analyze the chemistry of the soils at the surface and at depth in order to better understand the history of the water, the biohabitability of the soil, the availability of chemical energy sources, and the general geochemistry of the site.

The TECP, shown in Fig. 9.2, was mounted near the end of the 2.3-m robotic arm (RA) and was inserted either into the soil or held aloft in the atmosphere. It consisted of a single electronics box, fitted with four needle-like probes, which were inserted into the soil. The TECP was included on Phoenix to conduct in situ measurements to characterize the processes that control the distribution and exchange of H_2O between the atmosphere and subsurface and to observe the occurrence of liquid H_2O, both as thin films in the soil, or as atmospheric vapor. It was designed to specifically measure relative humidity, H_2O vapor abundance, temperature, dielectric

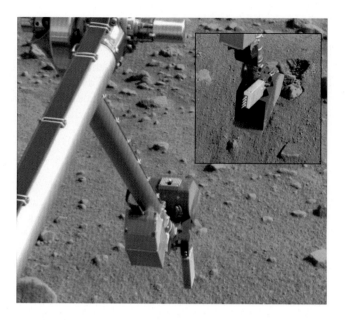

FIGURE 9.2

The RA with the TECP mounted near the end consisting of an electronics box fitted with four needle-like probes (insert) for placement into the soil. The TECP characterized the processes that control the distribution and exchange of H_2O between the atmosphere and subsurface by measuring relative humidity, H_2O vapor abundance, temperature, dielectric permittivity, thermal conductivity, and volumetric heat capacity of the soil. *Image Credit: NASA/JPL.*

permittivity, thermal conductivity, and volumetric heat capacity of the soil (Zent et al., 2009). Combined with data from the meteorology mast air temperature sensors, measurement of atmospheric saturation was also possible.

In addition, data from five other Phoenix instruments not directly aimed at volatiles measurements, the Optical (OM) and Atomic Force (AFM) microscopes (Hecht et al., 2008), the Surface Stereo Imager (SSI) (Lemmon et al., 2008; Smith et al., 2001), the Meteorological Station (Taylor et al., 2008), and the Light Detection and Ranging (LIDAR) system for atmospheric measurements (Whiteway et al., 2008), also provided results with significant implications for volatiles.

9.2 SULFUR

Mars is at least an order of magnitude more enriched in sulfur than Earth, and the apparent paucity of carbonate minerals generally implies a geochemical environment that is dependent on the sulfur cycle (Gaillard et al., 2012). Detection of sulfates through both orbital (Bishop et al., 2005; Gendrin, 2005; Langevin et al., 2005; Bibring et al., 2006; Arvidson et al., 2007; Mangold et al., 2008;

Calvin et al., 2009; Massé et al., 2012) and landed missions is evidence of a global distribution of sulfates both as a component of the global dust and concentrated in discrete locations (Clark et al., 1976; Bell et al., 2000; Madden et al., 2004; Wang et al., 2006; Squyres et al., 2006; Kounaves et al., 2010b; McAdam et al., 2014; Ming et al., 2014). Two likely pathways in the sulfur cycle on Mars that may produce globally distributed sulfates are the dissolution of volcanic SO_2 in water forming sulfuric acid, which rapidly weathers silicate minerals and precipitates sulfate bearing minerals, and the oxidation of iron sulfides as they are transported from reducing conditions of the mantle to oxidizing surface conditions (Gaillard et al., 2012).

9.2.1 SULFATES IN THE NORTHERN PLAINS

The initial indication that sulfur may be present in globally distributed soil came from the detection of approximately equivalent amounts of sulfur from both landing sites during the Viking missions (Clark et al., 1976). While evidence of a global dust with homogenous composition across the surface of Mars has been proposed and is widely supported (Clark et al., 1982; McSween Jr. and Keil, 2000; Yen et al., 2005; Berger et al., 2016), the sulfates present in the northern plains have distinct differences from other locations on Mars and are not likely to have originated from this dust. The circumpolar dune fields surround the north polar cap and include the largest gypsum deposits observed on Mars, initially detected by the OMEGA instrument on the Mars Express orbiter (Langevin et al., 2005). Gypsum deposits have since been detected throughout the north polar cap region, originating from the sediment layers embedded within the Basal Unit and North Polar Layered Deposits that comprise the north polar ice cap (Fishbaugh et al., 2007; Massé et al., 2010, 2012). Additionally, while iron- and Mg-sulfates are present throughout the midlatitudes, reflecting sulfate formation under oxidizing and arid conditions (Bishop et al., 2005; Arvidson et al., 2007; Mangold et al., 2008), no evidence for Fe-sulfate phases has been observed at high latitudes. The abundance of gypsum and lack of Mg- and Fe-sulfate phases surrounding the north pole suggests that these sulfates were formed through the reaction between SO_2 and water, rather than the oxidation of mantle-derived FeS, as may be the case at other locations.

9.2.2 ORIGIN OF SULFATES AT THE PHOENIX SITE

The location of the Phoenix landing site, ~ 20 km west of the Heimdal crater at the intersection of the VB and the Scandia Plains with Alba Mons in the south and the polar ice cap to the north (Heet et al., 2009), provided a source of both water and volcanic material for in situ formation of sulfates. These landforms were later excavated by the impact that formed Heimdal crater and, ultimately, the ejecta blanket that comprises the material at the Phoenix landing site. The Scandia Plains unit was formed during the Late Hesperian and is composed of volcanic sediments derived from Alba Mons, while the VB unit with its complex and varied landforms was subject to multiple depositional episodes resulting from outflow channels throughout the Hesperian. The intersection of these units that underlie Heimdal crater resulted in a high density of knobby geomorphic features and mesas that are composed of a volatile rich sedimentary mantle emplaced through possible mud volcanism and modified by differential erosion due to the loosely consolidated nature of the mantle (Tanaka et al., 2011). These products form the base features of the Phoenix landing site soil with possible alteration products resulting from the 600 Myr history of the blanket (Heet et al., 2009) and possible contributions from global dust.

9.2.3 DETECTION OF SULFATES AT THE PHOENIX SITE

Previous analysis from the Viking landing sites suggests that the sulfur content in the Martian soil is uniformly high at around 3 wt%, with likely phases comprised of Fe-sulfides and Na-, Mg-, Ca-, and/or Fe-sulfates (Clark et al., 1976). The WCL on Phoenix, using a $BaCl_2$ titration method to detect soluble sulfate (SO_4^{2-}), reported an equivalent concentration of ~ 1.3 (± 0.5) wt% SO_4^{2-} in the soil. This soluble portion of the soil sulfates is primarily attributed to a hydrated $MgSO_4 \cdot nH_2O$ phase (Kounaves et al., 2010a) based on geochemical models using the WCL measurements, thus limiting the sulfur phases in the local soil (and a significant portion of the globally distributed fines) to predominantly insoluble sulfur compounds. The Phoenix results also exclude the possibility of a soluble Fe-sulfate phase, as the presence of soluble iron in the WCL sample solution would have damaged or been detected by several sensors, which was not observed in any of the WCL analyses. However, despite the expectation and confirmation of sulfur mineralogy, the TEGA instrument suite failed to detect the evolution of any SO_2 during sample heating. As shown in Fig. 9.3, most S-bearing minerals would have evolved SO_2 in the range of 25 °C−1000 °C, the exception being the alkali metal sulfates (i.e., Na, K), and $CaSO_4$, which are thermally stable against SO_2 evolution but not against dehydration at these temperatures. Although the sulfate measurements made by Phoenix were subject to some uncertainty relating to instrument and

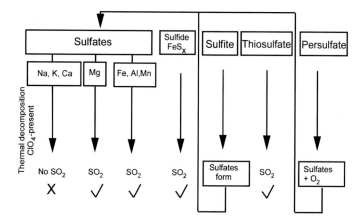

FIGURE 9.3

Flow diagram showing the expected behaviors of S-bearing phases for TEGA in the presence of a ClO_4^-. Upon heating to 1000 °C, there was no SO_2 detected in any soils analyzed at the Phoenix landing site. Thus the likely candidates for insoluble SO_4^{2-} would be limited to Ca-sulfates such as gypsum, anhydrite, or bassanite. The flow from the sulfite and persulfate columns to the sulfates column suggests that the sulfates that form during the thermal decomposition of sulfites and persulfates have thermal properties similar to the sulfates. X = SO_2 not evolved; √ = SO_2 evolved. *From Golden et al. (2009).*

protocol limitations, the combined WCL (Kounaves et al., 2010a, 2010b) and TEGA (Golden et al., 2009) results constrain the identity of any sulfate in the local soil and global dust to primarily insoluble and thermally stable Ca-sulfate, with small amounts of soluble Mg-sulfate phases.

9.3 CARBON

9.3.1 ISOTOPIC MEASUREMENTS OF ATMOSPHERIC CO_2

Isotopic fractionation patterns record the historical record of many geological processes including the interaction of volatile elements with aqueous and crustal material. Stable isotopes will become enriched or depleted in predictable ways depending on the dominant processes in which they are involved. The stable isotopic composition of atmospheric carbon dioxide (CO_2) is of particular significance due to its high partial pressure in the Martian atmosphere, and its participation in important planetary processes including atmospheric loss, volcanic degassing, and carbonate formation through interaction with surficial waters. Therefore, the isotopic composition of CO_2 in the Martian atmosphere holds an important key to understanding both past and current geological processes on Mars.

The TEGA magnetic-sector MS recorded masses 44, 45, and 46, which include all stable isotopes of carbon and oxygen, and were referenced against Vienna Pee Dee belemnite (VPDB) and Vienna standard mean ocean water (VSMOW) in accordance with typical terrestrial measurements. The results revealed an isotopic composition of $\delta^{13}C_{VPDB} = -2.5‰ \pm 4.3‰$ and $\delta^{18}O_{VSMOW} = 31.0‰ \pm 5.7‰$ (Niles et al., 2010), notably lacking in ^{13}C enrichments that would be expected if the dominant fractionation process in the Martian atmosphere was attributed to atmospheric loss. Instead, these results indicated that atmospheric carbon reservoirs maintain an active equilibrium with a source of ^{12}C and/or sink for ^{13}C. Comparison with isotopic measurements of carbonates and magmatic CO_2 derived from studies of Martian meteorites (Jull et al., 1997) suggests that the atmospheric carbon isotopic composition in communication with the northern plains can be explained by a mechanism incorporating ^{13}C depletion through carbonate formation and/or ^{12}C enrichment through equilibration with magmatic CO_2 and volcanic outgassing (Niles et al., 2010).

However, considering the effects of these mechanisms on $\delta^{18}O$ values suggests a less-active contribution from magmatic CO_2, as equilibrium exchange of oxygen under high temperature equilibrium conditions with molten silicate material would result in release of CO_2 bearing an oxygen fractionation pattern similar to the host rock. The ^{18}O measurements from silicate matrices of Martian meteorites indicate that Martian silicates have a $\delta^{18}O$ closer to 4‰ (Clayton and Mayeda, 1988), which is far more depleted than the observed ^{18}O in atmospheric CO_2. Alternatively, terrestrial studies have shown that at temperatures as low as 0 °C, the equilibration fractionation between CO_2 and water for ^{18}O is ~45‰ (Bottinga and Craig, 1968), which is in much better agreement with observed Martian atmospheric $\delta^{18}O$ values. Therefore, while mantle degassing may serve as a mechanism to dilute atmospheric ^{13}C values (Niles et al., 2010), the primary mechanism driving fractionation of atmospheric CO_2, and likely most currently active interaction of CO_2 with the Martian surface, is equilibration with low temperature water and the precipitation of carbonate phases.

9.3.2 COMPARING THE VOLATILE CARBON RESULTS OF THE NORTHERN PLAINS WITH EQUATORIAL REGIONS

The isotopic measurements reported by Phoenix agree with those measured at other locations in the northern plains by the Viking II lander (Nier et al., 1976), but differ markedly from those taken in the equatorial region by the Curiosity rover, which showed a $\sim 46\permil$ enrichment in ^{13}C (Webster et al., 2013), supporting the domination of atmospheric loss in isotope fractionation. While intuition may suggest atmospheric isotope ratios should be mostly homogenous across the surface of Mars, the collective evidence and inconsistency in isotopic measurements between regions suggest that localized variation exists, likely in response to seasonal cycles and geological setting.

The Phoenix mission occurred at the onset of the polar summer, after the retreat of the northern CO_2 caps. The seasonal cycles of CO_2 sublimation in close and regular proximity to subsurface ice and water vapor may serve to deplete the local atmosphere in ^{13}C through carbonate precipitation. Over the course of the Martian year and/or with distance from the polar caps, this effect would be diminished as atmospheric mixing dominates. The isotopic measurements observed at Gale crater by the Curiosity rover would, therefore, be expected to show an atmospheric enrichment in ^{13}C due to limited present day interactions between local sources of H_2O and CO_2.

9.3.3 DETECTION OF CARBONATES IN THE PHOENIX SOIL

Carbonates were detected at the Phoenix landing site by TEGA and identified as calcium rich phases at a concentration of 3−5 wt% (Boynton et al., 2009). The TEGA analyses identified two clear endothermic peaks, one in the range of 725 °C−820 °C and a second from 860 °C to 980 °C as well as two CO_2 releases, one coinciding with the 725 °C−820 °C endothermic transition, and a lower temperature release from 400 °C to 680 °C (Boynton et al., 2009). The correlation between the endothermic transition and CO_2 release occurring between 725 °C and 820 °C was attributed to the decomposition of a calcium-rich carbonate phase based on laboratory experiments (Sutter et al., 2012) with additional support from the WCL analyses.

Carbonate-rich solutions are characterized by a slightly alkaline pH and the ability to buffer the solution against pH changes upon acid addition. The WCL cells on Phoenix, which were designed to analyze the chemistry of a soil/water mixture, included pH sensors and an acid addition experiment to determine the buffering capacity of the soil. Upon soil addition, the pH was measured as 7.7 ± 0.5, which is consistent with a solution saturated with calcium carbonate under WCL operating conditions. Similarly, upon addition of 2-nitrobenzoic acid to the WCL solution no pH change was observed, confirming the presence of an acid buffering component in the soil (Kounaves et al., 2010b).

While the WCL results and laboratory-based TEGA results confirm the presence of a calcium rich carbonate phase, the identity of this phase and the origin of the remaining CO_2 release and endothermic peak remain uncertain. Initially, it was postulated that the lower temperature CO_2 release might be due to the decomposition of a magnesium or iron carbonate phase at a concentration too low to generate an endothermic response above the background level of the instrument (Sutter et al., 2012), with no assignment given to the higher temperature endothermic peak (Boynton et al., 2009). However, more recent laboratory experiments show that a reaction between hydrochloric acid, water vapor, and calcium carbonate, generated upon dehydration and thermal decomposition of hydrated magnesium perchlorate in the presence of calcium carbonate, will release CO_2 in the temperature range observed during the TEGA experiments. This reaction has also been suggested as a possible source for the unidentified high temperature endothermic peaks in connection with the decomposition of $Ca-Cl-O$ phases generated during the reaction (Cannon et al., 2012).

9.3.4 ORIGINS AND IMPLICATIONS OF CARBONATES AT THE PHOENIX LANDING SITE

Hypotheses for the origin of the detected carbonates at the Phoenix landing site include both inherited material and active in situ formation through thin films of water (Sutter et al., 2012). Two primary sources are cited as candidates for inherited carbonates: the VB material comprising the ejecta blanket that forms the Phoenix landing site and eolian deposition of magnesium carbonates, which have been detected at 2−5 wt% in the Martian dust (Bandfield et al., 2003). VB is believed to have formed during the Hesperian period from aqueous deposition of sediments prior to freezing and sublimation of standing water (Kreslavsky and Head, 2002). Therefore, material excavated from the VB region during the impact that formed the Phoenix landing site should contain some amount of carbonates as remnants of this aqueous activity and may serve as a source for carbonates at the Phoenix site.

The alternative source for the carbonates at the Phoenix landing site, as eolian deposition of magnesium carbonates, is supported by evidence from the microscopy experiments performed during the Phoenix mission. Optical microscopy was used to categorize general particle types at the Phoenix landing site and revealed a soil component comprised of a mixture of fines (< 10 μm) with similar spectral properties to globally observed airborne dust, and silt-sand-sized grains attributed to variable source regions (Goetz et al., 2010).

These results suggest that the mineral grains at the Phoenix landing site were, at least in part, formed in another location and deposited at the Phoenix site over time. Magnesium carbonates deposited with global dust may have undergone replacement of the magnesium with calcium through dissolution and reprecipitation on the surface of soil grains in the presence of thin films of water generated from subsurface ice (Sutter et al., 2012).

9.4 CHLORINE

The present levels and forms of chlorine in Martian soil are important indicators of the planet's aqueous geochemical history, and of volcanic emissions of chlorine-bearing gases. Planet-wide transport of soluble chloride (Cl^-) and localized enrichment of chloride-bearing salts are clear indicators of active weathering and transport processes involving liquid water. Orbital measurements by the Mars Odyssey Gamma Ray Spectrometer (GRS) have shown Cl to be widely distributed at the surface, predominantly at concentrations between 0.3 and 0.7 wt% (Keller et al., 2006). In situ measurements of chlorine (Cl) on Mars before the Phoenix mission were made by five different spacecraft, at varying locations over the planet's surface. The results from the two Viking Mars landers using X-ray fluorescence (Clark et al., 1982; Clark and Van Hart, 1981), the Pathfinder rover (Bell et al., 2000), and the Mars exploration rovers (MER) Opportunity and Spirit and Mars Science Laboratory Curiosity (Clark et al., 2005; Gellert et al., 2006; McLennan et al., 2014), using an alpha particle X-ray spectrometer, show similar levels of Cl across Mars' surface and at relatively higher concentrations than typically found in terrestrial or lunar materials. Interpretation and modeling of these results led to the conclusion that the chlorine was mostly in the form of $CaCl_2$ or $MgCl_2$, although not to the same degree for all samples and also depending on location and whether in soil or rock (Clark et al., 2005).

9.4.1 DETECTION OF CHLORINE IN THE PHOENIX SOIL

Each of the four WCL cells contained three different methods for detection of chlorine, an ISE sensor for chloride (Cl^-), an ISE sensor for perchlorate (ClO_4^-), and a silver-disk electrode for the chronopotentiometric (CP) determination of Cl^- (Kounaves et al., 2009). Chloride was determined independently by both the Cl^- ISE and CP in all three of the soil samples at an average concentration of 0.47 ± 0.13 mM Cl^-, equivalent to ~ 0.04 wt% Cl in the soil. Perchlorate was measured by the ClO_4^- ISE in three samples at similar concentrations with an average of 2.6 ± 0.10 mM ClO_4^-, or equivalent to ~ 0.22 wt% Cl in the soil (Kounaves et al., 2010b).

One interesting aspect of the above result is the close agreement between the total WCL-measured Cl concentration in the Phoenix site soil (~ 0.26 wt% Cl) and that extrapolated for the site by the Odyssey GRS of ~ 0.3 wt% Cl (Keller et al., 2006). This result lends strong support for the global accuracy of the GRS data, both in terms of the concentration and global distribution of chlorine (Clark and Kounaves, 2015). More importantly, and depending on the production and distribution processes, one might expect a global correlation between the concentration of Cl and ClO_4^- with the exact ratio providing insight into these processes. The question is, whether the distribution of chlorine between chloride and ClO_4^- (or other oxychlorines) is a constant fraction of total chlorine in various global Martian materials.

The Phoenix WCL results also showed the Cl to be predominantly in the ClO_4^- form, at a ratio for the Cl_{ClO4}^- to Cl_{Cl}^- of almost 6:1. However, the ratio of total oxychlorine species to chloride may be even larger because the ClO_4^- ISE was 10^3 times more sensitive to ClO_4^- than chlorate (ClO_3^-), and the signal for ClO_3^- would have been completely obscured. In conjunction with the WCL electrical conductivity measurements, it is possible to estimate that as much as 1–2 mM ClO_3^- may have also been present in the soil. On Earth, ClO_4^- and ClO_3^- are typically found together, and in many cases, the ClO_3^- is at a higher concentration (Jackson et al., 2015). Even though the Cl cycle on Mars is not yet fully understood, similar or identical production mechanisms and pathways are possible for Mars, and thus similar concentration ratios could be present.

9.4.2 THE MARTIAN GLOBAL OXYCHLORINE CYCLE

The terrestrial production of ClO_4^- predominantly involves the oxidation of chlorine containing species via UV activated ozone in the upper troposphere (Catling et al., 2010). However, this is not a viable production mechanism on Mars, especially in recent times, due to the lack of atmospheric chlorine (Smith et al., 2014). It has been recently shown that ClO_4^- and ClO_3^- can be produced photochemically under Martian conditions via UV irradiation of Cl-bearing minerals, without atmospheric or aqueous chlorine, with SiO_2 and metal-oxides acting as photocatalysts and generating

O_2^- radicals from O_2 that then react with chloride (Carrier and Kounaves, 2015). The in situ photochemical production of oxychlorines would most likely result in their presence as a noncrystalline (amorphous) phase. Such a mechanism is supported by the extremely rapid dissolution of the perchlorate salts on addition to the Phoenix WCL (Kounaves et al., 2010b) and by the nondetection of crystalline oxychlorine phases with the Chemistry and Mineralogy (CheMin) X-ray diffraction instrument (Vaniman et al., 2014) though they would be at the detection limit of CheMin (>3 wt%) (Bish et al., 2013). During the production of ClO_4^-, even more highly oxidizing intermediates, such as hypochlorite (ClO^-), chlorine dioxide gas (ClO_2), and radicals such as $^\bullet OCl$, $^\bullet Cl$, and $^\bullet OH$, can also be produced (Catling et al., 2010; Schuttlefield et al., 2011; Kang et al., 2008). Highly reactive oxychlorines and radicals can also be produced from the radiolysis of perchlorates by the action of cosmic gamma and X-rays (Quinn et al., 2013). These highly reactive intermediates from either of these pathways would be capable of destroying or altering organic molecules through oxidation and/or chlorination (Kounaves et al., 2014a). Taken together, along with laboratory experiments, the evidence points to a potentially global presence of strong oxidants (Clark and Kounaves, 2015) in the Martian soil and a complex ongoing oxychlorine production and destruction cycle as shown in Fig. 9.4.

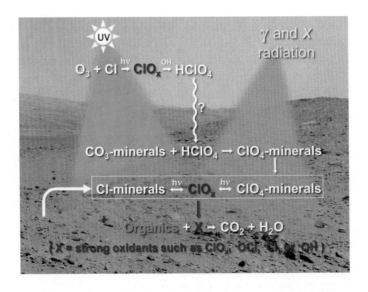

FIGURE 9.4

Possible processes and pathways for the production and destruction of perchlorates and the accompanying intermediary species produced. Strongly oxidizing species such as ClO^- and ClO_2^- are generated from Cl-bearing minerals by the action of UV and from ClO_4^- minerals by action of cosmic γ and X-rays. Taken together, the evidence points to a global presence of strong oxidants throughout the Martian soil and a complex ongoing oxychlorine production and destruction cycle. Oxychlorine compounds are thus potentially ubiquitous and ever present. *From Kounaves et al. (2014a).*

9.5 WATER

One of the objectives of the Phoenix mission was to verify the presence of near-surface H_2O ice deposits surrounding the northern polar cap as far south as 60° N. The reported estimates of the ice table depth in the 60−70° N latitudes derived from gamma ray and neutron spectroscopy (Boynton et al., 2002; Feldman et al., 2002), thermal emission spectroscopy and imaging (Titus and Prettyman, 2007; Putzig and Mellon, 2007; Bandfield, 2007), and thermodynamic ice stability theories (Mellon et al., 2004; Mellon and Jakosky, 1995), ranged from 2 to 18 cm. On the other hand, models of ice table depth in the region-D/Box-1 area, where Phoenix landed, predicted depths of about 2 to 6 cm, with an average depth of 4.0 ± 1.3 cm (Mellon et al., 2008). Variations, due to soil density, subsurface rock distribution, topography, or large buried boulders, did allow for a deeper ice table in some localized areas. Models also predicted ice-rich permafrost overlaid by ice-free soil with a sharp interface if the ice was in diffusive equilibrium with its environment. However, estimates of the soil:ice ratio suggested that the subsurface ice content exceeded the available soil pore space and made it difficult to reconcile on the basis of vapor diffusion alone.

On sol 5 of the mission the robotic arm camera (RAC) was pointed under the lander, revealing a shallow (~5 cm depth) ice table that had been excavated by the thrusters during landing (Fig. 9.5A). After over 12 trenches and 31 samples, similar ice depths were found in the center of the polygons, but icy soil was not encountered in the troughs down to the maximum excavation of 18 cm (Arvidson et al., 2009). During excavation in the Dodo-Goldilocks trench on sol 20, several

FIGURE 9.5

(A) Image taken under the lander by the RAC on sol 5 of the mission revealed a shallow ~5 cm depth ice table excavated by the thrusters during landing. Similar ice depths were found in the center of the polygons but icy soil was not encountered in the troughs down to the maximum excavation of 18 cm. (B) During sol 20, the excavation in the Dodo-Goldilocks trench revealed several 1−2 cm chunks of relatively pure ice that had been scraped lose from the trench, and which sublimated by sol 24 without leaving any residue. *Image Credit: NASA/JPL.*

1−2 cm chunks of relatively pure ice were scraped lose from the trench bottom (Fig. 9.5B). This ice had totally sublimated by sol 24, without any observable trace of residue (Cull et al., 2010). Even though pore ice had been predicted by thermodynamic models (Mellon et al., 2008), the presence of apparently pure H_2O ice would require a liquid H_2O or brine phase. This result supports the Odyssey GRS findings implying that ice concentrations exceed pore ice (Boynton et al., 2002; Feldman et al., 2007) and implies that the ice table was not in diffusive equilibrium with the atmosphere. Factors in addition to vapor diffusion may be involved. The segregated ice (Fig. 9.5B) and cemented soil clods are indicators of a possibly wet environment at some point in the past at this location, though this wet period could be anywhere within the past 5−500 Myr (Kreslavsky et al., 2008).

The presence of current or past liquid H_2O films in the soil was investigated by both the TECP and TEGA instruments. Results from the TECP indicated that during the daytime, atmospheric H_2O averaged 1.8 Pa ($R_H < 5\%$) but disappeared at night with R_H increasing to $> 95\%$, indicating that adsorption on the soil played a major role in removing H_2O from the atmosphere (Zent et al., 2010, 2016). An increase in permittivity recorded by the TECP on sol 70, which coincided with surface frost and a decrease in atmospheric water vapor, was modeled by laboratory experiments and shown to match an increase in adsorbed H_2O from ~ 1 to 3 monolayers in an analog soil, an amount an order of magnitude larger than inferred to have precipitated during the night (Stillman and Grimm, 2011). This observation was in line with that made by the SSI measurement of the LIDAR beam-scatter, which was used to estimate the ice−water content (IWC) of the near surface fog as 1.7 ± 1.0 mg/m^3 and, which compared to air aloft, was inhibited when IWC $> 30 \pm 24$ mg/m^3. When integrated over the night time, this represents up to 2.5 pr-μm (6% of the total water column) of diurnal adsorption by the soil at the Phoenix site (Moores and Schuerger, 2011).

The determination by TEGA of 3%−5% $CaCO_3$ in the soil implied its formation in a wet environment (Boynton et al., 2009). The presence of subsurface ice might provide a mechanism for in situ carbonate formation from dissolution of atmospheric CO_2 during periods when liquid water was temporarily stable. This may have occurred primarily during obliquity cycles when average temperatures in the northern plains were higher, or may be part of a slow but active hydrology (Kreslavsky et al., 2008). The diurnal cycles observed by the TECP could form thin films of water capable of participating in in situ carbonate formation. Combining the detection of carbonate and possible formation mechanisms with the isotopic measurements of atmospheric CO_2 taken by TEGA has led to hypotheses involving an active low-temperature hydrology and carbonate formation, and may provide insight into the history of water at the northern plains of Mars.

9.5.1 PHOENIX-BASED INDICATORS OF PROLONGED SEVERE ARIDITY AT THE NORTHERN PLAINS

The Phoenix WCL found that the soil at the landing site contained about 0.6 wt% ClO_4^- (Kounaves et al., 2010b). This discovery is important in terms of H_2O because the ClO_4^- ion's high solubility, distribution, chemical forms, and interactions with water can reveal much about the aqueous history of the planet. The WCL analyses indicated that the possible ClO_4^- parent salt counter cations could be Mg^{2+}, Ca^{2+}, Na^+, K^+, or $Fe^{2/3+}$, either singly or as a mixture. For a variety of reasons, Na^+, K^+, and $Fe^{2/3+}$ were considered as unlikely candidates (Kounaves et al., 2010b). The presence of $Ca(ClO_4)_2$ was also considered unlikely because highly insoluble calcium-sulfates and -carbonates would serve as sinks for the Ca^{2+} rather than $Ca(ClO_4)_2$, leaving $Mg(ClO_4)_2$ as the dominant ClO_4^- parent salt (Toner et al., 2015a, 2015b). Thus, if liquid water was present, there should be no $Ca(ClO_4)_2$ since the Ca^{2+} would have been precipitated by the excess soluble SO_4^{2-} as gypsum ($CaSO_4 \cdot 2H_2O$). However, refined analyses of the WCL Ca^{2+} sensor data taken on Mars and Earth laboratory experiments showed that the Phoenix soils contain a 3:2 ratio of $Ca(ClO_4)_2$ to $Mg(ClO_4)_2$ phases. Recent results from Curiosity also indicate that $Ca(ClO_4)_2$ provides the most reasonable match with the SAM data (Glavin et al., 2013). Even though the relationship between the duration of aridity and the presence of the ClO_4^- can only be accurately determined if the rate of ClO_4^- deposition is known, these results strongly support the conclusion that the Phoenix site has been an extremely arid environment, very likely since the formation of the Heimdal crater ejecta blanket ~ 600 Myr ago (Heet et al., 2009). This conclusion of course does not exclude the presence of current or recent liquid water at other locations on Mars where it appears to be indicated by such features as active gullies (Malin et al., 2006), dune and slope streaks (Kereszturi et al., 2010; Kreslavsky and Head, 2009), and seasonally recurring slope lineae (RSL) (McEwen et al., 2011). However, alternate mechanisms for formation of these features are possible and in general definitive evidence is still lacking (Núñez et al., 2016). It is also possible that at such locations $Ca(ClO_4)_2$ is not present. For example, data in support of such a hypothesis has been provided by the Compact Reconnaissance Imaging Spectrometer for Mars (CRISM) on the Mars Reconnaissance Orbiter (MRO) (Ojha et al., 2015). The spectral observations indicate the presence of hydrated salts at four locations in the seasons when RSL are present. The hydrated salts reported to be most consistent with the absorption features are $Mg(ClO_4)_2$, $Mg(ClO_3)_2$, and $Na(ClO_4)$, though the presence of both $Ca(ClO_4)_2$ and $Ca(ClO_3)_2$ could bring the limit of detection of either below that of the CRISM spectrometer. If liquid water is present at these RSL, the absence of $Ca(ClO_4)_2$ is exactly what would be predicted (Toner et al., 2015a, 2015b; Kounaves et al., 2014b).

Other Phoenix-based indicators of the extent of aqueous interaction with the soil are the OM and AFM microscopy observations of clay-sized particles and their fraction and size distribution (PSD) (Pike et al., 2011). Two size populations were identified: (1) large $20-200$ μm, mostly rounded grains and (2) small reddish fines, notably with a very low mass proportion for sizes < 2 μm. The latter represent the smallest scale formation processes and are indicative of a single production method for the particles < 11 μm (Pike et al., 2011). This value is significantly greater than that expected for aqueous clay formation, suggesting a primary formation mechanism of global eolian weathering under extremely arid conditions. The minor proportion of clay-sized particles implies that there has been < 5000 years of exposure to liquid H_2O (Pike et al., 2011) over the 600 Myr since the Heimdal crater ejecta blanket was formed (Heet et al., 2009).

The Phoenix results are also supported by more recent observations from Curiosity showing that the Rocknest regolith samples contain a mixture of both oxidized (ClO_4/ClO_3) and reduced (S^- or H-bearing) species as well minerals formed under alkaline (carbonates) and acidic (ferric sulfates) conditions (Archer et al., 2014). This intimate association implies a physical mixture in chemical disequilibrium that has never encountered sufficient water to reach chemical equilibrium and is consistent with formation by eolian processes in a cold and severely arid environment.

9.6 SUMMARY

Overall, the volatile compounds measured by Phoenix at the northern plains of Mars have produced new insights into the aqueous processes that have occurred in the past, and how these processes have evolved over time into their current state. The volatile sulfur compounds and their chemistry indicate an environment in which volcanic outgassing of SO_2 mixed with liquid water to produce precipitated minerals such as gypsum. Some of this water froze into a subsurface ice table of variable thickness that is now an active part of the hydrological cycle on Mars, while the remaining water likely sublimated and was lost through atmospheric escape, with little being retained as vapor in the present day thin atmosphere. Since the bulk of the atmospheric loss, CO_2 has interacted at low temperatures with temporarily stable liquid water producing the observed atmospheric enrichments that are maintained through seasonal cycles based on local geology. While the bulk of the carbonate minerals observed in the northern plains likely originates from earlier atmospheric−water interactions that precipitated the bulk of the insoluble $CaCO_3$ phases observed, this more recent atmospheric−water interaction may be the origin of the less abundant, soluble $MgCO_3$ phases. However, it is important to bear in mind that these more recent atmospheric−water interactions are still likely very limited and not sufficient to prevent the persistence of a state of chemical disequilibrium as indicated by the presence of a large proportion of calcium relative to magnesium perchlorates. Ultimately, the evidence provided by the volatiles measured at the northern plains by Phoenix suggests a once active hydrology that has ceased, and an extremely arid environment that has existed for at least the past 600 Myr.

REFERENCES

Archer, P.D., Franz, H.B., Sutter, B., Arevalo, R.D., Coll, P., Eigenbrode, J.L., et al., 2014. Abundances and implications of volatile-bearing species from evolved gas analysis of the Rocknest Aeolian deposit, Gale Crater, Mars. J. Geophys. Res. 119, 237−254.

Arvidson, R.E., Gendrin, A., Gondet, B., Langevin, Y., Mouelic, S. Le, Mangold, N., et al., 2007. Coupled ferric oxides and sulfates on the Martian surface. Science 317, 1206−1210.

Arvidson, R.E., Bonitz, R.G., Robinson, M.L., Carsten, J.L., Volpe, R.A., Trebi-Ollennu, A., et al., 2009. Results from the Mars Phoenix Lander Robotic Arm experiment. J. Geophys. Res. 114, E00E02. Available from: http://dx.doi.org/10.1029/2009JE003408.

Bandfield, J.L., 2007. High-resolution subsurface water-ice distributions on Mars. Nature 447, 64−67.

Bandfield, J.L., Glotch, T.D., Christensen, P.R., 2003. Spectroscopic identification of carbonate minerals in the Martian dust. Science 301, 1084−1087.

Bell, J.F., McSween, H.Y., Crisp, J.A., Morris, R.V., Murchie, S.L., Bridges, N.T., et al., 2000. Mineralogic and compositional properties of Martian soil and dust: results from Mars Pathfinder. J. Geophys. Res. 105, 1721−1755.

Berger, J.A., Schmidt, M.E., Gellert, R., Campbell, J.L., King, P.L., Flemming, R.L., et al., 2016. A global Mars dust composition refined by the alpha-particle X-ray spectrometer in Gale Crater. Geophys. Res. Lett. 43, 67−75.

Bibring, J.-P., Langevin, Y., Mustard, J.F., Poulet, F., Arvidson, R., Gendrin, A., et al., 2006. Global mineralogical and aqueous Mars history derived from OMEGA/Mars Express data. Science 312, 400−404.

Bish, D.L., Blake, D.F., Vaniman, D.T., Chipera, S.J., Morris, R.V., Ming, D.W., et al., 2013. X-ray diffraction results from Mars Science Laboratory: mineralogy of Rocknest at Gale Crater. Science 341, Available from: http://dx.doi.org/10.1126/science.1238932.

Bishop, J.L., Darby Dyar, M., Lane, M.D., Banfield, J.F., 2005. Spectral identification of hydrated sulfates on Mars and comparison with acidic environments on Earth. Int. J. Astrobiol. 3, 275–285.

Bottinga, Y., Craig, H., 1968. Oxygen isotope fractionation between CO_2 and water, and the isotopic composition of marine atmospheric CO_2. Earth Planet. Sci. Lett. 5, 285–295.

Boynton, W.V., Bailey, S.H., Hamara, D.K., Williams, M.S., Bode, R.C., Fitzgibbon, M.R., et al., 2001. Thermal and evolved gas analyzer: part of the Mars volatile and climate surveyor integrated payload. J. Geophys. Res. 106, 17683–17698.

Boynton, W.V., Feldman, W.C., Squyres, S.W., Prettyman, T.H., Brückner, J., Evans, L.G., et al., 2002. Distribution of hydrogen in the near surface of Mars: evidence for subsurface ice deposits. Science 297, 81–85.

Boynton, W.V., Ming, D.W., Kounaves, S.P., Young, S.M.M., Arvidson, R.E., Hecht, M.H., et al., 2009. Evidence for calcium carbonate at the Mars Phoenix landing site. Science 325, 61–64.

Calvin, W.M., Roach, L.H., Seelos, F.P., Seelos, K.D., Green, R.O., Murchie, S.L., et al., 2009. Compact Reconnaissance Imaging Spectrometer for Mars observations of northern Martian latitudes in summer. J. Geophys. Res. 114, E00D11. Available from: http://dx.doi.org/10.1029/2009JE003348.

Cannon, K.M., Sutter, B., Ming, D.W., Boynton, W.V., Quinn, R., 2012. Perchlorate induced low temperature carbonate decomposition in the Mars Phoenix Thermal and Evolved Gas Analzer (TEGA). Geophys. Res. Lett. 39, 2–6.

Carrier, B.L., Kounaves, S.P., 2015. The origins of perchlorate in the Martian soil. Geophys. Res. Lett. 42, 3739–3745.

Catling, D.C., Claire, M.W., Zahnle, K.J., Quinn, R.C., Clark, B.C., Hecht, M.H., et al., 2010. Atmospheric origins of perchlorate on Mars and in the Atacama. J. Geophys. Res. 115, E00E11. Available from: http://dx.doi.org/10.1029/2009JE003425.

Clark, B., Kounaves, S., 2016. Evidence for the distribution of perchlorates on Mars. Int. J. Astrobiol 15, 311–318. Available from: http://dx.doi.org/10.1017/S1473550415000385.

Clark, B.C., Van Hart, D.C., 1981. The salts of Mars. Icarus 45, 370–378.

Clark, B.C., Baird, A.K., Rose, H.J., Toulmin III, P., Keil, K., Castro, A.J., et al., 1976. Inorganic analyses of Martian surface samples at the Viking landing sites. Science 194, 1283–1288.

Clark, B.C., Baird, A.K., Weldon, R.J., Tsusaki, D.M., Schnabel, L., Candelaria, M.P., 1982. Chemical composition of Martian fines. J. Geophys. Res. Solid Earth 87, 10059–10067.

Clark, B.C., Morris, R.V., McLennan, S.M., Gellert, R., Jolliff, B., Knoll, A.H., et al., 2005. Chemistry and mineralogy of outcrops at Meridiani Planum. Earth Planet. Sci. Lett. 240, 73–94.

Clayton, R.N., Mayeda, T.K., 1988. Isotopic composition of carbonate in EETA 79001 and its relation to parent body volatiles. Geochim. Cosmochim. Acta 52, 925–927.

Cull, S., Arvidson, R.E., Mellon, M.T., Skemer, P., Shaw, A., Morris, R.V., 2010. Compositions of subsurface ices at the Mars Phoenix landing site. Geophys. Res. Lett. 37, L24203. Available from: http://dx.doi.org/10.1029/2010GL045372.

Feldman, W.C., Boynton, W.V., Tokar, R.L., Prettyman, T.H., Gasnault, O., Squyres, S.W., et al., 2002. Global distribution of neutrons from Mars: results from Mars Odyssey. Science 297, 75–78.

Feldman, W.C., Mellon, M.T., Gasnault, O., Diez, B., Elphic, R.C., Hagerty, J.J., et al., 2007. Vertical distribution of hydrogen at high northern latitudes on Mars: the Mars Odyssey neutron spectrometer. Geophys. Res. Lett. 34, L05201. Available from: http://dx.doi.org/10.1029/2006GL028936.

Fishbaugh, K.E., Poulet, F., Chevrier, V., Langevin, Y., Bibring, J.-P., 2007. On the origin of gypsum in the Mars north polar region. J. Geophys. Res. 112, E07002. Available from: http://dx.doi.org/10.1029/2006JE002862.

Gaillard, F., Michalski, J., Berger, G., McLennan, S.M., Scaillet, B., 2012. Geochemical reservoirs and timing of sulfur cycling on Mars. Space Sci. Rev. 174, 251–300.

Gellert, R., Rieder, R., Brückner, J., Clark, B.C., Dreibus, G., Klingelhöfer, G., et al., 2006. Alpha particle X-ray spectrometer (APXS): results from Gusev Crater and calibration report. J. Geophys. Res. 111, E02S05. Available from: http://dx.doi.org/10.1029/2005JE002555.

Gendrin, A., 2005. Sulfates in Martian layered terrains: the OMEGA/Mars Express view. Science 307, 1587–1591.

Glavin, D.P., Freissinet, C., Miller, K.E., Eigenbrode, J.L., Brunner, A.E., Buch, A., et al., 2013. Evidence for perchlorates and the origin of chlorinated hydrocarbons detected by SAM at the Rocknest Aeolian deposit in Gale Crater. J. Geophys. Res. 118, 1955–1973.

Goetz, W., Pike, W.T., Hviid, S.F., Madsen, M.B., Morris, R.V., Hecht, M.H., et al., 2010. Microscopy analysis of soils at the Phoenix landing site, Mars: classification of soil particles and description of their optical and magnetic properties. J. Geophys. Res. 115, E00E22. Available from: http://dx.doi.org/10.1029/2009JE003437.

Golden, D.C., Ming, D.W., Sutter, B., Clark, B.C., Morris, R.V., Boynton, W.V., et al., 2009. Sulfur mineralogy at the Mars Phoenix landing site. In: 40th Lunar and Planetary Science Conference. Lunar and Planetary Institute, Houston, Abstract #2319.

Hecht, M.H., Marshall, J., Pike, W.T., Staufer, U., Blaney, D., Braendlin, D., et al., 2008. Microscopy capabilities of the microscopy, electrochemistry, and conductivity analyzer. J. Geophys. Res. 113, E00A22. Available from: http://dx.doi.org/10.1029/2008JE003077.

Heet, T.L., Arvidson, R.E., Cull, S.C., Mellon, M.T., Seelos, K.D., 2009. Geomorphic and geologic settings of the Phoenix Lander mission landing site. J. Geophys. Res. 114, 1–19.

Hoffman, J.H., Chaney, R.C., Hammack, H., 2008. Phoenix Mars mission—the thermal evolved gas analyzer. J. Am. Soc. Mass Spectrom. 19, 1377–1383.

Jackson, W.A., Böhlke, J.K., Andraski, B.J., Fahlquist, L., Bexfield, L., Eckardt, F.D., et al., 2015. Global patterns and environmental controls of perchlorate and nitrate co-occurrence in arid and semi-arid environments. Geochim. Cosmochim. Acta 164, 502–522.

Jull, A.J.T., Eastoe, C.J., Cloudt, S., 1997. Isotopic composition of carbonates in the SNC meteorites, Allan Hills 84001 and Zagami. J. Geophys. Res. 102, 1663–1670.

Kang, N., Jackson, W.A., Dasgupta, P.K., Anderson, T.A., 2008. Perchlorate production by ozone oxidation of chloride in aqueous and dry systems. Sci. Total Environ. 405, 301–309.

Keller, J.M., Boynton, W.V., Karunatillake, S., Baker, V.R., Dohm, J.M., Evans, L.G., et al., 2006. Equatorial and midlatitude distribution of chlorine measured by Mars Odyssey GRS. J. Geophys. Res. 112, E03S08. Available from: http://dx.doi.org/10.1029/2006JE002679.

Kereszturi, A., Möhlmann, D., Berczi, S., Ganti, T., Horvath, A., Kuti, A., et al., 2010. Indications of brine related local seepage phenomena on the northern hemisphere of Mars. Icarus 207, 149–164.

Kounaves, S.P., Hecht, M.H., West, S.J., Morookian, J.-M., Young, S.M.M., Quinn, R., et al., 2009. The MECA Wet Chemistry Laboratory on the 2007 Phoenix Mars Scout Lander. J. Geophys. Res. 114, E00A19. Available from: http://dx.doi.org/10.1029/2008JE003084.

Kounaves, S.P., Hecht, M.H., Kapit, J., Quinn, R.C., Catling, D.C., Clark, B.C., et al., 2010a. Soluble sulfate in the Martian soil at the Phoenix landing site. Geophys. Res. Lett. 37, L09201. Available from: http://dx.doi.org/10.1029/2010GL042613.

Kounaves, S.P., Hecht, M.H., Kapit, J., Gospodinova, K., DeFlores, L., Quinn, R.C., et al., 2010b. Wet chemistry experiments on the 2007 Phoenix Mars Scout Lander mission: data analysis and results. J. Geophys. Res. 115, E00E10. Available from: http://dx.doi.org/10.1029/2009JE003424.

Kounaves, S.P., Carrier, B.L., O'Neil, G.D., Stroble, S.T., Claire, M.W., 2014a. Evidence of Martian perchlorate, chlorate, and nitrate in Mars meteorite EETA79001: implications for oxidants and organics. Icarus 229, 206–213.

Kounaves, S.P., Chaniotakis, N.A., Chevrier, V.F., Carrier, B.L., Folds, K.E., Hansen, V.M., et al., 2014b. Identification of the perchlorate parent salts at the Phoenix Mars landing site and possible implications. Icarus 232, 226–231.

Kreslavsky, M.A., Head, J.W., 2002. Fate of outflow channel effluents in the northern lowlands of Mars: the Vastitas Borealis Formation as a sublimation residue from frozen ponded bodies of water. J. Geophys. Res. 107, 5121–5128.

Kreslavsky, M.A., Head, J.W., 2009. Slope streaks on Mars: A new "wet" mechanism. Icarus 201, 517–527.

Kreslavsky, M.A., Head, J.W., Marchant, D.R., 2008. Periods of active permafrost layer formation during the geological history of Mars: implications for circum-polar and mid-latitude surface processes. Planet. Space Sci. 56, 289–302.

Langevin, Y., Poulet, F., Bibring, J.-P., Gondet, B., 2005. Sulfates in the north polar region of Mars detected by OMEGA/Mars Express. Science 307, 1584–1586.

Lemmon, M.T., Smith, P., Shinohara, C., Tanner, R., Woida, P., Shaw, A., et al., 2008. The Phoenix Surface Stereo Imager (SSI) investigation. In: 39th Lunar and Planetary Science Conference, Houston TX, Abstract #2156.

Madden, M.E.E., Bodnar, R.J., Rimstidt, J.D., 2004. Jarosite as an indicator of water-limited chemical weathering on Mars. Nature 431, 821–823.

Malin, M.C., Edgett, K.S., Posiolova, L.V., Mccolley, S.M., Noe Dobrea, E.Z., 2006. Rate and contemporary gully. Science 314, 1573–1577.

Mangold, N., Gendrin, A., Gondet, B., LeMouelic, S., Quantin, C., Ansan, V., et al., 2008. Spectral and geological study of the sulfate-rich region of West Candor Chasma, Mars. Icarus 194, 519–543.

Massé, M., Bourgeois, O., Le Mouélic, S., Verpoorter, C., Le Deit, L., Bibring, J.P., 2010. Martian polar and circum-polar sulfate-bearing deposits: sublimation tills derived from the North Polar Cap. Icarus 209, 434–451.

Massé, M., Bourgeois, O., Le Mouélic, S., Verpoorter, C., Spiga, A., Le Deit, L., 2012. Wide distribution and glacial origin of polar gypsum on Mars. Earth Planet. Sci. Lett. 317-318, 44–55.

McAdam, A.C., Franz, H.B., Sutter, B., Archer Jr., P.D., Freissinet, C., Eigenbrode, J.L., et al., 2014. Sulfur-bearing phases detected by evolved gas analysis of the Rocknest Aeolian deposit, Gale Crater, Mars. J. Geophys. Res. 119, 373–393.

McEwen, A.S., Ojha, L., Dundas, C.M., Mattson, S.S., Byrne, S., Wray, J.J., et al., 2011. Seasonal flows on warm Martian slopes. Science 333, 740–743.

McLennan, S.M., Anderson, R.B., Bell, J.F., Bridges, J.C., Calef, F., Campbell, J.L., et al., 2014. Elemental geochemistry of sedimentary rocks at Yellowknife Bay, Gale crater, Mars. Science 343, Available from: http://dx.doi.org/10.1126/science.1244734.

McSween Jr., H.Y., Keil, K., 2000. Mixing relationships in the Martian regolith and the composition of globally homogeneous dust. Geochim. Cosmochim. Acta 64, 2155–2166.

Mellon, M.T., Jakosky, B.M., 1995. The distribution and behavior of Martian ground ice during past and present epochs. J. Geophys. Res. 100, 11781–11799.

Mellon, M.T., Feldman, W.C., Prettyman, T.H., 2004. The presence and stability of ground ice in the southern hemisphere of Mars. Icarus 169, 324–340. Available from: http://dx.doi.org/10.1029/2007JE003067.

Mellon, M.T., Boynton, W.V., Feldman, W.C., Arvidson, R.E., Titus, T.N., Bandfield, J.L., et al., 2008. A pre-landing assessment of the ice table depth and ground ice characteristics in Martian permafrost at the Phoenix landing site. J. Geophys. Res. 113, E00A25. Available from: http://dx.doi.org/10.1029/2007JE003067.

Ming, D.W., Archer, P.D., Glavin, D.P., Eigenbrode, J.L., Franz, H.B., Sutter, B., et al., 2014. Volatile and organic compositions of sedimentary rocks in Yellowknife Bay, Gale Crater, Mars. Science 343, Available from: http://dx.doi.org/10.1126/science.1245267.

Moores, J.E., Schuerger, A.C., 2011. UV degradation of accreted organics on Mars: IDP longevity, surface reservoir of organics, and relevance to the detection of methane in the atmosphere. J. Geophys. Res. 117, L04203. Available from: http://dx.doi.org/10.1029/2012JE004060.

Nier, A.O., Mcelroy, M.B., Yung, Y.L., 1976. Isotopic composition of the Martian atmosphere. Science 194, 68−70.

Niles, P.B., Boynton, W.V., Hoffman, J.H., Ming, D.W., Hamara, D., 2010. Stable isotope measurements of Martian atmospheric CO_2 at the Phoenix landing site. Science 329, 1334−1338.

Núñez, J.I., Barnouin, O.S., Murchie, S.L., Seelos, F.P., McGovern, J.A., Seelos, K.D., et al., 2016. New insights into gully formation on Mars: constraints from composition as seen by MRO/CRISM. Geophys. Res. Lett. Available from: http://dx.doi.org/10.1002/2016GL068956.

Ojha, L., Wilhelm, M.B., Murchie, S.L., McEwen, A.S., Wray, J.J., Hanley, J., et al., 2015. Spectral evidence for hydrated salts in recurring slope lineae on Mars. Nat. Geosci. 8, 829−832.

Pike, W.T., Staufer, U., Hecht, M.H., Goetz, W., Parrat, D., Sykulska-Lawrence, H., et al., 2011. Quantification of the dry history of the Martian soil inferred from in situ microscopy. Geophys. Res. Lett. 38, L24201. Available from: http://dx.doi.org/10.1029/2011GL049896.

Putzig, N.E., Mellon, M.T., 2007. Apparent thermal inertia and the surface heterogeneity of Mars. Icarus 191, 68−94.

Quinn, R.C., Martucci, H.F., Miller, S.R., Bryson, C.E., Grunthaner, F.J., Grunthaner, P.J., 2013. Perchlorate radiolysis on Mars and the origin of Martian soil reactivity. Astrobiology 13, 515−520.

Schuttlefield, J.D., Sambur, J.B., Gelwicks, M., Eggleston, C.M., Parkinson, B.A., 2011. Photooxidation of chloride by oxide minerals: implications for perchlorate on Mars. J. Am. Chem. Soc. 133, 17521−17523.

Smith, M.L., Claire, M.W., Catling, D.C., Zahnle, K.J., 2014. The formation of sulfate, nitrate and perchlorate salts in the Martian atmosphere. Icarus 231, 51−64.

Smith, P.H., Reynolds, R., Weinberg, J., Friedman, T., Lemmon, M.T., Tanner, R., et al., 2001. The MVACS Surface Stereo Imager on Mars polar lander. J. Geophys. Res. 106, 17589−17607.

Smith, P.H., Tamppari, L., Arvidson, R.E., Bass, D., Blaney, D., Boynton, W., et al., 2008. Introduction to special section on the Phoenix mission: landing site characterization experiments, mission overviews, and expected science. J. Geophys. Res. 113, E00A18. Available from: http://dx.doi.org/10.1029/2008JE003083.

Smith, P.H., Tamppari, L.K., Arvidson, R.E., Bass, D., Blaney, D., Boynton, W.V., et al., 2009. H_2O at the Phoenix landing site. Science 58, 58−61.

Squyres, S.W., Knoll, A.H., Arvidson, R.E., Clark, B.C., Grotzinger, J.P., Jolliff, B.L., et al., 2006. Two years at Meridiani Planum: results from the opportunity rover. Science 313, 1403−1407.

Stillman, D.E., Grimm, R.E., 2011. Dielectric signatures of adsorbed and salty liquid water at the Phoenix landing site, Mars. J. Geophys. Res. 116, E09005. Available from: http://dx.doi.org/10.1029/2011JE003838.

Sutter, B., Boynton, W.V., Ming, D.W., Niles, P.B., Morris, R.V., Golden, D.C., et al., 2012. The detection of carbonate in the Martian soil at the Phoenix landing site: a laboratory investigation and comparison with the thermal and evolved gas analyzer (TEGA) data. Icarus 218, 290−296.

Tanaka, K.L., Fortezzo, C.M., Hayward, R.K., Rodriguez, J.A.P., Skinner, J.A., 2011. History of plains resurfacing in the Scandia region of Mars. Planet. Space Sci. 59, 1128−1142.

Taylor, P.A., Catling, D.C., Daly, M., Dickinson, C.S., Gunnlaugsson, H.P., Harri, A.-M., et al., 2008. Temperature, pressure, and wind instrumentation in the Phoenix meteorological package. J. Geophys. Res. 113, E00A10. Available from: http://dx.doi.org/10.1029/2007JE003015.

Titus, T.N., and Prettyman, T.H., 2007. Thermal inertia characterization of the proposed Phoenix landing site. In: Proc. 38th Lunar Planet Sci. Conf., Abstract #2088.

Toner, J.D., Catling, D.C., Light, B., 2015a. A revised Pitzer model for low-temperature soluble salt assemblages at the Phoenix site, Mars. Geochim. Cosmochim. Acta 166, 327−343.

Toner, J.D., Catling, D.C., Light, B., 2015b. Modeling salt precipitation from brines on Mars: evaporation versus freezing origin for soil salts. Icarus 250, 451−461.

Vaniman, D.T., Bish, D.L., Ming, D.W., Bristow, T.F., Morris, R.V., Blake, D.F., et al., 2014. Mineralogy of a mudstone at Yellowknife Bay, Gale Crater, Mars. Science 343, Available from: http://dx.doi.org/10.1126/science.1243480.

Wang, A., Haskin, L.A., Squyres, S.W., Jolliff, B.L., Crumpler, L., Gellert, R., et al., 2006. Sulfate deposition in subsurface regolith in Gusev Crater, Mars. J. Geophys. Res. 111, E02S17. Available from: http://dx.doi.org/10.1029/2005JE002513.

Webster, C.R., Mahaffy, P.R., Flesch, G.J., Niles, P.B., Jones, J.H., Leshin, L.A., et al., 2013. Isotope ratios of H, C and O in CO_2 and H_2O of the Martian atmosphere. Science 341, 260–264.

Whiteway, J., Daly, M., Carswell, A., Duck, T., Dickinson, C., Komguem, L., et al., 2008. Lidar on the Phoenix mission to Mars. J. Geophys. Res. 113, E00A08. Available from: http://dx.doi.org/10.1029/2007JE003002.

Yen, A.S., Gellert, R., Schröder, C., Morris, R.V., Bell III, J.F., Knudson, A.T., et al., 2005. An integrated view of the chemistry and mineralogy of Martian soils. Nature 436, 49–54.

Zent, A.P., Hecht, M.H., Cobos, D.R., Campbell, G.S., Campbell, C.S., Cardell, G., et al., 2009. Thermal and Electrical Conductivity Probe (TECP) for Phoenix. J. Geophys. Res. 114, E00A27. Available from: http://dx.doi.org/10.1029/2007JE003052.

Zent, A.P., Hecht, M.H., Cobos, D.R., Wood, S.E., Hudson, T.L., Milkovich, S.M., et al., 2010. Initial results from the Thermal and Electrical Conductivity Probe (TECP) on Phoenix. J. Geophys. Res. 115, E00E14. Available from: http://dx.doi.org/10.1029/2009JE003420.

Zent, A.P., Hecht, M.H., Hudson, T.L., Wood, S.E., Chevrier, V.F., 2016. A revised calibration function and results for the Phoenix mission TECP relative humidity sensor. J. Geophys. Res. 121, 626–651.

MARS EXPLORATION ROVER OPPORTUNITY: WATER AND OTHER VOLATILES ON ANCIENT MARS

10

Bradley L. Jolliff[1], David W. Mittlefehldt[2], William H. Farrand[3], Andrew H. Knoll[4], Scott M. McLennan[5], and Ralf Gellert[6]

[1]*Washington University in St. Louis, St. Louis, MO, United States* [2]*NASA/Johnson Space Center, Houston, TX, United States* [3]*Space Science Institute, Boulder, CO, United States* [4]*Harvard University, Cambridge, MA, United States* [5]*State University of New York at Stony Brook, Stony Brook, NY, United States* [6]*University of Guelph, Guelph, ON, Canada*

10.1 INTRODUCTION

"Follow the water!" was the mantra of the Mars Exploration Program when the Mars exploration rovers (MERs) arrived at the Red Planet in 2004. Meridiani Planum was selected as the landing site for MER-B Opportunity owing to the detection of hematite (α-Fe_2O_3) from orbit by the Thermal Emission Spectrometer (TES) on Mars Global Surveyor (MGS) (Christensen et al., 2001). Most known mechanisms of forming hematite over large spatial scales involve water, and orbital images indicated exposure of a hematite-bearing unit at the top of a thick, regionally distributed layered sequence. It was also known at the time, from Mars Odyssey gamma-ray spectrometer results, that Meridiani Planum contained a relatively high concentration of hydrogen in near-surface deposits (Feldman et al., 2004), likely in the form of water of hydration. The story of volatiles along Opportunity's journey is one of discovery of the past water action and aqueous processes on Mars, manifested in many ways along the rover's long and continuing path.

Opportunity landed in the 22-m-diameter Eagle crater on Meridiani Planum on January 24, 2004, scoring a geologic hole in one by a lucky bounce of its airbag landing mechanism (Squyres et al., 2004a). Opportunity's primary scientific objective was to explore a site where water may once have been present, to determine the role of water in the formation and alteration of the Martian surface, and to assess whether past environmental conditions might have been conducive to life. When the mast was deployed and the cameras took their first images within Eagle crater, Opportunity observed a thin ledge of light-toned outcrop along the inside of the crater rim. That small exposure proved to be layered bedrock, sedimentary in origin, and host to a variety of geologic evidence for a watery past. The first mineral detection by the rover's analytical payload was of hematite in small spheroidal granules littering the surface of the crater floor. These observations, made at the outset of the mission, almost immediately satisfied the primary objective of *finding the water*, or at least evidence of its presence on ancient Mars.

Volatiles in the Martian Crust. DOI: http://dx.doi.org/10.1016/B978-0-12-804191-8.00010-6

Opportunity explored Eagle crater for 56 sols (Martian days) before leaving the crater to venture onto the surrounding plains (Table 10.1). Following that auspicious beginning, Opportunity began a journey that would traverse many kilometers and visit craters of different (and generally increasing) size as a means of exploring not only the breadth but also the depth of the deposits of Meridiani Planum. The story of water begins with the discovery of ancient sulfate-rich evaporite deposits, formed initially by aqueous alteration of basalt, and includes the discovery of evidence for diagenesis in a groundwater-saturated environment, with occasional breaching of the surface by the groundwater table in a playa-like setting. The mineralogy is evocative of low-pH aqueous alteration (Morris et al., 2005) or reflects circumneutral initial aqueous alteration and reducing groundwater that later became oxidized through interaction in the near-surface environment from O_2 or photo oxidation, which would have led to low pH conditions (e.g., Hurowitz et al., 2010).

Opportunity explored the 150-m-diameter Endurance crater, the 350-m Erebus crater, and the 750-m Victoria crater, all of which expose sedimentary rocks of the sulfate-rich Burns formation. Opportunity then turned its sights toward Endeavour crater, ~ 22 km in diameter and 22 km (traverse distance) south of Victoria (Fig. 10.1). Endeavour crater is Noachian in age and thus older than the Meridiani sediments that Opportunity had explored to that time. As such, rocks exposed along its raised rim were expected to retain evidence of earlier and possibly different conditions relating to past Martian environments. Opportunity did indeed complete the traverse from Victoria to Endeavour crater, stopping along the way to examine several sites including the 100-m-diameter Santa Maria crater, and then began exploring exposures along the rim of Endeavour at a rim segment informally named Cape York ca. sol 2681 (August 2011). As part of the investigation of Cape York,

Table 10.1 Dates and Duration of Exploration Campaigns Described in This Chapter

Earth Date	Sol	Activity	Odometry (m) at End
Jan 24, 2004–Mar 21, 2004	001–056	Eagle crater	155
Mar 22, 2004–May 01, 2004	057–095	Eagle to Endurance	984
May 02, 2004–Dec 11, 2004	096–314	Endurance crater	1779
Dec 12, 2004–Nov 02, 2005	315–632	Endurance to Erebus	6418
Nov 03, 2005–Mar 12, 2006	633–758	Erebus crater	6701
Mar 13, 2006–Sep 26, 2006	759–951	Erebus to Victoria	9275
Sep 27, 2006–Aug 27, 2008	952–1634	Victoria crater	11,778
Aug 28, 2008–Dec 21, 2010	1619–2452	Victoria to Santa Maria	26,460
Dec 22, 2010–Mar 24, 2011	2453–2546	Santa Maria crater	26,709
Mar 25, 2011–Aug 08, 2011	2547–2681	Santa Maria to Endeavour	33,485
Aug 09, 2011–May 23, 2013	2682–3315	Cape York	36,027
Sep 05, 2012–May 12, 2013	3063–3307	Matijevic Hill	35,655
May 24, 2013–Aug 08, 2013	3316–3391	Cape York to Solander Point	38,182
Aug 09, 2013–Dec 04, 2013	3392–3506	Solander Point to Cook Haven	38,681
Dec 05, 2013–Mar 07, 2014	3507–3598	Cook Haven, Murray Ridge	38,748
Mar 08, 2014–Feb 13, 2015	3599–3933	Cook Haven to Marathon Valley	42,044

Detailed activity summaries in Squyres et al. (2004) and Arvidson et al. (2006, 2011).

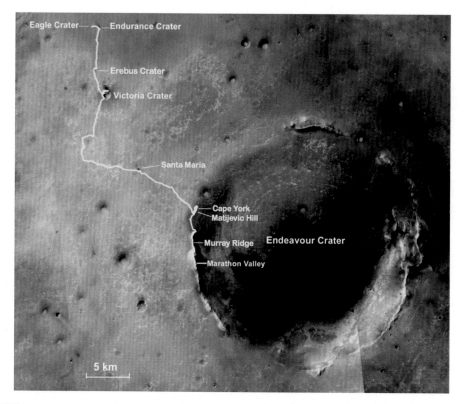

FIGURE 10.1

Opportunity's Traverse route through sol 3933 (42.04 km) to the beginning of its exploration of Marathon Valley. Traverse base map from the Context Imager on Mars Reconnaissance Orbiter. Traverse map was made at the New Mexico Museum of Natural History and Science, Albuquerque. *Image Credit: NASA/JPL-Caltech/MSSS/NMMNHS.*

Opportunity followed mineralogical clues in the form of orbital spectral reflectance signatures to locate ancient sediments bearing evidence of smectites (Arvidson et al., 2014). Exposed in the rim of Endeavour crater, these materials, dubbed the Matijevic formation, did indeed provide evidence of a different and more ancient environment, with new evidence of early aqueous processes.

On the Cape Tribulation rim segment, further south along the western rim of Endeavour, Opportunity encountered ancient fractures that host chemical and multispectral evidence of significant alteration and element mobility, indicative of water–rock interactions along fractures related to the formation of Endeavour crater (Arvidson et al., 2016). One area of special interest was Cook Haven, where Opportunity discovered Mn-rich coatings on overturned rocks some 660 m north of the stunning Endeavour crater overlook at Pillinger Point (ca. sol 3664). After completing a marathon distance (42.195 km, sol 3968), Opportunity arrived at a location named Marathon Valley where it investigated rim deposits to characterize the surface where orbital spectral observations indicated the presence of water-bearing smectite minerals. At the time of this writing (June 2017), Opportunity was continuing its traverse along Endeavour's western rim, approaching Cape Byron. In the following sections, the observations noted above are described in detail. This summary ends at the point of beginning the exploration at Marathon Valley, on approximately sol 4000 (April 24, 2015).

10.2 OPPORTUNITY INSTRUMENT PAYLOAD

The Opportunity Athena science payload (Squyres et al., 2004a) includes two instruments mounted on a 1.5-m-tall mast, a multispectral Panoramic Camera (430–1010 nm), Pancam (Bell et al., 2003), and a miniature thermal emission spectrometer, mini-TES (Christensen et al., 2003). The payload also includes a 5-degree-of-freedom robotic arm (Squyres et al., 2003), called the Instrument Deployment Device (IDD), housing four in situ instruments. The arm carries (1) an alpha particle X-ray spectrometer, APXS (Rieder et al., 2003), which measures elemental abundances of rocks and soils; (2) a Mössbauer Spectrometer (Klingelhöfer et al., 2003), which determines mineralogy and oxidation state of Fe-bearing phases; (3) a Microscopic Imager, MI (Herkenhoff et al., 2004), which takes high-resolution images (31 µm per pixel) of rock and soil surfaces; and (4) a Rock Abrasion Tool, RAT (Gorevan et al., 2003), which can remove up to ~5 mm of material over a circular area 45 mm in diameter. Motor currents of the RAT also provide a measure of resistance and thus of rock hardness (Arvidson et al., 2004; Herkenhoff et al., 2008a; Thomson et al., 2013). The mini-TES was no longer used after sol 2257 owing to temperature-related instrument degradation and dust accumulation on its mirrors (Arvidson et al., 2011); the Mössbauer Spectrometer was no longer used after sol 2900 owing to decay of its [57]Co radioactive source (Squyres et al., 2012).

10.3 GEOLOGIC CONTEXT OF MERIDIANI PLANUM

Observations from instruments on several orbiters provided data on the Meridiani Planum deposits prior to the arrival of Opportunity, including Mars Global Surveyor (MGS) Thermal Emission Spectrometer (TES; Christensen et al., 2000), Mars Orbiter Camera (MOC; Malin et al., 1992), Mars Express (MEX) Observatoire pour la Minéralogie, l'Eau, les Glaces et l'Activité (OMEGA; Bibring et al., 2005), and Mars Odyssey Thermal Emission Imaging System (THEMIS; Christensen et al., 2004a). The Mars Reconnaissance Orbiter (MRO) joined the fleet of orbiters in August 2006 (Zurek and Smrekar, 2007). Mapping based on orbital data shows that the oldest geologic unit in the region and underlying the Opportunity landing site is a highly eroded Noachian crater unit, which is overlain unconformably by a sequence of etched Late Noachian to Early Hesperian units (Flahaut et al., 2015; Hynek and Achille, 2017). Apparent ease of erosion and stratigraphic relationships in this region led Hynek and Phillips (2008) to conclude that a fluctuating groundwater table coupled with an eolian environment (Squyres and Knoll, 2005) or deposition of pyroclastic ash (McCollom and Hynek, 2005) were the most likely processes operating during formation of the etched units.

The hematite-bearing deposits of Meridiani Planum unconformably overlie the heavily cratered and eroded terrain, and bury ancient channel systems. Orbital reflectance spectra over the region are dominated by pyroxene, plagioclase feldspar, crystalline hematite, and nanophase iron oxide dust signatures (MEX OMEGA, Bibring et al., 2005). Analyses from the MGS TES

(Christensen et al., 2000, 2001), Mars Odyssey THEMIS (Christensen and Ruff, 2004), and MGS MOC (e.g., Edgett and Malin, 2002; Malin, 2005) also provided key information for geological context such as terrain morphology, thermal inertia, and mineralogy.

Meridiani Planum was initially identified as a hematite-bearing deposit and related to an origin involving water by means of MGS TES data (Christensen et al., 2000, 2001; Christensen and Ruff, 2004). Basaltic mineralogy (plagioclase and pyroxene) and alteration products were also inferred from the TES spectra (Arvidson et al., 2003). Morphologically, the Meridiani plains appear to consist of patches of low thermal-inertia and light-toned rock outcrop interspersed with areas of dark-toned dunes atop a thick section of layered deposits (Arvidson et al., 2003). Where the Meridiani plains thin, several hundred km to the northwest, north, and northeast of the Opportunity landing site, the etched terrain crops out from beneath the plains and exhibits layering and spectral evidence of hydrated sulfates, including kieserite and polyhydrated Mg-sulfate (Hynek et al., 2002; Arvidson et al., 2006).

Some 50 km to the south, the Meridiani plains appear to cover or merge with the highly eroded crater unit, which indicates that the hematite-bearing plains deposits are younger than the channel-forming fluvial activity reflected in the highly eroded terrain (Hynek and Achille, 2017). Such an age relationship is consistent with crater size-frequency distribution analysis (Arvidson et al., 2006; Hynek and Achille, 2017). Some of the larger impact craters in the area are partially buried by the Meridiani Planum deposits, such as the $\sim$20-km-diameter Endeavour crater located some 20 km southeast of the landing site (Arvidson et al., 2006), which is currently being explored by Opportunity. The crater size-frequency distribution data indicate relative ages for these surfaces that are Late Noachian to Early Hesperian; thus, Meridiani plains materials may have been deposited at the end of the early period of fluvial activity on Mars (Arvidson et al., 2006; Hynek and Achille, 2017).

Integrated analysis of available visible and near-IR hyperspectral data reveals a mineralogically more complex stratigraphy for the etched terrain, consisting of alternating sulfate and clay-rich units (Flahaut et al., 2015). The uppermost hematite-bearing, sulfate-rich unit is underlain by a unit that grades from sulfate-rich to clay-rich at the top of the sequence. This stratigraphy is important because these units require surficial water as well as groundwater, and they may all be exposed (and/or mixed) in the rim deposits of Endeavour crater. Flahaut et al. (2015) concluded that most hydrated minerals in the etched terrain at Meridiani Planum, dating to the Early Hesperian, could have formed at more neutral pH than indicated by the mineralogy of the upper, sulfate-rich sediments (see below). They also pointed out the similarity of the sulfate—clay—sulfate sequence to sediments observed by the Curiosity rover in Gale crater.

10.4 EAGLE CRATER AND FIRST DISCOVERIES OF AN AQUEOUS PETROGENESIS

Opportunity landed in the center of Eagle crater (Fig. 10.1), about 10 m from a ca. half meter thick outcrop of layered bedrock exposed in the slope of the crater wall. The first analysis of crater-filling regolith adjacent to the lander (Tarmac, sol 11; Table 10.2) showed it to contain dark, olivine-bearing, fine-grained basaltic sand; fine-grained dust; and small spherules, typically $\sim$1 to 6 mm in diameter (Fig. 10.2; Calvin et al., 2008). The small spherules became known informally as "blueberries" owing to their blue color in a common false-color Pancam scheme used to distinguish

Table 10.2 Chemical Compositions of Selected Meridiani Planum Soils and Dust

Sol Feature Target	011 Egress Soil Tarmac		023 Hematite Slope Hema2		060 Mont Blanc Les Hauches		237 Auk RAT	
Type	Soil	±	Soil	±	Soil/Dust	±	basaltic soil	±
APXS (wt%)								
SiO$_2$	46.3	0.47	42.7	0.45	45.3	0.29	48.8	0.47
TiO$_2$	1.04	0.07	0.78	0.07	1.02	0.07	0.85	0.07
Al$_2$O$_3$	9.26	0.13	8.59	0.13	9.22	0.09	10.41	0.13
Cr$_2$O$_3$	0.45	0.04	0.30	0.04	0.33	0.03	0.28	0.03
FeO(T)	18.8	0.13	24.8	0.17	17.6	0.07	15.9	0.11
MnO	0.37	0.01	0.31	0.01	0.34	0.01	0.35	0.01
MgO	7.58	0.12	7.50	0.12	7.63	0.08	6.90	0.11
CaO	7.31	0.06	6.13	0.06	6.59	0.04	7.38	0.06
Na$_2$O	1.83	0.28	2.12	0.26	2.24	0.19	2.39	0.24
K$_2$O	0.47	0.06	0.43	0.06	0.48	0.06	0.59	0.06
P$_2$O$_5$	0.83	0.08	0.81	0.08	0.94	0.07	0.84	0.07
SO$_3$	4.99	0.08	4.77	0.08	7.34	0.07	4.56	0.07
Cl	0.63	0.02	0.68	0.02	0.79	0.01	0.58	0.01
APXS (ppm)								
Ni	423	57	633	64	470	42	323	46
Zn	241	20	312	24	404	14	178	14
Br	32	17	37	19	26	14	21	16
Mössbauer (area percentage for component subspectra)								
Olivine	39		29		28		38	
Pyroxene	37		30		32		40	
Np Fe^{3+} Ox	14		13		30		14	
Jarosite	0		0		0		0	
Fe3D3	0		0		0		0	
Magnetite	6		6		5		6	
Hematite	4		22		5		1	
Fe^{2+}/Fe(T)	0.78		0.61		0.61		0.80	

APXS data: original: Rieder et al. (2004); updated in NASA Planetary Data System (PDS).
Errors (± values) represent precision (i.e., repeatability) 2-sigma statistical errors. Expected accuracy given in the MER calibration paper (Gellert et al., 2006) as relative deviations during laboratory calibration are as follows: Si 3%, Ti 20%, Al 7%, Cr 19%, Fe 7%, Mn 8%, Mg 14%, Ca 7%, Na 11%, K 15%, P 15%, S 15%, Cl 30%, Ni 16%, Zn 16%, Br 20%. See Gellert et al. (2015) for additional discussion of uncertainties.
Mössbauer data for Fe-bearing phase proportions and Fe(2)/Fe(T), from Morris et al. (2006).

various mineral components. Their true color is gray, consistent with specular hematite. Mini-TES data revealed a strong infrared signature of hematite, which correlated spatially with spherule abundance (Christensen et al., 2004b). The hematitic composition of Meridiani spherules was unambiguously confirmed with the Mössbauer Spectrometer (Klingelhöfer et al., 2004).

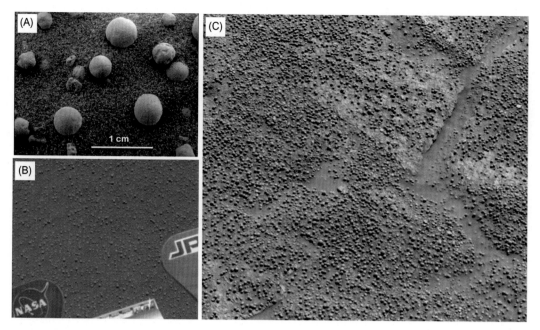

FIGURE 10.2

(A) MI image of soil Wisdom in Eagle crater (sol 14 MI sequence ID P2933). Image credit: NASA/JPL/USGS. (B) Soil near outcrop, littered with spherule lag deposit. Pancam approximate true color (Sol 014, sequence ID p2549). See Savransky and Bell (2004) and Bell et al. (2006) for the construction of Pancam approximate true color images. Rover deck components for scale. (C) Spherules on fractured outcrop pavement outside of Endurance crater, sol 114, sequence ID P2295, false color, filters: L257; L2, L5, and L7 correspond to wavelengths of 753, 535, and 432 nm. *(C) Image credit: D. Savransky and J. Bell/JPL/NASA/Cornell/ASU (Bell et al., 2003, 2006).*

The source of the hematitic spherules became clear as Opportunity began to explore Eagle crater outcrops. The first rocks visited by the rover were located along the light-colored, layered outcrop exposed in the low crater rim northeast of the landing point (Fig. 10.3). Observations at rock target Stone Mountain revealed that the spherules are embedded in the outcrop; because they are harder than the host rock, they weather out and form a lag deposit at the adjacent soil surface. Microscopic images revealed mm-scale laminations in the rocks and a texture dominated by well-sorted and moderately rounded sand grains (Herkenhoff et al., 2004), features collectively suggestive of grain transport by wind (Herkenhoff et al., 2008b). Chemical compositions measured with the APXS showed the rocks to have high concentrations of sulfur, with S — reported as SO_3 — as high as 20–25 wt% (Rieder et al., 2004; Table 10.3; Fig. 10.4). Deconvolution of mini-TES spectral data for mineralogy (Glotch et al., 2006) yielded Mg-, Fe-, and Ca-sulfates, plagioclase feldspar, nontronite, and hematite, plus strong evidence for a significant component of amorphous silica. All exposures of the sulfate sandstone showed evidence in Mössbauer spectra for jarosite, a hydrated ferric-iron sulfate [most likely $(Na,K)(Fe^{3+},Al)_3(SO_4)_2(OH)_6$], which may contribute as much as ~ 2 wt% H_2O to the outcrop matrix (Klingelhöfer et al., 2004). High concentrations of Cl and Br in APXS data (Rieder et al., 2004) further indicate the evaporitic nature of the sediments,

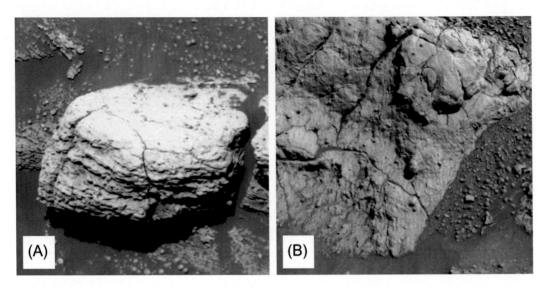

FIGURE 10.3

(A) Stone Mountain, in Eagle crater; concretions (blue) are several millimeters in diameter. (Sol 013, Sequence ID P2376, false color, filters: L257.) (B) Outcrop segment Cathedral Dome, showing purplish and tan or buff rock coloration. (Sol 033, Sequence ID P2589, false color, filters: L257.) *Image credits: D. Savransky and J. Bell/ JPL/NASA/ Cornell/ASU (Bell et al., 2003, 2006).*

although rock textures show clear evidence of sediment reworking prior to lithification. Use of the RAT on these outcrop materials showed the rocks to be extremely soft, consistent with their abundant sulfate mineralogy (Arvidson et al., 2004), and to exhibit brick-red grinding dust (Fig. 10.5) with a Pancam-measured spectrum consistent with red hematite produced by the comminution of coarse-grained gray hematite to finer grain sizes (Squyres et al., 2004b).

Geochemical data clearly indicate that outcrop sulfates originated by evaporation from a brine, but this does not necessarily mean deposition in a watery environment. In Eagle crater, however, an outcrop segment named Last Chance was found to exhibit cm-scale festoon cross-lamination (Grotzinger et al., 2005; Fig. 10.6), characterized by concave-up geometry. The festoon bed sets represent a ripple geometry that develops only in subaqueous flows on Earth, thus these features were interpreted by Grotzinger et al. (2005) as strong evidence for past subaqueous sediment transport. Modeling of bedforms produced by briny fluids with high ionic strength and viscosity implied by the evaporite mineralogy support the interpretation that ripples with the characteristics of those observed by Opportunity could have formed by subaqueous brine flow on the Martian surface (Lamb et al., 2012). As discussed in the next section, textures and m-scale crossbedding elsewhere also speak to eolian transport of ancient Meridiani sands, so the overall environment of deposition appears to have been an arid interdune—playa—sabkha setting, with a groundwater table that occasionally intersected and breached the surface (Grotzinger et al., 2005; McLennan and Grotzinger, 2008).

Table 10.3 Compositions of Burns Formation Sulfate Sandstones (RAT Abraded)

Sol	036	087	139	145	147	149	153	155	162	195	220
Feature	Guadalupe	Golf	Tennessee	Kentucky	Virginia	Ontario	Manitoba	Manitoba	Millstone	Axel	Escher
Target	Eagle	Pilbara	Vols	Cobble Hill2	Virginia	London	Grindstone	Kettlestone	Dramensf.	Heiberg	Kirchner
Location		Fram	Endurance	Endurance							Endurance
APXS (wt%)											
SiO$_2$	36.2	34.7	35.0	35.9	36.9	36.4	38.0	36.2	37.6	37.9	36.5
TiO$_2$	0.65	0.76	0.79	0.71	0.84	0.74	0.83	0.80	0.75	0.77	0.75
Al$_2$O$_3$	5.85	5.82	5.87	5.90	6.32	5.99	6.36	5.85	6.20	6.52	6.06
Cr$_2$O$_3$	0.17	0.19	0.20	0.18	0.21	0.20	0.19	0.20	0.21	0.23	0.18
FeO(T)	14.8	15.7	15.7	14.7	15.5	14.5	14.8	15.2	15.8	17.7	15.7
MnO	0.30	0.36	0.32	0.33	0.39	0.31	0.33	0.33	0.31	0.37	0.24
MgO	8.45	8.63	8.38	9.20	9.00	9.14	8.38	8.63	7.41	6.81	8.37
CaO	4.91	4.82	5.03	4.72	4.43	4.85	4.64	4.85	5.11	5.01	5.00
Na$_2$O	1.66	1.50	1.36	1.54	1.83	1.64	1.70	1.45	1.58	1.86	1.63
K$_2$O	0.53	0.50	0.58	0.57	0.60	0.57	0.58	0.55	0.59	0.60	0.57
P$_2$O$_5$	0.97	0.97	1.03	1.05	1.07	1.11	1.07	1.03	1.17	1.01	1.01
SO$_3$	24.9	25.2	24.9	24.4	22.1	23.7	21.5	23.0	21.1	19.3	23.0
Cl	0.50	0.66	0.65	0.65	0.60	0.72	1.45	1.75	1.98	1.69	0.78
APXS (ppm)											
Ni	589	634	679	618	664	638	604	644	616	933	564
Zn	324	526	533	371	381	357	319	346	437	499	314
Br	30	33	35	54	74	27	39	19	11	10	425
Mössbauer											
Sol	035	0.88	140	144	148	150	152	154	163	196	219
Target	King3	Golf	Vols	CobbleHill	Virginia	London	Grindstone	Kettlestone	Dramensf.	Aktineq3	Kirchner
Area for component subspectra (% of Fe-bearing phases)											
Olivine	1	0	1	1	2	1	2	2	1	2	1
Pyroxene	9	10	14	15	16	18	17	15	21	13	15
Jarosite	38	33	29	28	28	27	28	29	27	30	30
Fe3D3	16	20	18	20	19	19	20	21	18	14	20
Hem	36	37	39	35	35	35	34	32	33	41	35
Fe(2)/Fe(T)	0.10	0.10	0.15	0.16	0.18	0.19	0.19	0.18	0.22	0.15	0.16

(Continued)

Table 10.3 Compositions of Burns Formation Sulfate Sandstones (RAT Abraded) Continued

| Sol | 696 | 893 | 1182 | 1316 | 1375 | 1388 | 1481 | 1843 | 2515 | 3383 | Burns formation sandstone RAT-ground | |
| Feature | Ted | Baltra | Cercedilla | Steno | Smith | Lyell | Gilbert | Penryhn | Luis-DeTorres | Black-Shoulder | | |
Location	Erebus	Plains	Victoria	Victoria	Victoria	Victoria	Victoria	Plains	Santa Maria	Botany Bay	Average[a]	Std. Dev.
APXS (wt%)												
SiO_2	34.6	35.0	39.1	34.4	37.2	35.7	40.7	33.8	36.1	36.5	36.2	1.81
TiO_2	0.71	0.66	0.84	0.71	0.83	0.84	0.90	0.70	0.82	0.79	0.77	0.07
Al_2O_3	5.40	5.54	6.97	5.32	6.10	5.80	6.68	5.71	5.81	5.95	5.95	0.46
Cr_2O_3	0.18	0.17	0.23	0.18	0.19	0.19	0.22	0.26	0.20	0.18	0.20	0.02
FeO(T)	15.8	15.9	16.1	16.1	17.3	17.0	18.2	22.0	16.4	16.5	16.2	1.55
MnO	0.32	0.34	0.21	0.33	0.29	0.29	0.34	0.25	0.32	0.23	0.31	0.05
MgO	7.08	7.36	7.14	7.23	7.27	7.82	5.99	6.92	7.89	8.15	7.84	0.85
CaO	5.74	5.12	5.88	5.62	4.47	4.28	4.46	4.96	4.63	4.28	4.94	0.47
Na_2O	1.51	1.67	2.03	1.46	1.70	1.47	1.94	1.76	1.67	1.78	1.64	0.18
K_2O	0.55	0.54	0.66	0.52	0.58	0.58	0.63	0.50	0.59	0.64	0.57	0.04
P_2O_5	1.01	1.02	1.04	1.00	1.03	1.04	1.07	0.97	1.06	1.02	1.04	0.05
SO_3	26.5	25.8	19.1	26.4	22.0	23.4	17.4	21.4	23.8	22.4	23.2	2.66
Cl	0.46	0.73	0.57	0.68	0.84	1.43	1.41	0.69	0.54	1.60	0.95	0.48
APXS (ppm)												
Ni	537	568	371	523	624	603	520	730	644	612	614	101
Zn	554	488	273	507	777	455	553	391	480	294	437	117
Br	182	303	297	22	113	57	23	79	232	40	98	115

[a]Average and standard deviation values computed for the compositions given in the table.
Source of APXS data: NASA Planetary Data System; Mössbauer data from Morris et al. (2006). Dramensf. = Dramensfjord.

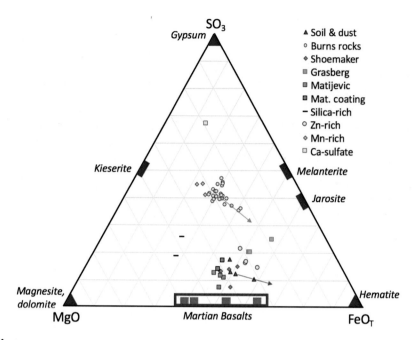

FIGURE 10.4

Mole proportions SO_3–MgO–FeO_T for abraded and unabraded Burns formation outcrop and soils, and representative compositions for Shoemaker, Grasberg, and Matijevic formation rocks. *Arrows* show trends of hematite enrichment for Meridiani soils and Burns formation rocks. All data points are for compositions included in the tables. Also shown are compositions of various end-member minerals and Martian basaltic rocks (Martian meteorites, Gusev and Meridiani basalts). *After McLennan et al. (2005), with permission from Elsevier.*

10.5 EVIDENCE OF DIAGENESIS

10.5.1 CEMENTS AND VUGS

Outcrop rocks in Eagle and Endurance craters are composed of moderately rounded sand grains (0.2–1 mm) that form mm-scale laminations commonly a single grain thick. The sand grains are sulfates and fine-grained siliciclastic materials reworked from an unknown source, likely a playa-lake setting (McLennan et al., 2005; Grotzinger et al., 2006). In MI images, these grains are held together by precipitated cements and some areas appear to be recrystallized (Herkenhoff et al., 2004, 2008b). Centimeter-sized vugs of tabular to discoidal shape (Fig. 10.7) are interpreted as crystal molds. By comparison to molds of sulfate euhedra in terrestrial sediments, these are interpreted as records of early diagenetic growth and subsequent dissolution of highly soluble evaporate crystals (Squyres et al., 2004b, 2006a,b; Herkenhoff et al., 2004; McLennan et al., 2005). Because most of the sediment preserved its overall texture (laminations) and Mg-sulfate was fairly abundant,

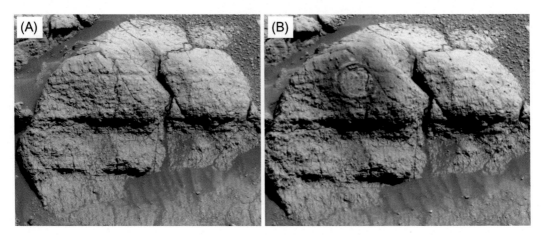

FIGURE 10.5

(A) Pancam image of outcrop crater portion of the El Capitan outcrop in Eagle crater showing bluish concretions embedded in laminated outcrop rock. Also seen in this exposure are "bird track" mineral casts formed by selective dissolution of lenticular mineral grains. (Sol 027, sequence ID P2387, filters: L257). (B) Same outcrop after use of the RAT to grind through the surface (sol 037). Grind-circle diameter is 4.5 cm. *Image credits: D. Savransky and J. Bell/JPL/NASA/ Cornell/ASU (Bell et al., 2003, 2006).*

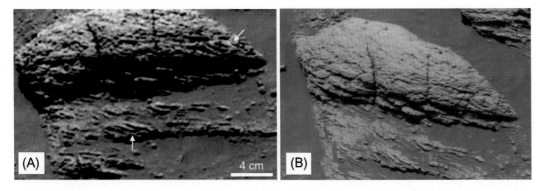

FIGURE 10.6

(A) Last Chance, in Eagle crater. *Lower arrow* shows climbing cross laminae, with flow inferred to have been from left to right. *Upper arrow* illustrates festoon cross lamination. (Sol 017, portion of false color image, sequence ID P2261, filters L257). (B) Last Chance imaged at a different angle. (Sol 040, portion of Pancam image, R1 channel, sequence ID P2541). Rock is ~30 cm across. Image Credits NASA/JPL/Cornell. *Part (A) from Grotzinger et al., 2005, reproduced with permission from Elsevier.*

FIGURE 10.7

Close-up views of rock surface taken with the Microscopic Imager showing crystal mold porosity and spherules in the rock matrix and relative to mm-scale laminations. (A) Portion of MI mosaic of rock target Diogenes taken on sol 125 outside of Endurance crater, showing randomly oriented, blade-shaped vugs that cut across laminae. (B) MI taken on Sol 29 in Eagle crater. View is about 3 cm wide. *After Herkenhoff et al. (2008b), their Fig. 12, reproduced with permission from John Wiley and Sons.*

the molds probably represented a mineral at least as soluble, if not more so, than Mg-sulfate. The relative timing of diagenetic features observed and interpreted for Burns formation rocks is summarized in Fig. 10.8.

10.5.2 CONCRETIONS

Textural, mineralogical, and chemical features of the "blueberry" spherules indicate that they are concretions, formed during diagenesis in a groundwater-saturated environment (McLennan et al., 2005; Chan et al., 2004, 2005). Observed in outcrops at many locations (Calvin et al., 2008), the spherules are dispersed throughout the bedrock and occur in a narrow size range, typically 1−6 mm in diameter (Figs. 10.2, 10.3), although spherules observed in some locations south of Endurance crater and between Victoria and Endeavour craters tend to be smaller, that is < 1 to 3 mm (Jones et al., 2014). Their distribution is statistically overdispersed, that is the distance between adjacent spherules is more regular than a random distribution (McLennan et al., 2005). Judging by analyses of surfaces ground by the RAT, the spherules constitute about 1%−3% of the volume of the rocks in which they occur (McLennan et al., 2005; Jolliff and McLennan, 2006). They are harder than the sulfate-rich matrix in which they are embedded, thus as the outcrop rock

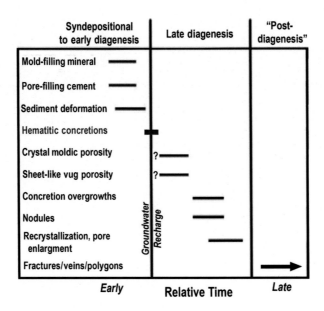

FIGURE 10.8

Schematic diagram showing relative timing of diagenetic features observed in Burns formation rocks. The boundary between early and late diagenesis is arbitrarily set at the formation of the concretions because this might have been a relatively rapid event associated with a chemically distinct groundwater recharge event. *After McLennan et al., 2005 their Fig. 4, with permission from Elsevier.*

weathers by eolian abrasion, the spherules remain and form a resistant lag deposit. In outcrop exposures abraded by the RAT, cross sections of spherules observed with the MI show them to be fine grained, homogeneous, and generally lacking internal structure; however, interpretation of mini-TES data suggests that there may be interior platy or radial concentric structure that is of too fine grain size or has too little contrast to be observed by the MI (Calvin et al., 2008). Either way, these characteristics are consistent with diffusion-limited growth of diagenetic concretions (McLennan et al., 2005; Calvin et al., 2008). Measurements of concretion shape demonstrate that the spherules are only very slightly oblate (elongated parallel to lamination with aspect ratios averaging ~1.06), suggesting that fluid transport was slow relative to (likely fairly rapid) concretion growth rate (McLennan et al., 2005; Sefton-Nash and Catling, 2008).

Measurements made using the Pancam, mini-TES, and Mössbauer Spectrometer confirm a hematite-rich mineralogy for the concretions (Bell et al., 2006; Christensen et al. 2004b; Klingelhöfer et al. 2004). Analyses of APXS and Mössbauer data indicate that the spherules contain >75% and perhaps as much as 90 wt% or more hematite (McLennan et al., 2005; Christensen et al., 2004b). Chemical data from the APXS indicate that the concretions also contain a siliciclastic component as might be expected for nodules growing within sediments, but the exact composition of the spherules is difficult to determine owing to contributions from dust, soil, and rock to most analyses of spherule-bearing targets (Jolliff et al., 2005). On the other hand, analyses of mini-TES spectra suggest that there may be no intrinsic siliciclastic component to the spherules (Calvin et al., 2008). In many instances, these spherules are encased in secondary cements, suggesting continued diagenesis and recrystallization of salts during or after concretion growth (Calvin et al., 2008).

10.6 BURNS FORMATION AND THE FURTHER CHARACTERIZATION OF MERIDIANI PLANUM BEDROCK

The small ledge of sulfate sandstone in Eagle crater provided a good initial representation of Meridiani Planum bedrock, but a fuller appreciation of these rocks came only with the investigation of thicker sections lining larger craters along Opportunity's traverse. These thicker sections record both a greater stratigraphic thickness and a wider range of sedimentary facies, sharpening our sense of regional environments at the time of Meridiani bedrock deposition.

10.6.1 ENDURANCE CRATER

In Endurance crater, located ∼800 m east of Eagle crater, about 7 m of outcrop stratigraphy were investigated between sols 134 and 314 (Table 10.1). A spectacular exposure of bedded sulfate sandstone in the crater rim was named Burns Cliff, after the late Roger Burns, who predicted some of the key geochemical and mineralogical features discovered by Opportunity (Burns, 1987; Squyres and Knoll, 2005). Grotzinger et al. (2005) mapped three units on the basis of stratigraphic and sedimentological characteristics and related these units to changes in the environment of deposition. The lower unit contains high-angle cross beds, representing a migrating eolian dune formation. This unit is truncated by an erosional contact with the overlying middle unit (interpreted as a deflation surface), characterized by planar laminations, low-angle cross beds, and extensive recrystallization that obscures lamination in places. The middle unit is interpreted to be a sand sheet deposit that accumulated as eolian ripples migrated across a nearly flat-lying sand surface. The contact between the middle and upper units appears to be diagenetic in origin, which is supported by enhanced chemical alteration at this contact (Cino et al., 2017). The lower part of the upper unit is also dominated by planar laminations and low-angle crossbedding, but within the upper extents of the upper unit, the texture grades into wavy bedding and small-scale festoon cross lamination, similar to the sedimentary features seen in Eagle crater outcrop (Squyres et al., 2004b, 2006a,b). The upper unit was interpreted to be correlative with the unit exposed in Eagle crater and to have resulted from ripple migration induced by surface flow of liquid water over sand (Grotzinger et al., 2005; Squyres and Knoll, 2005). The three units exposed in Burns Cliff thus appear to represent a "wetting upward" succession from dry eolian dunes in the lower unit to wet interdune deposits in the upper part of the upper unit where brine pooled on the surface between dunes. The facies motif in Burns Cliff is similar to those found in modern arid environments where transient playa lakes episodically develop among eolian sands (Grotzinger et al., 2005). Effects of groundwater interaction with accumulating sediments are seen in recrystallization textures and in the formation of the hematite-rich, spherical concretions.

Excellent outcrop exposures along the sloping inner wall of Endurance crater made it possible to use the rover's in situ analytical capabilities to investigate compositional and textural variations within parts of the exposed middle and upper units of Burns Cliff. APXS analyses of RAT-ground surfaces revealed a systematic change in chemistry with position in the sequence. Systematic and correlated variations such as between MgO and SO_3, and between SiO_2, Al_2O_3, Na_2O, K_2O, and P_2O_5 reflect systematic variations in mineral abundances in different parts of these units and are consistent with nonuniform deposition or with subsequent migration of mobile salt components, dominated by Mg-sulfates (Clark et al., 2005). Variations in the concentrations of Cl and Br provide further evidence of diagenetic changes and movement of sulfates and chlorides. Variations in the Cl/Br ratio within the outcrop at both Eagle and Endurance craters may indicate fractionation of Cl- and Br-salts by evaporative processes (Squyres et al., 2004b) or differential mobility facilitated by freezing-point depression of brines (Clark et al., 2005).

Within Endurance crater, the hematite-rich spherules are observed to occur throughout the exposed sequence of rock units. The sequence includes subaqueous (upper) as well and eolian (lower) dune and sand-sheet facies, yet all facies have spherules. The distribution is unaffected by facies boundaries; thus, the spherules must have formed after deposition of the entire sequence, during diagenesis, consistent with interpretation of the spherules as concretions (McLennan et al., 2005; Fig. 10.8). In outcrop, the spherules do not concentrate preferentially along bedding planes, indicating that they did not form as impact spherules or volcanic accretionary lapilli (e.g., Fralick et al., 2012).

Geologically, recent changes are also observed in Endurance crater. Bedrock exposed in the interior floor of the crater exhibits polygonal fractures. One large rock named Wopmay exhibits prominent desiccation fractures (Fig. 10.9) that resulted from volume change on dehydration of hydrated minerals, likely sulfates, since the time of formation of Endurance crater.

Wopmay, sol 251
Endurance crater

FIGURE 10.9

Desiccation fractures in a rock named Wopmay in Endurance crater (sol 251). The rock is ~1.6 m in length. The polygonal fracture pattern on this rock forms a three-dimensional structure and results from volume decrease caused by desiccation. (Sol 251, false color image mosaic, sequence ID P2432, filters L257). *Image credits: D. Savransky and J. Bell/JPL/NASA/ Cornell/ASU (Bell et al., 2003, 2006).*

10.6.2 **EREBUS CRATER**

Erebus crater, located approximately 4 km south of Endurance crater (Fig. 10.1), is a degraded ~350 m diameter crater with very low relief on its rim, but conspicuous exposures of bedrock on its eastern and western interior rim slopes. Sulfate sandstones crop out as a bedrock pavement in numerous locations between Endurance and Erebus craters, and around the rim of Erebus, and in situ analyses were done on some of these rocks to test for compositional variation of the Burns formation. These analyses essentially confirmed lithologic similarity to the earlier-analyzed rocks (Table 10.3). The main findings in and around Erebus contributing to the water story include primary sedimentary textures within the crater-rim outcrop exposures that indicate deposition under aqueous conditions (Grotzinger et al., 2006). Erebus was explored approximately between sols 633 and 758 (November 2005 to March 2006; Table 10.1).

10.6.2.1 *Trough cross-lamination*

Observations in Erebus crater at the Olympia outcrop showed an abundance of cm-scale trough cross-lamination occurrences, intimately associated with other facies characterized by probable desiccation cracking and soft-sediment deformation (Grotzinger et al., 2006). The cross-lamination texture is better expressed in the rocks at Erebus crater than at Eagle crater, and probable sediment desiccation cracks were observed.

Grotzinger et al. (2006) argued that trough cross-lamination as observed in rocks of the Olympia outcrop (Fig. 10.10), as well as similar rocks seen at Eagle and Endurance craters, most likely formed in shallow, subaqueous flows with moderate current velocities. Such an environment would form sinuous-crested ripples in fine- to medium-grained sand (e.g., Southard, 1973). A trough-shaped or festoon geometry is the diagnostic attribute of such ripples as exposed in cuts transverse to flow (Rubin, 1987). The presence of trough cross-lamination in the Burns formation at Eagle, Endurance, and Erebus craters provides compelling evidence for ancient surface water flow at Meridiani Planum.

10.6.2.2 *Shrinkage cracks*

Zones of soft-sediment deformation, including "curl-up" structures that show clear evidence of plastic deformation, were also observed in the exposed rocks at Erebus crater (Fig. 10.11). The cracks were interpreted by Grotzinger et al. (2006) as shrinkage cracks formed during desiccation of wet sediments. In terrestrial environments, intermittent wetting and drying of laminated sediments results in development of narrow but deep cracks that in plan view form polygons. Owing to the depth of the cracks and the effects of multiple desiccation events, these features can take a prismatic shape that is diagnostic of their environment of deposition. Such environments include tidal flats and interdune depressions that experience alternating wetting (expansion) and desiccation (shrinkage) intervals. Curl-up structures result from upward deflection of laminae when an exposed desiccating layer undergoes more evaporation than the protected underside of the same layer. Curl-up structures are a common feature of desiccated terrestrial sediments (Hardie and Garrett, 1977; Allen, 1982). These types of sedimentary features, seen in rocks exposed by Eagle, Endurance, and Erebus craters, reflect intermittent wetting and transient flooding of interdune depressions during groundwater movement within surface sediments with a measure of cohesion and in a fluctuating water table (Grotzinger et al., 2006).

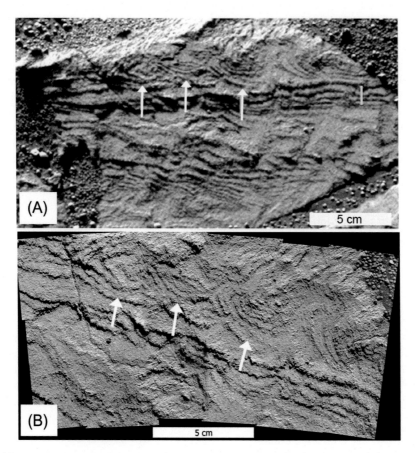

FIGURE 10.10

Trough (or festoon) cross-lamination in Overgaard outcrop, Erebus crater. (A) Cross-stratified lower unit, overlain by planar-laminated middle unit (indicated by *yellow bar*), and trough cross-laminated upper unit. *White arrows* point to three distinct troughs, as indicated by basal truncation and concave-upward geometry. Pancam image acquired on sol 716, sequence ID p2593 (low light), 432 nm filter. *Image credit: NASA/JPL/Cornell/ASU.* (B) Microscope imager (MI) mosaic of middle and upper units, highlighting trough cross-lamination. Mosaic constructed from MI images obtained on sols 721 and 723. *After Grotzinger et al. (2006), their Fig. 2 and Herkenhoff et al., 2008b, their Fig. 15. Reproduced with permission from the Geological Society of America and John Wiley and Sons (Journal of Geophysical Research).*

10.6.3 VICTORIA CRATER

Along the traverse from Endurance to Victoria, elevation increases by about 30 m, yet the regional outcrop is approximately flat-lying. Thus, on the way to Victoria crater, outcrop materials likely represent increasingly higher (younger) parts of the stratigraphic section, such that the Victoria crater section would be entirely higher than the section exposed at Endurance and Eagle craters but

FIGURE 10.11

Deformation cracks interpreted to be penecontemporaneous, resembling prism cracks. Those shown above are typically narrow and cut across lamination for more than 5 cm. (A) Lower Overgaard truncated laminae adjacent to cracks (*small arrows*) are deflected upward along crack margins, and cracks have lateral spacing of several cm. This "super-resolution" image was acquired on sol 698 using the Pancam 482 nm filter. (B) Skull Valley penecontemporaneous cracks (*small arrows*) cut across lamination, some oblique to bedding. Laminae are deflected upward along crack margins. Termination of prominent crack in center of rock at discrete bedding plane and truncation of upward-deflected laminae along discrete bedding planes in center and upper parts of rock (*large arrows* with "T"). This super-resolution Pancam image was acquired on sol 713 using the 482 nm filter. *After Grotzinger et al., 2006, their Fig. 5., reproduced with permission from the Geological Society of America.*

might overlap that preserved at Erebus crater (Edgar et al., 2012). Opportunity reached Victoria crater on sol 952 (September 2006; Table 10.1). The ~750-m crater has a serrated rim with sharp, steep promontories separated by more rounded, less steep alcoves (Squyres et al., 2009). By traversing along the northern rim of the crater, Opportunity imaged exposed sedimentary and impact structures on the promontory faces. Entering one of the shallower alcoves (Duck Bay), Opportunity was able to traverse and analyze exposed outcrop inside the rim of the crater. Within the crater rim ~10 m of vertical section of the Burns formation is exposed, and an analytical traverse similar to that done in Endurance crater was completed.

Rocks exposed at the rim and along the interior section of Victoria crater are essentially the same Burns formation sulfate sandstones as seen all along the traverse from Eagle to Victoria (Squyres et al., 2009; Edgar et al., 2012). As Opportunity approached Victoria, the size of spherules in outcrop was observed to decrease (Calvin et al., 2008). The variation in spherule size may document regional or stratigraphic variation in the intensity of groundwater processes after sand deposition (Calvin et al., 2008; Squyres et al., 2009). On entering the "smooth" annulus of material surrounding Victoria, larger spherules were observed in the soils, likely left behind by the weathering of ejecta from Victoria crater. Spherules of similar size (~6 mm) were observed in a rock at the rim (Cercedilla), likely ejected from deep within the crater. The chemical composition of RAT-ground Cercedilla (Table 10.3) is similar to rock compositions from the deepest part of Endurance crater and distinct from others analyzed in Victoria, suggesting that the unit may be correlative with the lowermost unit examined at Endurance, and further implying correlation of the diagenetic processes experienced by rocks separated by a lateral distance of over 7 km (from Eagle and Endurance to Victoria).

Decameter-scale cross stratification is displayed prominently on the steep promontory outcrop faces along the rim of Victoria crater. Bedding that is concordant with m-scale sets of trough cross-beds reflects eolian dune migration (Squyres et al., 2009; Hayes et al., 2011; Edgar et al., 2012). Evidence for diagenesis includes crosscutting fracture fills and preferentially eroded bedding, and weathering patterns that suggest differential cementation.

Within the interior of the Victoria Crater rim at Duck Bay, rock units with true thicknesses of 0.7−2 m and totaling at least 3 m were analyzed with the IDD instruments and imaged with the Pancam (see Edgar et al., 2012). These rocks exhibit similarities in chemistry (APXS), fine-scale textures (MI), and color (Pancam spectra) to the sequence of bedrock units analyzed in Endurance crater. In both craters, the rocks exhibit downslope (and down-section) decreases in S, Fe, and Mg, with corresponding Al and Si enrichment, and both also show a sharp discontinuity in Cl concentrations. Despite these similarities, differences in the detailed compositions make it unlikely that the sequences in rocks of the two craters are stratigraphically correlative. More likely, the similarities reflect recurring depositional or diagenetic processes, or more recent near-surface alteration associated with present-day surface exposure (e.g., Amundson et al., 2008) and possibly related to the impact process that could mobilize fluids to alter the near-surface. Regardless, exploration of Burns formation rocks at Victoria crater showed that the depositional and diagenetic processes documented at Eagle, Endurance, and Erebus craters likely occurred widely across the Meridiani Planum region (Squyres et al., 2009).

10.6.4 SANTA MARIA CRATER

During its 3-year traverse to Endeavour crater, Opportunity visited another significant, fresh exposure of the Burns formation at Santa Maria crater (Fig. 10.1, Table 10.1). This 100-m-diameter crater is ~11 to 17 m deep and exposes a lower stratigraphic level of the Burns formation than previously explored (Watters et al., 2011). The crater was investigated entirely by traversing the rim, imaging the crater interior and surroundings, and examining ejecta materials at and beyond its rim. Significant IDD campaigns were conducted at rock targets Ruiz Garcia and Luis DeTorres. The latter was analyzed with the APXS following a RAT grind and shown to have a composition very similar to average Burns formation rocks (Table 10.3).

Edgar et al. (2014) conducted a detailed study of sedimentary textures of ejecta blocks from Santa Maria and reported evidence for two modes of deposition for the sulfate-rich sediments. First, as seen elsewhere in Burns formation rocks, they reported sandstone textures with abundant wind-ripple laminations inferred to be consistent with eolian deposition processes followed by diagenesis. Second, they reported massive, fine-grained, and nodular textures, with sulfate sand-grain sizes smaller than could be resolved with the MI (i.e., $< 100 \, \mu m$ or several MI pixels). These rocks were interpreted to be the first mudstones observed in the Burns formation. Edgar et al. inferred formation of these sediments in a transient evaporitic lake as part of a playa lacustrine setting.

10.7 GENERAL OBSERVATIONS FROM OPPORTUNITY'S TRAVERSE FROM EAGLE CRATER TO SANTA MARIA CRATER

10.7.1 BURNS FORMATION ROCKS

Sedimentary features of the Burns formation reflect the presence of water, and in complementary fashion Burns formation mineralogy also records rock-water interactions on ancient Mars. Taken together, Mössbauer, APXS, mini-TES, and Pancam data indicate that the outcrop rocks throughout the Burns formation contain the following main components: silicate minerals, sulfate salts, amorphous silica, and oxidized-iron-bearing phases, especially hematite. On a S- and Cl-free basis, the chemical composition of the sediments is similar to that of an olivine basalt. Mössbauer spectra indicate the presence of pyroxene, and correlated variations between Si, Al, and alkalis suggest the presence of feldspar as relics of basaltic mineralogy (Clark et al., 2005). Olivine, however, although present in basaltic sands at the surface of Meridiani Planum, is not detected in the Burns formation rocks and thus is absent from the basaltic, siliciclastic component. Moreover, high concentrations of S and Cl require that a significant amount of major cations including Mg, Fe, and Ca be bound into sulfates and chlorides (Clark et al., 2005; Fig. 10.4). Combined with mini-TES data (Christensen et al., 2004b) cation abundances measured by APXS require that Mg-sulfates must be the dominant sulfate component, with subordinate Ca-sulfate and jarosite (Klingelhöfer et al., 2004). Hematite, also detected by the Mössbauer Spectrometer, occurs in two forms, hematite within the concretions and fine-grained hematite distributed within the outcrop. Mössbauer spectra also provide evidence of an alteration phase or combination of phases contributing to a spectral signature attributed to octahedrally coordinated Fe^{3+} (Fe3D3—Tables 10.2 and 10.3), which may represent a poorly crystalline Fe-oxide, Fe-oxyhydroxide, or Fe-sulfate (Klingelhöfer et al., 2004; Morris and Klingelhöfer, 2008). The identification of jarosite indicates that the fluid from which the jarosite precipitated was acidic (pH $\sim 2-4$; Burns and Fisher, 1990; Hurowitz et al., 2010); however, it remains unclear whether all of the minerals present formed in such a low-pH environment. If the oxidation of Fe was important in generating the low pH (Tosca et al., 2008b; Sefton-Nash and Catling, 2008; Hurowitz et al., 2010), then it is likely that the concretions formed in low pH as well.

The mineralogy and chemical compositions of altered materials detected by the MER analysis of Burns formation rocks suggests initial alteration in an acid-sulfate environment. Ming et al. (2008) discussed three possible chemical pathways for such alteration: (1) oxidative aqueous alteration of sulfide-bearing ultramafic igneous rocks; (2) sulfuric acid alteration of basaltic materials; and (3) acid-fog alteration of basalt or materials, for example, sediments, derived from basalt. Under such acidic conditions, the weathering cycle would be dominated by sulfates, not carbonates. Moreover, neutral or alkaline waters would be expected to form phyllosilicates, e.g., smectites. As indicated above, the S- and Cl-free composition of Burns formation sandstones is essentially that of an olivine basalt, indicating that the initial alteration of basalt was largely isochemical, with mobility mainly of Fe and Mg, and involved a low water-to-rock ratio (Hurowitz et al., 2006; Ming et al., 2008).

As an alternative to acid-sulfate alteration as described above, the acidity may have come from Fe oxidation related to reducing groundwaters that came into contact with the near surface environment (Hurowitz et al., 2010). An initial circumneutral weathering environment is also consistent with the best estimates of the chemical-index-of-alteration relationships of the siliciclastic materials (Cino et al., 2017).

Tosca et al. (2005) and Hurowitz et al. (2006) reported the results of experimental acid-sulfate weathering of Martian-like basalt and showed that the derived fluids are rich in Mg, Fe, Ca, $SiO_2(aq)$, and SO_4. Alumina is less mobile, thus an altered basaltic residue has a high ratio of Al_2O_3 to the sum of $MgO + FeO + CaO$ such that Burns formation rocks that have different proportions of altered siliciclastic residue and Mg-, Ca-, and Fe-sulfate components lie along a mixing line (Fig. 10.12). As a result, evaporite mineralogy formed from alteration solutions of such weathering at the Martian surface would be dominated by these components. The oxidation state of Fe is important and acid-sulfate weathering can produce both Fe^{2+} and Fe^{3+}. Thus, evaporites could include variably hydrated Mg sulfates (kieserite to epsomite), Ca sulfates (gypsum or anhydrite), and an Fe^{2+} sulfate such as melanterite or an Fe^{3+} sulfate such as jarosite. Among the major basaltic minerals, olivine is especially susceptible to alteration/dissolution (Tosca et al., 2005; McLennan and Grotzinger, 2008). Following the formation of evaporites, diagenetic transformation of jarosite to goethite is favored. Initially, melanterite is favored by slow oxidation rates, but melanterite is not indicated by Mössbauer spectra. Poorly crystalline goethite may have reacted to form hematite concretions, and moldic porosity may have formed by oxidation and destabilization of melanterite (e.g., McLennan et al., 2005; Sefton-Nash and Catling, 2008), with oxidation of $Fe^{2+}(aq)$ by reaction with atmospheric O_2 and/or photo-oxidized by UV radiation at the Martian surface (Hurowitz et al., 2010).

Burns formation evaporites required water for their formation, but these minerals also tell us something about the chemical activity of water as Meridiani brines matured. Making conservative assumptions about the composition of originally relatively dilute solutions, observed and inferred evaporates provide a quantitative estimate of water activity, basically the percentage of water molecules transiently bound to ions and, therefore, the percentage of water molecules available to do the chemical work of water (Tosca et al., 2008a). This matters to astrobiologists because water activity strictly limits the distribution of organisms on Earth. Mg-sulfates in the Burns formation imply a water activity tolerated by only a few organisms on Earth, while chloride minerals suggest water activities lower than any known to sustain any life that we know of (Tosca et al., 2008a). More generally, experimental results indicate microbial activity is inhibited by solutions of high ionic strength (Fox-Powell et al., 2016).

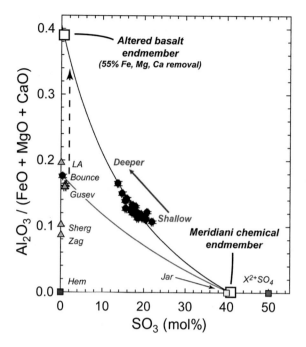

FIGURE 10.12

Molar $Al_2O_3/(FeO_T + MgO + CaO)$ vs SO_3 for Burns formation rocks (*black circles*). Also shown are the compositions of several Martian basalts, Zagami (Zag), Shergotty (Sherg), Los Angeles (LA) (basaltic shergottite meteorites); Bounce Rock (Bounce), a basaltic erratic—likely impact ejecta, several Gusev crater basalts (Gusev), the compositions of minerals observed within the Burns formation, hematite (Hem) and jarosite (Jar), and a composition derived by removing 55% of the divalent cations (Ca, Mg, and Fe) from typical Martian basalt to illustrate the effects of chemical weathering. *From Squyres et al., 2006b, their Fig. 3, reproduced with permission from the American Association for the Advancement of Science.*

Minerals such as jarosite and (across the planet, at Gusev crater) amorphous silica also inform us about the persistence of waters since the time of Meridiani deposition. On Earth, jarosite and amorphous silica seldom persist in sediments older than about a million years; for example, jarosite is unlikely to persist under aqueous condition on Mars for more than 1 Myr and likely less than 100,000 yr (Elwood Madden et al., 2009). Therefore, their presence in Late Noachian to Hesperian rocks suggests that Mars surface rocks have had limited contact with water since they formed (Tosca and Knoll, 2009). Complementing inferences of low rates of surface erosion (Golombek et al., 2006) and slow chemical alteration of exposed surfaces (Knoll et al., 2008), it appears that, on average, the Martian surface has been extremely dry for a long time.

The body of evidence from all of Opportunity's instruments indicates aqueous alteration of an olivine-basalt precursor, likely by acid-sulfate weathering involving dissolution of olivine, Fe-Ti oxides, phosphates, and to some extent, pyroxene. This alteration was followed by evaporation of ion-rich waters under acidic (sulfate-rich) and oxidizing conditions to produce the Burns formation sediments (Tosca et al., 2005; Ming et al., 2008; McLennan and Grotzinger, 2008). In short, Burns sedimentary rocks document a Martian surface that was arid, oxidizing and, at least periodically, acidic.

10.7.2 FORMATION OF CONCRETIONS

The hematitic concretions reflect formation in a complex and reactive Fe- and Mg-rich, acidic evaporite system. The resulting evaporitic mineralogy is highly variable as the stabilities of many of the minerals are sensitive to pH, oxidation state, ionic strength, and composition of groundwater.

In their modeling work, Tosca et al. (2005) considered the effects of fresh-water recharge on evaporites precipitated from fluids derived by interaction with olivine-bearing basalts and found that goethite or hematite could result from the breakdown of jarosite, for example,

$$(H_3O)Fe_3(SO_4)_2(OH)_6 \Rightarrow 3FeOOH + 2SO_4^{2-} + H_2O + 4H^+$$

or by oxidation of melanterite, which is a candidate highly soluble evaporite mineral, for example,

$$FeSO_4 \cdot 7H_2O + 0.25O_2 \Rightarrow FeOOH + SO_4^{2-} + 5.5H_2O + 2H^+$$

The likelihood of such processes has been largely confirmed by subsequent modeling (Sefton-Nash and Catling, 2008) and experimental (Tosca et al., 2008b; Zhao and McLennan, 2013) results.

Hematitic concretions in terrestrial sandstones attest to the physical aspects of such a model (Chan et al., 2004, 2005, 2012), and observed paragenesis in Pleistocene to modern sediments of the Rio Tinto, Spain, show its chemical plausibility; at Rio Tinto, modern sediments are rich in jarosite, schwertmannite, and (locally) melanterite, but terraces hundreds of years old contain only goethite, and those older than a million years contain hematite (Fernandez-Remolar et al., 2005).

In summary, diagenetic hematitic concretions embedded within the outcrops appear to have formed relatively rapidly under stagnant to very slow fluid migration. Formation of hematite by breakdown of jarosite during basin recharge by relatively dilute groundwater (Tosca et al., 2005) or oxidation of ferrous sulfates, such as melanterite, provide plausible mechanisms for their formation. Pathways involving schwertmannite ($Fe_8O_8(OH)_6(SO_4)$) and goethite (α-FeOOH) as intermediate products in the generation of hematite are also possible and are supported by Ni and Zn partitioning and abundances in Meridiani sediments (Zhao and McLennan, 2013).

10.7.3 SCALE OF THE BURNS FORMATION

The aqueous history recorded in rocks of the Burns formation appears to be characteristic of a large geographic region. Arvidson et al. (2011) summarized the Burns formation rocks examined by Opportunity as part of a regional-scale sedimentary deposit that is discontinuously present across most of Meridiani Planum, covering several hundred thousand square km and draped unconformably on dissected cratered terrain (the highly eroded crater unit of Hynek and Achille, 2017). The impact-crater size frequency distribution indicates a Late Noachian or Early Hesperian age (Arvidson et al., 2006; Hynek and Achille, 2017). Preexisting craters are evident and show partial burial by the sedimentary deposits, including the ~ 20-km-diameter Endeavour crater, discussed in the next section. Another large, nearby crater, Bopolu (~ 19 km diameter), located southwest of Endeavour crater, postdates the deposition of the sulfate rich sedimentary deposits (Hynek and Achille, 2017). Its ejecta exhibit a basaltic signature in remote sensing data and lack evidence of hematite (Christensen et al., 2001). Bopolu and other rayed craters that lack the hematite signature must have formed after the hematite spherules of the Burns formation became concentrated on the surface of Meridiani Planum (Golombek et al., 2010). The Burns formation rocks thus appear to have formed at a time when an early, wet Martian surface in this region was in the process of drying up (e.g., Hurowitz et al., 2010). One hypothesis posits that groundwater was sourced from surrounding highlands, perhaps the Tharsis region to the west, and that the Burns formation sulfate sandstones accumulated during episodes of rising groundwater (Andrews-Hanna et al., 2007, 2010; Arvidson et al., 2011). A lack of evidence for channel development or other large-scale water erosion after their formation suggests that these deposits formed at a time of transition in the history of water on Mars to the dry, arid conditions that have characterized Mars for much of its history since the Noachian.

10.7.4 EVIDENCE OF LATER WATER ACTIVITY ON MERIDIANI PLANUM: VENEERS, RINDS, AND FRACTURE FILLS

Veneers and thicker rinds that coat outcrop surfaces and partially cemented fracture fills or "fins" perpendicular to bedding document relatively late-stage alteration of ancient sedimentary rocks at Meridiani Planum. It is clear that the original sediments represented by Burns formation sulfate sandstones were relatively water rich. Perhaps shortly after their formation, these sediments would have been prone to compaction and, likely, loss of some of their less-tightly bound water through fractures and remobilization of salts would be expected. Much later processes, such as impact cratering, would also tend to expose rocks, form fractures in target bedrock, and remobilize groundwater or bound water. Exposed and near-surface rock would undergo desiccation as the Martian atmosphere thinned and the surface dried out. In more recent geologic times, variations in obliquity could also occasionally deposit frost and transient water vapor that would react with exposed surfaces (Knoll et al., 2008). The following paragraphs summarize rover observations relating to some of these kinds of effects.

10.7.4.1 Veneers

The chemistry of the buff-colored (in false-color rendition) veneers reflects multiple processes at work since establishment of the current Meridiani Planum surface. Burns formation rocks exposed at the plains' surface exhibit a systematic range of color differences when viewed in calibrated visible and near-infrared wavelength Pancam images. Most of the outcrop appears tinted either pale buff yellow or purple (Farrand et al., 2007; Knoll et al., 2008). Buff-colored surfaces ("HFS" spectral class of Farrand et al., 2007) tend to be flat or of low profile whereas purplish surfaces ("LFS" spectral class of Farrand et al., 2007) tend to protrude relative to local surface topography, and these differently colored surfaces have distinct multispectral signatures that are discernable in Pancam 11-band spectra (Fig. 10.13). Abrasion by the RAT shows that the buff-colored surfaces are thin veneers, less than a mm thick, and that beneath the veneers, the rock color is purplish to reddish. Most likely, the buff-colored veneer indicates some combination of dust coating and surface alteration (Farrand et al., 2007).

Dust on rock surfaces and the penetration depths of radiation for APXS ($\leq 10-50\,\mu m$) and Mössbauer Spectrometer (to several mm depth) measurements complicate direct comparisons to Pancam data for surface veneers. However, gentle brushing with the RAT to whisk away dust and reveal rock surfaces, followed by chemical analysis with the APXS, does reveal systematic chemical differences from deeper rock interiors accessed by grinding into rocks several mm with the RAT. Chemical data from the APXS indicate that brushed rock surfaces have higher Si and Al, but lower S than ground interiors of the same rock targets, and higher concentrations of Ti, Na, and Cl (Table 10.4; Knoll et al., 2008). These differences are consistent with exterior weathering that preferentially removed the softer and more chemically reactive sulfate minerals, leaving the exposed surface slightly enriched in silicate material. Increased concentrations of Cl, on the other hand, indicate

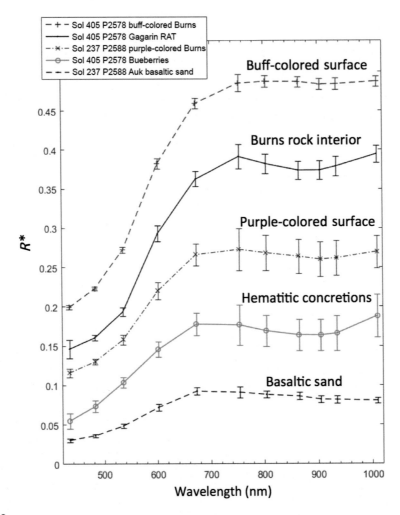

FIGURE 10.13

Pancam spectra of Burns formation rock surfaces (buff and purple, as discussed in the text) compared to rock interior (Gagarin RAT; see also Table 10.4), concretions (Sol 405 Blueberries), and basaltic sand (Sol 237Auk). R^* is the relative reflectance (defined as I/F divided by the cosine of the incidence angle, e.g., Bell et al., 2006; Farrand et al., 2007).

the possibility of mineral precipitation, perhaps mediated by thin, transient films of water. Precipitation might include NaCl, KCl, and some form of phosphate or P adsorbed onto ferric oxides. Chlorides might also be leached from soil or dust and then precipitated on the outcrop surface from a saturated film of water (Knoll et al., 2008). Competing processes of chemical alteration, perhaps at times affected by water beneath blanketing soils, and sandblasting of exposed outcrop surfaces by wind-driven eolian abrasion, determine the occurrence and distribution of veneers.

Table 10.4 Compositions (APXS) of Veneers, Rinds, and Fracture Fills

Sol	401	403	556	560	558	175
Feature	Gagarin	Gagarin	Fruitbasket	Fruitbasket	Fruitbasket	Hoghead Arnold Ziffel
Target Type	Rock Surface	Rock Interior	LemonRind Rind	LemonRind Rind	Strawberry No rind	Fracture fill
Preparation	Brushed	Ground 6 mm	Brushed	Ground 1.5 mm	Ground 3 mm	Flake, as is
APXS (wt%)						
SiO$_2$	38.3	32.6	40.2	35.1	32.8	41.5
TiO$_2$	0.78	0.68	0.89	0.75	0.72	0.83
Al$_2$O$_3$	6.71	4.90	7.49	6.17	5.72	7.89
Cr$_2$O$_3$	0.17	0.17	0.24	0.19	0.19	0.22
FeO(T)	15.2	15.9	17.0	16.0	15.8	15.7
MnO	0.31	0.35	0.37	0.38	0.39	0.25
MgO	7.25	7.33	7.07	7.83	8.09	8.29
CaO	5.36	5.78	6.07	5.20	5.13	5.45
Na$_2$O	1.79	1.35	2.22	2.02	1.57	2.11
K$_2$O	0.57	0.51	0.55	0.54	0.50	0.56
P$_2$O$_5$	1.03	1.07	1.07	1.05	0.99	0.99
SO$_3$	21.5	28.6	15.3	23.1	27.4	14.8
Cl	0.92	0.61	1.49	1.54	0.57	1.26
APXS (ppm)						
Ni	574	585	525	508	504	465
Zn	405	436	474	457	563	388
Br	67	54	67	67	84	108

"Preparation" refers to treatment of the surface with the RAT prior to APXS analysis.
APXS data from NASA PDS.

10.7.4.2 Rinds

Darker and thicker rinds, up to 8 mm thick, form conspicuous coatings that are distinct from veneers. In some cases, sedimentary textures are observed to be continuous between rinds and underlying rock; thus, the rinds represent an alteration surface. Pancam spectra and APXS chemistry, however, show that the rinds are distinct from the bedrock on which they occur. Rinds are observed in numerous locations but were studied in detail at an outcrop exposure near Erebus crater, where a thick rind target was analyzed relative to the interior of the host rock. Chemical trends were found to be similar to those found for veneers, and the processes of concentration were likewise interpreted to be similar to those proposed for veneers, namely concentration of siliciclastic residue, limited removal of sulfate, and precipitation of NaCl (Knoll et al., 2008). Rinds are observed to occur on blocks disturbed by impact craters with orientations showing formation both before and after crater formation, and they are older than polygonal cracks. Such relationships indicate formation well after the diagenesis that affected all Burns formation rocks, but before or contemporaneously with the polygonal fracture systems common at the present-day surface of Meridiani.

Formation of rinds, like veneers, may have been favored by partial soil cover, with formation at the rock—soil interface where limited fluids were afforded at least some protection from the dry and cold atmosphere. These rinds may represent one or more persistent or especially aggressive chemical weathering episodes (Knoll et al., 2008).

10.7.4.3 Fracture fills

Thin, linear, erosionally resistant, fin-like features occur commonly in Burns formation outcrop expanses. Like rinds and veneers, they are spectrally distinct from adjacent bedrock. They tend to be developed at the edges of fractures but do not completely fill the present-day fractures where they occur. Notably, they are not spatially related to the common polygonal fractures in Burns formation outcrop. On the basis of limited APXS analyses, they appear to share chemical characteristics, when compared to adjacent rocks, similar to rinds and veneers, with enriched Si and Al, but depleted S. They are highly variable in Cl and Br, which can be either enriched or depleted compared to adjacent rocks. These characteristics suggest formation closer to when fractures first formed by the action of brines acting to cement intraclastic material derived from adjacent rocks, and that the fractures later widened and became filled with basaltic soil and dust such as occur on the present-day surface. Similar features are observed at White Sands National Monument (Chavdarian and Sumner, 2006) where they are thought to have formed by sulfate dehydration followed by cement precipitation from brines that migrated through the cracks, and that were subsequently exposed by erosion. However, the lack of an association with polygonal fractures at Meridiani suggests that sulfate dehydration alone is not the source of the brines that formed the fracture fillings. That they are seen in outcrop pavement associated with impact craters may imply a relationship to sulfate dehydration and mobilization of fluids following impact crater formation.

All indications seem to be that erosion and chemical alteration rates have been very slow at Meridiani Planum since the current surface rocks were exposed and have perhaps been so for the past several billion years (Golombek et al., 2006). The occurrence of veneers, rinds, and fracture filling features reflect water-related activity in the past; however, this activity was rare or operated at exceedingly slow rates and perhaps was confined to transient events since the Meridiani rocks were exposed on the plains (Knoll et al., 2008).

10.8 SUMMARY OF EVIDENCE SUPPORTING SEDIMENTARY/AQUEOUS PROCESSES IN ROCKS OF MERIDIANI PLANUM

- The Opportunity landing site and surrounding Meridiani Plains occur on flat-lying sedimentary rocks that were produced by transport and deposition of sulfate-rich sand.
- Chemical aqueous alteration of basalts, followed by evaporation and oxidation, produced sulfate salts that fed the Meridiani sulfate-rich sand. The Meridiani bedrocks are a mixture of chemical and siliciclastic sediments with a complex diagenetic history, at least periodically at low pH,

including groundwater-influenced cementation that produced the Burns formation sulfate sandstones.
- Festoon crossbedding provides strong evidence of shallow-water subaqueous reworking of sediments.
- Occurrence of mudstones at Santa Maria crater suggests a playa lacustrine depositional setting for parts of the Burns formation.
- Cements, vugs, and crystal molds provide evidence of dissolution, precipitation, and recrystallization in a briny groundwater solution.
- Spherical hematite-rich concretions provide further evidence of diagenetic reaction in a stagnant or very low-flow groundwater environment.
- The environmental conditions that these sediments record are episodic groundwater recharge and inundation by shallow surface water, followed by evaporation, exposure, and desiccation.
- Later in Martian history, long after diagenesis, exposure of Meridiani rocks at the surface produced surface alteration that records occasional and perhaps transient interactions with water or water vapor.

10.9 ENDEAVOUR CRATER

As indicated above, Endeavour crater predates the Burns formation rocks and has been interpreted as Late Noachian in age (Squyres et al., 2012; Arvidson et al., 2014; Crumpler et al., 2015; Grant et al., 2016). As such, rocks exposed in the rim of Endeavour crater might be expected to contain evidence of more ancient water action on Mars than the rocks of Meridiani Planum. Following the exploration of Victoria crater, and owing to the increasingly apparent longevity and mechanical health of Opportunity (and the occasional dust-cleaning event to boost solar array efficiency), Endeavour crater was selected as the next long-range target for exploration (Arvidson et al., 2011). The rover emerged from Victoria crater on sol 1634 (August 2008; Table 10.1) and began its long traverse toward the rim of Endeavour crater, with one additional crater campaign to study the Burns formation at Santa Maria crater. Nearly 3 years later in August 2011, Opportunity reached the rim of Endeavour at a segment of the crater rim named Cape York for an initial investigation of the rim deposits.

Rocks of the Burns formation (rock target Gibraltar, sol 2669) were observed almost all the way to the contact of the Meridiani plains, thinning to feather edge along a morphologically prominent annulus surrounding Cape York. The annulus, composed of different material than the Burns formation, was christened the Grasberg formation (Squyres et al., 2012; Crumpler et al., 2015). The bedrock that cores the resistant materials of Cape York proved to be impact-melt breccia produced by Endeavour crater and was named the Shoemaker formation. Materials of both the Grasberg and the Shoemaker formations differ significantly from rocks of the Burns formation, but are still relatively rich in S, averaging about 6 wt% SO_3, and in Cl, averaging just under 1 wt% Cl (Table 10.5). As Opportunity explored Cape York, two features in particular became evident relating to water action. First is the common occurrence of Ca-sulfate veins in outcrop pavement of the Grasberg and Shoemaker formations, and second is the occurrence of the Matijevic formation rocks, which were signaled by orbital remote sensing, keying on the spectral signature of smectites (Arvidson et al., 2014). At the date of this writing, the Matijevic formation has only been observed at Cape York.

Table 10.5 Compositions of Representative Shoemaker and Grasberg Formation Rocks

Formation	Shoemaker					Grasberg		
Sol	2713	2722	2801	2920	3463	2990	3006	3022
Feature	Salisbury 1	Salisbury 1	Boesmans-kop	Amboy 4	Spinifex	Grasberg 1	Grasberg 3	Mons-Cupri
Preparation	As-is	Ground	Brushed	As-is	Brushed	As-is	Brushed	As-is
APXS (wt%)								
SiO_2	44.5	45.5	45.6	46.0	45.7	44.4	44.3	45.3
TiO_2	1.00	1.09	1.03	1.07	1.04	0.97	0.92	0.97
Al_2O_3	9.00	8.82	9.52	9.97	8.76	8.21	7.36	8.40
Cr_2O_3	0.24	0.25	0.25	0.19	0.22	0.23	0.23	0.25
$FeO(T)$	19.2	20.1	17.6	17.1	17.6	19.6	20.3	19.2
MnO	0.45	0.48	0.41	0.48	0.78	0.22	0.19	0.18
MgO	7.37	8.81	8.95	7.53	8.85	6.03	3.63	5.73
CaO	6.76	6.77	5.75	7.48	6.18	5.71	6.02	5.91
Na_2O	2.43	2.74	2.39	2.33	2.32	2.28	2.49	2.22
K_2O	0.49	0.41	0.51	0.50	0.70	0.63	0.71	0.71
P_2O_5	1.02	1.00	1.21	1.11	1.18	1.09	1.23	1.20
SO_3	6.16	3.09	5.58	5.41	5.52	8.49	9.75	8.12
Cl	1.31	0.87	0.99	0.74	0.95	1.88	2.60	1.71
APXS (ppm)								
Ni	474	482	615	306	537	425	444	470
Zn	283	246	350	194	460	863	935	558
Br	78	124	153	103	706	448	540	286

"Preparation" refers to treatment of the surface with the RAT prior to APXS analysis.
APXS data from NASA PDS.

10.9.1 GYPSUM VEINS

Cape York is surrounded by a low, broad topographic bench that grades from the interior of Cape York to the surrounding plains and the Burns formation (Crumpler et al., 2015). The bench materials overlie the Noachian Endeavour impact breccias that form the lower slopes and ridge of Cape York. A prominent feature of the bench is the common occurrence of light-toned veins several cm in width and of variable length that cut across the bench sediments (Fig. 10.14). Such veins are abundant in the inner parts of the bench, but also occur within outcrops of the onlapping Burns formation sandstone. APXS measurements at one location show that the veins are rich in Ca-sulfate, and Pancam spectra provide a best match to gypsum ($CaSO_4 \cdot 2H_2O$) on the basis of a drop in reflectance from 934 to 1009 nm, which corresponds to a water overtone feature in gypsum (Farrand et al., 2013, 2016). Anhydrite ($CaSO_4$) lacks a hydration band and bassanite ($CaSO_4 \cdot 0.5H_2O$) has a different and weaker hydration band (Rice et al., 2010).

The gypsum veins are also significant because their occurrence constrains conditions of formation to precipitation from relatively low-temperature and high-water-activity H_2O in fractures

FIGURE 10.14

Homestake gypsum vein in the Grasberg formation. Sol 2769 Pancam image sequence ID P2574, false color, filters: L257. *Image credits: D. Savransky and J. Bell/JPL/NASA/ Cornell/ASU (Bell et al., 2003, 2006).*

possibly related to extension associated with sediment compaction and dewatering (Squyres et al., 2012). These veins and their host rocks formed at a time of transition between an earlier style surficial geology characterized by phyllosilicate alteration of basaltic crust to the later evaporation of salt-rich groundwater that formed the sulfate-rich sandstones of the Burns formation. Evidence for the phyllosilicate alteration is described next.

10.9.2 CAPE YORK AND THE MATIJEVIC HILL DEPOSITS

Hyperspectral data from the Compact Reconnaissance Imaging Spectrometer for Mars (CRISM) on MRO indicated the presence of Fe^{3+}-rich smectites in a small area on the inboard (east) side of Cape York. This area, named Matijevic Hill, was investigated in detail by Opportunity during sols 3063–3307 (Table 10.1). The rocks exposed where CRISM detected the smectite signature differ significantly from the Shoemaker breccia that composes the bulk of Cape York (Table 10.6). The rocks include soft, fine-grained sandstones that are distinct from those of the Burns formation, for example containing only 2–5 wt% SO_3. The specific energy of RAT grinds into this sandstone is very low, indicative of rock as soft as the Burns sulfate-rich sandstones (Arvidson et al., 2014). These rocks lie stratigraphically below the Shoemaker impact breccia and thus appear to represent a preimpact target rock, which is thus likely of Noachian age.

The chemical compositions of these rocks are also generally basaltic, meaning that the alteration that produced them did not fractionate the compositions significantly from the precursor basalts. These rocks contain spherules, but these spherules are distinct from those of the Burns sandstones; they are not hematite rich (Arvidson et al., 2014) and in fact have compositions similar to encompassing sandstones (Table 10.6). Some are distinctly zoned, and their distribution in the sandstones is overdispersed, supporting a concretionary origin. Along one stratigraphic horizon, however, they are highly concentrated, and in this location (e.g., Kirkwood and Sturgeon River) their origin or mode of concentration or accumulation continues to be debated. The Matijevic sandstones also

Table 10.6 Chemical Compositions of Selected Matijevic Formation Rocks

Sol	3078	3087	3094	3150	3067	3209	3239	3305
Feature	Whitewat.-Lake, Rock	Whitewat.-Lake, Rock	Whitewat.-Lake, Rind	Sandcherry Rind	Kirkwood Spherule-Rich Rock	Fullerton3 Spherule-Poor Rock	Lihir Boxwork Alteration	Espérance Boxwork Alteration
Target	Azilda2	Azilda2	Chelmsf. 2	Sandcherry				
Type								
Preparation	Brushed	Ground	Brushed	Ground	Brushed	Brushed	As-is	Ground
APXS (wt%)								
SiO_2	50.3	51.3	46.3	44.9	49.2	50.1	58.4	62.5
TiO_2	0.90	0.88	0.87	0.91	0.79	0.96	1.16	0.93
Al_2O_3	10.5	10.6	9.46	9.36	9.88	10.5	12.9	15.4
Cr_2O_3	0.24	0.23	0.26	0.26	0.31	0.29	0.32	0.34
FeO(T)	15.4	15.4	15.8	16.6	16.8	14.8	5.80	4.43
MnO	0.35	0.37	0.61	0.43	0.23	0.28	0.16	0.19
MgO	7.40	7.97	7.70	8.70	8.45	8.22	5.89	4.73
CaO	6.07	5.90	6.78	6.85	5.04	5.81	4.03	2.14
Na_2O	2.48	2.46	2.49	3.00	2.44	2.25	1.66	2.25
K_2O	0.30	0.28	0.36	0.32	0.50	0.33	0.37	0.24
P_2O_5	1.39	1.49	1.22	1.26	0.73	0.89	1.19	1.14
SO_3	3.78	2.45	6.62	5.69	4.51	4.64	6.25	3.28
Cl	0.73	0.53	1.36	1.47	1.08	0.85	1.58	2.32
APXS (ppm)								
Ni	898	931	815	840	873	738	644	622
Zn	153	141	248	361	134	176	304	238
Br	128	50	176	302	114	159	114	35

Notes: Whitewat., Whitewater; Chelmsf. 2, Chelmsford2; Esperance, Esperance6 (see Clark et al., 2016). "Preparation" refers to treatment of the surface with the RAT prior to APXS analysis.
APXS data from NASA PDS.

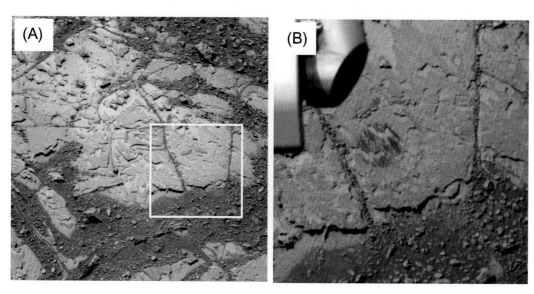

FIGURE 10.15

Rock Sandcherry—exposed on Matijevic Hill, exhibiting eroded remnants of a veneer or rind. (A) Sol 3135, sequence ID P2538, L257; (B) Sandcherry rind brushed with the RAT, Sol 3144 sequence ID P2540, L257. *White box* in (A) shows area of (B). *Image credits: D. Savransky and J. Bell/JPL/NASA/ Cornell/ASU (Bell et al., 2003, 2006).*

contain abundant Ca-sulfate veinlets as well as prominent but discontinuous rinds (Fig. 10.15), reflecting postdeposition aqueous alteration processes, with rinds having formed prior to the Endeavour impact and Ca-sulfate veinlets having formed from fluid circulation either before or after the impact. The rinds are interpreted to be the host of Fe^{3+} smectites detected in CRISM data and formed by reaction of mildly acidic waters with the basaltic sandstones (Arvidson et al., 2014).

One location on Matijevic Hill was found that exhibits strong chemical fractionation and a clear association with phyllosilicate alteration. The feature, named "Espérance," occurs along a boxwork-like fracture system characterized by planar fins of bright material and vertical laminae parallel to approximately orthogonal joints (Fig. 10.16; Arvidson et al., 2014). Chemical analysis of this material, including RAT ground portions, revealed the lowest FeO(T) and CaO (4.4 and 2.1 wt%, respectively), and highest SiO_2 and Al_2O_3 (62.5 and 15.4 wt%, respectively) of any materials analyzed by Opportunity at Meridiani Planum (Arvidson et al., 2014; Clark et al., 2016). The chemical compositions are consistent with the presence of a hydrous phyllosilicate, and Pancam spectra suggest the presence of a hydrated silica phase. Compositional data indicate formation of Al-rich smectite by means of aqueous leaching along the boxwork fracture network. Owing to the field of regard of the APXS, the trend indicated in Fig. 10.16 is interpreted as a mixing line between bedrock (sedimentary, but of basaltic composition) and highly altered boxwork vein material rather than as a progressive alteration trend.

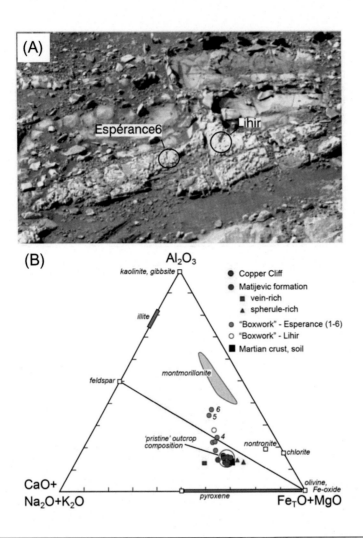

FIGURE 10.16

(A) Pancam false-color view, sol 3230, showing boxwork structure and areas examined in detail. *Circles* indicate the targets Espérance6 (which was abraded with the RAT) and Lihir. Approximate scale across the scene is 70 cm. (B) Ternary plot of mole fraction $Al_2O_3-(CaO + Na_2O + K_2O)-(FeO_T + MgO)$ (after Nesbitt and Wilson, 1992; Hurowitz et al., 2006; McLennan and Grotzinger, 2008) for selected rocks from Matijevic Hill and other materials. From Arvidson et al. (2014), their Fig. 11, reproduced with permission from the American Association for the Advancement of Science. Mineral compositions based on idealized stoichiometry, the field for montmorillonite based on structural formulae of 25 natural montmorillonites (Wolters et al., 2009), and average Martian crust and soils from Taylor and McLennan (2009).

10.9.3 COATED ROCKS (COOK HAVEN, MURRAY RIDGE)—OR AQUEOUS MOBILIZATION OF Mn AND Zn

Rocks exposed at several locations along the rim of Endeavour crater contain specific elemental enrichments that provide chemical evidence of significant aqueous alteration. One such rock, named Tisdale2, exposed by a small impact crater at Cape York, exhibits the highest concentrations of Zn measured by Opportunity. Another example includes several rocks found along Murray Ridge on Cape Tribulation that were dislodged by the rover and whose undersides exhibit coatings with strong enrichments of Mn.

10.9.3.1 Tisdale2 and Zn enrichment

Tisdale (Tisdale2) was among the first rocks to be analyzed at Cape York. The rock is a fresh ejecta block excavated from the $\sim 20\,m$ diameter Odyssey crater on the southern tip of Cape York (Squyres et al., 2012). Its Zn concentration ranges from ~ 700 to over 6000 ppm (0.78 wt% ZnO), averaging ~ 2500 ppm Zn over six analyses of unbrushed rock. These concentrations compare to ~ 300 ppm $\pm$ 40 (one standard deviation) for 21 other Shoemaker breccia analyses. The Zn enrichment in Tisdale2 could represent a deeper variant of the Shoemaker breccias because of its excavation by Odyssey crater, but its unique enrichment compared to all other Shoemaker breccias analyzed at Cape York suggests that it is a localized signature. Tisdale has a distinctive bright coating seen in Pancam spectra (Fig. 10.17), which indicates oxidation in the form of Fe^{3+} enrichment for the coating (Farrand et al., 2016); however, the highest Zn concentration is not on the top

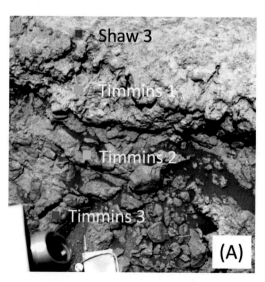
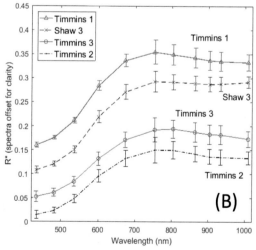

FIGURE 10.17

Rock Tisdale2—ejected from Odyssey crater. (A) Pancam image of rock Tisdale2 from sol 2697 sequence ID P2530 L256 false color. (B) Pancam 13-filter spectra of areas noted in (A) corresponding to APXS analyses in Table 10.7. Each spectrum is offset by 0.05 units for clarity.

surface, but is instead on the broken side of the rock below the surface. On Earth, enrichments of Zn are commonly associated with hydrothermal settings where Zn is deposited with S as a sulfide (e.g., Sverjensky, 1986), and in Tisdale2, the highest Zn content corresponds to the highest S content, but otherwise there is no correlation of S and Zn. Moreover, in this case the S is likely oxidized. The Zn concentrations correlate to some extent with P, and the highest Zn concentration corresponds to over 3% P_2O_5 (Table 10.7), which is one of the highest P concentrations in rocks analyzed by Opportunity. Squyres et al. (2012) concluded that this case of Zn enrichment represents a highly localized and heterogeneous secondary deposition of Zn, possibly associated with sulfate and/or phosphate, from an aqueous fluid flowing along a fracture and associated with hydrothermal alteration induced by the Endeavour impact.

10.9.3.2 Cook Haven rocks and Mn enrichment

While traversing Cape Tribulation along Murray Ridge, Opportunity approached a location dubbed Cook Haven. At this site, a serendipitous interaction between one of the rover wheels and a loose rock within a fracture zone caused several small rocks to flip, exposing their undersides, which attracted attention owing to their bright and unusual coloration. Several of these rocks, including Pinnacle Island and Stuart Island, and associated soil were investigated in detail and found to have high concentrations of Mn and S in coatings (Table 10.7). Pancam, MI, and APXS analyses of these coatings (Figs. 10.18, 10.19) were deconvolved to determine that the coatings include thin deposits of dark Mn oxides overlying sulfate minerals (Arvidson et al., 2016). The occurrence of these coated rocks in a fracture zone approximately radial to Endeavour crater suggests a relationship involving transportation of soluble cations and anions, including reduced, dissolved Mn, in fluids that migrated along fractures to zones where interaction with an oxidizing environment, perhaps near-surface in the form of O_2, led to precipitation of sulfates followed by one or more Mn oxide minerals. Location on the undersides of these rocks provided protection from physical erosion. Thin Ca-sulfate veins are also observed in the area, providing further evidence of fluid mobility and local mineralization.

10.9.4 EXPLORING MARATHON VALLEY AND LOOKING FORWARD

The use of CRISM data to guide Opportunity's search for evidence of phyllosilicate alteration in the Noachian materials of Endeavour crater was established and validated by the exploration of Matijevic Hill at Cape York. Exploration of the next rim segment to the south, Cape Tribulation, revealed additional exposures of Shoemaker formation impact breccias and other materials such as the Mn-rich coated rocks described above. The promise of further areas of phyllosilicate alteration lay to the south in discrete locations along the ridge of Cape Tribulation (Fox et al., 2014; Crumpler et al., 2015). Marathon Valley, so-named because it lies just beyond the distance of a marathon from Eagle crater, is one such site, where CRISM data signaled the presence of smectite alteration (Wray et al., 2009; Crumpler et al., 2016; Fox et al., 2016).

At the time of this writing (June 2017), Opportunity continues its exploration of the western rim of Endeavour crater, following various lines of evidence from orbital data, both from CRISM spectral data and from HiRISE terrain morphology, for the action of water associated with the Noachian crater rim deposits and radial fractures, and possibly evidence for ancient surface flow of water or brine. Thus far, evidence from the surface exploration by Opportunity along the Endeavour crater rim indicates that early Mars hosted different styles of active aqueous interaction and alteration of

Table 10.7 Chemical Compositions of Rocks Altered by Cation Mobilization in Aqueous Solutions

	Zn-rich				Mn-rich			Ca-rich
Sol	**2694**	**2695**	**2696**	**2702**	**3551**	**3564**	**3577**	**2764**
Feature	Tisdale2	Tisdale2	Tisdale2	Tisdale2	Pinnacle-Island3	Pinnacle-Island5	Stuart-Island4	Homestake gypsum vein
Target	Timmins1	Timmins2	Timmins3	Shaw3	Cook Haven	Cook Haven	Cook Haven	Cape York
Location	Cape York	Cape York	Cape York	Cape York	Unbrushed	Unbrushed	Unbrushed	Cape York
Preparation	Unbrushed	Unbrushed	Unbrushed	Unbrushed	Unbrushed	Unbrushed	Unbrushed	Unbrushed
APXS (wt%)								
SiO_2	42.6	46.0	45.4	45.6	18.1	20.1	25.6	25.4
TiO_2	0.99	1.05	1.05	1.01	0.44	0.58	0.65	0.29
Al_2O_3	8.86	9.97	10.10	8.16	3.48	3.68	4.87	4.78
Cr_2O_3	0.16	0.27	0.23	0.27	0.10	0.10	0.16	0.15
$FeO(T)$	17.6	18.0	18.8	21.4	15.1	15.5	16.8	6.05
MnO	0.38	0.38	0.23	0.46	**3.48**	**3.35**	**3.37**	0.17
MgO	6.20	6.04	6.19	5.90	13.0	11.5	11.7	4.77
CaO	7.13	6.78	5.88	4.83	7.66	8.26	4.85	**22.0**
Na_2O	1.84	2.16	2.54	2.09	0.86	0.84	0.86	1.63
K_2O	0.43	0.50	0.53	0.62	0.14	0.13	0.28	0.28
P_2O_5	3.14	1.22	1.20	2.01	2.37	2.44	1.40	0.71
SO_3	8.57	6.01	6.50	5.89	34.5	32.7	29.0	32.7
Cl	1.23	0.93	1.00	1.27	0.66	0.65	0.33	1.02
APXS (ppm)								
Ni	950	1405	2030	1005	1001	736	1022	21
Zn	**6267**	**1798**	**710**	**2314**	155	116	231	126
Br	779	722	377	1470	334	269	77	77

"Preparation" refers to treatment of the surface with the RAT prior to APXS analysis.
APXS data from NASA PDS.
Highlighted, underlined values indicate the significant elemental enrichments.

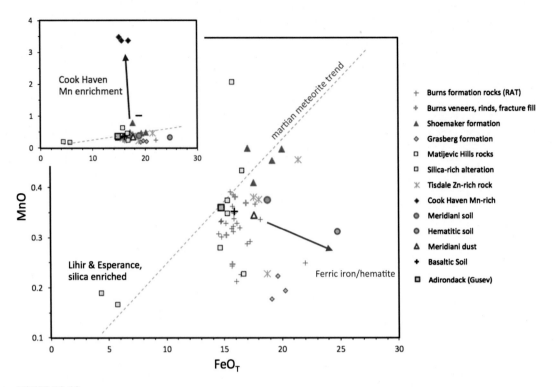

FIGURE 10.18

Mn as MnO vs Fe as FeO for compositions given in tables in this paper. The Martian meteorite trend is based on compositions of 25 representative Martian meteorites and is mainly an igneous trend, reflecting minimal chemical alteration and occurrence of Fe and Mn primarily in the divalent state. The *trend line* is constrained to the origin. Strong enrichment of Mn in the Cook Haven rocks (Table 10.7) and enrichment of Fe (as ferric iron) in other rocks reflect differential oxidation associated with alteration, fluid transport, and precipitation.

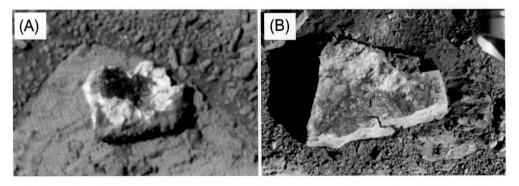

FIGURE 10.19

Coated rocks at Cook Haven that exhibit Mn and sulfate enrichment. (A) Pinnacle Island—sol 3541 sequence ID P2595 L257 false color. The rock is ~3.5 cm wide. (B) Stuart Island—sol 3567 sequence ID P2539 L257 false color. The rock is ~12 cm wide. *Image credits: D. Savransky and J. Bell/JPL/NASA/ Cornell/ASU (Bell et al., 2003, 2006).*

crustal materials, at the surface and in subsurface environments, over an extended period of time from the Noachian to Early Hesperian. Thus, terrain explored by Opportunity includes rocks that span the phyllosian to theiikian alteration eras, which are thought to be separated by a period of global climate change on Mars, plus diagenesis in the siderikian era (Bibring et al., 2006).

ACKNOWLEDGMENTS

The authors gratefully acknowledge support of the NASA Mars Exploration Rovers Project. We appreciate the superb work of the rover operations team at the Jet Propulsion Laboratory, California Institute of Technology, and NASA for continuing support of the Mars Exploration Rovers Mission. We also acknowledge and thank the MER instrument teams and the science operations working group for their dedicated efforts to deliver results from this fantastic mission.

REFERENCES

Allen, J.R.L., 1982. Sedimentary Structures, Their Character and Physical Basis, vol. 2. Elsevier, Amsterdam, 663 p.

Amundson, R., Ewing, S., Dietrich, W., Sutter, B., Owen, J., Chadwick, O., et al., 2008. On the in situ aqueous alteration of soils on Mars. Geochim. Cosmochim. Acta 72, 3845–3864.

Andrews-Hanna, J.C., Phillips, R.J., Zuber, M.T., 2007. Meridiani Planum and the global hydrology of Mars. Nature 446, 163–166.

Andrews-Hanna, J.C., Zuber, M.T., Arvidson, R.E., Wiseman, S.M., 2010. Early Mars hydrology: Meridiani playa deposits and the sedimentary record of Arabia Terra. J. Geophys. Res. 115, E06002. Available from: http://dx.doi.org/10.1029/2009JE003485.

Arvidson, R.E., Seelos, F.P., Deal, K.S., Koeppen, W.C., Snider, N.O., Kieniewicz, J.M., et al., 2003. Mantled and exhumed terrains in Terra Meridiani, Mars. J. Geophys. Res. 108, 8073. Available from: http://dx.doi.org/10.1029/2002JE001982.

Arvidson, R.E., Anderson, R.C., Bartlett, P., Bell III, J.F., Blaney, D., Christensen, P.R., et al., 2004. Localization and physical properties experiments conducted by Spirit at Gusev Crater. Science 305, 821–824.

Arvidson, R.E., Poulet, F., Morris, R.V., Bibring, J.-P., Bell III, J.F., Squyres, S.S., et al., 2006. Nature and origin of the hematite-bearing plains of Terra Meridiani based on analyses of orbital and Mars Exploration rover data sets. J. Geophys. Res. 111, E12S08. Available from: http://dx.doi.org/10.1029/2006JE002728.

Arvidson, R.E., Ashley, J.W., Bell III, J.F., Chojnacki, M., Cohen, B., Economou, T.E., et al., 2011. Opportunity Mars Rover Mission: overview and selected results from purgatory ripple to traverses to Endeavour Crater. J. Geophys. Res. 116, E00F15. Available from: http://dx.doi.org/10.1029/2010JE003746.

Arvidson, R.E., Squyres, S.W., Bell III, J.F., Catalano, J.G., Clark, B.C., Crumpler, L.S., et al., 2014. Ancient aqueous environments at Endeavour Crater, Mars. Science 343, 441–448.

Arvidson, R.E., Squyres, S.W., Morris, R.V., Knoll, A.H., Gellert, R., Clark, B.C., et al., 2016. High concentrations of manganese and sulfur in deposits on Murray Ridge, Endeavour Crater, Mars. Am. Mineral. 101, 1389–1405.

Bell III, J.F., Squyres, S.W., Herkenhoff, K.E., Maki, J.N., Arneson, H.M., Brown, D., et al., 2003. Mars Exploration Rover Athena Panoramic Camera (Pancam) investigation. J. Geophys. Res. 108 (E12), 8063. Available from: http://dx.doi.org/10.1029/2003JE002070.

Bell III, J.F., Joseph, J.R., Sohl-Dickstein, J., Arneson, H., Johnson, M., Lemmon, M., et al., 2006. In-flight calibration of the Mars Exploration Rover Panoramic Camera instrument. J. Geophys. Res. 111, E02S03. Available from: http://dx.doi.org/10.1029/2005JE002444.

Bibring, J.-P., Langevin, Y., Gendrin, A., Gondet, B., Poulet, F., Berthe, M., et al., 2005. Mars surface diversity as revealed by the OMEGA/Mars Express observations. Science 307, 1576–1581.

Bibring, J.-P., et al., 2006. Global mineralogical and aqueous Mars history derived from OMEGA/Mars Express data. Science 312, 400–404.

Burns, R.G., 1987. Ferric sulfates on Mars. J. Geophys. Res. 92, E570–E574.

Burns, R.G., Fisher, D.S., 1990. Iron–sulfur mineralogy of Mars-magmatic evolution and chemical-weathering products. J. Geophys. Res. 95, 14,415–14,421.

Calvin, W.M., Shoffner, J.D., Johnson, J.R., Knoll, A.H., Pocock, J.M., Squyres, S., et al., 2008. Hematite Spherules at Meridiani: results from MI, mini-TES and Pancam. J. Geophys. Res. 113, 1–27, E12S37.

Chan, M.A., Beitler, B., Parry, W.T., Ormö, J., Komatsu, G., 2004. A possible terrestrial analogue for haematite concretions on Mars. Nature 429, 731–734.

Chan, M.A., Bowen, B.B., Parry, W.T., Ormo, J., Komatsu, G., 2005. Red rock and red planet diagenesis: comparisons of Earth and Mars concretions. GSA Today 15, 4–10.

Chan, M.A., Potter, S.L., Boewen, B.B., Parry, W.T., Barge, L.M., Seiler, W., et al., 2012. Characteristics of terrestrial ferric oxide concretions and implications for Mars. In: Grotzinger, J.P., Milliken, R.E. (Eds.), Sedimentary Geology of Mars, SEPM Spec. Paper 102, pp. 253–270.

Chavdarian, G.V., Sumner, D.Y., 2006. Cracks and fins in sulfate sand: evidence for recent mineral-atmospheric water cycling in Meridiani Planum outcrops? Geology 34, 229–232.

Christensen, P.R., Ruff, S.W., 2004. Formation of the hematite-bearing unit in Meridiani Planum: evidence for deposition in standing water. J. Geophys. Res. 109, E08003. Available from: http://dx.doi.org/10.1029/2003JE002233.

Christensen, P.R., Bandfield, J.L., Clark, R.N., Edgett, K.S., Hamilton, V.E., Hoefen, T., et al., 2000. Detection of crystalline hematite mineralization on Mars by the Thermal Emission Spectrometer: evidence for near-surface water. J. Geophys. Res. 105, 9623–9642.

Christensen, P.R., Morris, R.V., Lane, M.D., Bandfield, J.L., Malin, M.C., 2001. Global mapping of Martian hematite deposits: remnants of water-driven processes on early Mars. J. Geophys. Res. 106, 23,873–23,886.

Christensen, P.R., Mehall, G.L., Silverman, S.H., Anwar, S., Cannon, G., Gorelick, N., et al., 2003. Miniature Thermal Emission Spectrometer for the Mars Exploration Rovers. J. Geophys. Res. 108, 8064.

Christensen, P.R., Jakosky, B.M., Kieffer, H.H., Malin, M.C., Kenneth Nealson Jr., H.Y.M., Mehall, G.L., et al., 2004a. The Thermal Emission Imaging System (THEMIS) for the Mars 2001 Odyssey mission. Space Sci. Rev. 110, 85–130.

Christensen, P.R., Wyatt, M.B., Glotch, T.D., Rogers, A.D., Anwar, S., Arvidson, R.E., et al., 2004b. Mineral compositions and abundances at the Meridiani Planum site from the mini-TES experiment on the Opportunity rover. Science 306, 1733–1739.

Cino, C.D., Dehouck, E., McLennan, S.M., 2017. Geochemical constraints on the presence of clay minerals in the Burns formation, Meridiani Planum, Mars. Icarus 281, 137–150.

Clark, B.C., Morris, R.V., McLennan, S.M., Gellert, R., Jolliff, B., Knoll, A.H., et al., 2005. Chemistry and mineralogy of outcrops at Meridiani Planum. Earth Planet. Sci. Lett. 240, 73–94.

Clark, B.C., Morris, R.V., Herkenhoff, K.E., Farrand, W.H., Gellert, R., Jolliff, B.L., et al., 2016. Esperance: multiple episodes of aqueous alteration involving fracture fills and coatings at Matijevic Hill, Mars. Am. Mineral 101, 1515–1526.

Crumpler, L.S., Arvidson, R.E., Bell, J., Clark, B.C., Cohen, B.A., Farrand, W.H., et al., 2015. Context of ancient aqueous environments on Mars from in situ geologic mapping at Endeavour Crater. J. Geophys. Res. 120, 538–569.

Crumpler, L.S., Arvidson, R.E., Mittlefehldt, D.W., Jolliff, B.L., Farrand, W.H., Fox, V., et al., 2016. Opportunity, geologic and structural context of aqueous alteration in Noachian outcrops, Marathon Valley and rim of Endeavour Crater. Lunar Planet. Sci. 46, abstract #2272.

Edgar, L.A., Grotzinger, J.P., Hayes, A.G., Rubin, D.M., Squyres, S.W., Bell, J.F., et al., 2012. Stratigraphic architecture of bedrock reference section, Victoria Crater, Meridiani Planum, Mars. In: Grotzinger, J.P., Milliken, R.E. (Eds.), Sedimentary Geology of Mars. SEPM Spec. Paper 102, pp. 195–209.

Edgar, L.A., Grotzinger, J.P., Bell, J.F., Hurowitz, J.A., 2014. Hypothesis for the origin of the fine-grained sedimentary rocks at Santa Maria Crater, Meridiani Planum. Icarus 234, 36–44.

Edgett, K.S., Malin, M.C., 2002. Martian sedimentary rock stratigraphy: outcrops and interbedded craters of northwest Sinus Meridiani and southwest Arabia Terra. Geophys. Res. Lett. 29, 2179. Available from: http://dx.doi.org/10.1029/2002GL016515.

Elwood Madden, M.E., Madden, A.S., Rimstidt, J.D., 2009. How long was Meridiani Planum wet? Applying a jarosite stopwatch to determine the duration of aqueous diagenesis. Geology 37, 635–638.

Farrand, W.H., Bell III, J.F., Johnson, J.R., Jolliff, B.L., Knoll, A.H., McLennan, S.M., et al., 2007. Visible and near-infrared multispectral analysis of rocks at Meridiani Planum, Mars, by the Mars Exploration Rover Opportunity. J. Geophys. Res. 112, E06S02. Available from: http://dx.doi.org/10.1029/2006JE002773.

Farrand, W.H., Bell III, J.F., Johnson, J.R., Rice, M.S., Hurowitz, J.A., 2013. VNIR multispectral observations of rocks at Cape York, Endeavour Crater, Mars by the Opportunity rover's Pancam. Icarus 225, 709–725.

Farrand, W.H., Johnson, J.R., Rice, M.R., Wang, A., Bell III, J.F., 2016. VNIR multispectral observations of aqueous alteration materials by the Pancams on the Spirit and Opportunity Mars Exploration Rovers. Am. Mineral 101, 2005–2019.

Feldman, W.C., Prettyman, T.H., Maurice, S., Plaut, J.J., Bish, D.L., Vaniman, D.T., et al., 2004. Global distribution of near-surface hydrogen on Mars. J. Geophys. Res. 109, E09006. Available from: http://dx.doi.org/10.1029/2003JE002160.

Fernandez-Remolar, D.C., Morris, R., Gruener, J.E., Amils, R., Knoll, A.H., 2005. The Rio Tinto Basin: mineralogy, sedimentary geobiology, and implications for interpretation of outcrop rocks at Meridiani Planum. Earth Planet. Sci. Lett. 240, 149–167.

Flahaut, J., Carter, J., Poulet, F., Bibring, J.-P., van Westrenen, W., Davies, G.R., et al., 2015. Embedded clays and sulfates in Meridiani Planum, Mars. Icarus 248, 269–288.

Fox, V., Arvidson, R., Murchie, S., Squyres, S., 2014. Smectite detections at Murray Ridge and Cape Tribulation, Mars, from along-track oversampled CRISM observations. In: Abstract for The Eighth International Conference on Mars, July 2014, abstract #1154.

Fox, V.K., Arvidson, R.E., Guinness, E.A., McLennan, S.M., Catalano, J.G., Murchie, S.L., et al., 2016. Smectite deposits in Marathon Valley, Endeavour Crater, Mars, identified using CRISM hyperspectral reflectance data. Geophys. Res. Lett. 43, 4885–4892.

Fox-Powell, M.G., Hallsworth, J.E., Cousins, C.R., Cockell, C.S., 2016. Ionic strength is a barrier to the habitability of Mars. Astrobiology 16, 427–442.

Fralick, P., Grotzinger, J.P., Edgar, L., 2012. Potential recognition of accretionary lapilli in distal impact deposits on Mars: a facies analog provided by the 1.85 Ga Sudbury impact deposit. In: Grotzinger, J.P., Milliken, R.E. (Eds.), Sedimentary Geology of Mars. SEPM Spec. Paper 102, pp. 211–227.

Gellert, R., Clark III, B.C., 2015. In situ compositional measurements of rocks and soils with the alpha particle X-ray spectrometer on NASA's Mars rovers. Elements 11, 39–44.

Gellert, R., Rieder, R., Brückner, J., Clark, B.C., Dreibus, G., Klingelhöfer, G., et al., 2006. Alpha particle X-ray spectrometer (APXS): results from Gusev Crater and calibration report. J. Geophys. Res. 111, 1–32.

Glotch, T.D., Bandfield, J.L., Christensen, P.R., Calvin, W.M., McLennan, S.M., Clark, B.C., et al., 2006. Mineralogy of the light-toned outcrop at Meridiani Planum as seen by the Miniature Thermal Emission Spectrometer and implications for its formation. J. Geophys. Res. 111, E12S03. Available from: http://dx.doi.org/10.1029/2005JE002672.

Golombek, M.P., Grant, J.A., Crumpler, L.S., Greeley, R., Arvidson, R.E., Bell, J.F., et al., 2006. Erosion rates at the Mars Exploration Rover landing sites and long-term climate change on Mars. J. Geophys. Res. 111, E12S10. Available from: http://dx.doi.org/10.1029/2006JE002754.

Golombek, M.P., Robinson, K., McEwen, A., Bridges, N., Ivanov, B., Tornabene, L., et al., 2010. Constraints on ripple migration at Meridiani Planum from Opportunity and HiRISE observations of fresh craters. J. Geophys. Res. 115, E00F08. Available from: http://dx.doi.org/10.1029/2010JE003628.

Gorevan, S.P., Myrick, T., Davis, K., Chau, J.J., Bartlett, P., Mukherjee, S., et al., 2003. Rock Abrasion Tool: Mars Exploration Rover mission. J. Geophys. Res. 108, 8068. Available from: http://dx.doi.org/10.1029/2003JE002061.

Grant, J.A., Parker, T.J., Crumpler, L.S., Wilson, S.A., Golombek, M.P., Mittlefehldt, D.W., 2016. The degradational history of Endeavour Crater, Mars. Icarus 280, 22−36.

Grotzinger, J.P., Arvidson, R.E., Bell III, J.F., Calvin, W., Clark, B.C., Fike, D.A., et al., 2005. Stratigraphy and sedimentology of a dry to wet eolian depositional system, Burns formation, Meridiani Planum, Mars. Earth Planet. Sci. Lett. 240, 11−72.

Grotzinger, J., Bell III, J., Herkenhoff, K., Johnson, J., Knoll, A., McCartney, E., et al., 2006. Sedimentary textures formed by aqueous processes, Erebus Crater, Meridiani Planum, Mars. Geology 34, 1085−1088.

Hardie, L., Garrett, P., 1977. General environmental setting. In: Hardie (Ed.), Sedimentation of the Modern Carbonate Tidal Flats of Northwest Andros Island, Bahamas. The Johns Hopkins University Press, Baltimore, MD, pp. 12−49.

Hayes, A.G., Grotzinger, J.P., Edgar, L.A., Squyres, S.W., Watters, W.A., Sohl-Dickstein, J., 2011. Reconstruction of eolian bedforms and paleocurrents from cross-bedded strata at Victoria Crater, Meridiani Planum, Mars. J. Geophys. Res. 116, E00F21. Available from: http://dx.doi.org/10.1029/2010JE003688.

Herkenhoff, K.E., Squyres, S.W., Arvidson, R., Bass, D.S., Bell III, J.F., Bertelsen, P., et al., 2004. Evidence from Opportunity's microscopic imager for water on Meridiani Planum. Science 306, 1727−1730.

Herkenhoff, K.E., Golombek, M.P., Guinness, E.A., Johnson, J.B., Kusack, A., Richter, L., et al., 2008a. In situ observations of the physical properties of the Martian surface. In: Bell, J. (Ed.), The Martian Surface: Composition, Mineralogy, and Physical Properties. Cambridge University Press, Cambridge, UK and New York, pp. 451−467.

Herkenhoff, K.E., Grotzinger, J., Knoll, A.H., McLennan, S.M., Weitz, C., et al., 2008b. Surface processes recorded by rocks and soils on Meridiani Planum, Mars: microscopic Imager observations during Opportunity's first three extended missions. J. Geophys. Res. 113, E12S32. Available from: http://dx.doi.org/10.1029/2008JE003100.

Hurowitz, J.A., McLennan, S.M., Tosca, N.J., Arvidson, R.E., Michalski, J.R., Ming, D.W., et al., 2006. In-situ and experimental evidence for acidic weathering on Mars. J. Geophys. Res. 111, E02S19. Available from: http://dx.doi.org/10.1029/2005JE002515.

Hurowitz, J., Fischer, W.W., Tosca, N.J., Milliken, R.E., 2010. Origin of acidic surface waters and the evolution of atmospheric chemistry on early Mars. Nat. Geosci. 3, 323−326.

Hynek, B.M., Di Achille, G., 2017. Geologic map of Meridiani Planum, Mars (ver. 1.1, April 2017): U.S. Geological Survey Scientific Investigations Map 3356, pamphlet 9 p., scale 1:2,000,000.

Hynek, B.M., Phillips, R.J., 2008. The stratigraphy of Meridiani Planum, Mars, and implications for the layered deposits' origin. Earth Planet. Sci. Lett. 274, 214−220.

Hynek, B.M., Arvidson, R.E., Phillips, R.J., 2002. Geologic setting and origin of Terra Meridiani hematite deposit on Mars. J. Geophys. Res. 107, 5088. Available from: http://dx.doi.org/10.1029/2002JE001891.

Jolliff, B.L., the Athena Science Team, 2005. Composition of Meridiani hematite-rich spherules: a mass-balance mixing-model approach. Lunar Planet. Sci 36, abstract #2269.

Jolliff, B.L., McLennan, S.M., the Athena Science Team, 2006. Evidence for water at Meridiani. Elements 2, 163−167.

Jones, D., Calvin, W.M., DeSouza, P., Jolliff, B.L., 2014. Martian spherule size distribution between Eagle and Endurance Craters from Opportunity rover MI images. Lunar Planet. Sci 45, abstract #2026.

Klingelhöfer, G., Morris, R.V., Bernhardt, B., Rodionov, D., de Souza Jr., P.A., Squyres, S.W., et al., 2003. Athena MIMOS II Mössbauer Spectrometer investigation. J. Geophys. Res. 108, 8067. Available from: http://dx.doi.org/10.1029/2003JE002138.

Klingelhöfer, G., Morris, R.V., Bernhardt, B., Schröder, C., Rodionov, D.S., de Souza Jr., P.A., et al., 2004. Jarosite and Hematite at Meridiani Planum from Opportunity's Mössbauer Spectrometer. Science 306, 1740−1745.

Knoll, A.H., Jolliff, B.L., Farrand, W.H., Bell III, J.F., Clark, B.C., Gellert, R., et al., 2008. Veneers, rinds, and fracture fills: relatively late alteration of sedimentary rocks at Meridiani Planum, Mars. J. Geophys. Res. 113, E06S16. Available from: http://dx.doi.org/10.1029/2007JE002949.

Lamb, M.P., Grotzinger, J.P., Southard, J.B., Tosca, N.J., 2012. Were aqueous ripples on Mars formed by flowing brines? In: Grotzinger, J.P., Milliken, R.E. (Eds.), Sedimentary Geology of Mars. SEPM Spec. Paper 102, pp. 139−150.

Malin, M.C., 2005. Hidden in plain sight: finding Martian landers. Sky Telescope 110, 42−46.

Malin, M.C., Danielson, G.E., Ingersoll, A.P., Masursky, H., Veverka, J., Ravine, M.A., et al., 1992. Mars Observer Camera. J. Geophys. Res. 97, 7699−7718.

McCollom, T.M., Hynek, B.M., 2005. A volcanic environment for bedrock diagenesis at Meridiani Planum on Mars. Nature 438, 1129−1131.

McLennan, S.M., Grotzinger, J.P., 2008. The sedimentary rock cycle of Mars. In: Bell III, J.F. (Ed.), The Martian Surface: Composition, Mineralogy, and Physical Properties. Cambridge University Press, Cambridge, UK and New York, pp. 541−577.

McLennan, S.M., Bell III, J.F., Calvin, W.M., Christensen, P.R., Clark, B.C., de Souza, P.A., et al., 2005. Provenance and diagenesis of the evaporite-bearing Burns formation, Meridiani Planum, Mars. Earth Planet. Sci. Lett. 240, 95−121.

Ming, D.W., Morris, R.V., Clark, B.C., 2008. Aqueous alteration on Mars. In: Bell III, J.F. (Ed.), The Martian Surface: Composition, Mineralogy, and Physical Properties. Cambridge University Press, Cambridge, UK and New York, pp. 519−540.

Morris, R.V., Klingelhöfer, G., 2008. Iron mineralogy and aqueous alteration on Mars from the MER Mössbauer Spectrometers. In: Bell III, J.F. (Ed.), The Martian Surface: Composition, Mineralogy, and Physical Properties. Cambridge University Press, Cambridge, UK and New York, pp. 339−365.

Morris, R.V., Ming, D.W., Graff, T.G., Arvidson, R.E., Bell III, J.F., Squyres, S.W., et al., 2005. Hematite spherules in basaltic tephra altered under aqueous, acid-sulfate conditions on Mauna Kea volcano, Hawaii: possible clues for the occurrence of hematite-rich spherules in the Burns formation at Meridiani Planum, Mars. Earth Planet. Sci. Lett. 240, 168−178.

Morris, R.V., Klingelhöfer, G., Schröder, C., Rodionov, D.S., Yen, A., Ming, D.W., et al., 2006. Mössbauer mineralogy of rock, soil, and dust at Meridiani Planum, Mars: Opportunity's journey across sulfate-rich outcrop, basaltic sand and dust, and hematite lag deposits. J. Geophys. Res 111, E12S15. Available from: http://dx.doi.org/10.1029/2006JE002791.

Nesbitt, H.W., Wilson, R.E., 1992. Recent chemical weathering of basalts. Am. J. Sci. 292, 740−777.

Rice, M., Bell III, J.F., Cloutis, E., Wang, A., Ruff, S., Craig, M., et al., 2010. Silica-rich deposits and hydrated minerals at Gusev Crater, Mars: Vis−NIR spectral characterization and regional mapping. Icarus 205, 375−395.

Rieder, R., Gellert, R., Brückner, J., Klingelhöfer, G., Dreibus, G., Yen, A., et al., 2003. The new Athena alpha particle X-ray spectrometer for the Mars Exploration Rovers. J. Geophys. Res. 108, 8066. Available from: http://dx.doi.org/10.1029/2003JE002150.

Rieder, R., Gellert, R., Anderson, R.C., Brückner, J., Clark, B.C., Dreibus, G., et al., 2004. Chemistry of rocks and soils at Meridiani Planum from the alpha particle X-ray spectrometer. Science 306, 1746−1749.

Rubin, D.M., 1987. Cross-bedding, Bedforms, and Paleocurrents. Society of Economic Paleontologists and Mineralogists, Tulsa, Oklahoma, 187 p.

Savransky, D., Bell III, J.F., 2004. True color and chromaticity of the Martian surface and sky from Mars Exploration Rover Pancam observations. In: EOS Trans. AGU, abstract #P21A-0197.

Sefton-Nash, E., Catling, D.C., 2008. Hematitic concretions at Meridiani Planum, Mars: their growth timescale and possible relationship with iron sulfates. Earth Planet. Sci. Lett. 269, 365−375.

Southard, J.B., 1973. Representation of bed configuration in depth-velocity-size diagrams. J. Sediment. Petrol. 41, 903–915.

Squyres, S.W., Knoll, A.H., 2005. Sedimentary rocks at Meridiani Planum: origin, diagenesis, and implications for life on Mars. Earth Planet. Sci. Lett. 240, 1–10.

Squyres, S.W., Arvidson, R.E., Baumgartner, E.T., Bell III, J.F., Christensen, P.R., Gorevan, S., et al., 2003. Athena Mars rover science investigation. J. Geophys. Res. 108, 8062. Available from: http://dx.doi.org/10.1029/2003JE002121.

Squyres, S.W., Arvidson, R.E., Bell III, J.F., Brückner, J., Cabrol, N.A., Calvin, W., et al., 2004a. The Opportunity Rover's Athena science investigation at Meridiani Planum, Mars. Science 306, 1698–1703.

Squyres, S.W., Grotzinger, J.P., Arvidson, R.E., Bell III, J.F., Calvin, W., Christensen, P.R., et al., 2004b. In situ evidence for an ancient aqueous environment at Meridiani Planum, Mars. Science 306, 1709–1714.

Squyres, S.W., Arvidson, R.E., Bollen, D., Bell III, J.F., Brückner, J., Cabrol, N.A., 2006a. Overview of the Opportunity Mars Exploration Rover mission to Meridiani Planum: Eagle Crater to Purgatory Ripple. J. Geophys. Res. 111, E12S12. Available from: http://dx.doi.org/10.1029/2006JE002771.

Squyres, S.W., Knoll, A.H., Arvidson, R.E., Clark, B.C., Grotzinger, J.P., Jolliff, B.L., et al., 2006b. Two years at Meridiani Planum: results from the Opportunity rover. Science 313, 1403–1407.

Squyres, S.W., Knoll, A.H., Arvidson, R.E., Ashley, J.W., Bell III, J.F., Calvin, W.M., et al., 2009. Exploration of Victoria Crater by the Mars Rover Opportunity. Science 324 (5930), 1058–1061.

Squyres, S.W., Arvidson, R.E., Bell, J.F., Calef, F., Clark, B.C., Cohen, B.A., et al., 2012. Ancient impact and aqueous processes at Endeavour Crater, Mars. Science 336, 570–576.

Sverjensky, D.A., 1986. Genesis of Mississippi Valley-type lead-zinc deposits. Ann. Rev. Earth Planet. Sci. 14, 177–199.

Taylor, S.R., McLennan, S.M., 2009. Planetary Crusts: Their Composition, Origin and Evolution. Cambridge University Press, Cambridge, 378 pp.

Thomson, B.J., Bridges, N.T., Cohen, J., Hurowitz, J.A., Lennon, A., Paulsen, G., et al., 2013. Estimating rock compressive strength from Rock Abrasion Tool (RAT) grinds. J. Geophys. Res. 118, 1233–1244.

Tosca, N.J., Knoll, A.H., 2009. Juvenile chemical sediments and the long term persistence of water at the surface of Mars. Earth Planet. Sci. Lett. 286, 379–386.

Tosca, N.J., McLennan, S.M., Clark, B.C., Grotzinger, J.P., Hurowitz, J.A., Knoll, A.H., et al., 2005. Geochemical modeling of evaporation processes on Mars: insight from the sedimentary record at Meridiani Planum. Earth Planet. Sci. Lett. 240, 122–148.

Tosca, N., Knoll, A.H., McLennan, S., 2008a. Water activity and the challenge for life on early Mars. Science 320, 1204–1207.

Tosca, N.J., McLennan, S.M., Dyar, M.D., Sklute, E.C., Michel, F.M., 2008b. Fe oxidation processes at Meridiani Planum and implications for secondary Fe mineralogy on Mars. J. Geophys. Res. 113, E05005. Available from: http://dx.doi.org/10.1029/2007JE003019.

Watters, W.A., Bell III, J.F., Calef, F., Golombek, M., Grant, J., Hayes, A., et al., 2011. Structure and morphology of Santa Maria Crater, Meridiani Planum, Mars. Lunar Planet. Sci. 42, 2586.

Wolters, F., Lagaly, G., Kahr, G., Nueeshch, R., Emmerich, K., 2009. A comprehensive characterization of dioctahedral smectites. Clays and Clay Miner. 57, 115–133.

Wray, J.J., Noe Dobrea, E.Z., Arvidson, R.E., Wiseman, S.M., Squyres, S.W., McEwen, A.S., et al., 2009. Phyllosilicates and sulfates at Endeavour Crater, Meridiani Planum, Mars. Geophys. Res. Lett. 36, L21201. Available from: http://dx.doi.org/10.1029/2009GL040734.

Zhao, Y.-Y.S., McLennan, S.M., 2013. Behavior of Ni, Zn and Cr during low temperature aqueous Fe oxidation processes on Mars. Geochim. Cosmochim. Acta 109, 365–383.

Zurek, R.W., Smrekar, S.E., 2007. An overview of the Mars Reconnaissance Orbiter (MRO) science mission. J. Geophys. Res. 110, E05S01. Available from: http://dx.doi.org/10.1029/2006JE002701.

ALTERATION PROCESSES IN GUSEV CRATER, MARS: VOLATILE/MOBILE ELEMENT CONTENTS OF ROCKS AND SOILS DETERMINED BY THE SPIRIT ROVER

David W. Mittlefehldt[1], Ralf Gellert[2], Douglas W. Ming[1], and Albert S. Yen[3]

[1]*NASA/Johnson Space Center, Houston, TX, United States* [2]*University of Guelph, Guelph, ON, Canada*
[3]*Jet Propulsion Laboratory, California Institute of Technology, Pasadena, CA, United States*

11.1 INTRODUCTION

The Mars exploration rover Spirit landed in Gusev crater on January 4, 2004 and explored for 2209 sols (Mars day; about 2.7% longer than an Earth day). She finally went silent on sol 2210 after having been entrapped in loose soil for 324 sols with her solar panels at an inconvenient angle with respect to the oncoming winter's lower Sun angle. During her explorations of Gusev crater, Spirit traversed 7730 m of diverse geological terranes. The instruments of Spirit's Athena payload (Squyres et al., 2003) were used to investigate the rocks and soils of Gusev crater. On a mast 1.5 m above the rover deck were the Panoramic Camera (Pancam; Bell et al., 2003) and the mirror for the Miniature Thermal Emission Spectrometer that was housed within the rover body (Mini-TES; Christensen et al., 2003). On the Instrument Deployment Device (IDD, a.k.a. the Arm) were the Alpha Particle X-ray Spectrometer (APXS; Rieder et al., 2003), the MIMOS II Mössbauer spectrometer (MB; Klingelhöfer et al., 2003), the microscopic imager (MI; Herkenhoff et al., 2003), and the rock abrasion tool (RAT; Gorevan et al., 2003). The science instruments were supported by imagery from the engineering cameras—Navigation Cameras (Navcam) on the mast, and front and rear Hazard Avoidance Cameras (Hazcam) on the rover body below the deck (Maki et al., 2003).

The main goal of the Mars exploration rover mission, still being pursued by Opportunity in Meridiani Planum, is to explore sites on Mars where water may once have been present, to evaluate past environmental conditions at those sites and to assess the suitability of those conditions for life (Squyres et al., 2003). Gusev crater was selected as the landing site because of prior orbital imaging and remote sensing that suggested that Gusev crater might once have hosted a large body of standing water (Golombek et al., 2003). Gusev crater is a flat-floored, 158-km diameter Noachian-aged crater on the Southern Highlands near the dichotomy boundary with the Northern Lowlands. The

Volatiles in the Martian Crust. DOI: http://dx.doi.org/10.1016/B978-0-12-804191-8.00011-8

southern rim of the crater is breached by Ma'adim Vallis, an approximately 900-km-long valley system with morphological features suggestive of flowing water having been present in the past. The lower northern rim of Gusev crater was thought to have provided an outlet for the water, but the crater would have been substantially filled. A set of mesas at the mouth of Ma'adim Vallis were thought to be erosional remnants of deltaic sediments. The flat floor of the crater is of Early Hesperian age and was thought to represent sediment fill, which might be as thick as ~ 900 m in the center of the crater (Carter et al., 2001). However, it was recognized that some morphologic features of the crater plains, particularly wrinkle-ridge-like features, suggested that it could be capped by a thin volcanic layer. Nevertheless, it was thought that craters on the plains could have excavated lake sediments from beneath a volcanic cover and made them available for study, albeit without proper geologic context (Golombek et al., 2003).

The focus of this chapter is on the chemical data returned by Spirit's APXS instrument that show the compositional effects of various alteration processes on the rocks and soils in Gusev crater. These compositional data are put in geological and mineralogical context using interpretations gleaned from data from the other instruments of the Athena Science Package. Compositional data used here are reported in several publications, with later reports sometimes including the data presented earlier. All data used here are published in Arvidson et al. (2010), Gellert et al. (2006), Ming et al. (2008a), and Schmidt et al. (2009). All of the Spirit APXS data are available on the NASA Planetary Data System website.

The rocks and soils in Gusev crater are divided into classes using bulk compositions determined on targets abraded by the RAT (Squyres et al., 2006). However, because of the hardness of many of the rocks, the cutting heads became worn down, precluding further use of the abrasion tool after sol 416. For this reason, rock classes defined later in the mission are based on brushed or naturally cleaned surfaces (Ming et al., 2008a). Although the team tried to analyze untreated surfaces that were as clean as possible, some airfall dust and/or eolian sand was present to varying degrees. The data used here are as quantified by the APXS without any attempt to subtract a dust/soil component. Mössbauer-derived iron mineralogy, Pancam 13 filter (13f) spectra, and Mini-TES spectra allow for more complete characterization of rock and soil class definitions and classification of remotely sensed targets (Farrand et al., 2006, 2008; Morris et al., 2006, 2008; Ruff et al., 2006, 2007).

Table 11.1 lists the rock and soil classes defined in Gusev crater, gives their average Mineralogical Alteration Index (MAI), and their average concentrations of the more labile elements that are the main focus of this chapter: SO_3, Cl, Zn, and Br. These elements show the greatest variations associated with alteration. We refer to the more labile elements here under the general rubric of volatile/mobile elements because their variations in the rocks and soils, coupled with the documented or inferred alteration mineralogy, indicate that these elements were vapor- and/or fluid-mobile during alteration. The MAI is defined by Morris et al. (2006, 2008) as the sum of the percent of total Fe in five alteration phases: nanophase ferric oxide, hematite, goethite, Fe-sulfate, FeS_2. [A doublet with uniquely low isomer shift in MB spectra was assigned to FeS_2 (Morris et al., 2008). It is found only in the Fuzzy Smith target, and therefore, its inclusion among the list of alteration phases has no impact on the calculation of MAI of any of the other rocks and soils discussed here.] Subsequent to Morris et al. (2008), a MB spectral feature in the Comanche targets originally assigned to pyroxene was determined to be due to Fe-bearing carbonate alteration phases (Morris et al., 2010). The MAI of Comanche in Table 11.1 has been recalculated to include carbonate among the alteration phases.

The averages in Table 11.1 are based on the same targets as used by Ming et al. (2006, 2008a) for compositions, and Morris et al. (2008) for MAI. The only exception to this is for the Paso-Robles-class soils for which a large number of APXS targets were analyzed later in the mission (Arvidson et al.,

Table 11.1 Summary of Rock and Soil Classes Defined by Compositions from Gusev Crater, with Average Mineralogic Alteration Index (MAI) and Volatile/Mobile Element Contents

Class	Location	MAI[a]	n[b]	SO₃ wt%	Cl wt%	Zn μg/g	Ge[c] μg/g	Br μg/g	n[b]
Rocks									
Adirondack	Cratered plains	12	8	1.68	0.31	121	b.d.	84	5
Clovis	West Spur	63	7	5.30	1.57	117	b.d.	506	17
Wishstone	Husband Hill	38	4	2.75	0.58	79	b.d.	45	6
Watchtower	Husband Hill	64	8	4.88	1.16	114	b.d.	257	14
Peace	Husband Hill	14	2	9.23	1.08	169	b.d.	185	6
Backstay	Husband Hill	15	1	1.82	0.40	272	b.d.	21	2
Independence	Husband Hill	38	2	4.34	0.70	274	b.d.	63	5
Descartes	Husband Hill	50	1	7.52	1.21	215	b.d.	161	4
Irvine	Husband Hill	7	3	2.37	0.46	299	b.d.	93	2
Algonquin	Husband Hill	15	3	4.80	1.06	187	b.d.	100	5
Comanche	Husband Hill	51	1	3.91	0.65	149	b.d.	154	2
Barnhill	Home Plate	27	10	4.57	1.27	507	41	166	13
Fuzzy Smith	Home Plate	63	1	3.39	0.63	679	191	21	1
Elizabeth Mahon	Eastern Valley	30	3	3.70	0.50	507	b.d.	56	6
Halley	Low Ridge	52	7	7.01	0.72	1090	b.d.	70	9
Montalva	Low Ridge	84	1	5.64	1.00	608	b.d.	76	2
Everett	Eastern Valley	18	2	3.40	1.43	1160	b.d.	234	2
Good Question	Eastern Valley	38	1	4.34	1.53	1030	b.d.	245	1
Torquas	Mitcheltree Ridge	35	1	4.85	1.25	1750	54	295	1
Soils									
Laguna	Widespread	21	40	5.55	0.65	260	b.d.	39	26
Boroughs	Widespread	34	3	10.7	0.85	345	b.d.	119	4
Paso Robles	Columbia Hills	76	5	30.8	0.41	153	b.d.	127	13
Gertrude Weise	Eastern Valley	30	2	1.05	0.15	278	b.d.	15	3
Eileen Dean	Eastern Valley	17	1	3.14	1.66	1070	b.d.	122	2

[a]Average MAI rolled up to the class level, calculated from data in Morris et al. (2008).
[b]Number of measurements averaged.
[c]Only Ge measurements that are ≥4 times the measurement precision are averaged (see Section 11.2); b.d., below detection.

2010); no MB analyses were made on these later targets. Germanium possibly shows variations associated with alteration, but the data are close to the detection limit. Only 28% of the analyses have Ge contents greater than or equal to four times the measurement *precision* (see the next section), and only three classes have reliable Ge averages (Table 11.1). For this reason, Ge is not a major focus of the discussion. Two rock classes, Descartes and Fuzzy Smith, are listed in Table 11.1 for completeness but are not discussed in this chapter. The Descartes class shows only limited geochemical evidence for alteration. Fuzzy Smith is a single erratic that lacks geological context.

Table 11.2 gives the averages and standard deviations for all elements in all the rock and soil classes listed in Table 11.1. Ming et al. (2006, Table 1, 2008a, Table 3) describe the distinguishing compositional characteristics of the defined classes in relation to the Adirondack-class

Table 11.2 Average Compositions and Standard Deviations of Rock and Soil Classes Defined from Gusev Crater

Rocks

Class	n[a]		Na₂O wt%	MgO wt%	Al₂O₃ wt%	SiO₂ wt%	P₂O₅ wt%	SO₃ wt%	Cl wt%	K₂O wt%	CaO wt%	TiO₂ wt%	Cr₂O₃ wt%	MnO wt%	FeO wt%	Ni µg/g	Zn µg/g	Br µg/g
Adirondack	5	ave	2.65	9.89	10.6	45.9	0.63	1.68	0.31	0.15	7.87	0.58	0.58	0.41	18.7	201	121	84
		std	0.16	0.73	0.4	0.3	0.12	0.93	0.14	0.09	0.24	0.08	0.08	0.01	0.2	83	59	65
Clovis	17	ave	2.99	11.9	9.98	45.5	1.01	5.30	1.57	0.32	4.69	0.83	0.18	0.21	15.4	576	117	506
		std	0.42	2.1	0.85	1.6	0.10	2.21	0.49	0.11	0.95	0.06	0.03	0.06	1.4	80	45	290
Wishstone	6	ave	4.95	4.68	14.9	44.9	3.73	2.75	0.58	0.53	7.76	2.58	0.01	0.23	12.4	60	79	45
		std	0.38	0.82	0.8	1.4	1.54	0.99	0.12	0.02	1.15	0.43	0.01	0.02	0.7	29	18	24
Watchtower	14	ave	3.42	8.40	12.6	45.6	2.64	4.88	1.16	0.49	6.63	1.93	0.05	0.23	11.8	109	114	257
		std	0.37	0.82	1.1	1.4	0.69	1.07	0.30	0.13	0.57	0.25	0.04	0.04	1.0	52	49	74
Peace	6	ave	1.75	15.6	5.40	40.5	0.66	9.23	1.08	0.15	4.95	0.56	0.61	0.40	19.0	581	169	185
		std	1.35	4.4	2.41	2.6	0.26	2.10	0.27	0.12	0.25	0.12	0.08	0.04	0.8	102	85	65
Backstay	2	ave	3.99	8.30	13.1	49.4	1.34	1.82	0.40	1.02	6.04	0.93	0.16	0.25	13.2	210	272	21
		std	0.23	0.03	0.3	0.1	0.07	0.43	0.07	0.08	0.00	0.01	0.01	0.00	0.2	27	4	6
Independence	5	ave	2.48	5.98	15.8	52.5	2.50	4.34	0.70	0.80	5.60	1.31	1.17	0.13	6.47	1110	274	63
		std	0.67	1.93	2.6	1.8	0.83	1.46	0.26	0.23	1.41	0.43	1.46	0.03	1.60	650	231	24
Descartes	4	ave	3.20	9.37	10.4	46.1	1.42	7.52	1.21	0.64	5.46	0.98	0.16	0.22	13.2	410	215	161
		std	0.14	0.23	0.6	1.1	0.06	0.44	0.16	0.10	0.21	0.05	0.02	0.03	1.4	47	61	20
Irvine	2	ave	3.04	9.54	8.34	47.5	0.94	2.37	0.46	0.60	5.80	1.06	0.20	0.37	19.7	342	299	93
		std	0.52	1.55	0.08	0.6	0.04	0.00	0.01	0.11	0.32	0.00	0.00	0.01	0.7	75	97	124
Algonquin	5	ave	2.52	14.6	6.91	42.0	1.22	4.80	1.06	0.50	4.36	0.66	0.52	0.37	20.3	612	187	100
		std	0.55	4.4	1.72	1.8	0.94	0.49	0.28	0.32	1.37	0.31	0.27	0.01	0.7	272	39	47
Comanche	2	ave	1.77	19.5	4.95	41.6	0.59	3.91	0.65	0.14	2.97	0.38	0.69	0.43	22.2	934	149	154
		std	0.92	7.4	2.86	0.4	0.20	1.73	0.06	0.14	1.48	0.18	0.04	0.01	0.5	94	24	2
Barnhill	13	ave	3.04	9.62	9.49	46.1	1.12	4.57	1.27	0.42	6.29	0.95	0.37	0.33	16.3	381	507	166
		std	0.31	0.80	0.66	0.9	0.22	0.88	0.46	0.21	0.31	0.15	0.07	0.05	0.9	103	122	133
Fuzzy Smith	1	n/a	2.92	4.16	6.31	68.4	0.68	3.39	0.63	2.76	1.93	1.71	0.06	0.15	6.76	272	679	21
Elizabeth Mahon	6	ave	1.20	8.25	4.29	68.7	0.56	3.70	0.50	0.18	2.51	0.64	0.25	0.10	9.01	566	507	56
		std	0.60	4.05	1.22	5.0	0.09	0.18	0.03	0.10	0.74	0.10	0.02	0.02	3.58	489	336	27
Halley	9	ave	2.77	8.86	9.60	45.6	0.94	7.01	0.72	0.86	6.69	0.86	0.25	0.28	15.3	604	1090	70
		std	0.19	0.67	0.58	2.2	0.13	2.71	0.13	0.49	1.33	0.12	0.02	0.01	1.1	173	250	55
Montalva	2	ave	1.10	12.9	6.56	45.3	0.68	5.64	1.00	2.91	3.34	0.88	0.14	0.14	19.2	792	608	76
		std	0.05	0.2	0.05	0.1	0.01	0.67	0.03	0.08	0.30	0.06	0.01	0.02	0.4	32	55	8
Everett	2	ave	0.76	19.7	4.57	45.6	0.56	3.40	1.43	0.57	2.63	0.45	0.58	0.31	19.2	847	1160	234
		std	0.79	2.0	1.79	0.6	0.05	0.01	0.39	0.14	0.68	0.11	0.18	0.05	0.1	4	260	79
Good Question	1	n/a	1.96	10.8	5.99	53.9	0.72	4.34	1.53	0.42	3.85	0.73	0.48	0.20	14.8	996	1030	245
Torquas	1	n/a	1.93	12.6	8.82	43.8	1.08	4.85	1.25	1.78	4.85	0.98	0.24	0.32	16.9	1980	1750	295

Table 11.2 Average Compositions and Standard Deviations of Rock and Soil Classes Defined from Gusev Crater *Continued*

Class	n[a]		Na₂O wt%	MgO wt%	Al₂O₃ wt%	SiO₂ wt%	P₂O₅ wt%	SO₃ wt%	Cl wt%	K₂O wt%	CaO wt%	TiO₂ wt%	Cr₂O₃ wt%	MnO wt%	FeO wt%	Ni µg/g	Zn µg/g	Br µg/g
Soils																		
Laguna	26	ave	3.08	8.39	10.2	46.2	0.94	6.43	0.74	0.47	6.25	0.92	0.31	0.31	15.7	459	304	35
		std	0.21	0.22	0.6	0.8	0.07	0.86	0.10	0.02	0.15	0.11	0.03	0.02	0.6	95	77	11
Boroughs	4	ave	2.47	9.63	8.61	42.1	0.76	**10.7**	0.85	0.35	5.78	0.88	0.39	0.34	17.0	469	345	**119**
		std	0.07	0.66	0.59	2.3	0.05	2.6	0.17	0.01	0.05	0.02	0.03	0.01	0.4	28	94	76
Paso Robles	13	ave	***1.06***	***5.34***	***4.81***	**30.5**	**1.44**	**30.8**	***0.41***	***0.20***	6.21	***0.49***	0.46	***0.19***	**18.0**	580	**153**	127
		std	0.48	2.17	0.94	4.6	1.67	3.0	0.15	0.12	1.51	0.17	0.16	0.05	2.3	223	42	160
Gertrude Weise	3	ave	***0.32***	***2.18***	***1.77***	**90.6**	***0.31***	***1.05***	***0.15***	***0.01***	***0.71***	1.23	0.33	***0.03***	***1.32***	***154***	278	***15***
		std	0.01	0.29	0.10	0.5	0.01	0.06	0.01	0.00	0.06	0.03	0.02	0.00	0.22	3	6	3
Eileen Dean	2	ave	***1.36***	**15.8**	***6.08***	49.5	***0.67***	**3.14**	**1.66**	0.42	**3.82**	***0.50***	**0.53**	0.24	16.1	**717**	**1070**	**122**
		std	0.62	1.0	0.68	0.6	0.02	0.13	0.31	0.01	0.30	0.05	0.01	0.04	0.1	6	10	38

Key differences in rock classes from Adirondack-class basalts, and soil classes from Laguna-class soils, when high are indicated in bold font, and when low are indicated in bold-italic font.
[a]Number of measurements averaged.

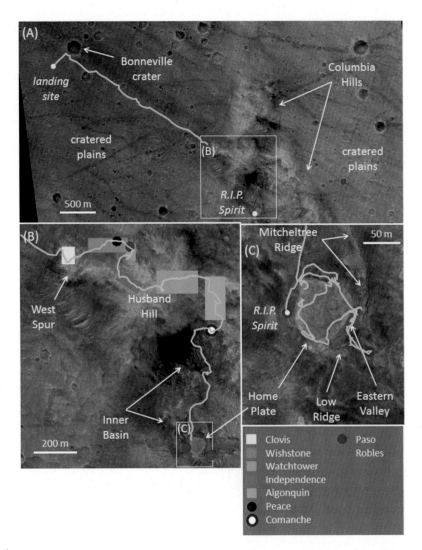

FIGURE 11.1

Location images from HiRISE image PSP_005456_1650_RED with schematic rover track (after Arvidson et al., 2006; 2008; 2010) showing details for the cratered plains (A), West Spur-Husband Hill-Inner Basin (B), and Home Plate-Eastern Valley (C) regions. Generalized locations of West Spur-Husband Hill rock classes (after McCoy et al., 2008), and Paso-Robles-class soils (after Arvidson et al., 2010; Yen et al., 2008).

basalts or the cratered plains soils. In Table 11.2, the key differences from these reference classes for each defined class are highlighted.

Geographic names used in this chapter include formally accepted names and informal designations used by the mission team to facilitate discussion of operations and science targets. Fig. 11.1 shows the region explored by Spirit, the geographic features discussed in this chapter, and the regions where some of the rock classes were encountered.

11.2 THE APXS DATASET

The APXS determines chemical compositions of rocks and soils using X-ray spectroscopy after irradiation with energetic alpha particles and X-rays. It therefore resembles a combination of the standard laboratory methods of X-ray fluorescence spectrometry (XRF) and particle induced X-ray emission spectrometry (Rieder et al., 2003). The typical sample field of view has a diameter of about 35 mm. Concentrations are extracted from the X-ray spectra using the empirical method described in Gellert et al. (2006). The areas of the characteristic peaks of each element are determined with a nonlinear least-squares-fit algorithm, and the peak areas are then quantified into elemental concentrations using the calibration sample set for MER, comprised of about 50 geological reference materials and additional simple chemical compounds (cf., Gellert et al., 2006; Rieder et al., 2003). For each element, the typical oxide—for example, SiO_2 for quantified Si, MgO for Mg—is assumed. The sum of all oxides is normalized to 100% to compensate for a variable stand-off distance, and Fe is reported as FeO. In the analysis model, self-absorption is taken into account using the assumption of a homogeneous, glass-like sample. This assumption is probably never correct and is the underlying reason for the lower accuracy compared to analyses of glass disks in standard XRF. The absorption of the emitted X-rays, especially for lower Z elements that come from depths of only a few micrometers, depends on the composition of the host phase. Of necessity, absorption corrections for the APXS data use the average sample composition.

The results are reported with uncertainties for each element that represent 2σ precision errors of the peak areas (e.g., Gellert et al., 2006; Ming et al., 2008a). Precision uncertainties are well suited to judge the similarity of samples rather than using the larger accuracy errors and can be used to group rocks by their similar composition. The rocks likely share a similar mineralogy, and therefore, any inaccurate corrections in the APXS analysis stemming from the microscopic heterogeneity would be minimized for these rocks. The validity of using precision error bars for comparing and grouping rocks in classes is justified by the near identical and consistent composition of fine-grained, homogeneous igneous rocks like the Adirondack basalts. The relative large accuracy error bars can be explained by the very different compositions of possible minerals. For example, two possible Cl-rich minerals include NaCl and $NaClO_4$, where the difference in oxygen causes substantial differences in the absorption cross-sections that are needed for accurate correction. Independent knowledge of the mineralogy of the targets would be required to improve the accuracy of analyses.

In addition to deriving elemental concentrations through analysis of the emitted X-rays, the Compton/Rayleigh backscattered peak ratio of a spectrum depends importantly on the abundances of light elements such as C and O (Campbell et al., 2008; Rieder et al., 2003). This ratio can be used to infer the abundance of light-element compounds (H_2O, CO_2, etc.) in cases where other data, for example, a Mini-TES spectra showing a bound water emission feature, can be used to constrain which light-element compounds are present.

11.3 GEOLOGICAL CONTEXT

Gusev crater was formed in materials interpreted to be of Noachian age. The immediate area surrounding Gusev crater on all but the southern rim is mapped as the Early Noachian highland unit representing undifferentiated impact, volcanic, fluvial, and basin materials, and the area of the southern rim is the Middle Noachian highland unit consisting similarly of undifferentiated materials (Tanaka et al., 2014). The age of Gusev crater is Early to Middle Noachian (Kuzmin et al., 2000), while the floor is of Early Hesperian age (Tanaka et al., 2014). This time span covers the alteration eras Phyllosian to the Theiikian, which are thought to be separated by a period of global climate change that occurred in the Late Noachian (Bibring et al., 2006).

The cratered plains proved to be of volcanic origin, not lacustrian as mapped prior to landing (Squyres et al., 2004). The bedrock is dark, fine-grained, olivine basalt (McSween et al., 2004, 2006a) covered by an impact-generated regolith (Golombek et al., 2006a). The basaltic layer must be at least 10 m thick, the depth plumbed by the largest craters near the rover traverse, as no other rock types are evident in the regolith (Golombek et al., 2006a; Squyres et al., 2004). Orbital observations indicate that most of the surface of Gusev crater consists of volcanic bedrock and regolith (Martínez-Alonso et al., 2005). The basaltic plains lap onto the Columbia Hills (Fig. 11.1) and thus are younger (Crumpler et al., 2005). Crater counting places the age of the volcanic plains at 3.65 Ga (Greeley et al., 2005; Parker et al., 2010).

Unlike the uniformity in lithology found on the cratered plains, rocks on the Columbia Hills are diverse in lithology (e.g., Gellert et al., 2006; Ming et al., 2006, 2008a; Morris et al., 2006, 2008), and distinct rock units are mapped (Crumpler et al., 2011). The rock units on the Columbia Hills have dips that are generally conformable with the local topography and are interpreted to be airfall materials, of either volcanic or impact in origin, that were draped over preexisting topography (Crumpler et al., 2011; McCoy et al., 2008). These include the rock units on West Spur, Husband Hill, and the Inner Basin that lies between Husband Hill and McCool Hill (Fig. 11.1). The substrate of these materials is not exposed, but the Columbia Hills are thought to be part of the ancient central peak or ring of the crater and to consist of breccias and deformed uplifted crustal materials, or possibly breccias formed on rims of overlapping smaller craters (Crumpler et al., 2011; McCoy et al., 2008; Parker et al., 2010). Stratigraphy thus indicates that the Columbia Hills core was first draped by a sequence of airfall units and then encroached upon by the olivine basalts of the plains. An alternative interpretation of the geology of the Columbia Hills is that they represent a Noachian-aged layered intrusion brought up as part of the central uplift resulting from formation of Gusev crater (Francis, 2011). However, this interpretation is based on compositions and iron mineralogy determined by the APXS and MB, and is not informed by geologic mapping, Pancam 13f or Mini-TES spectroscopy, or microscopic textures.

Home Plate is a quasicircular raised feature roughly 80 m in diameter and 1−2 m high located in the Inner Basin of the Columbia Hills (Fig. 11.1). The rocks that form Home Plate are layered, with a lower, coarser-grained parallel-layered unit and an upper, finer-grained, laminated, and crossbedded unit (Squyres et al., 2007). It is interpreted to be a remnant of a pyroclastic surge deposit draped over preexisting topography; the upper unit possibly was reworked by eolian processes (Lewis et al., 2008; Squyres et al., 2007). The lower contact of the basal unit is not exposed, and therefore, the substrate for the deposit remains unknown (Crumpler et al., 2010; Lewis et al., 2008). Thus, the age of Home Plate with respect to the units draping Husband Hill is not established based upon local stratigraphy. However, the draping units on Husband Hill are younger at lower elevations, and the Home Plate units are inferred to overlie the youngest unit on the eastern side of Husband Hill (Crumpler et al., 2011).

11.4 **CRATERED PLAINS**

The rocks that make up the upper >10 m of the cratered plains are olivine basalts of the Adirondack class (McSween et al., 2004, 2006a). The fine-fraction of the regolith developed on the basalts is in turn of basaltic composition and was used to define the Laguna class of soils in Gusev crater (Morris et al., 2006, 2008). Although the cratered plains of Gusev crater are surfaced by volcanics, observations from early in the mission showed that mobilization of some elements by fluids had occurred. The ages of these processes are difficult to pin down, but they are recent in a relative sense as they are observed on blocks in the ejecta blanket from Bonneville crater, the freshest crater on the plains in the region explored by Spirit (Golombek et al., 2006a). Because the cratered plains are Early Hesperian in age, alteration of rock and regolith on the surface occurred within the Theiikian alteration era (Bibring et al., 2006) or later; phyllosilicate-dominated mineralogy is not expected.

11.4.1 **ROCK COATINGS AND VEINS**

The basaltic rocks on the cratered plains exhibit variations in their Mini-TES spectra. These have been interpreted as indicating that four components are present, two of which are possible alteration features (Christensen et al., 2004). One is fairly resistant to removal by brushing and might be a coating or rind possibly composed of oxides. The other is bright and easily removed by brushing but is distinct from the light-toned airfall dust (Christensen et al., 2004). These components are not present in spectra of abraded interiors, suggesting they represent surface alteration. Imaging by the MI of partially abraded surfaces where the angle of the rock surface relative to the RAT plane results in a shallow, beveled abrasion through the surface show evidence for thin coatings consistent with the Mini-TES observations (Herkenhoff et al., 2004; McSween et al., 2006a).

Compositional results show systematic variations of some volatile/mobile elements related to rock surface treatment. Gellert et al. (2004) showed that S and Cl contents were systematically lower in abraded compared to brushed targets (Fig. 11.2A), and the same is true of Zn (Fig. 11.2B). Gellert et al. (2004) and McSween et al. (2004) noted that trends on variation diagrams, such as Mg vs S, are inconsistent with simple mixing of rock and soil. From this, these authors concluded that the high S-Cl component represented an alteration coating. Other volatile/mobile elements do not show systematic trends with surface treatment. For Adirondack and Humphrey, K_2O, Ni, Zn, Ge, and Br are at equal concentrations in brushed and abraded surfaces (Gellert et al., 2004, 2006). The Br contents of the abraded interior target of the rock Mazatzal is higher than on the untreated surface, but Ni, Zn, and Ge are lower (Gellert et al., 2004, 2006). The surface of Mazatzal is distinct from those of Adirondack and Humphrey in that the former holds mineralogical evidence for more extensive alteration. Brushed and abraded surfaces of Adirondack have identical low MAI (7−8) and fraction of total iron as Fe^{3+} (Fe^{3+}/Fe_T: 0.16−0.17), while Humphrey possibly has a slightly more altered brushed surface compared to interior: MAI 10 vs 7; Fe^{3+}/Fe_T 0.19 vs 0.15 (Morris et al., 2004, 2006). In contrast, these indicators for the brushed vs abraded surfaces of the Mazatzal target New York are MAI 32 vs 14; Fe^{3+}/Fe_T 0.36 vs 0.18 (Morris et al., 2004, 2006).

Gellert et al. (2004) inferred that the Br in the Mazatzal abraded target is hosted in the interior alteration zones and veins noted in MI images of these rocks (Herkenhoff et al., 2004). The Cl/Br

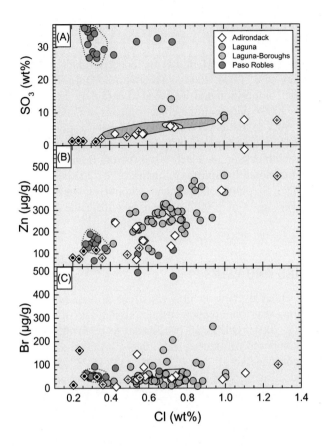

FIGURE 11.2

Volatile/mobile element contents of Adirondack-class rocks, and Laguna-class and Paso-Robles-class soils. *Yellow field*—trench soil targets in Laguna Hollow (see text); *gray field* (in A)—Laguna-class soils; *purple dotted outline*—Paso-Robles-class soils from Troy; Adirondack-class targets that were brushed (*plus sign*) or abraded (*dot*) are indicated.

ratios for these targets are low, and there is no correlation between Br and Cl (Fig. 11.2C), which might indicate formation from residual brines after Cl-rich salt crystallization (Gellert et al., 2004; Haskin et al., 2005). Hurowitz et al. (2006a) concluded that the compositional differences between brushed and abraded Adirondack-class basalt surfaces are consistent with a short period of alteration by acidic solutions at low water/rock ratios. The more soluble major element cations were partially removed from the surfaces, and evaporation of the residual brines resulted in enhancement in the acidic anion elements S and Cl in coatings.

The simplest interpretation of the results on basaltic ejecta blocks is that the alteration rinds and coatings were formed after the Bonneville impact had ejected the blocks to their current resting places. Imaging shows that the surface has deflated by up to 25 cm through removal of regolith fines and eolian abrasion, and that rocks now exposed were initially partially or completely buried in the ejecta deposit (Golombek et al., 2006b). Haskin et al. (2005) conclude that the alteration coatings were formed while the rocks were buried in the regolith under low water/rock ratio conditions. However, Hurowitz et al. (2006a) suggested that the coatings were rather formed in an acid-fog-like environment in which sulfuric acid aerosols collected on exposed rock surfaces and altered them. This could only have taken place after the rock surfaces were exposed by deflation and would have occurred more recently than the process envisioned by Haskin et al. (2005).

Close to Bonneville crater, there are spheroidally weathered and case-hardened rocks mixed in among the angular blocks. The former have been taken as evidence that fluid-mediated alteration might have occurred in the interiors of basalt flows and thus would have predated the Bonneville impact (Crumpler et al., 2005). Rinds and coating on angular blocks might represent preexisting planes of weakness along altered fractures in the basalt flow that influenced breakup of the rock during impact. Alternatively, formation of the unique spheroidally weathered and case-hardened morphologies could have occurred after deposition of angular basalt blocks in the Bonneville ejecta (Crumpler et al., 2005). However, this would not provide a straightforward explanation for why the unique morphologies are interspersed with fresh, angular blocks; the former scenario seems more plausible.

11.4.2 SOIL INDURATIONS

Laguna-class soils are subdivided by MB into several subclasses, but for most subclasses, the differences in iron mineralogy are gradational (Morris et al., 2006, 2008); these are treated together here (Table 11.1). The Laguna-class soils display a positive correlation between their SO_3 and Cl contents (Fig 11.2A). Comparison of lighter toned and darker toned soil targets shows that the former have higher SO_3 and Cl contents, consistent with mixing of bright airfall dust with basaltic sand (Yen et al., 2005). A general positive relationship also exists between Zn and Cl (Fig. 11.2B) suggesting two-component mixing might also be responsible for the Zn variation. Bromine is not correlated with Cl (Fig. 11.2C), and this has been attributed to mobilization by liquid water films that form on soil particles under climatic conditions that might be similar to those of present-day Mars (Yen et al., 2005). Bromine salts are typically more soluble than chlorine salts; Br would be expected to be more mobile than Cl under the low water/rock ratio conditions inferred.

Some soils also show evidence for mobilization of some elements at low temperature in the recent Martian environment. These are the Boroughs subclass soils (Table 11.1). Soil surfaces commonly display surface crusts that are a few mm thick indicating some induration of the soil particles (Arvidson et al., 2004). Three trenches in soils were made on the cratered plains using a wheel of Spirit as a trenching tool, and the APXS was then used to document compositional variations. One trench in Laguna Hollow was dug in drift sand on the margin of the ejecta deposit of Bonneville crater, while two others were dug in older regolith that lies outside the ejecta blankets of the larger, fresher craters (see Crumpler et al., 2005, Fig. 11.1). The Laguna Hollow trench soils have contents of S, Cl, and Br within ranges typical of surface soils (yellow field, Fig. 11.2). The soils from the other two trenches have higher S contents than Laguna-class soils (Haskin et al. 2005). The Big Hole soils are at the high-S end of the Laguna-class soil trends, exhibiting modestly higher S and Cl contents (Haskin et al., 2005). The two Boroughs targets have much higher S and Br contents than Laguna-class soils, but their Cl contents are typical of the latter (Haskin et al., 2005). The Boroughs trench was made furthest from large, fresh caters (see Crumpler et al., 2005). Because of their unique volatile/mobile element contents, the Big Hole and Boroughs soils are defined as the Boroughs subclass of the Laguna-class soils (Table 11.1; Morris et al., 2006, 2008). Haskin et al. (2005) explain the soil compositional data as indicating that brines within the regolith migrated upward following a thermal gradient, depositing a sequence of salts as water was evaporated from the surface.

The timing of these alteration processes cannot be determined. The cratered plains are Early Hesperian in age but Bonneville, the freshest crater, must be substantially younger than this. Because the rock coatings occur on blocks that were exhumed by deflation of Bonneville ejecta, formation of the coatings was occurring relatively late in the geological history of Gusev crater as discussed above. On the other hand, evaporative salts were formed in regolith older than Bonneville crater. Both of these alteration signatures suggest low water/rock ratios (Haskin et al., 2005), and it is likely that these alteration processes took place over an extended time period subsequent to development of regolith on the cratered plains; post-Early Hesperian.

11.5 WEST SPUR-HUSBAND HILL

The West Spur-Husband Hill region contains sundry rock classes, unlike the general uniformity of the basaltic rocks encountered on the cratered plains (Table 11.1). Most of these rocks drape the unexposed core of Husband Hill and are interpreted to be volcanic and impact-generated airfall deposits (Crumpler et al., 2011; Squyres et al., 2006). However, some linear boulder features composed of dark, fine-grained mafic rock (Backstay class) might represent mafic dike intrusions (McSween et al., 2006b). The rocks on the Husband Hill have textures that are distinct and more varied than the fine-grained, homogeneously textured, porphyritic basalts of the cratered plains (Herkenhoff et al., 2006). Using the APXS data, the rocks from the West Spur and Husband Hill region are divided into nine classes (Table 11.1). The generalized locations of the classes discussed here are shown on Fig. 11.1B.

11.5.1 VARIABLE STYLES AND DEGREES OF ALTERATION

Remote sensing and microscopic imaging of the rocks in the West Spur-Husband Hill region show that they are of varied texture and mineralogy, in part a result of variable degrees of alteration. Using spectral analysis and statistical techniques on Pancam 13f images, the rocks on West Spur-Husband Hill can be divided into 11 different spectral classes (Farrand et al., 2006, 2008). One of the spectral classes includes the Independence class, which has uniquely low FeO contents (Farrand et al., 2008; Ming et al., 2006); this class is discussed in Section 11.5.2. Another of the spectral classes is associated with the Comanche outcrop, which is discussed separately in Section 11.5.3. Some of the spectral variation is caused by differences in olivine and pyroxene abundances in the rocks and thus is related to variations in primary mineralogy (Farrand et al., 2008). The 535-nm band depth and the slope from 482 to 673 nm are related to the degree of oxidation of iron; greater band depths and/or steeper slopes indicate greater degrees of oxidation. The Pancam 13f observations show that the degree of oxidation of the rocks in this region is variable (Farrand et al., 2008).

Spectra determined by the Mini-TES also show variations among rocks from West Spur-Husband Hill. After correcting for various observational challenges, six spectral classes are identified among the rocks from West Spur up to approximately the summit region of Husband Hill (Ruff et al., 2006, 2007). Two of these are basaltic rocks that occur scattered on Husband Hill as a minor component of the rock suite. The other four classes are Clovis, Wishstone, Watchtower, and Peace (Table 11.1). The Peace class is unique among these in that it has an emissivity peak at roughly $1630 \, \mathrm{cm}^{-1}$ ($\sim 6.1 \, \mu\mathrm{m}$) that is interpreted to be due to bound molecular water, likely in a hydrated sulfate (Ruff et al., 2006). The Mini-TES spectra of the other three classes are well modeled as being dominated by primary igneous phases (some combination of plagioclase, pyroxene, olivine), basaltic glass with or without high-silica glass, and sulfate (Ruff et al., 2006, 2007). The Clovis-class rocks have the strongest signature of alteration phases with goethite and secondary silicates making up perhaps as much as 30% of the rock (Ruff et al., 2006, 2007). Although glass is required in mixing models to achieve a good fit to the Mini-TES spectra, this does not mean that the glass is pristine. At the time when the modeling was done, there was insufficient knowledge of the spectral characteristics that might be expected of altered glass. This issue is discussed at length in Ruff et al. (2006). In fact, the Mössbauer spectra, high MAI and Fe^{3+}/Fe_T of Clovis-class rocks, discussed below, indicate that Fe-bearing glass, pristine or not, is unlikely.

The compositional data have been used to divide the rocks from West Spur and Husband Hill into nine classes (Ming et al., 2006, 2008a; see Table 11.1). By combining rock compositions with Mössbauer-derived iron mineralogy and oxidation state, some of these classes have been further broken down into subclasses (Ming et al., 2006, 2008a,b; Morris et al., 2006, 2008). The discussion here is at the class level. As mentioned above, the Comanche subclass of the Algonquin class has a unique volatile-bearing mineralogy (Morris et al., 2010) and is discussed separately. The MB-derived iron mineralogy is divided into two broad groupings: primary igneous and alteration phases (Morris et al., 2006, 2008, 2010). The alteration phases are nanophase ferric oxide, hematite, goethite, Fe-sulfate, FeS_2, and Fe-bearing carbonates, and the primary igneous phases are olivine, pyroxene, ilmenite, chromite, and magnetite. Magnetite can also be formed during alteration, but is assumed to be primary (cf., Morris et al., 2006). Similarly, FeS_2 can be a minor/trace primary phase in Martian magmatic rocks (cf., McSween and Treiman, 1998), but because it occurs as a major iron-bearing phase in only one rock, Fuzzy Smith, that does not have a plausible magmatic composition (Table 11.2), it is considered an alteration phase (Morris et al., 2008). Table 11.1 includes the MAI for the rock and soil classes.

Mössbauer spectra show that three of the rock classes on Husband Hill are dominated by little-altered igneous phases: Algonquin, Backstay, and Irvine. The fractions of total iron as Fe^{3+} (Fe^{3+}/Fe_T) in Algonquin and Backstay are similar to that of the Adirondack class and their iron mineralogy predominately consists of igneous phases; the MAI of Algonquin and Backstay are 15 compared to an average for the Adirondack class of 11 (Table 11.1). The only iron-bearing alteration phases detected in Backstay are nanophase ferric oxide and hematite, and these are plausibly contained within a thin weathered zone (Morris et al., 2008). The Irvine class has higher Fe^{3+}/Fe_T than Adirondack class basalts, but that is because of a high proportion of magnetite, which is plausibly a magmatic phase. The Irvine class has an MAI of only 7 (Table 11.1) contributed by nanophase ferric oxide and hematite (Morris et al., 2008). Algonquin and Backstay targets have shallow 535-nm band depths in Pancam 13f images indicating a minor component of iron-oxide phases (Farrand et al., 2008), consistent with the MB results.

The compositions of the Backstay and Irvine classes, when recast as CIPW normative mineralogy following the assumptions of Ming et al. (2006), have high normative diopside, similar to the

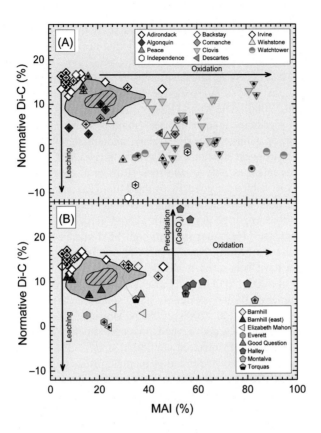

FIGURE 11.3

Normative diopside-corundum vs MAI for rocks from West Spur/Husband Hill (A) and the Home Plate/Eastern Valley (B) regions. *Gray field*—Laguna-class soils; *crosshatched* area—75% of Laguna-class soils; targets that were brushed (*plus sign*) or abraded (*dot*) are indicated. See text for explanation of y axis scale.

contents of abraded and brushed Adirondack-class basalts (Fig. 11.3A). A key assumption in the calculation is that the halogens and S were added as acidic anions and did not change the cation contents of the rocks except by dilution, that is, an acid-fog-type of alteration. High normative diopside (Di) is a hallmark of pristine basaltic rocks (Ca in excess of that needed to combine with Al in plagioclase), while alteration under high water/rock ratio conditions causing extensive leaching of the more soluble elements results in corundum normative (C) mineralogy (Al in excess of $Ca+Na+K$). Normative diopside and corundum cannot both occur in a recast bulk composition; positive values of Di-C in Fig. 11.3 indicate normative diopside, negative values indicate normative corundum. The low MAI and high normative diopside of the Backstay and Irvine classes thus indicate only minimal alteration occurred in these rocks. The Backstay and Irvine classes also have mean SO_3, Cl, and Br contents that are similar to those of Adirondack basalts, but their Zn contents

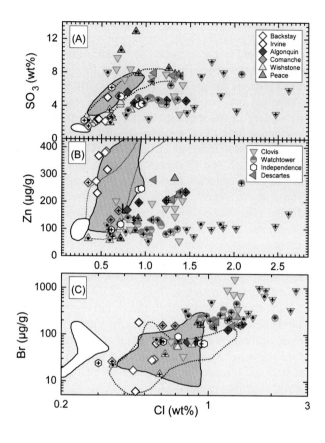

FIGURE 11.4

Volatile/mobile element contents of rock from West Spur and Husband Hill. *White field*—abraded Adirondack-class rocks; *gray field*—Laguna-class soils; *dotted outline*—untreated and brushed Adirondack-class rocks; targets that were brushed (*plus sign*) or abraded (*dot*) are indicated. Note that panel C is log-log scale.

are substantially higher (Table 11.1); note that the Adirondack class mean composition is based on abraded targets while those of Backstay and Irvine are not (cf., Ming et al., 2008a). The Backstay- and Irvine-class rocks fall within the fields for untreated and brushed Adirondack-class rocks for SO_3, Cl, Zn, and Br (Fig. 11.4). Thus, the volatile/mobile element contents of the Backstay and Irvine classes are consistent with inferences from spectroscopy and major element compositions that these rocks are minimally altered.

Algonquin-class rocks have lower normative diopside than do the Adirondack, Backstay, and Irvine classes (Fig. 11.3A). This does not necessarily indicate a higher degree of alteration for this class. The Algonquin class is more ultramafic in composition, and even pristine rocks would have lower normative diopside (cf., Mittlefehldt et al., 2006). A higher proportion of the bulk Fe of Algonquin-class targets is in olivine compared to the Adirondack class (Morris et al., 2008). The Algonquin class has higher SO_3 and Cl contents than the other three magmatic classes

(Table 11.1). Several of the Algonquin-class targets were brushed, and this operation generally resulted in an increase in Cl with a possible slight decrease in SO_3 (Fig. 11.4A). This change drives the compositions away from the field of Laguna-class soils indicating some removal of contaminating soil. The brushed targets also have generally higher Zn and Br contents (Fig. 11.4B and C).

Algonquin-class rocks have clastic textures composed of poorly sorted, clast-supported angular to more rounded grains, although one target is more massive (Yingst et al., 2007). Adirondack, Backstay, and Irvine class rocks are all massive, fine-grained basalts (McSween et al., 2004, 2006a,b; McSween et al., 2008). Penetrating, light-toned veins on abraded surfaces of Adirondack class basalts show late introduction of volatile/mobile elements along cracks (Gellert et al., 2004; McSween et al., 2004, 2006b). The generally higher SO_3 and Cl contents of Algonquin-class rocks might indicate that the more open texture allowed these elements to penetrate the clastic matrix as brines, which then precipitated out as salts. Thus, the Algonquin class could be cemented clastic rocks. Substantial alteration of primary phases did not occur during this process as indicated by the dominance of primary igneous minerals in the MB spectra.

There is variation in average clast size within the Algonquin class (Yingst et al., 2007), but all have the same SO_3 content (Fig. 11.4A) suggesting that, if the above scenario is correct, there was not a substantial clast-size-related variation in porosity within the class. There is some variation in Cl content among the brushed Algonquin targets (Fig. 11.4A), but this does not vary with average clast size. Because we did not abrade any Algonquin-class targets, the higher volatile/mobile element contents could simply indicate that a thin coating rich in these elements is present. The volatile/mobile element signature of the Algonquin class is consistent with low temperature interaction with brines as is inferred for the rock coatings and indurated soils on the cratered plains. However, an acid-fog environment such as posited for formation of coating on Adirondack-class basalt (Hurowitz et al., 2006a) could plausibly explain the data.

The process just outlined for the Algonquin class is similar to that inferred for formation of the Peace-class rocks. These latter are finely laminated clastic rocks composed mostly of fine granules a few hundred microns in size bound by a cementing agent (Squyres et al., 2006). The Pancam 13f spectra of Peace-class rocks are most similar to those of the Adirondack class. They have generally flat spectra in the 482 to 673-nm regions and no appreciable 535-nm absorption indicating a low degree of oxidation (Farrand et al., 2008). The iron mineralogy of Peace-class rocks is dominated by olivine, pyroxene, and magnetite; the MAI of this class is 14 with nanophase ferric oxide being the only alteration phase (Morris et al., 2008). Analyses by the APXS were done on untreated, brushed, and abraded Peace-class targets. The untreated surfaces have the highest halogen and Zn contents while the abraded targets have the lowest concentrations (Fig. 11.4). The untreated and brushed surfaces have essentially identical SO_3 contents, but the abraded interior targets have roughly 50% higher SO_3 (Fig. 11.4A). These are the highest SO_3 contents of all rocks from Gusev crater. The high SO_3 contents are coincident with high MgO contents of $\sim$20 wt%. The Peace-class rocks are thought to have formed through cementation by Ca- and Mg-sulfates of relatively unaltered ultramafic sands composed of igneous phases (Ming et al., 2006; Squyres et al., 2006). The CIPW normative mineralogy of Peace targets has similar diopside contents to those of the Adirondack class (Fig. 11.3A), but this is likely because of added Ca in the sulfate cement. The sulfates are thought to have precipitated from an aqueous solution that likely was not very acidic. The Mini-TES spectra show that some of these sulfates are hydrated. The cementation was plausibly a low temperature process.

Clovis- and Watchtower-class rocks are among the most altered on West Spur and Husband Hill (Ming et al., 2006, 2008b). They have class-average MAI of 63−64 (Table 11.1) and include different combinations of the alteration phases: nanophase ferric oxide, hematite, and goethite (Morris et al., 2008). Although glass is a component in the modeled Mini-TES spectra of Clovis-class rocks (Ruff et al., 2006, 2007), Mössbauer spectra do not provide evidence for Fe-bearing, basaltic composition glasses in these rocks (Morris et al., 2008). The degree of alteration is variable within these two classes. The range in MAI for individual Clovis-class targets is from 45 to 83, and for Watchtower-class rocks is from 37 to 94 (Morris et al., 2008). Many of the Clovis-class rocks have high normative diopside, although most of the abraded targets have very low Di-C (Fig. 11.3A). While this might suggest that only modest leaching occurred in some Clovis rocks, those with high Di-C also have the highest SO_3 and CaO contents. A simpler explanation is that $CaSO_4$ was added to some of the Clovis rocks during alteration. Clovis-class rocks were found only on West Spur of Husband Hill where they dominate (Fig. 11.1B). These rocks are poorly sorted clastic rocks composed of angular to rounded grains up to $\sim$1 cm in size (Herkenhoff et al., 2006; Squyres et al., 2006). Watchtower-class rocks are found only near the summit of Husband Hill (Fig. 11.1B). They are diverse in texture but are generally fine-grained and have knobby protuberances and sub-mm wide veins visible on brushed surfaces (Herkenhoff et al., 2006; Squyres et al., 2006). Watchtower-class rocks have low Di-C indicating a substantial loss of soluble elements during alteration that is not correlated with the degree of oxidation as indicated by the MAI (Fig. 11.3A). Because of similarities in some unusual compositional characteristics, for example, high P_2O_5 and TiO_2, the Wishstone class is thought to be genetically connected to the Watchtower class (e.g., Hurowitz et al., 2006b; Squyres et al., 2006). The Wishstone class has an MAI intermediate between those of the Adirondack class and the Clovis and Watchtower classes (Table 11.1; Morris et al., 2008), consistent with mineralogical modeling of Wishstone-class compositions (Ming et al., 2006). Wishstone-class rocks are poorly sorted clastic rocks containing angular to rounded grains set in a fine-grained matrix (Herkenhoff et al., 2006; Squyres et al., 2006).

The Clovis class has wide ranges in volatile/mobile element contents, even if only abraded targets are considered (Fig. 11.4). This is consistent with the wide range in MAI for these rocks that indicates quite differing degrees of alteration. Halogen element contents are higher in brushed targets, which plot away from the field of Laguna-class soils (Fig. 11.4C). This indicates that the compositions of untreated targets are mixtures of rock surfaces and partial soil cover. The SO_3 and Zn contents do not change much with brushing, indicating the Clovis class has similar contents of these elements to those of soils. Note that on average, the abraded Clovis-class rocks have lower halogen contents than do the brushed targets (Fig. 11.4C), and indeed, the highest halogen contents measured for rock targets in Gusev crater are on brushed Clovis-class rocks. This suggests that halogen-rich coatings are present. These are plausibly relatively recent additions to the surface enabled by films of water forming, mobilizing the most soluble phases and reprecipitating salts on surfaces. The abraded interiors of Clovis-class rocks have higher SO_3 and halogen element contents than do Adirondack-class basalts from the cratered plains. The volatile/mobile element contents of the Watchtower class are similar to those of the Clovis class, but the class displays less variability in these elements (Fig. 11.4). Only one Watchtower-class target was abraded, and it has the lowest SO_3 and Cl contents of the class. Wishstone-class rocks have on average lower contents of SO_3 and halogens than either the Clovis or Watchtower class, but there is no appreciable difference in Zn (Fig. 11.4). One of the two abraded Wishstone targets has volatile/mobile element contents approaching the low concentrations found for abraded Adirondack-class basalts. The major element compositions of Clovis-class rocks are not plausible Martian basalt compositions because they are corundum normative (Fig. 11.3A, see Ming et al., 2006). This suggests alteration of primary phases and removal of more soluble elements leaving an Al-enriched residue. Ming et al. (2006) modeled the compositions of Clovis-class rocks using Pancam, Mini-TES, and MB spectroscopy results to constrain the phases present. Based on the modeled secondary mineralogy, Ming et al. (2006) concluded that the class is mostly composed of secondary aluminosilicates, sulfates, and secondary ferric oxides.

Clovis-class rock outcrops are variable in morphology from massive with little structure evident to rocks with distinctive fine-scale, plane-parallel fabric and are draped over preexisting topography (Crumpler et al., 2011; Squyres et al., 2006). The Clovis class consists of poorly sorted clastic rocks with angular to rounded clasts ranging in size from the resolution limit of the MI ($\sim 100\,\mu m$) to roughly 1 cm (Herkenhoff et al., 2006; Squyres et al., 2006). Based on outcrop morphology, rock texture, mineralogy, and composition, the Clovis class is interpreted to be an airfall deposit of impact ejecta that was subsequently highly altered (Squyres et al., 2006). A scenario developed by Ming et al. (2006) for the alteration on West Spur that engendered the Clovis class is that SO_2-, HCl-, and HBr-rich volcanic vapors and/or fluids interacted with impact-ejecta deposits. Acidic solutions with a pH of 0−1 (e.g., Hurowitz et al., 2006a) diffused into the outcrops, reacted with the primary phases, and altered them to Ca-, Mg-, and other sulfates, halogen salts, iron oxides/oxyhydroxides, and secondary aluminosilicates. Although initially acidic, the solutions were substantially neutralized by basalt dissolution. The final solutions had a pH high enough to initiate hydrolysis of Fe to form Fe-oxides or oxyhydroxides. Sufficient water was present to facilitate removal of the more soluble elements leaving a corundum normative residue (Ming et al., 2006; Squyres et al., 2006). The temperature at which this alteration occurred is not well constrained, but Ming et al. (2006) note that the alteration phases they model for the Clovis class are similar to those produced by hydrolytic and sulfatetic alteration of basaltic tephra on Earth under hydrothermal conditions.

The compositions of the Wishstone and Watchtower classes show some similarities and are distinct from other rock classes on Husband Hill (e.g., Gellert et al., 2006; Ming et al., 2006). They are distinct from the Adirondack-class basalts in having higher Al, Ti, and P contents, and lower Cr contents (Table 11.2; Ming et al., 2006). The highest P_2O_5 contents measured among Gusev crater rock targets (4.5−5.2 wt%) are for some Watchtower- and Wishstone-class rocks. The most notable distinction between Wishstone and Watchtower rocks in the nonvolatile/nonmobile elements is higher MgO in the latter (Table 11.2; Ming et al., 2006). The compositions of these two classes can be modeled as representing a mixing relationship between a Wishstone-like end member and a component that has no identified lithological analog among the rocks on Husband Hill (Hurowitz et al., 2006b). The mixing relationships are interpreted by Hurowitz et al (2006b) with a model in which a Wishstone-like deposit interacted with an oxidized Mg-Zn-SO_3-Cl-Br rich fluid at a pH in the range 4−9. This resulted in alteration of the Wishstone-class protolith with concomitant enrichment in elements derived from the fluid in the Watchtower class (Fig. 11.5). Hurowitz et al. (2006b) did not discuss the possible thermal regime for this process, but Ming et al. (2006) suggested the alteration process was like that which produced the Clovis class.

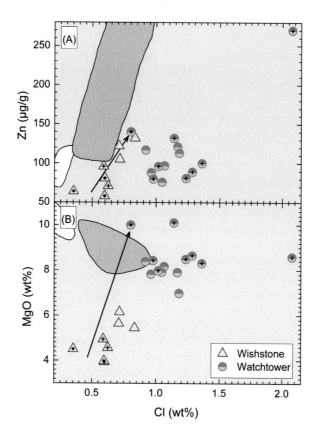

FIGURE 11.5

Zn vs Cl and MgO vs Cl for Wishstone and Watchtower classes from Husband Hill. *Arrows* show suggested mixing relationship for these classes based on abraded targets (see text). *White field*—abraded Adirondack-class rocks; *gray field*—Laguna-class soils; targets that were brushed (*plus sign*) or abraded (*dot*) are indicated.

11.5.2 INDEPENDENCE CLASS

Independence-class outcrops are located on the western flank of Husband Hill that was investigated as Spirit ascended toward the summit (Fig. 11.1B). Pancam 13f spectra of the Independence class show a "negative" 535-nm band depth, that is, the relative reflectance at 535 nm is greater than the value interpolated from the relative reflectances at 432 and 673 nm (Farrand et al., 2008). The absence of absorption at 535 nm indicates that crystalline ferric iron phases are not present. This is borne out by MB spectra that do not require hematite in order to model the data (Morris et al., 2008). The class has moderate MAI of 38 (Table 11.1), with nanophase ferric oxide as the only Fe-bearing alteration phase (Morris et al., 2008). Independence-class targets show a range in Mini-TES spectra that bespeaks a range in mineralogical make up. Modeling of the spectra indicates that an amorphous aluminosilicate component dominates them, but for some of the targets, the model fits were rather poor (Clark et al., 2007).

The compositions of Independence-class rocks are characterized by exceptionally low total Fe contents (3.8−8.1 wt% FeO compared to 18.7 wt% for the average Adirondack-class basalt, cf., Table 11.2) and high Al, Si, P, and Ti contents (Clark et al., 2007; Ming et al., 2008a). Independence-class rocks are corundum normative ((Fig. 11.3A), an indicator that their compositions have been changed from any potential magmatic precursor composition (cf., Ming et al., 2006, 2008b). The more Al-rich compositions plot in the field of altered terrestrial basalts on a ternary FeO_T + MgO, Al_2O_3, CaO + Na_2O + K_2O diagram that is used to describe the effects of extensive open hydrologic system alteration (cf., Nesbitt and Wilson, 1992). Nevertheless, the Mössbauer spectra demonstrate that primary igneous minerals—pyroxene, ilmenite, and/or chromite—contain 44%−68% of the total Fe in the rocks (Morris et al., 2008). Thus, the Independence class must contain highly altered material and relic igneous minerals (cf., Ming et al., 2008b). Clark et al. (2007) attempted to calculate the composition of that portion of one of the Independence-class rocks that remains after deconvolving the bulk analysis into known (MB mineralogy) or inferred (based on element−element correlations) phases. They found that the remainder amounted to about 80% of the rock and was rich in Al and Si with an Al/Si ratio like that of terrestrial montmorillonites. Modeling of the Mini-TES spectra does not support the presence of significant (>5%) crystalline phyllosilicates; thus, Clark et al. (2007) suggested that the alteration material was an amorphous form that either did not mature into crystalline phyllosilicate or was structurally destroyed after formation. These authors did not put tight constraints on the alteration process but did favor alteration by aqueous fluids with a high enough water/rock ratio to mobilize and leach more soluble elements leaving a corundum-normative residue. Ming et al. (2008a) concluded that the Independence class was leached under more neutral pH conditions than typical for Husband Hill alterations such that Al was enriched and a smectite-like composition was formed in an open hydrologic system.

11.5.3 COMANCHE AND CARBONATE FORMATION

Comanche is a rock outcrop on the eastern side of Husband Hill that was investigated as Spirit descended toward the Inner Basin (Fig. 11.1B). It was originally thought to be a lower member of a putative mafic-ultramafic sequence from that area (Mittlefehldt et al., 2006). In this sequence, compatible elements in igneous systems (Mg, Cr, Ni) increased and igneous incompatible elements (Al, P, Ca, Ti) decreased systematically with descent along the rover track (Ming et al., 2008a; Mittlefehldt et al., 2006) consistent with traversing down a layered mafic-ultramafic body. However, current understanding of the geology of the Columbia Hills shows that the stratigraphic sequence is inverted relative to topography; the ultramafic rocks are stratigraphically higher than the mafic rocks (Crumpler et al., 2011). The iron mineralogy of all the rocks is dominated by olivine + pyroxene and Comanche was found to have the lowest olivine/pyroxene ratio (Morris et al., 2008). Pancam 13f spectra are consistent with a higher pyroxene content in Comanche targets (Farrand et al., 2008). However, the octahedrally coordinated Fe^{2+} doublet identified as pyroxene in Comanche MB spectra has distinctive parameters that early on raised doubts about its mineralogical identification (Morris et al., 2008).

Subsequent effort to understand the MB spectra of Comanche showed that the unusual Fe^{2+} doublet is instead well matched by natural and synthetic Fe-bearing carbonates (Morris et al., 2010). Except as noted, the discussion that follows is taken from that paper. Mini-TES spectra of

Comanche-class outcrops are well fit as a mixture of only three phases: magnesian olivine; an amorphous silicate of basaltic composition; Mg-Fe carbonate making up ~34% of the scene. The APXS spectra of Comanche targets include photon scattering peaks that are due to light elements in the target. Analysis of these peaks, assuming all of the excess light element is C, results in 12 ± 5 wt% CO_2 in Comanche, which translates to 16–34 wt% of a carbonate phase. Using MB constraints on Fe speciation, the Comanche composition can be modeled as a three-component mix of ~41 wt% magnesian olivine, ~26 wt% Mg-Fe-carbonate, and ~33 wt% residual silica- and ferric-iron-rich material. The carbonates so calculated are similar in composition to those in Martian meteorite Allan Hills (ALH) 84001 (e.g., Mittlefehldt, 1994). By analogy with laboratory experiments that form carbonates like those in ALH 84001 (Golden et al., 2000), Morris et al. (2010) consider that the likely formation mechanism for the carbonates was at hydrothermal conditions with the solutions having near-neutral pH. Comanche targets have volatile/mobile element contents within the range of those of Laguna-class soils (Fig. 11.4) indicating that the carbonate enrichment was not accompanied by significant enrichments of other volatile/mobile elements.

The timing of the hydrothermal activity is not well constrained other than it is postdeposition of the Comanche rocks onto Husband Hill, and prior to flooding of Gusev crater by olivine basalts—that is, prior to Early Hesperian (cf., Crumpler et al., 2011). Morris et al. (2010) note that hydrothermal activity in the Home Plate region is dominated by acid−sulfate solutions. Following Bibring et al. (2006), this would suggest that carbonate formation in Comanche predates hydrothermal activity around Home Plate (discussed below), with the latter occurring after the global change in alteration conditions and the former before. Morris et al. (2010) suggest hydrothermal activity at Comanche occurred during the Noachian.

An alternative view of carbonate formation in Comanche has been offered by Ruff et al. (2014). The texture of Comanche consists of well-rounded, well-sorted grains in a cemented matrix (Yingst et al., 2007). The rounded grains are similar to accretionary lapilli found in terrestrial volcaniclastic deposits, and Comanche is interpreted to be a phreatomagmatic tuff deposit (Yingst et al., 2007). Ruff et al. (2014) interpret the textures of Algonquin and Comanche as being tephras, and the uniformity in inferred olivine compositions is consistent with formation in a single eruption event. The Mini-TES spectra of Comanche can be well modeled as a mixture of Algonquin rock, Mg-Fe-carbonates and amorphous silica. The best fit is achieved if several carbonates—end member magnesite and siderite, plus intermediate Mg-Mn-Fe-carbonates—are used. This is similar to the range of carbonate compositions found for Martian meteorite ALH 84001 (cf., Corrigan and Harvey, 2004; Mittlefehldt, 1994). Modeling of mildly acidic aqueous fluids in equilibrium with Algonquin rock at T below about 50 °C and buffered by a CO_2-rich atmosphere (pCO_2 of 100 mbar) results in fluid compositions rich in Fe, Mg, Si and bicarbonate. Solutions like these derived by leaching Algonquin-like precursors elsewhere could precipitate the observed suite of Mg-Fe-carbonates and amorphous silica present in Comanche. Ruff et al. (2014) note that rocks of similar morphology and thermophysical properties are scattered in the region surrounding the Columbia Hills, and they posit that these rocks represent remnants of a wide-spread Algonquin-like tephra layer. Periodic flooding of Gusev crater via Ma'adim Vallis generated ephemeral lakes that enabled carbonate formation in the Comanche outcrop. Algonquin-class outcrops were encountered at higher elevation on Husband Hill than was the Comanche outcrop (Fig. 11.1B, see Crumpler et al., 2011); the Ruff et al. (2014) scenario would suggest a shoreline was crossed during the rover traverse although one has not been identified in rover imagery.

11.6 INNER BASIN

The Columbia Hills are a collection of several hills with intervening topographic lows. The region between Husband Hill and McCool Hill to the southeast is referred to as the Inner Basin (Fig. 11.1). After summiting Husband Hill, Spirit was commanded to descend the eastern slope into the Inner Basin with a goal of reaching a light-toned, quasicircular feature visible in orbital imaging (Fig. 11.1). This feature, named Home Plate, was visible in Pancam imaging from the summit of Husband Hill. An extensive campaign of remote and in situ observations in the Inner Basin was done. Here, we focus on the Home Plate structure and kindred volcanics, and silica-rich deposits near Home Plate in a region referred to as the Eastern Valley (Fig. 11.1).

11.6.1 HOME PLATE—HYDROTHERMAL ALTERATION

The units forming Home Plate are interpreted to be hydrovolcanic deposits formed as a result of a phreatomagmatic explosion (Lewis et al., 2008; Squyres et al., 2007). This interpretation is based on the overall basaltic composition, microscopic textures that are consistent with granules being accretionary lapilli, high-volatile element contents consistent with condensation from volcanic vapors, and the presence of a bomb sag (Squyres et al., 2007). The explosion might have been triggered when basaltic magma encountered near-surface, briny groundwaters (Lewis et al., 2008; Squyres et al., 2007). The pyroclastic deposits were likely lithified penecontemporaneously with deposition, but if the upper unit was reworked by eolian processes as seems possible, then lithification of that unit occurred later (Lewis et al., 2008). The bomb sag is hosted in the coarser-grained lower unit (Squyres et al., 2007). Bomb sags form in pyroclastic deposits by soft-sediment deformation of ash with a high-intergranular water content (e.g., Fisher and Schmincke, 1984). The morphology of the deformation surrounding the Home Plate bomb sag is consistent with impact into water-saturated ash (Manga et al., 2012).

Remote sensing from orbit and by Spirit shows that there is a systematic difference in visible/near-infrared reflectance between the eastern and western sides of Home Plate (Schmidt et al., 2009). This difference is also established by MB spectra. Of the total Fe present, $\sim 17\%$ is in olivine, $\sim 23\%$ in pyroxene, $\sim 28\%$ in a nanophase ferric oxide, and on the order of 29% is contained in magnetite on the western side (Morris et al., 2008; Squyres et al., 2007). On the eastern side, these percentages are $\sim 1\%$ in olivine, $\sim 39\%$ in pyroxene, $\sim 9\%$ in nanophase ferric oxide, and $\sim 48\%$ in magnetite (Morris et al., 2008). Thus, rocks on the western side of Home Plate have generally higher MAI than those on the eastern side (Fig. 11.3B), and the oxidation state differs, with Fe^{3+} making up $\sim 53\%$ of the total iron on the western side but only $\sim 46\%$ on the eastern side (Morris et al., 2008). Mineralogy determined from Mini-TES spectra do not entirely match determinations from MB spectra largely due to differences in analysis volume (e.g., areal coverage, penetration; see Squyres et al., 2007), but nevertheless, Mini-TES-derived mineralogy shows systematic east-to-west differences (Schmidt et al., 2009).

The rocks on Home Plate have major element compositions that are similar to those of Irvine class basalts (cf., Table 11.2; Ming et al., 2008a; Schmidt et al., 2008; Squyres et al., 2007). These latter are massive to vesicular alkali-rich basalts (McSween et al., 2006b, 2008) that occur on Husband Hill and in the Inner Basin; those close to Home Plate are often vesicular. Some of the

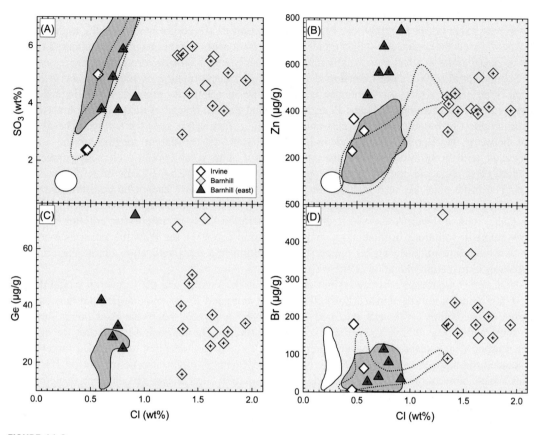

FIGURE 11.6

Volatile/mobile element contents of Barnhill-class and Irvine-class rocks. *White field*—abraded Adirondack-class rocks; *gray field*—Laguna-class soils; *dotted outline*—untreated and brushed Adirondack-class rocks; targets that were brushed are indicated with a *plus sign*.

Barnhill-class rocks from eastern Home Plate have MAI and normative mineralogy like those of Irvine-class basalts, but others have higher MAI and lower normative diopside indicating a greater degree of alteration (Fig. 11.3B). The Barnhill-class rocks from western Home Plate have higher MAI than Irvine-class basalts, but similar normative diopside contents consistent with oxidation but little chemical leaching (Fig. 11.3B). One of the compositional differences between the Irvine-class and Barnhill-class rocks is that the latter have generally higher contents of the volatile/mobile elements S, Cl, Zn, Ge, and Br (Fig. 11.6), although the ranges of their contents overlap to some degree. Germanium and Zn enrichments are observed in terrestrial hydrothermal deposits, and halogenated Ge species are mobilized in hydrothermal systems (Bernstein, 1985). The high Cl, Zn, and Ge contents of Barnhill-class rocks (Table 11.1, Fig. 11.6) are consistent with additions to volcanic ashes from hydrothermal sources. There are also differences in the contents of some of the volatile/mobile elements between the east and west sides (Ming et al., 2008a; Squyres et al., 2007). The rocks from eastern Home Plate have higher Si/Mg, Al, Zn, Ni, and K, while Cl and Br are higher in the west (Schmidt et al., 2009; cf., Fig. 11.6). Schmidt et al. (2008) interpreted the higher volatile/mobile element contents to indicate that Barnhill-class ashes are the magmatic equivalent of the Irvine-class basalts that had their volatile/mobile elements enhanced by hydrothermal processes.

Schmidt et al. (2008, 2009) developed a scenario for the formation of Barnhill-class rocks on Home Plate, from which this summary is based. The morphologic, textural, mineralogical, and compositional observations of Home Plate indicate that the ash layers resulted from a phreatomagmatic eruption of basaltic lava that encountered briny groundwaters near the surface. The distinction between the Barnhill class and Irvine-class basalts is the parent magma of the former interacted with groundwaters while the latter did not. The volatile/mobile element contents of Barnhill-class rocks are consistent with interaction with a Na-Mg-Zn-Ge-Cl-Br brine that was low in S. The brine plausibly also included Fe, Ca, and CO_2, but these cannot be constrained by the data. The low S content was a consequence of removal of acid-sulfate vapor from a boiling hydrothermal system at depth. Additional hydrothermal alteration likely occurred after eruption. Further, the systematic variations in mineralogy and composition across Home Plate are taken as evidence for two alteration regimes, higher temperature alteration and recrystallization of the eastern side and lower temperature alteration of the western side.

Note that comparisons between Irvine- and Barnhill-class rocks, and the formation model developed, are based on mostly brushed Barnhill-class and untreated Irvine-class targets and thus carry a degree of uncertainty. Although untreated surfaces were analyzed, the Irvine-class target surfaces are shiny and appear to have been abraded; alteration rinds might not be present. Earlier we showed that brushed Clovis-class rocks have higher halogen element contents than do abraded targets and concluded the halogen-rich coating likely are present. Coatings might also be present on Barnhill-class rocks, none of which were abraded. The substantially lower halogen element contents for brushed Barnhill-class targets on eastern Home Plate compared to those from the other parts of the structure (Fig. 11.6D) argue that coatings are not likely to be an important part of the story. Nevertheless, the model for hydrothermal enhancement of volatile/mobile elements in the rocks on Home Plate is not uniquely able to explain the APXS results.

Filiberto and Schwenzer (2013) did theoretical modeling of alteration at Home Plate using a thermochemical modeling package that has been applied to terrestrial hydrothermal alteration processes in mafic and ultramafic rocks. They used Barnhill-class target Fastball from the west side of Home Plate as their starting rock composition for the modeling. The modeling requires numerous assumptions that cannot be verified, but nevertheless, provides useful constraints on plausible alteration scenarios. The modeling shows that pyroxene would be consumed at moderate-to-high temperatures (150 °C−300 °C) for water/rock ratios >20 (Filiberto and Schwenzer, 2013). Pyroxene is a major host for Fe in all Barnhill-class rocks (Morris et al., 2008) and the FeO contents of Barnhill-class rocks are only slightly lower than those of Adirondack-class and Irvine-class basalts (Table 11.2). Together, these indicate that the alteration could not have occurred at high water/rock ratios. However, chlorite is a major alteration phase under these conditions (Filiberto and Schwenzer, 2013), but it is not identified in MB spectra of Barnhill-class targets (Morris et al., 2008). In general, alteration mineral assemblages that result from the thermochemical modeling do not match well with phases identified in Barnhill-class rocks. This likely indicates that some of the assumptions used in the modeling are invalid and/or that the real-life alteration process did not follow the ideal alteration pathways that the thermochemical modeling uses.

In the formation model developed by Schmidt et al. (2008), the low S/Cl ratios measured in Barnhill-class rocks are inferred to have resulted from exhalation of acid-sulfate vapors from a deep hydrothermal system. The Paso-Robles-class sulfate-rich soils (discussed in Section 11.7) at Tyrone are roughly 50-m east of Home Plate and the sulfate-rich soils at Troy are about 10 m of the western margin. This invites the interpretation that the alterations of all of these materials are related, but we do not have sufficient age constrains to evaluate this.

11.6.2 EASTERN VALLEY REGION—SILICA-RICH DEPOSITS; HEMATITE-RICH ROCKS

The Eastern Valley region includes the shallow depression along the eastern margin of Home Plate, with Low Ridge forming a boundary to the south and Mitcheltree Ridge to the east (Figs. 11.1 and 11.7a). Within this terrain, Spirit discovered a region of light-toned, generally nodular outcrops and associated light-toned soils. The nodular outcrops form low rises a few cm high extending for several m subparallel to the margin of Home Plate (see Crumpler et al., 2010). The outcrops are relatively tough and often were not crushed when overridden by the rover wheels. The nodular outcrops appear strataform, lying above a platy, buff-colored unit interpreted as an altered, volcanic ash layer (Fig. 11.7B; Arvidson et al., 2008; Ruff et al., 2011; Squyres et al., 2008). The platy unit is the lowest unit of the exposed stratigraphic section in the Eastern Valley region and occurs at the base of Low Ridge and Mitcheltree Ridge (Fig. 11.7C; Arvidson et al., 2008; Crumpler et al., 2010). The nodular outcrops appear to be laterally continuous and there is no evidence that they are localized by fractures or other morphologic features (Ruff et al., 2011). The light-toned, nodular outcrops are most commonly found in the Eastern Valley, but a morphologically similar outcrop was encountered along the northern margin of Home Plate. This latter occurrence is in a similar stratigraphic sequence—lying above the lowest exposed unit—but the lowest unit here is not the platy, buff-colored unit (Ruff et al., 2011). Because of the difference in lithology of the substrate to the nodular outcrops between the Eastern Valley and the northern Home Plate margin localities, it is possible that they were formed on a topographic surface rather than being truly strataform. Indeed, a preliminary geologic map and cross section suggests that the nodular outcrops might have been developed on a disconformable contact within the local stratigraphy (Crumpler et al., 2010).

The textures of the nodular rocks were described by Ruff et al. (2011). A characteristic of the nodular outcrops is an irregular surface composed of cm-sized, rounded, digitate protuberances that have an interior porous, sponge-like texture when exposed on broken surfaces. Some have a botryoidal texture with evidence for pitting or a porous nature. Two rocks were crushed by the rover wheels, and these have interior textures that are fragmental; they consist of poorly sorted mixtures of pebble-sized angular clasts in a matrix of angular to subrounded, intermediate to light-toned, sand-sized grains. This texture is thought to be detrital in origin. One of these crushed rocks, Innocent Bystander, contains some mm-sized, dark-toned clasts that show a finely vesicular, volcanic microtexture. The dark clasts typically have lighter-toned rims suggesting alteration in situ. The larger, light-toned grains in Innocent Bystander have fine-grained detrital textures, and the matrix consists of subrounded to rounded, light-toned, sand-sized grains. Many of these have cores of dark-toned grains, again suggesting in situ alteration or cementation.

FIGURE 11.7

(A) Locator image of Eastern Valley region showing locations of outcrop detail images (B-D), from HiRISE image PSP_005456_1650_RED. (B) Portion of Pancam sol 1165 false-color image (P2584) using the 432, 535 and 753 nm filters from within the Eastern Valley showing Elizabeth-Mahon-class nodular outcrops (n) lying above buff-colored, platy outcrops (b) thought to be correlative with the Halley class. (C) The Troll outcrop in a portion of Navcam mosaic from sols 1087-1089, site 128, position 213 showing stratigraphic relationships between the Montalva class (M) and Halley-class outcrop containing Raquelme (R); buff-colored, platy outcrops (b) are also present. (D) Pancam sol 1183 false-color image (P2599) using the 432, 535 and 753 nm filters from within the Eastern Valley showing Everett-class brushed-target ExamineThis_Slide (bright circle, E) lying below the sole Good-Question-class target (Q) analyzed by the APXS.

Mini-TES spectra of the nodular outcrops show two absorptions attributed to distortions of Si-O bonds, one specific to opaline silica (Squyres et al., 2008). After correcting the Mini-TES spectra to remove the effects of dust on the mirror, the spectrum of nodular rock target Clara Zaph4 shows the strong absorptions that are expected for opal-A, an amorphous, hydrated form of opaline silica

(Ruff et al., 2011). Although the Mini-TES can detect a feature due to bound water, this feature is greatly subdued in solid opal-A, and it was not present in the Clara Zaph4 spectrum (Ruff et al., 2011). The bound water feature is more pronounced in fine-grained materials, and an emission due to bound water is present in the Mini-TES spectrum of nearby silica-rich soil Gertrude Weise (Ruff et al., 2011). Pancam 13f spectra of silica-rich targets show negative slopes from 934 to 1009 nm (Wang et al., 2008) that is attributed to absorption associated with a water combination band (Rice et al., 2010). Pancam 13f spectra have been used to define a hydration signature, and 13f images of the light-toned nodular outcrops and the silica-rich soil show that these targets are hydrated (Rice et al., 2010).

Rocks of the platy, buff-colored unit were examined along Low Ridge off the southeastern side of Home Plate and on Mitcheltree Ridge to the east (Figs. 11.1 and 11.7C). The lower section on Low Ridge is composed of fine-grained, laminated rocks, while the upper section is a thinly bedded rock composed of round, coarse mm-sized grains (Arvidson et al., 2008). Pancam spectra for the platy outcrops on Low Ridge show a pronounced slope in the low wavelength range indicative of ferric iron (Arvidson et al., 2008; cf., Farrand et al., 2008) and a downturn from 934 to 1009 nm indicative of hydration (Rice et al., 2010). Mössbauer-derived iron mineralogy shows that the platy, buff-colored unit on Low Ridge is dominated by nanophase ferric oxide and hematite (Morris et al., 2008). However, the equivalent rock from the stratigraphic section on Mitcheltree Ridge, Torquas, has only about 4% of its total iron in hematite, 46% is in magnetite (Morris et al., 2008).

Rock and soil targets in the Eastern Valley region are divided into classes and subclasses based on composition and MB iron mineralogy (Ming et al., 2008a; Morris et al., 2008); the upper level classes are used here (Table 11.1). The nodular outcrops are Elizabeth Mahon class and include those with opaline silica and hydration signatures. The Good Question class is composed of only the eponymous target on a nodular outcrop; it lacks opaline silica and hydration signatures (Ruff et al., 2011). The Everett class is a light-toned, non-nodular outcrop that stratigraphically lies below the Good Question target (Fig. 11.7D). As mentioned above, the Elizabeth Mahon class nodular-outcrops are stratigraphically above buff-colored platy rocks (Fig. 11.7B). These latter are the Halley and Montalva classes. Montalva lies stratigraphically below a Halley class outcrop of the lithology of round, mm-sized grains (Fig. 11.7C; Arvidson et al., 2008). The Torquas-class target from Mitcheltree Ridge is finely layered rock composed of mm-sized rounded grains (Arvidson et al., 2008). Two soil classes have been defined in the Eastern Valley, the Gertrude-Weise class and the Eileen Dean class. The former includes soils with opaline silica and hydration signatures, and they have extreme SiO_2 contents, while the latter does not (Ruff et al., 2011). Eileen Dean was located near an Elizabeth Mahon class nodular-outcrop.

The geologic map and stratigraphy of the Eastern Valley region of Crumpler et al. (2010) are somewhat at odds with the rock classes defined by APXS and MB data. These authors subdivide the lower unit on Home Plate (B) into upper (B2) and lower (B1) members. These are equivalent to the coarser-grained, planar-bedded to massive lower Home Plate unit of Lewis et al. (2008) and Squyres et al. (2007). Crumpler et al. (2010) include the lithology from the upper section on Low Ridge that consists of a thinly bedded rock composed of round, coarse mm-sized grains as being the basal lithology of B1. This Low Ridge unit is the Graham Land subclass of the Halley class (Ming et al., 2008a). Thus, some Halley-class rocks are considered to be part of the Home Plate volcanic ashes by Crumpler et al. (2010). The valley floor outcrops (unit D) are described as being granular to nodular outcrops of light-toned materials (Crumpler et al., 2010), that is,

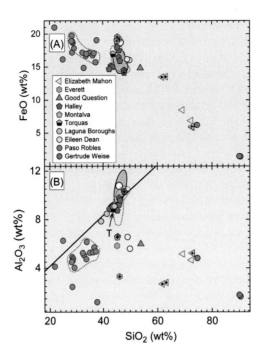

FIGURE 11.8

FeO and Al$_2$O$_3$ vs SiO$_2$ for rocks from the Eastern Valley region near Home Plate and various soil classes. *White field*—abraded Adirondack-class rocks; *gray field*—Laguna-class soils; *line*—Al/Si ratio of Al-poor Laguna-class soils; *purple dotted outline*—Paso-Robles-class soils from Troy; *red dotted outline*—Barnhill-class rocks; rock targets that were brushed (*plus sign*) or abraded by wheel scuff (*dot*) are indicated.

Elizabeth Mahon class rocks. Montalva (Fig. 11.7C) and the platy, buff-colored Halley-class rocks (see Ruff et al., 2011, Fig. 42) lie below the Graham Land subclass, with Elizabeth Mahon class outcrops lying above them. This would indicate that unit D of Crumpler et al. (2010) includes the Halley subclass of the Halley class (Ming et al., 2008a). A disconformity separates units D and B1 (Crumpler et al., 2010).

Halley-class rocks are essentially basaltic in composition but show some compositional distinctions compared to the olivine basalts of the cratered plains, particularly in higher contents of some minor, incompatible elements (P, K, Ti) and lower contents of minor, compatible elements (Cr, Mn) (Table 11.2; Ming et al., 2008a). Among major elements, a characteristic of the class is generally lower Al and Fe contents than those of the Adirondack-class basalts (Fig. 11.8). They are broadly similar in major element composition to the Barnhill class of Home Plate (Fig. 11.8). Ming et al. (2008a) noted that there is an excellent correlation between Ca and S in Halley-class targets that suggests variable amounts of Ca-sulfate in these rocks. Montalva differs from the Halley class in having very high K contents, the highest measured in Gusev crater, and very low Mn contents (Table 11.2; Ming et al., 2008a). The Fe content of Montalva is at the high end of the range of the Halley class and similar to Adirondack-class basalts, but its Al content is very low (Fig. 11.8).

Montalva is distinct from Barnhill-class rocks in major element composition (Fig. 11.8). Torquas also has higher P, K, and Ti compared to Adirondack-class basalts (Table 11.2), and is similar to the Halley class in Al, Si, and Fe (Fig. 11.8). In contrast, the Elizabeth Mahon class is far from basaltic in composition with high Si and low Al and Fe contents (Table 11.2; Fig. 11.8). The Gertrude-Weise-class soils take the compositional characteristics of the Elizabeth Mahon class to extremes, with ~90 wt% SiO_2 in some targets.

The Halley class has a higher MAI than does the Barnhill class but its normative diopside is similar, attesting that oxidation during alteration was not accompanied by substantial leaching (Table 11.1; Fig. 11.3B). The Halley-class targets with high normative diopside are those rich in Ca and SO_3 due to precipitation of $CaSO_4$ (Ming et al., 2008a). Torquas also has a major element composition similar to the Barnhill class (Fig. 11.8), but it does have higher MgO. It has a lower MAI and slightly lower normative diopside than the Halley class (Fig. 11.3B) and thus experienced a lower degree of oxidation. Because of the low Al content of Montalva (Fig. 11.8), the low normative diopside content cannot be used to evaluate possible leaching during alteration for this rock. The same is true for the silica-rich and Al-poor classes Elizabeth Mahon, Everett, and Good Question.

The volatile/mobile element contents of the Elizabeth Mahon, Halley, and Montalva classes are generally similar to those of the Barnhill class from the eastern side of Home Plate, and to those of Laguna-class soils, while Torquas is more similar to the noneastern Home Plate rocks (Fig. 11.9). Notable exceptions are the generally higher Zn contents of the Eastern Valley rocks and the higher S content of those Halley-class rocks that are enriched in $CaSO_4$ (Ming et al., 2008a). The highest Zn content measured in Gusev crater (2270 µg/g) was on a Halley-class target. The Everett, Good Question, and Eileen Dean classes are more similar to the noneastern Barnhill-class targets in volatile/mobile element contents, again with the notable exception of Zn (Fig. 11.9). The Gertrude-Weise-class soils, especially the two targets with the highest Si content, have generally low volatile/mobile element contents compared to other rocks and soils from the region.

The silica-rich rocks and soils in the Eastern Valley region are thought to have formed by hydrothermal processes connected to the volcanism that formed the Barnhill class ash layers (Ming et al., 2008a; Squyres et al., 2008). Squyres et al. (2008) noted that hydrothermal silica deposits can form under a variety of conditions. These authors favored a low pH, acid-sulfate process for the Eastern Valley because the rocks also show enrichments in Ti, which is insoluble in acidic solutions, and because nearby light-toned soils are rich in ferric sulfates (see section 11.7). The dominance of hematite in the iron mineralogy of the platy rocks just below the silica-rich classes could indicate aqueous alteration of a volcanic protolith (e.g., Ming et al., 2008a). However, formation of hematite by dry, thermal oxidation of preexisting iron-bearing oxides and silicates could also be responsible (Morris et al., 2008).

Ruff et al. (2011) have suggested that the silica-rich deposits are sinters formed by deposition from hydrothermal vapors/solutions rather than via acid-sulfate alteration of preexisting rocks. These authors find that the relatively low oxidation state (MAI 18−38 for the rock classes; 17−30 for the soil classes; Table 11.1), and low S contents (SO_3 3.4−4.3 wt% for the rock classes; 1.0−3.1 wt% for the soil classes; Table 11.1) argue against an acid-sulfate hydrothermal process. Ruff et al. (2011) conclude that the strataform nature of the deposits and macro- and microtextures favor an origin as amorphous opaline silica sinter precipitated from solutions. One possible scenario

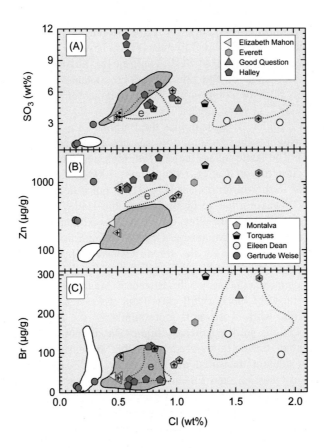

FIGURE 11.9

Volatile/mobile element contents of rocks from the Eastern Valley region. *White field*—abraded Adirondack-class rocks; *gray field*—Laguna-class soils; *red dotted outline*—Barnhill-class rocks; targets that were brushed (*plus sign*) or scuffed by the rover wheels (*dot*) are indicated. Note that Zn is plotted on a log scale.

for formation of the silica-rich deposits is as precipitates from thermal springs or geysers that had a near-neutral to alkaline pH.

Gertrude-Weise-class soils have the highest silica measured in Gusev crater, and are also richer in TiO_2 compared to outcrop rocks that are the substrate to the silica-rich nodular rocks in the Eastern Valley, the Halley, and Montalva classes. The coupled high SiO_2 and TiO_2 is taken as support for an acid-leaching origin for the silica-rich deposits in the Eastern Valley as these oxides are relatively insoluble under such conditions (Ming et al., 2008a; Morris et al., 2008; Squyres et al., 2008). Ruff et al. (2011) did not consider the high TiO_2 contents of Gertrude Weise to be compelling evidence for leaching by acidic fluids or vapors in a hydrothermal environment. Rather, they found that the low SO_3 contents were a compositional characteristic that argued against such a model as the fluids/vapors would have been S-rich. The SO_4^{2-} anion would be highly soluble in aqueous fluids and could have been flushed through the silica-rich deposits, as acknowledged by Ruff et al. (2011). Some members of the Halley-class rock substrate have coupled enrichments in CaO and SO_3 indicating mobilization of Ca-sulfates (Ming et al., 2008a). The approximately factor of two enrichment in SiO_2 in Gertrude Weise relative to the Halley and Montalva classes that are thought to represent prototlith compositions requires a high water/rock ratio to efficiently remove soluble cations, and this would do the same for anions. Thus, the low SO_3 contents of the silica-rich rocks and soils are not a compelling argument against an acidic hydrothermal alteration process for the origin of these materials.

The silica-rich deposits, the highly oxidized Halley and Montalva classes, and the much less altered Barnhill-class rocks from eastern Home Plate are all located within a region of $\sim 375 \ m^2$. They occupy only a few 10s of cm of a stratigraphic section (Crumpler et al., 2010). Their volatile/mobile element contents occupy much of the range for all rocks in Gusev crater (cf., Figs. 11.4 and 11.9). This attests to a highly localized and heterogeneous nature for alteration in the Eastern Valley region, indicating nearby and localized sources for the volatile/mobile elements, such as vents or fumaroles associated with volcanic activity. The alteration and compositional distribution is not consistent with large scale, equilibrium fluid processes.

11.7 SULFATE-RICH PASO-ROBLES-CLASS SOILS

Soils with extraordinarily high S contents were encountered at several locations once Spirit began exploring the Columbia Hills region. These soils are variously referred to as sulfate-rich soils (or fines) (e.g., Crumpler et al., 2011), light-toned salty soils (a name that includes other compositional types) (e.g., Wang et al., 2008), and Paso-Robles-class soils (e.g., Ming et al., 2006). The latter term is used here and will be restricted to soils with >26 wt% SO_3 (Table 11.1; Fig. 11.2A). The locations of the Paso-Robles-class soils interrogated by the APXS are shown on Fig. 11.1. The first of these that was analyzed by the entire Athena instrument package was the eponymous Paso Robles that was serendipitously discovered during the ascent of Husband Hill. An uphill drive faulted-out due to excessive wheel slip, and Hazcam images revealed a trench in soil that showed light-toned patches, prompting an extensive investigation. All Paso-Robles-class soils were similarly stumbled upon when Spirit's wheels unexpectedly dug into the regolith, and all were challenging for rover mobility. The last encountered is on the western side of Home Plate at Troy and proved to be a rover trap; it is the final resting place of Spirit (Fig. 11.1C; Arvidson et al., 2010).

Paso-Robles-class soils are light-toned in Navcam and Pancam images. Pancam multispectral images from each site show differences in color and spectra indicating cm-scale variations in mineralogy and differences between sites (Johnson et al., 2007; Wang et al., 2008). Modeling of Pancam 13f spectra have led to interpretations of the Paso-Robles-class soils as mixtures of typical Gusev soils with differing assemblages of sulfates dominated by ferric sulfates (Johnson et al., 2007). Deconvolution of Mini-TES spectra yields a best fit that includes $\sim 20\%$ ferric sulfates (Lane et al., 2007). An emission peak in Mini-TES spectra at 6.1 μm due to bound water is strongest in sulfate-rich soils in the region around Paso Robles (Kuzmin et al., 2006). The light-toned (sulfate-rich) fractions of Paso Robles have an enhanced bound water emission compared to dark-toned fractions (Yen et al., 2008). Mössbauer spectra of Paso-Robles-class soils are unique among Gusev soils in that the majority of their Fe content, on average 65%, is in ferric sulfates; other soils do not contain measurable ferric sulfate (Morris et al., 2006, 2008). Nanophase iron oxides, containing $\sim 15\%-30\%$ of the Fe in Laguna-class soils, are not detected by MB in Paso-Robles-class soils (Morris et al., 2006, 2008). The ratio of Compton to Rayleigh backscattered peaks in APXS spectra depends significantly on the abundance of light elements such as H and O. The spectra of Paso Robles soils have ratios of these backscattered peaks that are consistent with approximately 20 wt% bound water (Campbell et al, 2008).

All Paso-Robles-class soils are characterized by very high S contents, but compositions from different sites are strongly influenced by local bedrock compositions (Ming et al., 2006, 2008a,b; Yen et al., 2008). As discussed in Section 11.5.1, rocks on Husband Hill show varying degrees of alteration. An assessment of different measures of alteration of rocks and soils on Husband Hill consistently indicated that the type soil Paso Robles was among the most altered materials encountered in Gusev crater through sol 512 (Ming et al., 2006; Morris et al., 2006). The initial interpretation of the formation of Paso Robles was that it is a mixture of Laguna-class soils and salt deposits formed via the evaporation of aqueous solutions or brines whose chemistry was established by acid-sulfate alteration of local materials; fumarolic vents may have supplied the SO_2-rich vapors needed for this process (Ming et al., 2006).

Yen et al. (2008) discussed the origin of Paso-Robles-class soils from three sites, Paso Robles, Arad (a.k.a. Dead Sea), and Tyrone. What follows here was taken from that paper unless otherwise indicated. Paso Robles lies on the western side of Husband Hill, Arad lies on the eastern side in the Inner Basin, and Tyrone is on the east side of Home Plate (Fig. 11.1B and C). These three sites differ in elevation by 33 m. Tyrone occupies a local hollow, but the other two do not; Paso Robles is on a slope and Arad is at the base of a slope. The fourth Paso-Robles-class soil location is in a hollow at Troy, a mostly filled, 8-m-diameter crater (Fig. 11.1C; Arvidson et al., 2010). Local ponding of groundwater can be ruled out as a cause of accumulation of brines at Paso Robles and Arad, and thus seems likely to be inconsequential for formation of Paso-Robles-class soils more generally. With the exception of their high SO_3 content, the compositions of the soils are different and take on the compositional character of the local rocks. For example, Paso Robles, located near outcrops of P-rich Watchtower- and Wishstone-class outcrops, has the highest P_2O_5 content measured in Gusev crater (5.6 wt%). This indicates that solution compositions are significantly governed by reactions with nearby bedrock. Reconstruction of the compositional data into plausible mineralogical components shows that all are dominated by ferric sulfates, Mg-sulfates, and silica that are mixed with differing amounts of typical, Laguna-class soil. These phases are best modeled as sublimates from high-temperature vapors or precipitates from high-temperature fluids. The vapor/fluid was likely volcanic in origin because of the evidence for volcanism and associated hydrothermal activity recorded at Home Plate (Schmidt et al., 2008, 2009; Squyres et al., 2007) and the Eastern Valley (Ming et al., 2008a; Ruff et al., 2011; Squyres et al., 2008), and rocks interpreted as possibly being mafic dikes occur on Husband Hill (McSween et al., 2006b).

Paso-Robles-class soils are generally low in Cl and Zn relative to Laguna-class soils (Table 11.1; Fig. 11.2B). Those soils at Troy have lower Cl and Br contents than almost all other Paso Robles soils, but include the highest SO_3 contents for this class (Fig. 11.2). Thus, the sulfate-enrichment was not accompanied by volatile/mobile element enrichments more generally. Yen et al. (2008) showed that the compositions of Paso-Robles-class soils often indicate that they contain excess silica in addition to being sulfate-rich. This is shown in Fig. 11.8B where many of the Paso-Robles-class soil targets, including all of those from Troy, plot on the Si-rich side of the lowest Al/Si ratio observed for Laguna-class soils. Simple addition of sulfate cements, such as occurred in the Laguna-Boroughs-subclass soils, would drive soil compositions to lower Al and Si without changing their ratio (cf., Laguna Boroughs soils; Fig. 11.8B). The enrichments in S engendering ferric sulfates and the excess silica are consistent with an acid-sulfate-type alteration process that seems to have pervasively operated in the Columbia Hills region.

The ages of the processes responsible for the formation of Paso-Robles-class soils are not significantly constrained. They occur in regolith which must postdate the local bedrock, consistent with the influence of local geology on compositions of different light-toned soils. Paso-Robles-class soils were not encountered on the basaltic plains, which might indicate that the fumarolic/hydrothermal activity predated emplacement of the basaltic flows, but this could also be explained by a lack of such fumarolic/hydrothermal activity far from the Columbia Hills region. Tyrone and Troy are close to Home Plate and thus are plausibly coeval with the volcanism and hydrothermal alteration that engendered those ash layers (Schmidt et al., 2008, 2009).

11.8 SUMMARY: VOLATILE/MOBILE ELEMENTS AND ALTERATION IN GUSEV CARTER

The Mars exploration rover Athena science payload onboard Spirit investigated a variety of rocks and soils during its more than 3 Mars years of activity. Mineralogical characterization of the materials by Pancam, Mini-TES, and MB shows that different styles and degrees of alteration are recorded. The APXS quantified the compositions of the rocks and soils, and the data show wide ranges of volatile/mobile element contents. Enrichments in volatile/mobile elements in some of the rocks and soils in Gusev crater, especially on Husband Hill and in the Inner Basin, provide evidence constraining the aqueous alteration processes that occurred. In many cases, these elements were mobilized in vapor emissions or hydrothermal fluids associated with volcanic sources. They are also mobile in aqueous systems and were transported some distance and precipitated either as fluids evaporated, or as the intrinsic properties (e.g., P, T, pH) of solutions changed resulting in solubility limits being exceeded.

The most recent alteration process is recorded in rock coatings and soil indurations, and is most readily observed as correlations of SO_3 and Zn with Cl in Adirondack-class basalts and Laguna-class soils. The coatings and indurations can be understood as arising from thin films of water dissolving and mobilizing halogen and sulfate salts (and associated cations), which are then precipitated upon drying. This process occurs at very low water/rock ratios and might be occurring in the present or recent-past Martian climate. However, formation of rock coatings could have been caused by an acid-fog alteration process affecting exposed rock surfaces. This again is a low water/rock ratio process, and could have occurred within the recent past, as basaltic magmatism on Mars within the last few hundred million years is documented through studies of Martian meteorites. Crater statistics indicate that large-scale, caldera-style volcanism occurred up until ~ 150 Myr ago on Mars (Robbins et al., 2011), and plains-style basaltic volcanism was occurring within the last few tens of millions of years (Hauber et al., 2011).

Rocks and soils from the Columbia Hills region show that the most pervasive alteration processes in these older rocks was acid-sulfate alteration by fumarolic/hydrothermal activity, plausibly engendered by regional volcanic activity that predated flooding of the crater by the Adirondack-class plains-basalts. The water/rock ratios were variable, with high ratios leading to leaching and corundum-normative compositions, and low ratios resulting in compositions that have normative diopside contents like those in pristine basalts, but with high MAI. The variability in alteration conditions is reflected in the compositions of the rocks. The rocks and soils have varying volatile/mobile element contents engendered as fluid compositions changed in response to chemical interactions with the rocks, fluids separated into vapor and brine phases, and temperatures and pH changed. Thus, for example, the Clovis-Watchtower-Wishstone classes have a range in Cl contents that is not correlated with Zn, while the Halley class has positively correlated Zn-Cl enrichments. The Barnhill-class ash deposits have high Cl, Zn, and Ge contents, but neither cation is correlated with Cl. Even so, a hydrothermal origin for these elements in the Barnhill class is likely. The alteration processes in the Columbia Hills region might have reoccurred periodically and locally, but the geologic context does not tightly constrain this.

An extreme example of an acidic hydrothermal alteration process is provided by the silica-rich rocks and soils in the Eastern Valley. The most silica-rich of these—the Elizabeth Mahon class rocks and Gertrude-Weise-class soils—have low contents of S, Cl, Ge, and Br indicating the hydrothermal solutions flushed any volatile/mobile elements through the system along with the soluble cations from the protolith. An alternative scenario for the formation of these silica-rich deposits, precipitation from near-neutral waters, nevertheless had their origin mediated by volcanic-sourced heat. In either case, high/water rock is indicated.

One possible moderate-temperature, high water/rock ratio process that produced a unique mineralogy occurred in the region of Gusev crater explored by Spirit. The Comanche outcrop contains abundant Mg-Mn-Fe-carbonates, but is not enriched in other volatile/mobile elements. The alteration that produced the Comanche class might have been accomplished by near-neutral hydrothermal solutions, but an alternative scenario is alteration at moderate-to-low temperatures in an ephemeral lake.

11.8.1 GUSEV CRATER: PRELANDING PREDICTIONS; POSTLANDING REALITY

Gusev crater once hosted a crater-filling lake formed through discharge down Ma'adim Vallis (Golombek et al., 2003). Ma'adim Vallis is thought to have formed in a single, catastrophic flooding event around the time of the Noachian/Hesperian boundary when the rim of Eridania basin to the south was breached, unleashing the lake it contained (Irwin et al., 2004). Mesas at the mouth of Ma'adim Vallis that are considered to be remnants of deltaic sediments are several hundred meters higher in elevation than the region explored by Spirit, and thus the entire landing zone would have been submerged (Irwin et al., 2004). Failure to find lacustrian sediments on the oldest surface explored by Spirit, the Columbia Hills, is not critical evidence against a crater-lake model. The estimated volume of material eroded from Ma'adim Vallis is broadly compatible with the estimate of crater fill in Gusev (see Irwin et al., 2004), and there is no requirement then that the Columbia Hills should have been covered by thick, water-borne sediments. Thin, fine-grained lacustrian sediments that might have been deposited on the Columbia Hills from a quiet, 400-m-deep lake could have been eroded by subsequent eolian activity. Thus, lack of lacustrian sediments is not a fatal flaw for a model invoking a crater-filling lake based on the evidence in hand.

However, neither did Spirit find evidence consistent with widespread, low temperature, high water/rock ratio alteration of the rocks of the Columbia Hills. The alteration uncovered by Spirit was more typically highly localized and under temperature, pH and water/rock ratios inconsistent with those of a large body of standing water. Many of the observed alteration products—various sulfates and halogen salts—would have dissolved in a large lake. Taken together, the evidence suggests that localized alteration must have occurred subsequent to draining of the Gusev crater lake. If wide-spread alteration on the Columbia Hills did occur during the lake stage of Gusev geological history, evidence for it was overprinted by subsequent localized processes. Some of the scenarios invoked to explain alteration of rocks on the Columbia Hills require a source of groundwater. For example, Home Plate was formed as a result of interaction of magma with briny groundwaters. These groundwaters could plausibly have been remnants of the Gusev crater lake.

Finally, the alternate scenario for alteration of the Comanche outcrop and formation of carbonates is that it took place in a shallow, ephemeral lake (Ruff et al., 2014). Geological evidence supports formation of Ma'adim Vallis by a single catastrophic flood event that filled Gusev crater with water above a thick sediment pile (Irwin et al., 2004). If alteration of the Comanche outcrop occurred in an ephemeral lake, it either represented the last drying of the Gusev crater lake, or later recharge by subsequent, small-scale flooding events. Regardless, Comanche stands several m above the elevation of several of the sulfate-rich Paso-Robles-class soils, and following the reasoning given above, this would indicate alteration of Comanche before formation of the sulfate-rich soils.

ACKNOWLEDGMENTS

Rover operations described in this paper were conducted at the Jet Propulsion Laboratory, California Institute of Technology, under a contract with NASA. We thank L.S. Crumpler for helpful discussions regarding Eastern Valley geology and stratigraphy, but any inaccuracies presented here are ours alone. We thank the members of the MER project who enabled daily science observations at the Spirit landing site. We thank the reviewers T. J. McCoy and M. E. Schmidt, and the editor J. Filiberto for their efforts, which led to significant improvements in this chapter. The senior author is supported by the NASA Mars Exploration Rover Participating Scientist Program. The other DWM acknowledges support of the NASA Mars Fundamental Research Program and NASA Johnson Space Center.

REFERENCES

Arvidson, R., Bell, J., Bellutta, P., Cabrol, N., Catalano, J., Cohen, J., et al., 2010. Spirit Mars Rover Mission: overview and selected results from the northern Home Plate Winter Haven to the side of Scamander crater. J. Geophys. Res. 115, E00F03. Available from: https://doi.org/10.1029/2010JE003633.

Arvidson, R.E., Anderson, R., Bartlett, P., Bell, J., Blaney, D., Christensen, P., et al., 2004. Localization and physical properties experiments conducted by Spirit at Gusev Crater. Science 305, 821–824.

Arvidson, R.E., Ruff, S.W., Morris, R.V., Ming, D.W., Crumpler, L.S., Yen, A.S., et al., 2008. Spirit Mars Rover Mission to the Columbia Hills, Gusev Crater: mission overview and selected results from the Cumberland Ridge to Home Plate. J. Geophys. Res. 113, E12S33. Available from: https://doi.org/10.1029/2008JE003183.

Bell, J., Squyres, S., Herkenhoff, K., Maki, J., Arneson, H., Brown, D., et al., 2003. Mars Exploration Rover Athena Panoramic Camera (Pancam) investigation. J. Geophys. Res. 108 (E12), 8063. Available from: https://doi.org/10.1029/2003JE002070.

Bernstein, L.R., 1985. Germanium geochemistry and mineralogy. Geochim. Cosmochim. Acta 49, 2409–2422.

Bibring, J.-P., Langevin, Y., Mustard, J.F., Poulet, F., Arvidson, R., Gendrin, A., et al., 2006. Global mineralogical and aqueous Mars history derived from OMEGA/Mars Express data. Science 312, 400−404.

Campbell, J.L., Gellert, R., Lee, M., Mallett, C.L., Maxwell, J.A., O'Meara, J.M., 2008. Quantitative in situ determination of hydration of bright high-sulfate Martian soils. J. Geophys. Res.: Planets 113, E06S11. Available from: https://doi.org/10.1029/2007JE002959.

Carter, B. L., Frey, H., Sakimoto, S. E. H. and Roark, J., 2001. Constraints on Gusev Basin infill from the Mars Orbiter Laser Altimeter (MOLA) topography. In: 32nd Lunar and Planetary Science Conference. Lunar and Planetary Institute, Houston, Abstract #2042.

Christensen, P.R., Mehall, G.L., Silverman, S.H., Anwar, S., Cannon, G., Gorelick, N., et al., 2003. Miniature Thermal Emission Spectrometer for the Mars Exploration Rovers. J. Geophys. Res. 108 (E12), 8064. Available from: https://doi.org/10.1029/2003JE002117.

Christensen, P.R., Ruff, S., Fergason, R., Knudson, A., Anwar, S., Arvidson, R., et al., 2004. Initial results from the Mini-TES experiment in Gusev Crater from the Spirit Rover. Science 305, 837−842.

Clark, B.C., Arvidson, R.E., Gellert, R., Morris, R.V., Ming, D.W., Richter, L., et al., 2007. Evidence for montmorillonite or its compositional equivalent in Columbia Hills, Mars. J. Geophys. Res. 112, E06S01. Available from: https://doi.org/10.1029/2006JE002756.

Corrigan, C.M., Harvey, R.P., 2004. Multi-generational carbonate assemblages in Martian meteorite Allan Hills 84001: implications for nucleation, growth, and alteration. Meteorit. Planet. Sci. 39, 17−30.

Crumpler, L.S., Squyres, S., Arvidson, R., Bell, J., Blaney, D., Cabrol, N., et al., 2005. Mars Exploration Rover geologic traverse by the Spirit rover in the plains of Gusev Crater, Mars. Geology 33, 809−812.

Crumpler, L.S., Arvidson, R., Squyres, S., Yingst, A., McCoy, T., DesMarais, D., et al., 2010. Overview of the field geologic context of Mars Exploration Rover Spirit, Home Plate and Surroundings. In: 41st Lunar and Planetary Science Conference. Lunar and Planetary Institute, Houston, Abstract #2557.

Crumpler, L.S., Arvidson, R.E., Squyres, S.W., McCoy, T., Yingst, A., Ruff, S., et al., 2011. Field reconnaissance geologic mapping of the Columbia Hills, Mars, based on Mars Exploration Rover Spirit and MRO HiRISE observations. J. Geophys. Res. 116, E00F24. Available from: https://doi.org/10.1029/2010JE003749.

Farrand, W., Bell, J., Johnson, J., Squyres, S., Soderblom, J., Ming, D., 2006. Spectral variability among rocks in visible and near-infrared multispectral Pancam data collected at Gusev crater: examinations using spectral mixture analysis and related techniques. J. Geophys. Res. 111, E02S15. Available from: https://doi.org/10.1029/2005JE002495.

Farrand, W.H., Bell, J., Johnson, J.R., Arvidson, R.E., Crumpler, L.S., Hurowitz, J.A., et al., 2008. Rock spectral classes observed by the Spirit Rover's Pancam on the Gusev Crater Plains and in the Columbia Hills. J. Geophys. Res. 113, E12S38. Available from: https://doi.org/10.1029/2008JE003237.

Filiberto, J., Schwenzer, S.P., 2013. Alteration mineralogy of Home Plate and Columbia Hills—formation conditions in context to impact, volcanism, and fluvial activity. Meteorit. Planet. Sci. 48, 1937−1957. Available from: http://dx.doi.org/10.1111/maps.12207.

Fisher, R., Schmincke, H., 1984. Pyroclastic Rocks. Springer, Berlin, p. 472. Available from: http://dx.doi.org/10.1007/978-3-642-74864-6.

Francis, D., 2011. Columbia Hills—an exhumed layered igneous intrusion on Mars? Earth Planet. Sci. Lett. 310, 59−64.

Gellert, R., Rieder, R., Anderson, R., Brückner, J., Clark, B., Dreibus, G., et al., 2004. Chemistry of rocks and soils in Gusev Crater from the Alpha Particle X-ray Spectrometer. Science 305, 829−832.

Gellert, R., Rieder, R., Brückner, J., Clark, B.C., Dreibus, G., Klingelhöfer, G., et al., 2006. Alpha particle X-ray spectrometer (APXS): results from Gusev crater and calibration report. J. Geophys. Res. 111, E02S05. Available from: https://doi.org/10.1029/2005JE002555.

Golden, D.C., Ming, D.W., Schwandt, C.S., Morris, R.V., Yang, S.V., Lofgren, G.E., 2000. An experimental study on kinetically-driven precipitation of calcium—magnesium—iron carbonates from solution: implications for the low-temperature formation of carbonates in Martian meteorite Allan Hills 84001. Meteorit. Planet. Sci. 35, 457—465.

Golombek, M., Grant, J., Parker, T., Kass, D., Crisp, J., Squyres, S., et al., 2003. Selection of the Mars Exploration Rover landing sites. J. Geophys. Res. 108 (E12), 8072. Available from: https://doi.org/10.1029/2003JE002074.

Golombek, M., Crumpler, L., Grant, J., Greeley, R., Cabrol, N., Parker, T., et al., 2006a. Geology of the Gusev cratered plains from the Spirit rover transverse. J. Geophys. Res. 111, E02S07. Available from: https://doi.org/10.1029/2005JE002503.

Golombek, M., Grant, J., Crumpler, L., Greeley, R., Arvidson, R., Bell, J., et al., 2006b. Erosion rates at the Mars Exploration Rover landing sites and long-term climate change on Mars. J. Geophys. Res. 111, E12S10. Available from: https://doi.org/10.1029/2006JE002754.

Gorevan, S., Myrick, T., Davis, K., Chau, J., Bartlett, P., Mukherjee, S., et al., 2003. Rock abrasion tool: Mars Exploration Rover mission. J. Geophys. Res. 108 (E12), 8068. Available from: https://doi.org/10.1029/2003JE002061.

Greeley, R., Foing, B.H., McSween, H.Y., Neukum, G., Pinet, P., van Kan, M., et al., 2005. Fluid lava flows in Gusev crater, Mars. J. Geophys. Res. 110, E05008. Available from: https://doi.org/10.1029/2005JE002401.

Haskin, L.A., Wang, A., Jolliff, B.L., McSween, H.Y., Clark, B.C., Des Marais, D.J., et al., 2005. Water alteration of rocks and soils on Mars at the Spirit rover site in Gusev Crater. Nature 436, 66—69.

Hauber, E., Brož, P., Jagert, F., Jodłowski, P., Platz, T., 2011. Very recent and wide-spread basaltic volcanism on Mars. Geophys. Res. Lett. 38, L10201. Available from: http://dx.doi.org/10.1029/2011GL047310.

Herkenhoff, K., Squyres, S., Bell, J., Maki, J., Arneson, H., Bertelsen, P., et al., 2003. Athena Microscopic Imager investigation. J. Geophys. Res. 108 (E12), 8065. Available from: https://doi.org/10.1029/2003JE002076.

Herkenhoff, K., Squyres, S., Arvidson, R., Bass, D., Bell, J., Bertelsen, P., et al., 2004. Textures of the soils and rocks at Gusev Crater from Spirit's Microscopic Imager. Science 305, 824—826.

Herkenhoff, K.E., Squyres, S.W., Anderson, R., Archinal, B.A., Arvidson, R.E., Barrett, J.M., et al., 2006. Overview of the Microscopic Imager Investigation during Spirit's first 450 sols in Gusev crater. J. Geophys. Res. 111, E02S04. Available from: https://doi.org/10.1029/2005JE002574.

Hurowitz, J.A., McLennan, S.M., Tosca, N.J., Arvidson, R.E., Michalski, J.R., Ming, D.W., et al., 2006a. In situ and experimental evidence for acidic weathering of rocks and soils on Mars. J. Geophys. Res. 111, E02S19. Available from: https://doi.org/10.1029/2005JE002515.

Hurowitz, J.A., McLennan, S.M., McSween, H.Y., DeSouza, P.A., Klingelhöfer, G., 2006b. Mixing relationships and the effects of secondary alteration in the Wishstone and Watchtower Classes of Husband Hill, Gusev Crater, Mars. J. Geophys. Res. 111, E12S14. Available from: https://doi.org/10.1029/2006JE002795.

Irwin, R.P., Howard, A.D., Maxwell, T.A., 2004. Geomorphology of Ma'adim Vallis, Mars, and associated paleolake basins. J. Geophys. Res. 109, E12009. Available from: https://doi.org/10.1029/2004JE002287.

Johnson, J., Bell, J., Cloutis, E., Staid, M., Farrand, W., McCoy, T., et al., 2007. Mineralogic constraints on sulfur-rich soils from Pancam spectra at Gusev Crater, Mars. Geophys. Res. Lett. 34, L13202. Available from: https://doi.org/10.1029/2007GL029894.

Klingelhoefer, G., Morris, R.V., Bernhardt, B., Rodionov, D., De Souza, P., Squyres, S., et al., 2003. Athena MIMOS II Mössbauer spectrometer investigation. J. Geophys. Res. 108 (E12), 8067. Available from: https://doi.org/10.1029/2003JE002138.

Kuzmin, R.O., Greeley, R., Landheim, R., Cabrol, N. and Farmer, J., 2000. Geologic map of the MTM-15182 and MTM-15187 quadrangles. In: Gusev Crater-Ma'adim Vallis Region Mars: US Geological Survey Map, Geologic Investigation Series I-2666.

Kuzmin, R.O., Christensen, P.R., Ruff, S.W., Graff, T.G., Knudson, A.T., Zolotov, M.Y. et al., 2006. Spatial and temporal variations of bound water content in the Martian Soil within the Gusev Crater: preliminary results of the TES and Mini-TES data analysis. In: 37th Lunar and Planetary Science Conference. Lunar and Planetary Institute, Houston, Abstract #1673.

Lane, M.D., Bishop, J.L., Dyar, M.D., Parente, M., King, P.L. and Hyde, B.C., 2007. The ferric sulfate and ferric phosphate minerals in the light-toned Paso Robles rover track soils: a multi-instrument analysis. In: Seventh International Conference on Mars. Lunar and Planetary Institute, Houston, Abstract #3331.

Lewis, K.W., Aharonson, O., Grotzinger, J.P., Squyres, S.W., Bell, J.F., Crumpler, L.S., et al., 2008. Structure and stratigraphy of Home Plate from the Spirit Mars Exploration Rover. J. Geophys. Res. 113, E12S36. Available from: https://doi.org/10.1029/2007JE003025.

Maki, J., Bell, J., Herkenhoff, K., Squyres, S., Kiely, A., Klimesh, M., et al., 2003. Mars Exploration Rover Engineering Cameras. J. Geophys. Res. 108 (E12), 8071. Available from: https://doi.org/10.1029/2003JE002077.

Manga, M., Patel, A., Dufek, J., Kite, E.S., 2012. Wet surface and dense atmosphere on early Mars suggested by the bomb sag at Home Plate, Mars. Geophys. Res. Lett. 39, L01202. Available from: https://doi.org/10.1029/2011GL050192.

Martínez-Alonso, S., Jakosky, B.M., Mellon, M.T., Putzig, N.E., 2005. A volcanic interpretation of Gusev Crater surface materials from thermophysical, spectral, and morphological evidence. J. Geophys. Res. 110, E01003. Available from: https://doi.org/10.1029/2004JE002327.

McCoy, T., Sims, M., Schmidt, M., Edwards, L., Tornabene, L., Crumpler, L., et al., 2008. Structure, stratigraphy, and origin of Husband Hill, Columbia Hills, Gusev Crater, Mars. J. Geophys. Res. 113, E06S03. Available from: https://doi.org/10.1029/2007JE003041.

McSween, H.Y., Treiman, A.H., 1998. Martian meteorites. In: Papike, J.J. (Ed.), Planetary Materials, Reviews in Mineral, 36. Mineralogical Society of America, Washington, DC, pp. 6.1−6.53.

McSween, H.Y., Arvidson, R., Bell, J., Blaney, D., Cabrol, N., Christensen, P., et al., 2004. Basaltic rocks analyzed by the Spirit rover in Gusev Crater. Science 305, 842−845.

McSween, H.Y., Wyatt, M.B., Gellert, R., Bell, J.F., Morris, R.V., Herkenhoff, K.E., et al., 2006a. Characterization and petrologic interpretation of olivine-rich basalts at Gusev Crater, Mars. J. Geophys. Res.: Planets 111, E02S10. Available from: https://doi.org/10.1029/2005JE002477.

McSween, H.Y., Ruff, S.W., Morris, R.V., Bell, J.F., Herkenhoff, K., Gellert, R., et al., 2006b. Alkaline volcanic rocks from the Columbia Hills, Gusev Crater, Mars. J. Geophys. Res.: Planets 111, E09S91. Available from: https://doi.org/10.1029/2006JE002698.

McSween, H.Y., Ruff, S.W., Morris, R.V., Gellert, R., Klingelhöfer, G., Christensen, P.R., et al., 2008. Mineralogy of volcanic rocks in Gusev Crater, Mars: reconciling Mössbauer, Alpha Particle X-Ray Spectrometer, and Miniature Thermal Emission Spectrometer spectra. J. Geophys. Res.: Planets 113, E06S04.

Ming, D.W., Mittlefehldt, D.W., Morris, R.V., Golden, D.C., Gellert, R., Yen, A., et al., 2006. Geochemical and mineralogical indicators for aqueous processes in the Columbia Hills of Gusev Crater, Mars. J. Geophys. Res.: Planets 111, E02S12. Available from: https://doi.org/10.1029/2005JE002560.

Ming, D.W., Gellert, R., Morris, R.V., Arvidson, R.E., Brückner, J., Clark, B.C., et al., 2008a. Geochemical properties of rocks and soils in Gusev Crater, Mars: results of the Alpha Particle X-Ray Spectrometer from Cumberland Ridge to Home Plate. J. Geophys. Res.: Planets 113, E12S39. Available from: https://doi.org/10.1029/2008JE003195.

Ming, D.W., Morris, R.V., Clark, B.C., 2008b. Aqueous alteration on Mars. In: Bell III, J.F. (Ed.), The Martian Surface: Composition, Mineralogy, and Physical Properties. Cambridge University Press, Cambridge, UK, pp. 519−540.

Mittlefehldt, D.W., 1994. ALH84001, a cumulate orthopyroxenite member of the Martian meteorite clan. Meteoritics 29, 214−221.

Mittlefehldt, D.W., Gellert, R., McCoy, T., McSween, H.Y.J., Li, R. and Team, A.S., 2006. Possible Ni-rich Mafic-ultramafic magmatic sequence in the Columbia Hills: evidence from the Spirit Rover. In: 37th Lunar and Planetary Science Conference. Lunar and Planetary Institute, Houston, Abstract #1505.

Morris, R.V., Klingelhoefer, G., Bernhardt, B., Schröder, C., Rodionov, D.S., De Souza, P., et al., 2004. Mineralogy at Gusev Crater from the Mössbauer spectrometer on the Spirit Rover. Science 305, 833−836.

Morris, R.V., Klingelhoefer, G., Schröder, C., Rodionov, D.S., Yen, A., Ming, D.W., et al., 2006. Mössbauer mineralogy of rock, soil, and dust at Gusev crater, Mars: spirit's journey through weakly altered olivine basalt on the plains and pervasively altered basalt in the Columbia Hills. J. Geophys. Res. 111, E02S13. Available from: https://doi.org/10.1029/2005JE002584.

Morris, R.V., Klingelhöfer, G., Schröder, C., Fleischer, I., Ming, D.W., Yen, A.S., et al., 2008. Geochemical properties of rocks and soils in Gusev Crater, Mars: results of the alpha particle X-ray spectrometer from Cumberland Ridge to Home Plate. J. Geophys. Res. 113, E12S39. Available from: https://doi.org/10.1029/2008JE003195.

Morris, R.V., Ruff, S.W., Gellert, R., Ming, D.W., Arvidson, R.E., Clark, B.C., et al., 2010. Identification of carbonate-rich outcrops on Mars by the Spirit rover. Science 329, 421−424.

Nesbitt, H.W., Wilson, R.E., 1992. Recent chemical weathering of basalts. Am. J. Sci. 292, 740−777.

Parker, Mv.K., Zegers, T., Kneissl, T., Ivanov, B., Foing, B., Neukum, G., 2010. 3D structure of the Gusev Crater region. Earth Planet. Sci. Lett. 294, 411−423.

Rice, M., Bell, J., Cloutis, E., Wang, A., Ruff, S., Craig, M., et al., 2010. Silica-rich deposits and hydrated minerals at Gusev Crater, Mars: vis−NIR spectral characterization and regional mapping. Icarus 205, 375−395.

Rieder, R., Gellert, R., Brückner, J., Klingelhöfer, G., Dreibus, G., Yen, A., et al., 2003. The new Athena Alpha Particle X-ray Spectrometer for the Mars Exploration Rovers. J. Geophys. Res. 108 (E12), 8066. Available from: https://doi.org/10.1029/2003JE002150.

Robbins, S.J., Di Achille, G., Hynek, B.M., 2011. The volcanic history of Mars: high-resolution crater-based studies of the calderas of 20 volcanoes. Icarus 211, 1179−1203.

Ruff, S., Christensen, P., Blaney, D., Farrand, W., Johnson, J., Michalski, J., et al., 2006. The rocks of Gusev Crater as viewed by the Mini-TES instrument. J. Geophys. Res. 111, E12S18. Available from: https://doi.org/10.1029/2006JE002747.

Ruff, S., Christensen, P., Blaney, D., Farrand, W., Johnson, J., Michalski, J., et al., 2007. The rocks of Gusev Crater as viewed by the Mini-TES instrument. J. Geophys. Res. 111, E12S18. Available from: https://doi.org/10.1029/2006JE002747.

Ruff, S.W., Farmer, J.D., Calvin, W.M., Herkenhoff, K.E., Johnson, J.R., Morris, R.V., et al., 2011. Characteristics, distribution, origin, and significance of opaline silica observed by the Spirit rover in Gusev crater, Mars. J. Geophys. Res. 116, E00F23. Available from: https://doi.org/10.1029/2010JE003767.

Ruff, S.W., Niles, P.B., Alfano, F., Clarke, A.B., 2014. Evidence for a Noachian-aged ephemeral lake in Gusev Crater, Mars. Geology 42, 359−362.

Schmidt, M.E., Ruff, S.W., McCoy, T.J., Farrand, W.H., Johnson, J.R., Gellert, R., et al., 2008. Hydrothermal origin of halogens at Home Plate, Gusev Crater. J. Geophys. Res.: Planets 113, (1991−2012).

Schmidt, M.E., Farrand, W.H., Johnson, J.R., Schröder, C., Hurowitz, J.A., McCoy, T.J., et al., 2009. Spectral, mineralogical, and geochemical variations across Home Plate, Gusev Crater, Mars indicate high and low temperature alteration. Earth Planet. Sci. Lett. 281, 258−266.

Squyres, S.W., Arvidson, R.E., Baumgartner, E.T., Bell III, J.F., Christensen, P.R., Gorevan, S., et al., 2003. Athena Mars rover science investigation. J. Geophys. Res.: Planets 108, 8062. Available from: https://doi.org/10.1029/2003JE002121.

Squyres, S.W., Arvidson, R.E., Bell, J.F., Brückner, J., Cabrol, N.A., Calvin, W., et al., 2004. The Spirit Rover's Athena Science Investigation at Gusev Crater, Mars. Science 305, 794−799.

Squyres, S.W., Arvidson, R.E., Blaney, D.L., Clark, B.C., Crumpler, L., Farrand, W.H., et al., 2006. Rocks of the Columbia Hills. J. Geophys. Res.: Planets 111, E02S11. Available from: https://doi.org/10.1029/2005JE002562.

Squyres, S.W., Aharonson, O., Clark, B.C., Cohen, B.A., Crumpler, L., De Souza, P., et al., 2007. Pyroclastic activity at home plate in Gusev Crater, Mars. Science 316, 738−742.

Squyres, S.W., Arvidson, R.E., Ruff, S., Gellert, R., Morris, R., Ming, D., et al., 2008. Detection of silica-rich deposits on Mars. Science 320, 1063−1067.

Tanaka, K.L., Skinner, J.A., Jr., Dohm, J.M., Irwin, R.P., III, Kolb, E.J., Fortezzo, C.M., et al. (2014) Geologic map of Mars. U.S. Geological Survey Scientific Investigations Map 3292.

Wang, A., Bell, J., Li, R., Johnson, J., Farrand, W., Cloutis, E., et al., 2008. Light-toned salty soils and coexisting Si-rich species discovered by the Mars Exploration Rover Spirit in Columbia Hills. J. Geophys. Res. 113, E12S40. Available from: https://doi.org/10.1029/2008JE003126.

Yen, A.S., Gellert, R., Schröder, C., Morris, R.V., Bell, J.F., Knudson, A.T., et al., 2005. An integrated view of the chemistry and mineralogy of Martian soils. Nature 436, 49−54.

Yen, A.S., Morris, R.V., Clark, B.C., Gellert, R., Knudson, A.T., Squyres, S., et al., 2008. Hydrothermal processes at Gusev Crater: an evaluation of Paso-Robles-class soils. J. Geophys. Res.: Planets 113, E06S10. Available from: https://doi.org/10.1029/2007JE002978.

Yingst, R.A., Schmidt, M.E., Herkenhoff, K.E., Mittlefehldt, D.W. and Team, A.S., 2007. Linking home plate and algonquin class rocks through microtextural analysis: evidence for hydrovolcanism in the inner basin of Columbia Hills, Gusev Crater. In: Seventh International Conference on Mars. Lunar and Planetary Institute, Houston, Abstract #3296.

VOLATILE DETECTIONS IN GALE CRATER SEDIMENT AND SEDIMENTARY ROCK: RESULTS FROM THE MARS SCIENCE LABORATORY'S SAMPLE ANALYSIS AT MARS INSTRUMENT

12

Brad Sutter[1], Amy C. McAdam[2], and Paul R. Mahaffy[2]

[1]*Jacobs, NASA Johnson Space Center, Houston, TX, United States* [2]*NASA, Goddard Space Flight Center, Greenbelt, MD, United States*

12.1 INTRODUCTION

The Sample Analysis at Mars (SAM) instrument aboard the Mars Science Laboratory's Curiosity rover has provided a detailed volatile analysis of the eolian sediment and sedimentary rock of Gale crater, Mars (Glavin et al., 2013; Leshin et al., 2013; Archer et al., 2014; McAdam et al., 2014; Ming et al., 2014; Freissinet et al., 2015; Mahaffy et al., 2015; Stern et al., 2015; Sutter et al., 2017). The SAM instrument consists of a quadrupole mass spectrometer (QMS), a 6-column gas chromatograph, and a tunable laser spectrometer (TLS) that analyzes evolved gases derived from sample pyrolysis (Mahaffy et al., 2012). Notable detections from Gale crater materials include the first in situ detection of nitrates and organic carbon, which both have implications for Martian microbial habitability (Leshin et al., 2013; Ming et al., 2014; Freissinet et al., 2015; Stern et al., 2015, 2017; Sutter et al., 2017). Water (0.9−2.5 wt%) was detected in every sample from a variety of sources (e.g., adsorbed water, hydrated salts, phyllosilicates). Detections of SO_2 are linked to the presence of iron and magnesium sulfates and possibly iron sulfides (McAdam et al., 2014; Franz et al., 2016; Sutter et al., 2017). Oxygen and HCl detections are consistent with the presence of oxychlorine phases (e.g., chlorate and/or perchlorate) (Glavin et al., 2013; Ming et al., 2014; Sutter et al., 2017). Detections of CO_2 are consistent with organic-C and/or carbonate (Leshin et al., 2013; Ming et al., 2014) while coevolution of CO and CO_2 provides additional support for the presence of organic-C in Gale crater surface material (Sutter et al., 2017).

Isotopic analyses by SAM of gases evolved from solid samples have included the deuterium to hydrogen (*D/H*) ratio in water along with $\delta^{13}C$ in CO_2, $\delta^{37}Cl$ in HCl, $\delta^{34}S$ in SO_2 and ^3He, ^{21}Ne, ^{36}Ar, and ^{40}Ar (Leshin et al., 2013; Mahaffy et al., 2015; Farley et al., 2014, 2016). The C and H isotope ratios in evolved gases detected by SAM can be compared with these same ratios in

Volatiles in the Martian Crust. DOI: http://dx.doi.org/10.1016/B978-0-12-804191-8.00012-X

atmospheric CO_2 and H_2O, as well as ratios detected in Mars meteorites (Leshin, 2000; Greenwood et al., 2008; Usui et al., 2012, 2015; Kurokawa et al., 2014). In addition, SAM atmospheric measurements of CO_2, N_2, H_2O, and a variety of noble gases have secured values for $^{15}N/^{14}N$, $^{38}Ar/^{36}Ar$, $^{40}Ar/^{36}Ar$, as well as the numerous Kr and Xe isotopes in the Martian atmosphere. The substantial fractionation observed in atmospheric $\delta^{13}C$, $\delta^{18}O$, $\delta^{15}N$, and δ^2H are all signatures of atmospheric loss to space over time leaving the atmosphere enriched in the heavier isotopes (e.g., Webster et al., 2013; Mahaffy et al., 2013; Atreya et al., 2013), while the xenon isotopes are signatures of processes on early Mars (Conrad et al., 2016). These isotopic measurements in gases evolved from soils and rocks provide insight into a variety of processes including rock age and exposure age chronologies (Farley et al., 2014) in the case of the light evolved noble gases. Isotope fractionation effects of $\delta^{37}Cl$ (e.g., Farley et al., 2016) and $\delta^{34}S$ can also provide clues to understanding S and Cl redox cycling on Mars (Franz et al., 2016).

The remarkable aspects of this SAM data, when evaluated alongside mineralogical data from the CheMin X-ray diffraction instrument, are that Gale crater sediments consist of oxidized (sulfate, nitrate, oxychlorine) and reduced (sulfides, organics) phases along with mineralogies that suggest acidic (e.g., Fe-sulfate) and alkaline (e.g., phyllosilicate, carbonate, and apatite) conditions (e.g., Bish et al., 2013; Blake et al., 2013; Leshin et al., 2013; Ming et al., 2014; Vaniman et al., 2014; Morris et al., 2016; Rampe et al., 2017). This mix of reduced/oxidized and acidic/alkaline mineralogies demonstrates that Gale crater was exposed to varying geochemical conditions at likely differing time periods, represented by mineral inventories at different stratigraphic levels. The inferred past geochemical conditions may have been favorable for microbial life (e.g., Grotzinger et al., 2014), even those settings where acidic (Fe-sulfate) and alkaline (phyllosilicate, apatite) mineralogies coexist (Hurowitz et al., 2017; Rampe et al., 2017).

12.2 PREVIOUS IN SITU VOLATILE ANALYSES OF MARTIAN SURFACE MATERIAL

The previous Viking (1976) and Phoenix (2008) lander missions also sought to utilize thermal analytical techniques to evaluate the volatile content of Martian sediments with the goal of detecting organics (e.g., Biemann et al., 1977; Boynton et al., 2009; Hecht et al., 2009; Smith et al., 2009). Loose unconsolidated surface sediment from Chryse Planitia and Utopia Planitia were examined by the Viking Lander's gas chromatography/mass spectrometry (GCMS) instrument where samples underwent "flash" (1−8 s) pyrolysis to selected temperatures of up to 500 °C. Viking GCMS detected 1 wt% H_2O, 50−700 ppm CO_2 and background organics (Biemann et al., 1977). Proposed sources of water include adsorbed water, hydrated salts, and phyllosilicate water with adsorbed CO_2 offered as the only likely candidate source of CO_2 (Biemann et al., 1977). Subsequent laboratory analog work suggested oxidation of organics during Viking GCMS analysis could have been responsible for some of the CO_2 release (Navarro-Gonzalez et al., 2006), and that some of the detected chlorinated organics were the result of Martian perchlorate reacting with Martian organics (Navarro-González et al., 2010). These reinterpretations of the Viking data, however, have been challenged (Biemann, 2007; Biemann and Bada, 2011). The chlorinated organics detected by Viking have been suggested to be consistent with reactions of background organics with Martian oxychlorine (Sutter et al., 2016).

The thermal evolved gas analyzer (TEGA) instrument (Hoffman et al., 2008; Boynton et al., 2009; Hecht et al., 2009; Smith et al., 2009) aboard the Phoenix lander detected water, CO_2, and O_2 in northern polar surface material. Unlike the Viking GCMS analyses, which rapidly heated samples to select temperatures, TEGA heated samples from Mars ambient to 1000 °C at 20 °C/min. Furthermore, the TEGA instrument consisted of a scanning calorimeter connected to a mass spectrometer, which allowed for simultaneous thermal (endothermic vs exothermic) analysis with evolved gas analysis (Boynton et al., 2009). TEGA detected low-temperature water (295 °C−735 °C) that was attributed to dehydration of hydrated salts and/or dehydroxylation of goethite, phyllosilicates, and/or jarosite, while the high-temperature (>735 °C) water was attributed to dehydroxylation of phyllosilicates, serpentine, and/or amphiboles (Smith et al., 2009). Evolved CO_2 detected between 400 °C and 680 °C was consistent with combustion of organics and/or Fe-rich carbonate while the high-temperature (>725 °C) CO_2 release and corresponding endotherm release was attributed to Ca-rich carbonate (Boynton et al., 2009; Sutter et al., 2012). Wet Chemistry Laboratory (WCL) analysis of the Phoenix soil resulted in solution pH of 7.8, which is also consistent with the presence of carbonate (Hecht et al., 2009). Evolved O_2 detected between 325 °C and 625 °C was attributed to thermal decomposition of perchlorate (Hecht et al., 2009). Perchlorate in the Phoenix soil was confirmed by WCL (ion-selective electrode) soil solution analysis (Hecht et al., 2009; Kounaves et al., 2014; Toner et al., 2014). Other than possible detection of organic combustion to CO_2, no other evidence of organic-C detection by TEGA has been reported.

12.3 SAMPLE ANALYSIS AT MARS ANALYSES OBJECTIVES

The Curiosity rover's SAM instrument along with the Chemistry Mineralogy (CheMin) X-ray diffractometer (Bish et al., 2013; Blake et al., 2013; Ming et al., 2014; Vaniman et al., 2014; Treiman et al., 2015; Morris et al., 2016), Alpha Particle X-ray Spectrometer (APXS) (e.g., Blake et al., 2013; McLennan et al., 2014), Chemistry Camera (ChemCam) (e.g., Meslin et al., 2013; Rapin et al., 2016), and Dynamic Albedo of Neutrons (DAN) instrument (Mitrofanov et al., 2014) have all continued to significantly enhance our understanding of the volatile nature of Martian materials. The mobility provided by the Curiosity rover, in contrast to fixed landers, enables numerous samples to be analyzed in scoop or targeted drill locations along stratigraphic sections to reveal geochemical and mineralogical diversity. The remainder of this discussion will provide an assessment of key SAM detections of evolved gas (water, SO_2, NO, CO_2, CO, and O_2) from 11 separate sampling locations. Results of this work will provide insight into the mineralogy and geochemical history of Gale crater, Mars. The *D/H* ratio of water from a mudstone will also be evaluated in an effort to understand water loss on Mars.

12.4 GALE CRATER SAMPLES AND ANALYSIS METHODS
12.4.1 SAMPLES

The eleven samples analyzed by SAM were acquired in Gale crater along a 12-km traverse as of sol 1237 (Fig. 12.1). One eolian drift deposit at Rocknest (RN) (Fig. 12.2A) and an eolian

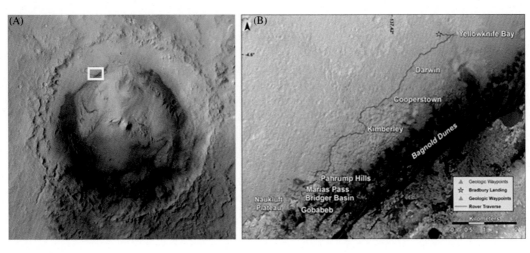

FIGURE 12.1

(A) Gale crater, Mars, where the white box indicates the location of the Mars Science Laboratory, Curiosity rover traverse. (B) The Mars Science Laboratory, Curiosity rover traverse. Rocknest was sampled $\sim$100 m to southwest of the John Klein and Cumberland sampling sites in Yellowknife Bay. Windjana was sampled at Kimberly. The Confidence Hills, Mojave, and Telegraph Peak samples were aquired in the Pahrump Hills region. The Buckskin sample was obtained near Marias Pass while the Big Sky and Greenhorn sediments were aquired in Bridger Basin. Gobabeb samples were scooped from the Bagnold dune field. *Image credit: NASA/JPL-Caltech/University of Arizona.*

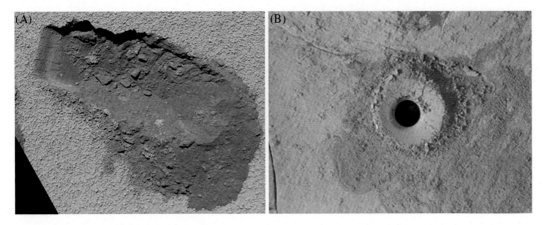

FIGURE 12.2

(A) Representative scooping of Martian eolian drift material at Rocknest. Rocknest scoop 3 width is 4 cm wide (MAHLI-0084MH0001120010100946E01, sol 84). (B) Representative drill hole sample aquired at Mojave 2 (MAHLI- 0882MH0003970010302481C00, sol 882). Drill hole diameter is 1.6 cm. *Image credit: NASA/JPL-Caltech/MSSS.*

dune sample at Gobabeb (GB) were acquired by the rover's scoop. SAM analysis of RN consisted of four subsamples that were analyzed individually by SAM. The reported contents for all evolved gases represent an average of the four RN subsamples of each evolved gas. Two subsamples of GB were acquired (GB1, GB2). Only GB1 contents are reported because an unknown volume of GB2 was delivered to SAM for analysis. The amount of scooped coarser grained (150 μm to 1 mm) GB2 sample delivered to SAM was unknown because sample delivery models used to calculate delivery amount were based on finer grained components ($<$ 150 μm) that encompassed all other samples analyzed by SAM. The sample delivery models for the coarse grained components have yet to be developed. The RN and GB samples were the first and last samples, respectively, to be analyzed by SAM between landing and sol 1237.

Nine sedimentary rock samples subsequent to RN were acquired with a drill (1.6 cm diameter, ∼5−6 cm deep hole) (Fig. 12.2B; Anderson et al., 2012). The acquisition order was John Klein (JK; mudstone), Cumberland (CB; mudstone), Windjana (WJ; sandstone), Confidence Hills (CH; mudstone), Mojave 2 (MJ; mudstone), Telegraph Peak (TP; mudstone), Buckskin (BK; mudstone), Big Sky (BS; sandstone), and Greenhorn (GH; sandstone). SAM analysis of JK encompassed four subsamples (JK1, JK2, JK3, and JK4). The JK1, JK2, and JK3 samples were exposed to a preanalysis heating procedure to minimize background contributions from N-methyl-N-(tert-butyldimethylsilyl)trifluoroacetamide MTBSTFA (Ming et al., 2014; Freissinet et al., 2015). Only JK4 is presented because it did not undergo a preanalysis heating procedure, which allowed the comparison of the JK4 results to RN and the other samples that were similarly prepared. The CB analyses consisted of seven subsamples of which CB1, CB2, CB3, and CB5 underwent analytical procedures similar to the RN samples and JK4. The remaining drilled samples and the GB dune material were heated with the same procedure as RN and JK4. Several CB1 evolved gas abundances were much less than CB2, CB3, and CB5, which may have been, for reasons not entirely clear, attributed to less sample being dumped into the cup relative to the other subsamples; thus, the CB1 subsample will not be discussed. The evolved gas release profiles for CB2, CB3, and CB5 are similar (Ming et al., 2014); thus, only CB2 profiles are presented. CB contents are reported as average of CB2, CB3, and CB5. All SAM-analyzed samples except GB2 were passed through a 150-μm sieve in the rover's sample handling system. The 150 μm to 1 mm fraction of GB2 sample was analyzed by SAM.

Samples were either drilled or scooped by the Sample Acquisition/Sample Processing and Handling (SA/SPaH) hardware and then delivered to SAM (Anderson et al., 2012). All samples were sieved ($<$ 150 μm) except one (GB2) (150 μm to 1 mm) and then fed into a portioning tube (76 mm^3), followed by delivery into one of the SAM quartz sample cups. Subsampling for the JK, CB, and GH materials consisted of samples derived from the same bulk sample in the SA/SPaH that were individually portioned and delivered to different SAM cups for separate analyses. Most samples consisted of single portion delivery to SAM except for JK3, WJ, BK, BS, GH, and GB1, which were all triple portions. Experimental tests done on Earth with analog materials before flight coupled with theoretical models determined that single and triple portions consisted of 45 ± 18 and 135 ± 31 mg (2σ standard deviation), respectively, for $<$ 150 μm grain sizes (Archer et al., 2014; Ming et al., 2014). Portion masses have yet to be determined for $>$ 150 μm grain sizes.

12.4.2 SAMPLE ANALYSIS AT MARS INSTRUMENT

The SAM instrument operates by heating samples in one of two ovens to $\sim 870\,°C$ (35 °C/min) under a helium purge (~ 0.8 sccm; 25 mbar) where a portion of evolved gases was sent for analysis by the QMS, which is known as SAM-evolved gas analyzer (SAM-EGA). The remaining portion of evolved gases was either analyzed by the GCMS or a TLS (Mahaffy et al., 2012). Evolved inorganic gases (e.g., H_2O, SO_2, NO, CO_2, CO, and O_2) from volatile bearing phases (e.g., minerals and amorphous materials) were released at characteristic temperatures that can be used to identify the presence of particular mineral phases. Organics were detected either by the SAM-EGA or by trapping a portion of the gas released from the sample for subsequent GCMS analysis. This work focuses on results on evaluating evolved gases as detected by the SAM-EGA.

A portion of the released gas stream may also be diverted to the TLS where two lasers scan rovibrational lines within a narrow wavelength region around 2.78 and 3.27 μm to specifically target CH_4, CO_2, and H_2O and their C, O, and H isotopes. The 81-pass Herriott cell with mirror separation of nearly 20 cm designed to provide sub-ppb volume mixing ratio measurements for atmospheric methane provides excellent sensitivity as well for evolved CH_4 and its isotopes. *D/H* in water can also be independently derived from the SAM-EGA data. The *D/H* ratios reported here were determined by SAM-EGA and SAM-TLS analyses.

12.4.3 MTBSTFA CONTRIBUTIONS TO CO_2, CO, AND NO

The SAM-GCMS derivatization agent MTBSTFA was determined to have contributed to the SAM-EGA detections of CO_2 and NO (e.g., Glavin et al., 2013; Ming et al., 2014; Freissinet et al., 2015; Stern et al., 2015). MTBSTFA was sealed in seven cups inside SAM that would undergo cup puncture to release MTBSTFA for a derivatization GCMS analysis. Thermal decomposition fragments of MTBSTFA unfortunately were detected by SAM-EGA during analysis of the first sample (RN) indicating at least one of the cups leaked MTBSTFA. Combustion of MTBSTFA by O_2 released from Martian oxychlorine was therefore presumed to have contributed to some of the evolved CO_2 and NO detected by SAM-EGA. Glavin et al. (2013) determined that 900 nmol carbon from combustion of MTBSTFA for each analysis, whether it was a single or triple portion, contributed to evolved CO_2 detections. Thus 900 nmol of carbon was used as worst-case scenario and was subtracted from each CO_2 content determination to provide a corrected CO_2 content.

The MTBSTFA contribution to the NO detected by SAM-EGA utilized the background correction method by Stern et al. (2015). MTBSTFA decomposes into nitrogen-free ion fragments that were detected by SAM-EGA that include Tert-butyldimethylsilanol (monosilylated H_2O, MSW) (*m/z* 75), 1,3-bis(1,1-dimethylethyl)-1,1,3,3-tetramethyldisiloxane (bisilylated H_2O, BSW) (*m/z* 147), tert-butyldimethylfluorosilane (TBDMS-F) (*m/z* 134), 2-methylpropene (C_4H_8) (*m/z* 41), and a contribution at *m/z* 15, either CH_4 or methylene ions (Stern et al., 2015; Supplementary Information). The detection of these fragment masses in the sample of interest indicates how much MTBSTFA decomposed. Each MTBSTFA molecule has one nitrogen and thus the amount of nitrogen presumably evolved as NO when MTBSTFA decomposed can be calculated as follows:

$$\text{MTBSTFA-nitrogen} = 2 \times BSW + MSW + TBDMS\text{-}F + C_4H_8 + 1/5 \times CH_4$$
$$\text{Corrected-nitrogen} = \text{Total nitorgen detected by SAM-EGA} - \text{MTBSTFA-nitrogen}$$

Contributions to evolved CO_2 from MTBSTFA combustion were estimated to range up to 36% of detected CO_2 for all samples except TP, which may have as much as 60% CO_2 derived from MTBSTFA (Sutter et al., 2017). MTBSTFA contributions to evolved NO ranged up to 58% for all samples except the CH sample, which could have as much as 89% of its NO from MTBSTFA (Stern et al., 2015; Sutter et al., 2017). MTBSTFA contributions to NO were high in some cases because so little NO evolved from the sample. The CO_2 carbon and NO nitrate contents are plotted as both uncorrected and corrected for comparative purposes (see Figs. 12.6 and 12.8).

12.4.4 O_2 CORRECTIONS FROM NITRATE DECOMPOSITION

Evolved oxygen used to calculate the amount of oxychlorine present must remove O_2 contributions from nitrate thermal decomposition (Ettarh and Galwey, 1996), for example,

$$2Ca(NO_3)_2 \rightarrow 2CaO + 4NO + 3O_2 \tag{12.1}$$

Nitrate contribution of O_2 was calculated by multiplying the MTBSTFA-corrected nitrate amount by 0.75 and subsequently subtracting this nitrate-oxygen from total oxygen evolved at temperatures consistent with oxychlorine thermal decomposition. Evolved oxygen above 600 °C from sulfate decomposition were not corrected for nitrate-oxygen as these temperatures were too high to have contributions from nitrate thermal decomposition. Between 1% and 8% of the total evolved O_2 could be derived from nitrate thermal decomposition. This O_2 correction was applied to the perchlorate (ClO_4) abundances presented in Fig. 12.5B.

12.4.5 ALPHA PROTON X-RAY SPECTROMETER

Total S contents (SO_3 wt%) of all samples were determined by the APXS and are compared to SAM detections of SO_2 (reported as SO_3 wt%). The APXS method, the instrument, and operation are detailed elsewhere (Gellert et al., 2006, 2015; Campbell et al., 2012; Schmidt et al., 2014; Berger et al., 2016; Thompson et al., 2016). Briefly, the scoop or drilled material was acquired by the rover's SA/SPaH system, which then passed portions of the sample through a sieve to SAM for analysis (Anderson et al., 2012). Sieved material similar to what was analyzed by SAM was dumped onto the Martian surface at a later date and subsequently analyzed by the APXS. This "post-sieve" sample APXS analysis containing total S content was then compared with the SAM-EGA S content. The APXS determines the elemental composition of the sample using X-ray spectroscopy by excitation with high energetic alpha-particles and X-rays from the internal ^{244}Cm sources. The method resembles a combination of the terrestrial standard methods of particle-induced X-ray emission (PIXE) and X-ray fluorescence (XRF). Sulfur is identified by its characteristic X-ray energy and quantified with small statistical uncertainty down to a few 1000 ppm abundance. The signal for S comes from about $5-10\,\mu m$ within the sample. The S unfortunately occurs within heterogeneous natural samples that can cause absorption effects of the S X-ray signal, which can skew abundance determinations. This results in an accuracy error ($\pm 15\%$ of measured S value) that is attributed to the APXS calibration, which assumes a homogeneous sample for the correction of self-absorption within the sample.

12.5 RESULTS/DISCUSSION

12.5.1 H_2O

Water (0.9 ± 0.3 to 2.2 ± 1.4 wt%) was detected in all samples and evolved over the entire SAM temperature range (Fig. 12.3). Possible sources of water include adsorbed water ($< \sim 200$ °C), phyllosilicate interlayer water (100 °C−300 °C) and hydrated salts (e.g., bassanite, perchlorate salts) (< 350 °C). Dehydroxylation of CheMin-detected jarosite in CH, MJ, and TP (Rampe et al., 2017) is consistent with evolved water peak between 450 °C and 500 °C in those samples. Phyllosilicates detected by CheMin (Ming et al., 2014. Vaniman et al., 2014; Treiman et al., 2015; Rampe et al., 2017) are likely responsible for water evolved between 650 °C and 800 °C in CB, JK, and MJ. Phyllosilicates were not detected but amorphous material was indicated by CheMin in BK, BS, and GH (Rampe et al., 2017; Yen et al., 2017). This suggests that inclusion water in the amorphous material is a possible source of the broad temperature water release above 500 °C in BK, BS, and GH.

The D/H ratio measured by SAM may suggest a significant loss of the Martian water inventory before 3 Ga if the D/H ratio in the modern atmosphere can be interpreted as the result of atmospheric escape over geologic time from an initial inventory. Some suggest the imprint of more recent events (e.g., ice-rich impacts, outflow floods) could affect Martian D/H ratios (Carr, 1990). The D/H ratio measured by the SAM-EGA and TLS on high temperature water releases (> 550 °C) is the most strongly bound water or hydroxyl groups in the CB drill sample of Yellowknife Bay phyllosilicates. This D/H ratio in the water reflects the Martian atmosphere D/H ratio when the phyllosilicates formed during the Hesperian era (> 3 Ga). This ratio (Mahaffy et al., 2015) of 3 times the standard mean ocean water (SMOW) is less than half that of the current Mars atmosphere (Villanueva et al., 2015). Evidence from a Martian meteorite suggests that the D/H ratio on pre-Noachian (4.5 Ga) Mars was slightly higher ($< 1.3 \times$) than SMOW (Usui et al., 2012). The higher ($3 \times$ SMOW) CB D/H value that is nevertheless less than half the value of the reservoir of water that can readily exchange with the current Mars atmosphere provides insight into the history of hydrogen escape to space over time. The D/H ratio of the Hesperian (> 3 Ga) CB phyllosilicates midway between more ancient and modern Mars is evidence for substantial loss of water with escape of hydrogen to space prior to the formation of the clays in these mudstones with continued loss of water after this time. Water released from samples at lower temperatures in SAM-EGA experiments shows a D/H ratio closer to the current Mars atmosphere suggesting partial exchange with the water in the current atmosphere.

12.5.2 SO_2

Most SO_2 was evolved above 500 °C in all samples with minor SO_2 evolution below 500 °C in CB and GB2 (Fig. 12.4A). Iron- and/or Mg-sulfate ranging from crystalline (e.g., jarosite) to amorphous sulfate, and adsorbed sulfate forms are candidate sources of SO_2 above 500 °C (McAdam et al., 2014). Minor sulfides oxidized to stable sulfates by O_2 derived from oxychlorine decomposition could contribute to some SO_2 evolved above 500 °C. The shape of the major SO_2 release profiles above 500 °C typically consists of at least two "peaks" within the broad temperature release (e.g., RN, CB, MJ), from ~ 500 °C to 700 °C and > 700 °C (Fig. 12.4A). The lower temperature

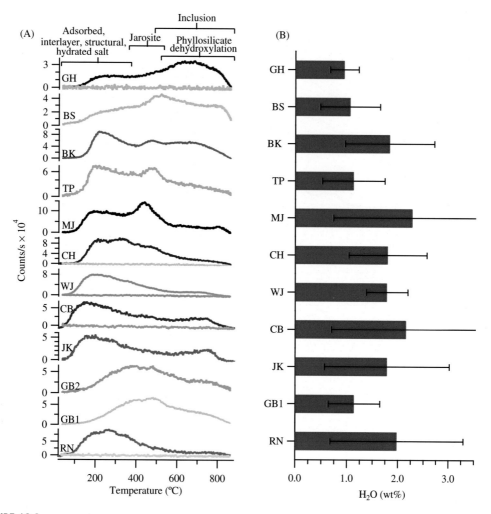

FIGURE 12.3

(A) Evolved water (m/z 20) (counts/s) vs temperature as detected by the SAM-EGA. Evolved water release is represented by m/z 20 isotopologue of water as m/z 18 saturated the detector. Lighter toned plots for RN, CB, CH, and GH refer to empty cup analyses that were run before their respective sample analysis. Temperature ranges for proposed sources of evolved water are noted with brackets. (B) Water content detected by the SAM-EGA.

peak is consistent with Fe-sulfate phases that typically decompose between 500 °C and 700 °C (e.g., McAdam et al., 2014). Magnesium sulfate typically decomposes at temperatures higher than those reached by SAM analyses but when mixed with other phases, the thermal decomposition temperature of Mg-sulfate can drop to within the SAM temperature range (Mu and Perlmutter, 1981a). This suggests that SO_2 peaks detected above 700 °C are candidates for $MgSO_4$.

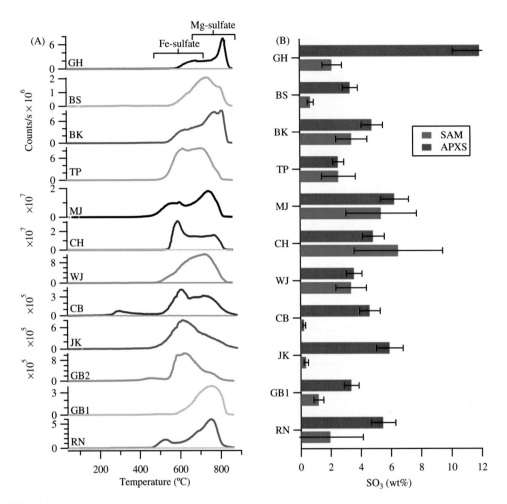

FIGURE 12.4

(A) Evolved SO_2 (m/z 64) (counts/s) vs temperature as detected by the SAM-EGA. Lighter toned plots for RN, CB, CH, and GH refer to empty cup analyses that were run before their respective sample analysis. Proposed sources of evolved SO_2 are noted with brackets. Y-axis scale 10^6 counts/s except where noted. (B) SO_2 content detected by the SAM-EGA converted to SO_3 and compared to APXS SO_3 content.

Aluminum-sulfate and/or organo-sulfur bearing phases are additional candidates for the evolved SO_2 (e.g., Mu and Perlmutter, 1981a; McAdam et al., 2014; Eigenbrode et al., 2015). There is no geochemical or mineralogical evidence to suggest that Al-sulfate is present, but similar to the Fe-sulfates detected by SAM and CheMin, Al-sulfates form under acidic conditions and could be present in low concentrations. The combustion of organic-S could contribute to evolved SO_2. The SAM-EGA detections of CO_2 and CO (discussed below) are consistent with the presence of organic C that could possess some S. Based on the levels of detected C (up to ~ 2400 μg C/g) and assuming a meteoritic C source (12.3 C/S wt% ratio Murchison meteorite), organic-S would contribute a small portion (0.05 wt% SO_3) of total S detected by the SAM-EGA.

Oxidation of sulfide directly to SO_2 could have contributed to the minor SO_2 evolutions in CB and GB2 below 500 °C. Pyrrhotite (FeS) ($\sim$ 1 wt%) was detected by CheMin in CB, while sulfides in GB2 were not detected but could occur at concentrations below CheMin detection limits. Oxychlorine decomposition occurring below 600 °C could have provided O_2 (Fig. 12.5A) for sulfide oxidation resulting in SO_2 evolution (Ming et al., 2014).

Evolved S_2 is an unlikely source of the *m/z* 64 detection attributed to SO_2. Photochemical models suggest that elemental sulfur (e.g., S_8) was an expected solid phase that mixed into the Martian surface when there was significant input of volcanic sulfur gases into the atmosphere (Sholes et al., 2016). The SAM-EGA detected no evidence for corresponding elemental sulfur S_8 (*m/z* 256) or S_4 (*m/z* 128). However, elemental S oxidation to sulfate by evolved O_2 from decomposing oxychlorine cannot be completely excluded as a possibility. Such a sulfate could have then thermally decomposed releasing the detected SO_2.

The disparity of the SAM and APXS-SO_3 contents demonstrate that more samples are dominated by Ca-sulfate than Fe- and Mg-sulfates. Seven samples (RN, GB, JK, CB, BK, BS, and GH) have SAM-SO_3 contents that are less than total APXS-SO_3 while only four samples (WJ, CH, MJ, and TP) have similar SAM and APXS-SO_3 contents (Fig. 12.4B). The inequality of SAM and APXS SO_3 contents is caused by the presence of Ca-sulfate, which is stable under SAM-oven temperatures and will not decompose to evolve SAM detectable SO_2. Only in samples that are dominated by Fe- and Mg- sulfate, APXS and SAM-SO_3 contents are similar. This is because Fe- and Mg-sulfate can decompose and evolve SO_2 under SAM operating temperatures (Sutter et al., 2017). When this inequality occurred between SAM and APXS, samples are typically observed to have bright veins that are enriched in Ca and S (e.g., Grotzinger et al., 2014; McLennan et al., 2014). The Ca-sulfate veins typically cross cut sediments in which they were present indicating that these veins postdate sediment deposition. The presence of more Ca-sulfate than Fe- and Mg-sulfate in the eolian RN and GB samples also suggest that Ca-sulfate could dominant regional and local and possibly global sulfate more so than Fe- and Mg- sulfate.

12.5.3 O_2

Oxygen releases (<600 °C) were detected in all samples suggesting the presence of oxychlorine phases (e.g., chlorate/perchlorate) (0.05 ± 0.01 to 1.05 ± 0.44 wt% ClO_4) (Fig. 12.5; Glavin et al., 2013; Leshin et al., 2013; Ming et al., 2014; Sutter et al., 2017). The oxychlorine cation as well as the oxychlorine species affects the oxychlorine decomposition temperature, which typically follows the order of Fe-perchlorate < Mg-chlorate < Ca-chlorate < Mg-perchlorate < Ca-perchlorate $\sim$ Na-chlorate < Na-perchlorate < K-chlorate < K-perchlorate (e.g., Glavin et al., 2013; Sutter et al., 2015). The lowest temperature O_2 release such as JK and CB could have more chlorate than perchlorate while the remaining Gale crater samples could consist of chlorate, perchlorate, or mixtures of the two. Further complicating the identification of oxychlorine species are the Gale crater Fe- phases (e.g., hematite, magnetite) that catalyze perchlorate and chlorate to lower thermal decomposition temperatures relative to their pure phases (Rudloff and Freeman, 1970; Sutter et al., 2015; Clark et al., 2016).

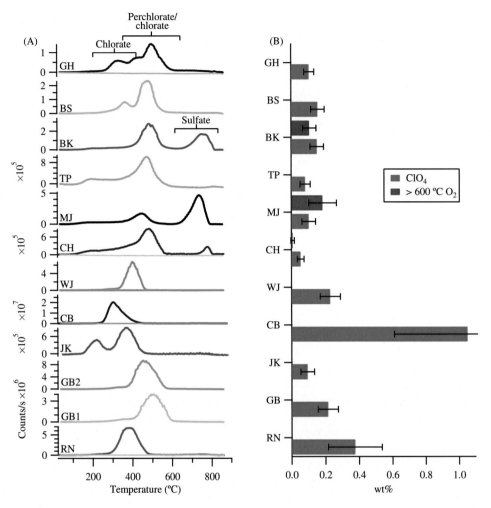

FIGURE 12.5

(A) Evolved O_2 (m/z 32) (counts/s) vs temperature as detected by the SAM-EGA. Lighter toned plots for RN, CB, CH, and GH refer to empty cup analyses that were run before their respective sample analysis. Proposed sources of evolved O_2 are noted with brackets. Y-axis scale is 10^6 counts/s except where noted. (B) O_2 content below 600 °C calculated as perchlorate (ClO_4) and calculated as O_2 above 600 °C (> 600 °C$-O_2$).

Oxygen releases detected above 600 °C in CH, MJ, and BK are not expected for oxychlorine phases and are instead attributed to the thermal decomposition of Fe- and/or Mg- sulfate phases that are proposed for these samples (Fig. 12.4A). Sulfate thermal decomposition results in the evolution of O_2 as demonstrated by the decomposition of Mg-sulfate (e.g., Scheidema and Taskinen, 2011).

$$2MgSO_4 \rightarrow 2MgO + 2SO_2 + O_2 \tag{12.2}$$

The oxychlorine contents detected in all samples (<0.4 wt% ClO_4) except CB (1.05 wt% ClO_4) (Fig. 12.5B) were at quantities below the CheMin X-ray diffraction instrument detection limits (1−3 wt%). The CB possesses oxychlorine abundance that should be detectable by CheMin. This suggests that either the CB oxychlorine phase is amorphous or that it consists of multiple perchlorate/chlorate salts that individually occur at contents below CheMin detection limits.

Variation in initial oxychlorine solution concentrations for each depositional episode or variation in postdepositional processes (e.g., leaching or evaporation) are possible reasons for oxychlorine content differences in the Gale sedimentary rock samples. The high oxychlorine-containing CB (~1.1 wt% ClO_4) samples occur in the topographically lowest portion of the traverse that may have accumulated oxychlorine through evaporative processes. The neighboring JK sample unexpectedly has much lower oxychlorine content than CB. This has been attributed to the postdepositional leaching along fractures that resulted in the loss of oxychlorine. The remaining sedimentary rock oxychlorine contents at stratigraphically higher locations were much lower than CB (<0.4 wt% ClO_4). The lower oxychlorine contents could reflect the range of initial oxychlorine solution concentrations that were responsible for providing oxychlorine to these sediments. Postdepositional leaching processes could alternatively have lowered any higher oxychlorine contents that were initially deposited in those sediments.

The detection of oxychlorine phases in the Gale crater samples adds further support to the view that oxychlorine phases are widespread across Mars. Perchlorate has been argued to be present at the Viking landing sites (Navarro-González et al., 2010; Sutter et al., 2016) and has been definitively detected at the Phoenix landing site (up to 0.6 wt% ClO_4) (Hecht et al., 2009; Kounaves et al., 2014) and select locations on Mars by orbital reflectance spectroscopy (Ojha et al., 2015). The widespread nature of oxychlorine is consistent with processes that involve the oxidation (photooxidation or radiolysis) of chloride either in the dust or in surface material (Catling et al., 2010; Carrier and Kounaves, 2015; Wilson et al., 2016).

12.5.4 CARBON (CO_2/CO)

Carbon dioxide (160 ± 248 to 2373 ± 820 μg CO_2-C/g; corrected for background CO_2) was detected over most of the SAM temperature range (Fig. 12.6), which was consistent with the presence of organic and inorganic C sources (Leshin et al., 2013; Glavin et al., 2013; Ming et al., 2014; Eigenbrode et al., 2015; Freissinet et al., 2015; Sutter et al., 2017). Organic sources of CO_2 include simple organic carbon (<300 °C), refractory macromolecular carbon (300 °C−600 °C), and trapped organics and magmatic carbon (>600 °C) (e.g., Campbell et al., 1980; Mu and Perlmutter, 1981b; Espitalié et al., 1984; Grady et al., 2002, 2004; Steele et al., 2012; Francois et al., 2016). Inorganic sources of CO_2 include adsorbed CO_2 (<250 °C), carbonates (>450 °C), and/or CO_2 inclusions in mineral or glass phases (>600 °C) (Hochstrasser and Antonini, 1972; Sutter et al., 2012; Manning et al., 2013).

Organic-C can be derived from exogenous and/or indigenous sources. Exogenous inputs of reduced and oxidized organic carbon from infall of meteorites and interplanetary dust particles (IDP) are expected for Mars (Flynn, 1996). High-temperature CO_2 releases are consistent with decomposition of organics trapped in minerals (e.g., Aubrey et al., 2006; Bowden and Parnell, 2007). Organics trapped within thermally decomposing sulfates were shown to evolve CO_2 (Francois et al., 2016). Martian magmatic carbon can be a source of abiotic indigenous organic-C,

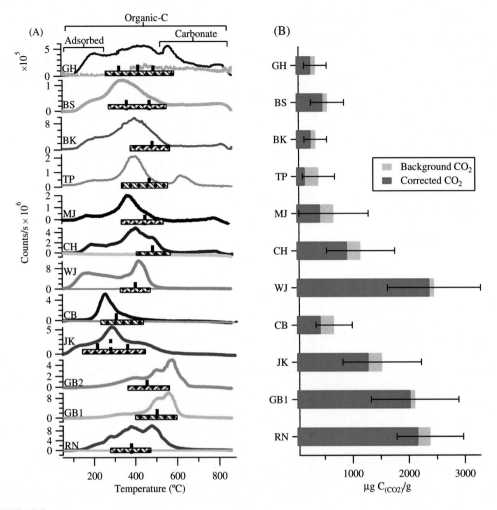

FIGURE 12.6

(A) Evolved CO_2 (m/z 44) (counts/s) vs temperature as detected by the SAM-EGA. Lighter toned plots for RN, CB, CH, and GH refer to empty cup analyses that were run before their respective sample analysis. Temperature ranges of proposed sources of evolved CO_2 are indicated in brackets. Y-axis scale is 10^6 counts/s except where noted. (B) Carbon content (μg $C_{(CO2)}/g$) as derived by evolved CO_2 release detected by the SAM-EGA. Darker bar refers to corrected CO_2 carbon content corrected for potential contributions of CO_2 from MTBSTFA represented by lighter toned bar.

which can occur as graphite, carbynes, hydrocarbons, and/or "amorphous" carbon (Mathez and Delaney, 1981). Magmatic carbon can occur in inclusions or vesicles of mineral or glass phases and along grain surfaces (Grady et al., 2004). Magmatic carbon has been detected in several Martian meteorites through pyrolysis and combustion and through Raman imaging (Leshin et al., 1996; Grady et al., 2004; Steele et al., 2007, 2012, 2016; Agee et al., 2013). Microbiological sources of indigenous organic-C are possible; however, unequivocal evidence for past or present microbiological activity on Mars has not been detected.

The detection of mostly oxygen-bearing C phases [CO_2 and CO (Figs. 12.6 and 12.7)] is consistent with the oxidation of initially reduced-C inputs on the Martian surface or combustion of carbon-bearing phases during analysis. Sources of O_2 for combustion are from oxychlorine decomposition. Low amounts of SAM-detectable reduced carbon phases (<300 ppb) were reported for the CB sample (Freissinet et al., 2015); however, all other samples have not resulted in the detection of reduced organic C above background levels below 450 °C. This suggests that initially deposited reduced organic-C could have been exposed to oxidative processes (e.g., photooxidation,

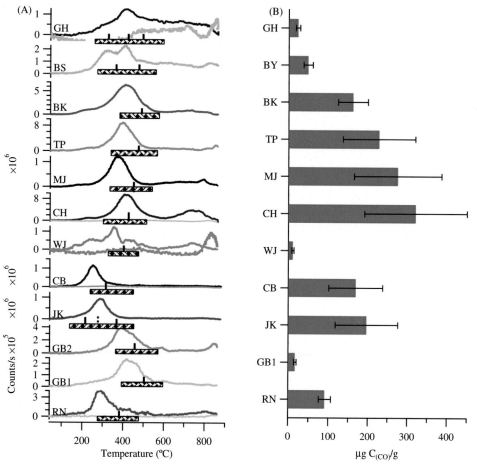

FIGURE 12.7

(A) Evolved CO (m/z 28) (counts/s) vs temperature as detected by the SAM-EGA. Lighter toned plots for RN, CB, CH, and GH refer to empty cup analyses that were run before their respective sample analysis. Y-axis scale is 10^5 counts/s except where noted. (B) CO carbon content (μg $C_{(CO)}$/g) detected by the SAM-EGA.

radiolysis) on the Martian surface (Benner et al., 2000; Pavlov et al., 2012; Eigenbrode et al., 2014; Applin et al., 2015) resulting in the formation of oxygen-bearing carboxylic acids [e.g., mellitic acid (RCOOH), acetate ($CH_3CO_2^-$), and oxalates ($C_2O_4^{2-}$)]. Thermal decomposition of these oxygen-bearing phases could explain the dominance of CO_2 and CO carbon detections by SAM. Another possibility is that evolved oxygen from decomposing oxychlorine and sulfates combusted reduced organic-C phases contributing to the detection of CO_2 and CO.

Inorganic sources of CO_2 include atmospherically adsorbed CO_2, inclusion CO_2, and/or carbonates. Adsorbed CO_2 on Martian material (e.g., Fanale and Cannon, 1974; Zent and Quinn, 1995) could contribute to CO_2 detected below 250 °C (Hochstrasser and Antonini, 1972; Jänchen, et al., 2009). Magmatic CO_2 gas inclusions within mineral or amorphous material can yield high-temperature CO_2 (>600 °C) releases during pyrolysis (e.g., Macpherson et al., 1999; Manning et al., 2013). Carbon dioxide evolved from 450 °C to 800 °C is consistent with contributions from carbonates (Leshin et al., 1996, 2013; Sutter et al., 2012; Cannon et al., 2012; Ming et al., 2014). The temperatures of the highest Gale CO_2 release peaks are consistent with Fe-rich carbonate for RN and CH, Mg-rich carbonate for TP, and Fe-Mg carbonate for GH and GB. The amounts of carbonates if present are low (<1 wt%) and below CheMin detection limits.

Carbon monoxide releases (11 ± 3 to 320 ± 130 µg CO-C/g) were detected in all samples and occurred over a wide temperature range (Fig. 12.7) and are consistent with the presence of organic-C. The majority of the CO releases occurred between 150 °C and 575 °C and were mostly coincident with at least one of the CO_2 release peaks, which suggests that CO and CO_2 releases were being derived from the same organic-C source. The CO content associated with CO_2 was always less than CO_2 (Fig. 12.7B). The presence of CO indicates that organic-C is present as CO evolution is not associated with adsorbed CO_2 or thermal decomposition of carbonate.

Background CO of an unknown source was detected in the blanks above 400 °C and especially above 750 °C where the most intense background CO releases were detected (Fig. 12.7A). The contribution from background CO between 400 °C and 700 °C was significant for GH and possibly WJ as these samples had low CO release intensities compared with the other samples. The background CO contributions above 750 °C were significant for GB, TP, and BS as CO release intensities were nearly similar for these samples and the blanks above 750 °C (Fig. 12.7A). Carbon contributions to CO from MTBSTFA were not likely as no evidence of MTBSTFA fragments in the blanks or samples was detected above 600 °C. There was no evidence that N_2 was the background source of m/z 28 because the m/z 14 peak normally associated with m/z 28 was not detected. Laboratory analog work will be required to understand nature of the CO background above 600 °C.

The loose unconsolidated eolian RN (2160 µg $C_{(CO2)}$/g) and GB (1680 µg $C_{(CO2)}$/g) samples had high carbon contents (Fig. 12.6B) indicating that these C contents could be widespread in locally and regionally and possibly globally loose surface fines. The CO_2-C contents are higher than what was detected in the loose unconsolidated material at the Viking 2 landing site (up to 191 µg CO_2-C/g). The maximum analytical temperature achieved by the Viking pyrolysis analyses was 500 °C, which required 8 s to achieve and was held there for 30 s (Biemann et al., 1977). This may have not been enough time to allow for as much CO_2 evolution as in SAM, which heated samples to higher temperatures (870 °C) and at much slower heating rate (35 °C/min). Nevertheless, the coarse grained (>100 µm) GB sediment consisted of mostly of >450 °C CO_2 releases that were consistent with carbonate but organic-C cannot be excluded. The finer grained (<150 µm) RN sample had mostly CO_2 releases below 450 °C, which was consistent with organic-C. These results suggest that the coarser grained eolian material consisted mostly of carbonate while the finer grained

($< 150\,\mu$m) material consisted of more organic-C. As discussed above, the amount of globally sourced material in RN was difficult to determine but the possibility exists that organic-C in surface fines could be prevalent all over Mars.

12.5.5 NO

The detection of evolved NO by SAM-EGA indicated the presence of nitrates (0.002–0.06 wt% NO_3 corrected for background) in all sediments (Fig. 12.8). Multiple peak NO releases suggests

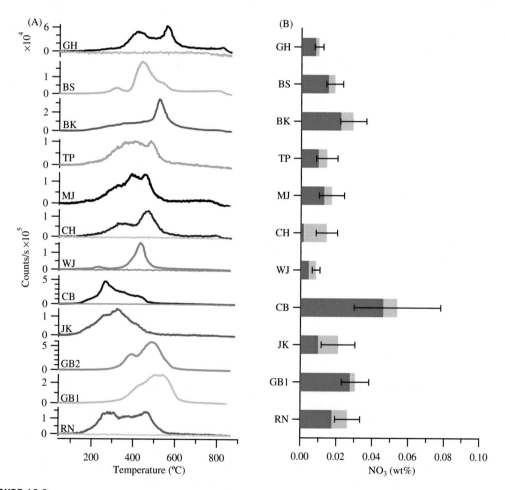

FIGURE 12.8

(A) Evolved NO (m/z 30) (counts/s) vs temperature as detected by the SAM-EGA. Lighter toned plots for RN, CB, CH, and GH refer to empty cup analyses that were run before their respective sample analysis. Y-axis scale is 10^5 counts/s except where noted. (B) Nitrate (NO_3^-) content (wt%) calcuated from evolved NO detected by the SAM-EGA. Darker bar refers to NO_3 content corrected for potential NO_3 contributions from MTBSTFA represented by lighter toned bar.

that multiple cation species may be associated with nitrates in the Martian sediments. Fe-bearing nitrates tend to evolve NO at lower temperatures followed by Mg-, Ca-, Na-, and K-bearing nitrates at correspondingly higher temperatures (e.g., Gordon and Campbell, 1955; Stern et al., 2015). Nitrate contents vary in Gale crater with highest contents in the CB and BK mudstone and eolian RN and GB samples while the CH mudstone and WJ sandstone had the lowest nitrate contents.

The nitrate content is variable throughout the Gale crater materials where evaporative processes and/or alteration fluid nitrate concentration can affect the final content of this highly soluble salt. Evaporative processes or postdepositional fluids enriched in nitrate are possible causes of relatively high nitrate contents in CB and BK. Nitrate and oxychlorine phases have similar solubilities suggesting the two might have accumulated in CB. Similar to oxychlorine, nitrate occurs in lower concentration in JK, which suggests that nitrate was leached by fluids depleted in nitrate that deposited the Ca-sulfate veins in JK. The accumulation of nitrate in heavily altered BK (e.g., Hurowitz et al., 2017; Rampe et al., 2017) is consistent with evaporation of postdepositional fluids enriched in nitrate. Lower nitrate contents in CH and WJ suggests that CH and WJ were leached of nitrate or were never exposed to evaporative environments and/or fluids enriched in nitrate.

The SAM detection of nitrate is significant as this is the first in situ Mars detection of nitrates. Similar to organic-C, the presence of nitrate in the eolian RN and GB samples suggests nitrate could be widespread on the surface of Mars. Following the grain size distribution discussion outlined in the carbon section for GB and RN samples, nitrate in those samples are likely derived from regional and/or local sources with possible sources derived from global dust. Noachian impacts (Manning et al., 2009) and atmospheric photochemical (Smith et al., 2014) processes are proposed to be the source of Martian nitrate and thus it would not be surprising if nitrate occurred globally.

12.6 SUMMARY

The SAM instrument results continue to build on the understanding of the volatile nature of Martian sediment and sedimentary rock that began with evolved gas analyses by the Viking and Phoenix lander missions. Significant SAM findings include the detections of nitrate, organic-C, oxychlorine phases, carbonate, Fe-sulfate, and Mg-sulfate (Table 12.1). Evolved H_2O was derived from multiple sources including adsorbed water, phyllosilicate interlayer and structural water, hydrated salts, jarosite-hydroxyls, and mineral/glass inclusion water. The *D/H* ratio of the Hesperian (>3 Ga) CB phyllosilicates midway between more ancient and modern Mars demonstrates that substantial amount of Martian water was lost with escape of hydrogen to space prior to the phyllosilicate formation in these mudstones with continued loss of water after this time. The first detections of nitrate and organic-C demonstrates that two major elements required for life are present in Martian materials suggesting that if life ever arose on Mars, N and organic-C may have been available to support life. The detection of oxychlorine in all Gale samples coupled with detections made elsewhere on Mars (e.g., Hecht et al., 2009; Ojha et al. 2015) indicates oxychlorine phases are widespread across Mars. Evolved SO_2 detected by SAM is consistent with Fe- and Mg-sulfates in all Gale crater samples while combined APXS-S and SAM-S content results

Table 12.1 Abundances of Water and Nitrate (NO₃) as Determined from Evolved Gases Detected by the SAM-EGA

Sample Type	Sample	H_2O wt%	NO_3 wt%	Sulfate	OC	CO_3	OxyCl
Eolian	RN	2.0 (1.3)	0.02 (0.01)	Ca > Fe, Mg	x	x	x
	GB1	1.1 (0.5)	0.03 (0.01)	Ca > Fe, Mg	x	x	x
Mudstone	JK	1.8 (1.2)	0.01 (0.01)	Ca ≫ Fe, Mg	x		x
	CB	2.5 (1.6)	0.06 (0.03)	Ca ≫ Fe, Mg	x		x
Sandstone	WJ	1.8 (0.4)	0.005 (0.002)	Fe, Mg ≫ Ca	x		x
Mudstone	CH	1.8 (0.8)	0.002 (0.007)	Fe, Mg ≫ Ca	x		x
	MJ2	2.3 (1.5)	0.01 (0.01)	Fe, Mg ≫ Ca	x		x
	TP	1.1 (0.6)	0.01 (0.01)	Fe, Mg ≫ Ca	x	x	x
	BK	1.8 (0.9)	0.02 (0.01)	Fe, Mg > Ca	x		x
Sandstone	BY	1.1 (0.6)	0.02 (0.01)	Ca ≫ Fe, Mg	x		x
	GH2	0.9 (0.3)	0.01 (0.003)	Ca ≫ Fe, Mg	x	x	x

Samples where the SAM-EGA analysis results have yielded detection of evolved gases consistent with presence of organic carbon (OC), carbonate (CO₃) and oxychlorine (OxyCl) species are indicated by "x." Relative abundances of Ca-sulfate vs Fe- and Mg-sulfates are indicated and are based on combined SAM-EGA, APXS, and CheMin analyses.
From Blake et al. (2013), Bish et al. (2013), Vaniman et al. (2014), Treiman et al. (2015), Rampe et al. (2017), and Yen et al. (2017).

indicated that Ca-sulfate has a significant presence as well. The detection of these sulfates in Gale crater continues to add to the growing evidence that sulfates are prevalent across Mars (e.g., Langevin et al., 2005; Morris et al., 2006; Mangold et al., 2008; Murchie et al., 2009). The presence of acidic conditions forming Fe-sulfates are difficult to reconcile when coexisting with samples containing CO_2 releases consistent with alkaline carbonate (e.g., RN, GB, and TP). Furthermore, all samples possess oxidized phases like nitrate, oxychlorine, and sulfate but some samples (e.g., JK and CB) have evidence for possessing reduced Fe-sulfides (Vaniman et al., 2014) and reduced hydrocarbons (Freissinet et al., 2015). The coexistence of acidic and alkaline mineralogies and oxidized and reduced mineralogies in the same samples suggests that Gale crater sedimentary rocks have been exposed to a complicated authigenic and diagenetic history that involved temporal and possibly spatial variations in pH, redox chemistry, and salt concentration.

REFERENCES

Agee, C.B., et al., 2013. Unique meteorite from Early Amazonian Mars: water-rich basaltic breccia Northwest Africa 7034. Science 339, 780–785.

Anderson, R.C., et al., 2012. Collecting samples in Gale Crater, Mars: an overview of the Mars Science Laboratory Sample Acquisition, Sample Processing and Handling System. Space Sci Rev. 170, 57–75.

Applin, D.M., Izawa, M.R.M., Cloutis, E.A., Goltz, D., Johnson, J.R., 2015. Oxalate minerals on Mars? Earth Planet. Sci. Lett. 420, 127–139.

Archer Jr., P.D., et al., 2014. Abundances and implications of volatile-bearing species from evolved gas analysis of the Rocknest aeolian deposit, Gale Crater, Mars. J. Geophys. Res. Planets 119, 237–254.

Atreya, S.K., et al., 2013. Primordial argon isotope fractionation in the atmosphere of Mars measured by the SAM instrument on Curiosity and implications for atmospheric loss. Geophys Res. Lett. 40, 5605–5609.

Aubrey, A., Cleaves, H.J., Chalmers, J.H., Skelley, A.M., Mathies, R.A., Grunthaner, F.J., et al., 2006. Sulfate minerals and organic compounds on Mars. Geology 34, 357–360.

Benner, S.A., Devine, K.G., Matveeva, L.N., Powell, D.H., 2000. The missing organic molecules on Mars. Proc. Natl. Acad. Sci. 97, 2425–2430.

Berger, J.A., et al., 2016. A global Mars dust composition refined by the Alpha-Particle X-ray Spectrometer in Gale Crater. Geophys. Res. Lett. 43. Available from: http://dx.doi.org/10.1002/2015GL066675.

Biemann, K., 2007. On the ability of the Viking gas chromatograph-mass spectrometer to detect organic matter. Proc. Natl. Acad. Sci. 104, 10310–10313.

Biemann, K., Bada, J.L., 2011. Comment on "Reanalysis of the Viking results suggests perchlorate and organics at midlatitudes on Mars" by Rafael Navarro-González et al. J. Geophys. Res. 116, E12001. Available from: http://dx.doi.org/10.1029/2011JE003869.

Biemann, K., et al., 1977. The search for organic substances and inorganic volatile compounds in the surface of Mars. J. Geophys Res. 82, 4641–4657.

Bish, D.L., et al., 2013. X-ray diffraction results from Mars Science Laboratory: mineralogy of Rocknest at Gale Crater. Science 341. Available from: http://dx.doi.org/10.1126/science.1238932.

Blake, D.F., et al., 2013. Curiosity at Gale Crater, Mars: characterization and analysis of the Rocknest Sand Shadow. Science 341. Available from: http://dx.doi.org/10.1126/science.1239505.

Bowden, S.A., Parnell, J., 2007. Intracrystalline lipids within sulfates from the Haughton Impact Structure: implications for survival of lipids on Mars. Icarus 187, 422–429.

Boynton, W.V., et al., 2009. Evidence for calcium carbonate at the Mars Phoenix Landing site. Science 325, 61–64.

Carr, M.H., 1990. D/H on Mars: effects of floods, volcanism, impacts and polar processes. Icarus 87, 210–227.

Carrier, B.L., Kounaves, S.P., 2015. The origins of perchlorate in the Martian soil. Geophys. Res. Lett. 42. Available from: https://doi.org/10.1002/2015GL064290.

Catling, D.C., Claire, M.W., Zahnle, K.J., Quinn, R.C., Clark, B.C., Hecht, M.H., et al., 2010. Atmospheric origins of perchlorate on Mars and in the Atacama. J. Geophys. Res. 115, E00E11. Available from: http://dx.doi.org/10.1029/2009JE003425.

Conrad, P.G., et al., 2016. In situ measurement of atmospheric krypton and xenon on Mars with Mars Science Laboratory. Earth Planet. Sci. Lett. 454, 1–9.

Campbell, J.H., Gallegos, G., Gregg, M., 1980. Gas evolution during oil shale pyrolysis. 1: Nonisothermal rate measurements. Fuel 59, 718–726.

Campbell, J.L., Perrett, G.M., Gellert, R., Andrushenko, S.M., Boyd, N.L., Maxwell, J.A., et al., 2012. Calibration of the Mars Science Laboratory alpha particle X-ray spectrometer. Space Sci. Rev. 170, 319–340.

Cannon, K.M., Sutter, B., Ming, D.W., Boynton, W.V., Quinn, R., 2012. Perchlorate induced low temperature carbonate decomposition in the Mars Phoenix Thermal and Evolved Gas Analyzer (TEGA). Geophys. Res. Lett. 39, L13203. Available from: http://dx.doi.org/10.1029/2012GL051952.

Clark, J.V., et al., 2016. The investigation of chlorate/iron-phase mixtures as a possible source of oxygen and chlorine detected by the Sample Analysis at Mars (SAM) instrument in Gale Crater. 47th Lunar Planet Sci Conf. Abstract #1537.

Eigenbrode, J.L., et al., 2014. Decarboxylation of carbon compounds as potential source for CO2 and CO observed by SAM at Yellowknife Bay, Gale Crater, Mars. 45th Lunar Planet. Sci. Conf. Abstract #1605.

Eigenbrode, J.L., et al., 2015. Evidence of refractory organic matter preserved in the mudstones of Yellowknife Bay and the Murray Formations. American Geophysics Union, Fall Meeting, #79168 San Francisco, CA, December 14−18.

Espitalié, J., Senga Makadi, K., Trichet, J., 1984. Role of the mineral matrix during kerogen pyrolysis. Org. Geochem. 6, 365−382.

Ettarh, C., Galwey, A.K., 1996. A kinetic and mechanistic study of the thermal decomposition of calcium nitrate. Thermochim. Acta 288, 203−219.

Fanale, F.P., Cannon, W.A., 1974. Exchange of adsorbed H_2O and CO_2 between the regolith and atmosphere of Mars caused by changes in surface insolation. J. Geophys. Res. 79, 3397−3402.

Farley, K.A., et al., 2014. In-situ radiometric and exposure age eating of the Martian surface. Science 343. Available from: http://dx.doi.org/10.1126/science.1247166.

Farley, K.A., et al., 2016. Light and variable $^{37}Cl/^{35}Cl$ ratios in rocks from Gale Crater, Mars: possible signature of perchlorate. Earth Planet. Sci. Lett. 438, 14−24.

Flynn, G.J., 1996. The delivery of organic matter from asteroids and comets to the early surface of Mars. Earth Moon Planets 72, 469−474.

Francois, P., Szopa, C., Buch, A., Coll, P., McAdam, A.C., Mahaffy, P.R., et al., 2016. Magnesium sulfate as a key mineral for the detection of organic molecules on Mars using pyrolysis. J. Geophys. Res. Planets 121. Available from: http://dx.doi.org/10.1002/2015JE004884.

Franz, H.B., et al., 2016. Large sulfur isotope fractionations in Martian sediments. Nat. Geosci. 10, 658−662.

Freissinet, C., et al., 2015. Organic molecules in the Sheepbed Mudstone, Gale Crater, Mars. J. Geophys. Res. Planets 120, 495−514.

Gellert, R., et al., 2006. Alpha Particle X-Ray Spectrometer (APXS): results from Gusev Crater and calibration report. J. Geophys. Res. 111, E02S05. Available from: http://dx.doi.org/10.1029/2005JE002555.

Gellert, R., et al., 2015. Chemical evidence for an aqueous history at Pahrump, Gale Crater, Mars, as seen by the APXS. 46th Lunar Planet. Sci. Conf. Abstract #1855.

Glavin, D.P., et al., 2013. Evidence for perchlorates and the origin of chlorinated hydrocarbons detected by SAM at the Rocknest aeolian deposit in Gale Crater. J. Geophys. Res. Planets 118, 1955−1973.

Gordon, S., Campbell, C., 1955. Differential thermal analysis of inorganic compounds: nitrates and perchlorates of alkali and alkaline Earth groups and subgroups. Anal. Chem. 27, 1102−1109.

Grady, M.M., Verchovsky, A.B., Franchi, I.A., Wright, I.P., Pillinger, C.T., 2002. Light element geochemistry of the Tagish Lake CI2 chondrite: comparison with CII and CM2 meteorites. Meteor. Planet. Sci. 37, 713−735.

Grady, M.M., Verchovsky, A.V., Wright, I.P., 2004. Magmatic carbon in Martian meteorites: attempts to constrain the carbon cycle on Mars. Int. J. Astrobio. 3, 117−124.

Greenwood, J.P., Itoh, S., Sakamoto, N., Vicenzi, E.P., Yurimoto, H., 2008. Hydrogen isotope evidence for loss of water from Mars through time. Geophys. Res. Lett. 35, L05203. Available from: http://dx.doi.org/10.1029/2007GL032721.

Grotzinger, J.P., et al., 2014. A habitable fluvio-lacustrine environment at Yellowknife Bay, Gale Crater, Mars. Science 343. Available from: http://dx.doi.org/10.1126/science.1242777.

Hecht, M.H., et al., 2009. Detection of perchlorate and the soluble chemistry of the Martian soil at the Phoenix Lander site. Science 325, 64−67.

Hochstrasser, G., Antonini, J.F., 1972. Surface states of pristine silica surfaces I. ESR studies of Es' dangling bonds and of CO_2-adsorbed radicals. Surf. Sci 32, 644−664.

Hoffman, J.H., Chaney, R.C., Hammack, H., 2008. Phoenix Mars Mission—the Thermal Evolved Gas Analyzer. J. Am. Soc. Mass. Spectrom. 19, 1377−1383.

Hurowitz, J.A., et al., 2017. Redox stratification of an ancient lake in Gale Crater, Mars. Science. Available from: http://dx.doi.org/10.1126/science.aah6849.

Jänchen, J., Morris, R.V., Bish, D.L., Janssen, M., Hellwig, U., 2009. The H_2O and CO_2 adsorption properties of phyllosilicate-poor palagonitic dust and smectites under Martian environmental conditions. Icarus 200, 463–467.

Kounaves, S.P., et al., 2014. Identification of the perchlorate parent salts at the Phoenix Mars landing site and possible implications. Icarus 232, 226–231.

Kurokawa, H., Sato, M., Ushioda, M., Matsuyama, T., Moriwaki, R., Dohm, J.M., et al., 2014. Evolution of water reservoirs on Mars: constraints from hydrogen isotopes in Martian meteorites. Earth Planet. Sci. Lett. 394, 179–185.

Langevin, Y., Poulet, F., Bibring, J.P., Gondet, B., 2005. Sulfates in the north polar region of Mars by OMEGA/Mars Express. Science 307, 1584–1586.

Leshin, L.A., Epstien, S., Stopler, E.M., 1996. Hydrogen isotope geochemistry of SNC meteorites. Geochim. Cosmochim. Acta 60, 2635–2650.

Leshin, L.A., 2000. Insights into Martian water reservoirs from analyses of Martian meteorite QUE94201. Geophys. Res. Lett. 27, 2017–2020.

Leshin, L.A., et al., 2013. Volatile, isotope, and organic analysis of Martian fines with the Mars Curiosity Rover. Science 341. Available from: http://dx.doi.org/10.1126/science.1238937.

Macpherson, C.G., Hilton, D.R., Newman, S., Mattey, D.P., 1999. CO_2, $^{13}C/^{12}C$ and H_2O variability in natural basaltic glasses: a study comparing stepped heating and FTIR spectroscopic techniques. Geochim. Cosmochim. Acta 63, 1805–1813.

Mahaffy, P.R., et al., 2012. The Sample Analysis at Mars investigation and instrument suite. Space Sci. Rev. 170, 401–478.

Mahaffy, P.R., et al., 2013. Abundance and isotopic composition of gases in the Martian atmosphere from the Curiosity Rover. Science 341, 263–266.

Mahaffy, P.R., et al., 2015. The imprint of atmospheric evolution in the D/H of Hesperian clay minerals on Mars. Science 347, 412–414.

Mangold, N., et al., 2008. Spectral and geological study of the sulfate-rich region of West Candor Chasma, Mars. Icarus 194, 519–543.

Manning, C.E., Shock, E.L., Sverjensky, D.A., 2013. The chemistry of carbon in aqueous fluids at crustal and upper-mantle conditions: experimental and theoretical constraints. Rev. Miner. Geochim. 75, 109–148.

Manning, C.V., Zahnle, K.J., McKay, C.P., 2009. Impact processing of nitrogen on early Mars. Icarus 199, 273–285.

Mathez, E.A., Delaney, J.R., 1981. The nature and distribution of carbon in submarine basalts and peridotite nodules. Earth Planet. Sci. Lett. 56, 217–232.

McAdam, A.C., et al., 2014. Sulfur-bearing phases detected by evolved gas analysis of the Rocknest aeolian deposit, Gale Crater, Mars. J. Geophys. Res. Planets 119, 373–393.

McLennan, S.M., et al., 2014. Elemental geochemistry of sedimentary rocks at Yellowknife Bay, Gale Crater, Mars. Science 343. Available from: http://dx.doi.org/10.1126/science.1244734.

Meslin, P.Y., et al., 2013. Soil diversity and hydration as observed by ChemCam at Gale Crater, Mars. Science 341. Available from: http://dx.doi.org/10.1126/science.1238670.

Ming, D.W., et al., 2014. Volatile and organic compositions of sedimentary rocks in Yellowknife Bay, Gale Crater, Mars. Science 343. Available from: http://dx.doi.org/10.1126/science.1245267.

Minitti, M.E., et al., 2013. MAHLI at the Rocknest sand shadow: science and science-enabling activities. J. Geophys. Res. Planets 118. Available from: http://dx.doi.org/10.1002/2013JE004426.

Mitrofanov, I.G., et al., 2014. Water and chlorine content in the Martian soil along the first 900 m of the Curiosity Rover traverse as estimated by the DAN instrument. J. Geophys. Res. Planet. 119, 1579–1596.

Morris, R.V., et al., 2006. Mössbauer mineralogy of rock, soil, and dust at Meridiani Planum, Mars: Opportunity's journey across sulfate-rich outcrop, basaltic sand and dust, and hematite lag deposits. J. Geophys. Res. 111, E12S15. Available from: http://dx.doi.org/10.1029/2006JE002791.

Morris, R.V., et al., 2016. Silicic volcanism on Mars evidenced by tridymite in high-SiO_2 sedimentary rock at Gale Crater. Proc. Natl Acad. Sci. 113, 7071−7076.

Mu, J., Perlmutter, D.D., 1981a. Thermal decomposition of inorganic sulfates and their hydrates. Ind. Eng. Chem. Process Des. Dev. 20, 640−646.

Mu, J., Perlmutter, D.D., 1981b. Thermal decomposition of carbonates, carboxylates, oxalates, acetates, formats, and hydroxides. Thermochim. Acta 49, 207−218.

Murchie, S.L., et al., 2009. A synthesis of Martian aqueous mineralogy after 1 Mars year of observations from the Mars Reconnaissance Orbiter. J. Geophys. Res. 114, E00D06. Available from: http://dx.doi.org/10.1029/2009JE003342.

Navarro-Gonzalez, R., et al., 2006. The limitations on organic detection in Mars-like soils by thermal volatilization-gas chromatography-MS and their implications for the Viking results. Proc. Natl Acad. Sci. 103, 16089−16094.

Navarro-González, R., Vargas, E., de la Rosa, J., Raga, A.C., McKay, C.P., 2010. Reanalysis of the Viking results suggests perchlorate and organics at midlatitudes on Mars. J. Geophys. Res. 115, E12010. Available from: http://dx.doi.org/10.1029/2010JE003599.

Ojha, L., Wilhelm, M.B., Murchie, S.L., McEwen, A.S., Wray, J.J., Hanley, J., et al., 2015. Spectral evidence for hydrated salts in recurring slope lineae on Mars. Nat. Geosci. 8, 829−832.

Pavlov, A.A., Vasilyev, G., Ostryakov, V.M., Pavlov, A.K., Mahaffy, P., 2012. Degradation of the organic molecules in the shallow subsurface of Mars due to irradiation by cosmic rays. Geophys. Res. Lett. 39, L13202. Available from: http://dx.doi.org/10.1029/2012GL052166.

Rampe, E.B., et al., 2017. Mineralogy of an ancient lacustrine mudstone succession from the Murray formation, Gale Crater, Mars. Earth Planet. Sci. Lett. Available from: http://dx.doi.org/10.1016/j.epsl.2017.04.021.

Rapin, W., et al., 2016. Hydration state of calcium sulfates in Gale Crater, Mars: identification of bassanite veins, Earth Planet. Sci. Lett. 452. pp. 197−205.

Rudloff, W.K., Freeman, E.S., 1970. Catalytic effect of metal oxides on thermal decomposition reactions. II. The catalytic effect of metal oxides on the thermal decomposition of potassium chlorate and potassium perchlorate as detected by thermal analysis methods. J. Phys. Chem. 74, 3317−3324.

Scheidema, M.N., Taskinen, P., 2011. Decomposition thermodynamics of magnesium sulfate. Ind. Eng. Chem. Res. 50, 9550−9556.

Schmidt, M.E., et al., 2014. Geochemical diversity in first rocks examined by the Curiosity Rover in Gale Crater: evidence for and significance of an alkali and volatile-rich igneous source. J. Geophys. Res. Planets 119, 64−81.

Sholes, S.F., Smith, M.L., Claire, M.W., Zahnle, K.J., Catling, D.C., 2016. Anoxic atmospheres on early Mars driven by volcanism: implications for past environments and life. Icarus. Available from: http://dx.doi.org/10.1016/j.icarus.2017.02.022

Sklute, E.C., Jensen, H.B., Rogers, A.D., Reeder, R.J., 2015. Morphological, structural, and spectral characteristics of amorphous iron sulfates. J. Geophys. Res. Planets 120, 809−830.

Smith, P.H., et al., 2009. H_2O at the Phoenix landing site. Science 325, 58−61.

Smith, M.L., Claire, M.W., Catling, D.C., Zahnle, K.J., 2014. The formation of sulfate, nitrate and perchlorate salts in the Martian atmosphere. Icarus 231, 51−64.

Steele, A., Fries, M.D., Amundson, H.E.F., Mysen, B.O., Fogel, M.L., Schweizer, M., et al., 2007. Comprehensive imaging and Raman spectroscopy of carbonate globules from Martian meteorite ALH 84001 and a terrestrial analogue from Svalbard. Meteoritics Planet. Sci. 42, 1549−1566.

Steele, A., et al., 2012. A reduced organic carbon component in Martian basalts. Science 337, 212−215.

Steele, A., McCubbin, F.M., Fries, M.D., 2016. The provenance, formation, and implications of reduced carbon phases in Martian meteorites. Meteoritics Planet. Sci. Available from: http://dx.doi.org/10.1111/maps.12670.

Stern, J.C., Sutter, B., Jackson, W.A., Navarro-González, R., McKay, C.P., Ming, D.W., et al., 2017. The nitrate/(per)chlorate relationship on Mars. Geophys. Res. Lett. 44. Available from: http://dx.doi.org/10.1002/2016GL072199.

Stern, J.S., et al., 2015. Evidence for indigenous nitrogen in sedimentary and aeolian deposits from the Curiosity Rover investigations at Gale Crater, Mars. Proc. Natl. Acad. Sci. 112, 4245−4250.

Sutter, B., Boynton, W.V., Ming, D.W., Niles, P.B., Morris, R.V., Golden, D.C., et al., 2012. The detection of carbonate in the Martian soil at the Phoenix Landing site: a laboratory investigation and comparison with the Thermal and Evolved Gas Analyzer (TEGA) data. Icarus 218, 290−296.

Sutter, B., et al., 2015. The investigation of perchlorate/iron phase mixtures as a possible source of oxygen detected by the Sample Analysis at Mars (SAM) instrument in Gale Crater, Mars. 46th Lunar Planet. Sci. Conf. Abstract #2137.

Sutter, B., Quinn, R.C., Archer, P.D., Glavin, D.P., Glotch, T.D., Kounaves, S.P., et al., 2016. Measurements of oxychlorine species on Mars. Int. J. Astrobio. Available from: http://dx.doi.org/10.1017/S1473550416000057.

Sutter, B., et al., 2017. Evolved gas analyses of sedimentary rocks and eolian sediment in Gale Crater, Mars: results of the Curiosity Rover's Sample Analysis at Mars (SAM) instrument from Yellowknife Bay to the Namib Dune. J. Geophys. Res. Planets. 122, 2574−2609.

Thompson, L.M., et al., 2016. Potassium-rich sandstones within the Gale impact crater, Mars: the APXS perspective, J. Geophys Res. Planets 121, 1981−2003.

Toner, J.D., Catling, D.C., Light, B., 2014. Soluble salts at the Phoenix Lander site, Mars: a reanalysis of the Wet Chemistry Laboratory data. Geochim. Cosmochim. Acta 136, 142−168.

Treiman, A.H., et al., 2015. Mineralogy, provenance, and diagenesis of a potassic basaltic sandstone on Mars: CheMin X-ray diffraction of the Windjana sample (Kimberley area, Gale Crater). J. Geophys. Res. Planets 121, 75−106.

Usui, T., Alexander, C.M.O.D., Wang, J., Simon, J.I., Jones, J.H., 2012. Origin of water and mantle-crust interactions on Mars inferred from hydrogen isotopes and volatile element abundances of olivine-hosted melt inclusions of primitive shergottites. Earth Planet. Sci. Lett. 357, 119−129.

Usui, T., Alexander, C.M.O.D., Wang, J., Simon, J.I., Jones, J.H., 2015. Meteoritic evidence for a previously unrecognized hydrogen reservoir on Mars. Earth Planet. Sci. Lett. 410, 140−151.

Vaniman, D.T., et al., 2014. Mineralogy of a mudstone at Yellowknife Bay, Gale Crater, Mars. Science 343. Available from: http://dx.doi.org/10.1126/science.1243480.

Villanueva, G.L., Mumma, M.J., Novak, R.E., Käufl, H.U., Hartogh, P., Encrenaz, T., et al., 2015. Strong water isotopic anomalies in the Martian atmosphere: probing current and ancient reservoirs. Science 348, 218−221.

Webster, C.R.M., et al., 2013. Isotope ratios of H, C, and O in CO_2 and H_2O of the Martian atmosphere. Science 341, 260−263.

Wilson, E.H., Atreya, S.K., Kaiser, R.I., Mahaffy, P.R., 2016. Perchlorate formation on Mars through surface radiolysis-initiated atmospheric chemistry: a potential mechanism. J. Geophys. Res. Planets 121, 1472−1487.

Yen, A.S., et al., 2017. Multiple stages of aqueous alteration along fractures in mudstone and sandstone strata in Gale Crater, Mars. Earth Planet Sci. Lett. Available from: http://dx.doi.org/10.1016/j.epsl.2017.04.033.

Zent, A.P., Quinn, R.C., 1995. Simultaneous adsorption of CO_2 and H_2O under Mars-like conditions and application to the evolution of the Martian climate. J. Geophys. Res. Planets 100, 5341−5349.

CONCLUSIONS AND IMPLICATIONS FOR HABITABILITY OF THE MARTIAN CRUST

Justin Filiberto[1,2], Karen Olsson-Francis[2], and Susanne P. Schwenzer[2]

[1]*Southern Illinois University, Carbondale, IL, United States* [2]*The Open University, Milton Keynes, United Kingdom*

In the introduction of this book, we briefly explored the history of the exploration of Mars, where we discussed a major shift in the scientific thinking and public perception of the nature of the surface of Mars as new missions explored the Red Planet. The first observations of the surface sparked speculations of Mars being inhabited by intelligent beings, inspiring fantasy, science fiction, and, of course, further exploration. The turn came when Mariner 4 returned images of a lunar-like landscape, a monotonous basaltic world with many impact craters but no water, and therefore no prospects for life to find a habitable niche. The authors of the chapters of this book show how different our understanding of Mars is today from the historical Mariner 4 perspective!

To begin with, Mars is not the monotonous basaltic world that was initially revealed. Magmatic rocks of basaltic compositions do account for a large proportion of the planet's rocks, but with increasing numbers of observations the complexity of the planet shows up in the data. This diversity also shows up in the volatile content and inventory of the Martian interior with the Martian mantle being heterogeneous in terms of volatile elements and their isotopic signatures. Petrologic, halogen, noble gas, and isotopic data show that different reservoirs likely exist within Mars. Petrologic evidence (Chapter 2: Volatiles in Martian Magmas and the Interior: Inputs of Volatiles Into the Crust and Atmosphere) from Martian meteorites, surface basalts, and orbital measurements suggest that volatile elements (specifically water, halogens, and to some extent even sulfur) are distributed heterogeneously in the crust and mantle of Mars. Further, noble gas data (Chapter 3: Noble Gases in Martian Meteorites: Budget, Sources, Sinks, and Processes) have shown that interior reservoirs with different ^{244}Pu-fission history exist, but that the ^{40}Ar-concentrations might also be inhomogeneously distributed in the Martian interior. This might be surprising, if only for the lack of plate tectonics on Mars, but has been shown in different data sets. Possible reasons for the different reservoirs include distribution from primary differentiation during magma ocean solidification, fractionation during magma genesis and/or differentiation, loss from magmas during eruption at low pressure, and redistribution in the crust by low-temperature aqueous processes. Remelting of crustal rocks during basin-forming impacts could have played an additional role. Hydrogen isotopes of Martian meteorites, telescopic observations, and MSL Curiosity measurements at Gale crater

Volatiles in the Martian Crust. DOI: http://dx.doi.org/10.1016/B978-0-12-804191-8.00019-2

(Chapter 4: Hydrogen Reservoirs in Mars as Revealed by Martian Meteorites) show that there are three hydrogen reservoirs on Mars: Primordial water in the Martian interior, atmospheric water, and crustal water. All three reservoirs have different concentrations and isotopic signatures. The crustal water reservoir, which is the most important for habitability and our discussion in this book, is the most uncertain and likely varies with location, depth in the crust, and presumably geologic time. This reveals a more diverse world than is seen in the Martian meteorites alone. It is this diversity in reservoirs, which then forms the basis for further diversification through exogenous geologic processes.

Magmatic rocks on Mars undergo alteration in very similar ways than their counterparts on Earth. Hydrothermal alteration around volcanic activity, due to impact heating, as well as weathering, has all been confirmed on Mars, not least through finding evidence for both high-temperature magmatic hydrothermal activity, as well as lower temperature alteration in the Martian meteorites. These meteorites specifically show evidence for degassing of high-temperature (700 °C), chlorine-rich, water-poor fluids (Chapter 2: Volatiles in Martian Magmas and the Interior: Inputs of Volatiles Into the Crust and Atmosphere). Such a high-temperature, chlorine-rich, water-poor fluid would be well beyond the limits of known terrestrial organisms. Halophiles (salt-loving organisms) can thrive in NaCl-rich systems but only at significantly lower temperatures (DasSarma, 2006). However, during cooling of the fluid, the system could have reached a temperature that was potentially habitable for halophiles (Rothschild and Mancinelli, 2001; DasSarma, 2006).

The nakhlite meteorites also contain evidence for low-temperature vein-type alteration, which is attributed to impact-generated hydrothermal processes (Chapter 5: Carbonates on Mars). The carbonates in the nakhlites are composed of calcite, dolomite-ankerite, and magnesite-siderite. The carbonates formed as hydrothermal fluids flowed through cracks within the nakhlite cumulate pile. This introduced CO_2-rich fluids that dissolved parts of the nakhlite rocks and precipitated the carbonates (Chapter 5: Carbonates on Mars). In the veinlets in nakhlites, fractionated noble gases have been found, with their nature still under scrutiny (Chapter 3: Noble Gases in Martian Meteorites: Budget, Sources, Sinks, and Processes). Are they associated with the alteration, adsorbed onto wall rock and subsequently incorporated into the magma, or even a climate or seasonal signal? While the Mars Science Laboratory (MSL) rover has measured noble gas isotopes in the atmosphere, elemental ratios of the heavy noble gases in the current atmosphere are still outstanding, and solids have not yet been measured for noble gases by the mission (Chapter 3: Noble Gases in Martian Meteorites: Budget, Sources, Sinks, and Processes). Formation of the veinlets themselves, though, has been ascribed to short-lived, low-temperature hydrothermal activity of circumneutral to alkaline fluids. Thermochemical modeling suggests that the nakhlite hydrothermal brines contain elements essential for life, for example, potassium, calcium, and sodium (Bridges and Schwenzer, 2012). Conditions like those are habitable on Earth, in the same way as clay formation in pedogenic processes creates niches for terrestrial bacteria.

Further, alteration in Martian meteorites appears as carbonate globules in ALH 84001. ALH 84001 is a unique Martian meteorite being an ancient sample composed almost entirely of orthopyroxene (Chapter 5: Carbonates on Mars). ALH 84001 is also unique because it contains carbonate globules that have received significant attention for containing supposed evidence of ancient Martian life (McKay et al., 1996) (Chapter 5: Carbonates on Mars). However, the hypothesis that ALH 84001 contains evidence of ancient life is not well accepted by the scientific community and has received significant contrary evidence (Anders et al., 1996; Bradley et al., 1998; Treiman,

2003; Treiman and Essene, 2011). Compositions of the carbonates are somewhat different than those in the nakhlites and range continuously from calcite cores, through dolomite-ankerite to siderite-magnesite, and to nearly pure magnesite rims. Formation of the carbonates likely occurred at low temperatures (near 18 °C) from a Ca-, Fe-, Mg-, and of course C-rich fluid. The conditions required for the formation of carbonates may also be deemed habitable. Low temperature, pH 4–5, key bioessential elements, and a rich inorganic carbon source, which could be used by autotrophic (produce complex organic molecules from inorganic molecules) microorganisms, are all conducive to life (Cockell, 2014). Carbonates have also been analyzed on the Martian surface, which will be discussed in more detail below.

One of the prevailing aspects of Mars that has been revealed by all spacecraft, Martian meteorite and telescopic investigations is that Mars is a sulfur-rich planet (Chapter 6: Sulfur on Mars From the Atmosphere to the Core). A sulfur-rich crust was noted all the way back to the Viking landers, and reconfirmed by all of the later rover in situ investigations (Chapters 6, 9–12). In fact, one of the first minerals analyzed at Meridiani Planum was jarosite, a potassium- and iron-bearing hydrous sulfate that forms from acidic waters (Chapter 6: Sulfur on Mars From the Atmosphere to the Core; and Chapter 10: Mars Exploration Rover Opportunity: Water and Other Volatiles on Ancient Mars). Sulfur is an especially important element because of its several redox states. The oxidation of reduced sulfur compounds, such as sulfide, elemental sulfur, and thiosulfate, is a process by which chemolithotrophic microorganisms can produce energy (for a review see Ghosh and Dam, 2009). Sulfur can also be used in anaerobic respiration and some microorganisms called sulfate-reducing bacteria reduce sulfate, sulfite, thiosulfate, or elemental sulfur (for review see Muyzer and Stams, 2008). Sulfur may also form sulfuric acid, which enhances silicate dissolution (Tosca et al., 2004; Zolotov and Mironenko, 2007). Its presence, therefore, increases the amount of dissolved species available for biologic processes, but challenges them with an oxidizing, acidic environment (e.g., Drever, 1997). Therefore, the widespread availability of sulfur in the Martian crust could have provided both an energy source and nutrients for microbes (Chapter 6: Sulfur on Mars From the Atmosphere to the Core).

Looking at ancient Mars, the evidence for water activity is overwhelming: Fluvial valleys, outflow channels, fans, deltas, paleolakes, lobate craters, possible signs of glaciation, and even potential evidence for seas and oceans have all been detected from orbit, while conglomerates and mudstones, grain size sorting, cross bedding, and rounding of pebbles have been observed from the surface (Chapter 7: The Hydrology of Mars Including a Potential Cryosphere). Yet, the question of whether there have been oceans and seas on Mars is just one of the highly debated ones, with mineralogical evidence seemingly contradicting the possibility but orbital radar data finding conceivable evidence (Chapter 7: The Hydrology of Mars Including a Potential Cryosphere). In contrast to potentially contradicting the ocean hypothesis, mineralogical evidence clearly supports water–rock interaction, with evidence for both alkaline to circumneutral and acidic conditions present in the time–location space. The detection of hydrated minerals—ranging from phyllosilicates to opaline silica and zeolites—through orbital instruments highlights the widespread, yet diverse nature of the alteration environments; and hydrated minerals could store between 15 and as much as 1000 global equivalent layers (GEL) of water (Chapter 8: Sequestration of Volatiles in the Martian Crust Through Hydrated Minerals: A Significant Planetary Reservoir of Water). Notably, all but one of the observed hydrous minerals are indicative of a low-temperature formation. Prehnite is the only exception, and the lack of amphiboles in orbital observations is most puzzling (Chapter 8:

Sequestration of Volatiles in the Martian Crust Through Hydrated Minerals: A Significant Planetary Reservoir of Water). This lack of amphiboles is in stark contrast to Earth, where amphiboles are common in alteration and magmatic mineral assemblages. The more evidence is returned from meteorites, orbiters, landers, and rovers, the more complex the history of water on Mars becomes. To fully understand the Martian habitability through time and between locations, more detailed investigations are required, especially discerning mineral assemblages from out-of-equilibrium co-detections, and, of course, the association of mineralogy with geomorphology must be understood much better.

In contrast to Noachian Mars, liquid water is currently not stable on the surface of Mars but ground ice and water-ice in the polar caps have been evidenced from orbiters and landers. Like water, ice is not stable at the surface outside the polar caps, for example, the Phoenix landing site where it evaporated after excavation within 24 hours (Chapter 9: Volatiles Measured by the Phoenix Lander at the Northern Plains of Mars). There is, however, an active water cycle driven by diurnal adsorption and desorption of as much as 6% of the total water column (Phoenix landing site, Chapter 9: Volatiles Measured by the Phoenix Lander at the Northern Plains of Mars). The polar regions are covered by layered deposits, whereby the North polar cap has a remarkable young surface, either due to deposition or reworking of material. The north polar layered deposit is estimated to contain as much as 7.5 GEL of water, while in the south a smaller cap is exposed but an extensive masked layered deposit exists, accounting for as much as 9.5 GEL of water stored as ice (Chapter 7: The Hydrology of Mars Including a Potential Cryosphere). Orbital geomorphological evidence, including visible ice in fresh craters, and lander evidence (Phoenix, Chapter 9: Volatiles Measured by the Phoenix Lander at the Northern Plains of Mars) for ground ice outside the polar layered deposits add up to 2.6 GEL to the amount of water currently stored as ice. Despite the frozen state of most ice, and the low surface pressures and temperatures making water unstable at the surface under most circumstances, transient contemporary activity is observed: Gullies, periglacial landforms, and recurring slope lineae are all considered water-driven features (Chapter 7: The Hydrology of Mars Including a Potential Cryosphere), which need to be much better understood in order to enable the interpretation of the Martian environmental conditions today.

The Mars Exploration Rovers Spirit (Chapter 11: Alteration Processes in Gusev Crater, Mars: Volatile/Mobile Element Contents of Rocks and Soils Determined by the Spirit Rover) and Opportunity (Chapter 10: Mars Exploration Rover Opportunity: Water and Other Volatiles on Ancient Mars) landed in January 2004 to investigate the potential for evidence of ancient water in the Martian crust (Squyres et al., 2004a,b, 2007). Meridiani Planum was selected for the Opportunity landing site due to the detection of hematite from orbit (Golombek et al., 2003, 2006). Importantly, most formation models for hematite require water, therefore this site was chosen to follow the water (Golombek et al., 2003, 2006), and in fact hematite was the first mineral analyzed by Opportunity in Eagle crater as small spherical grains (nicknamed blueberries) (Chapter 10: Mars Exploration Rover Opportunity: Water and Other Volatiles on Ancient Mars). Opportunity went on to do a detailed analysis of the Burns Formation at Meridiani Planum where it detected jarosite (the Fe-K sulfate mentioned above). The sequence of the Burns Formation at Eagle, Endurance, Erebus, and Victoria craters revealed ancient sulfate-rich evaporite deposits, formed initially by aqueous, acidic alteration of basalt. Opportunity then drove to Endeavor crater where it is currently exploring an older sequence of rocks suggestive of being deposited in more neutral waters, as exemplified by the discovery of a hydrous-mineral smectite. Although microorganisms can grow in acidic conditions, growth is limited to acidophilic (acid-loving) microorganisms. In contrast, pH will not be a limiting factor in neutral waters, therefore increasing the possibility of habitability.

MER Spirit landed at Gusev crater to investigate ancient lake sediments (Golombek et al., 2003, 2006), but after landing and exploring the region around the landing site revealed a basaltic lava flow with no evidence for lake sediments (McSween et al., 2006). Instead, Spirit landed on an ancient basaltic lava flow (Chapter 11: Alteration Processes in Gusev crater, Mars: Volatile/Mobile Element Contents of Rocks and Soils Determined by the Spirit Rover). It then drove to the Columbia Hills where it analyzed more evolved magmatic compositions and carbonate-bearing rocks. Comanche is a carbonate-bearing rock outcrop on the eastern side of Husband Hill in the Columbia Hills. The carbonates (based on chemistry and comparison with the carbonates in ALH 84001) likely formed from near-neutral pH waters during hydrothermal conditions. Finally, Spirit drove to the Home Plate pyroclastic deposit, which was deposited in water-saturated muddy sediments, as revealed by a bomb sag (Squyres et al., 2006). To the east of the Home Plate outcrop, in the Eastern Valley region, light-toned, nodular outcrops and light-toned soils consisting of opaline silica were discovered. The opaline silica-bearing rocks and soils are thought to have formed by hydrothermal processes connected to the volcanism at Home Plate.

The most recent arrival on the Martian surface is the Mars Science Laboratory rover Curiosity. Some few sols after landing, in fact, in the excavation of the exhaust engines and in the rocks encountered along the first drives, rounded pebbles were observed, which are now known as part of the fan deposit sequence in the area. Williams et al. (2013) first described the conglomerate, which confirmed the presence of sustained liquid water activity at the time of the fan deposit formation about 4 Ga ago. Subsequently, mudstones, siltstones, and complex sequences interpreted to be lake bed sediments were investigated (Vaniman et al., 2014; Grotzinger et al., 2014, 2015). Most importantly for habitability, a wide range of volatiles were found and investigated in the rock samples by the SAM instrument (Chapter 12: Volatile Detections in Gale Crater Sediment and Sedimentary Rock: Results From the Mars Science Laboratory's Sample Analysis at Mars Instrument). Water was detected in every sample, with a maximum abundance of 2.5 wt%, and even more importantly, never before in situ measured species—nitrogen and organic carbon—were found and studied (Chapter 12: Volatile Detections in Gale Crater Sediment and Sedimentary Rock: Results From the Mars Science Laboratory's Sample Analysis at Mars Instrument). Oxygen and hydrochloric acid releases confirm the presence of perchlorates (Chapter 12: Volatile Detections in Gale Crater Sediment and Sedimentary Rock: Results From the Mars Science Laboratory's Sample Analysis at Mars Instrument), the infamous substances unexpectedly interfering with the Viking experiments some 40 years ago. This wide variety of species indicates both reducing and oxidizing conditions, as well as alkaline, circumneutral, and acidic conditions, and thus confirms the mineralogical evidence from sulfides (reducing conditions) and sulfates and hematite (oxidizing conditions), as well as jarosite (acidic conditions) and phyllosilicates (circumneutral to alkaline conditions). SAM D/H measurements confirm observations discussed earlier of atmosphere loss before 3 Ga (Chapter 12: Volatile Detections in Gale Crater Sediment and Sedimentary Rock: Results From the Mars Science Laboratory's Sample Analysis at Mars Instrument). Curiosity and Opportunity will continue their investigations of the Martian surface and atmosphere for—hopefully—many, many more sols to come, while six orbiters are currently studying (or about to start their science phase) Mars from orbit: Mars Odyssey, Mars Reconnaissance Orbiter, MAVEN (all NASA), Mars Express, Mars Trace Gas Orbiter (both ESA), and Mangalyaan (ISRO). As evidence adds, we will understand better and better how the Martian surface, subsurface, and atmosphere interacted over time, which processes shaped the landscape, which physical conditions prevailed, and what chemical substances dominated the environments.

Based on our growing understanding of the mineralogy and geochemistry of Mars, there has been large speculation regarding habitability of the planet (past and present). Liquid water, nutrients and an energy source are required for life on Earth, and so their presence is central to theories about when, where, and under what conditions life may have existed. Evidence presented in Chapter 7, The Hydrology of Mars Including a Potential Cryosphere, demonstrates that large ancient fluvial systems existed on ancient Mars with speculation that within these systems, the conditions existed that could support life (Grotzinger et al., 2014). Another independent strand of evidence for a more hospitable past is impact-generated hydrothermal systems, which may have provided habitable conditions even during periods of very cold climate (Abramov and Kring, 2005; Schwenzer and Kring, 2009). Original speculation regarding life on Mars centered around extremophilic microorganisms that could survive in extreme environments. However, with the growing geological data, the concept of habitable environments has changed and the consensus is that certain environments on ancient Mars were less extreme than originally thought, for example, mesophilic temperature and circumneutral pH, which are conducive for life. Using rover-based instrumentation on the surface of Mars (MSL and in future ExoMars and Mars 2020) and in future sample return missions, we will be able to address the question whether Mars was once habitable.

REFERENCES

Abramov, O., Kring, D.A., 2005. Impact-induced hydrothermal activity on early Mars. J. Geophys. Res. 110 (20), E12S09. Available from: http://dx.doi.org/10.1029/2005JE002453.

Anders, E., Shearer, C.K., Papike, J.J., Bell, J.F., Clemett, S.J., Zare, R.N., et al., 1996. Evaluating the evidence for past life on Mars. Science 274, 2119–2125.

Bradley, J.P., McSween Jr., H., Harvey, R.P., 1998. Epitaxial growth of nanophase magnetite in Martian meteorite Allan Hills 84001: implications for biogenic mineralization. MAPS 33, 765–773.

Bridges, J.C., Schwenzer, S.P., 2012. The nakhlite hydrothermal brine. EPSL 359–360, 117–123.

Cockell, C.S., 2014. Trajectories of martian habitability. Astrobiology 14 (2), 182–203. Available from: http://dx.doi.org/10.1089/ast.2013.1106.

DasSarma, S., 2006. Extreme halophiles are models for astrobiology. Microbe 1, 120–126.

Drever, J.I., 1997. The Geochemistry of Natural Waters: Surface and Groundwater Environments. Prentice Hall, Upper Saddle River, NJ.

Ghosh, W., Dam, B., 2009. Biochemistry and molecular biology of lithotrophic sulfur oxidation by taxonomically and ecologically diverse bacteria and archaea. Microbiol. Rev. 33, 999–1043.

Golombek, M.P., Grant, J.A., Parker, T.J., Kass, D.M., Crisp, J.A., Squyres, S.W., et al., 2003. Selection of the Mars Exploration Rover landing sites. J. Geophys. Res. 108. Available from: http://dx.doi.org/10.1029/2003JE002074.

Golombek, M.P., Crumpler, L., Grant, J.A., Greeley, R., Cabrol, N.A., Parker, T., et al., 2006. Geology of the Gusev cratered plains from the Spirit rover transverse. J. Geophys. Res 111, E02S07. Available from: http://dx.doi.org/10.1029/2005JE002503.

Grotzinger, J.P., Sumner, D.Y., Kah, L.C., Stack, K., Gupta, S., Edgar, L., et al., 2014. A habitable fluvio-lacustrine environment at Yellowknife Bay, Gale Crater, Mars. Science 343. Available from: http://dx.doi.org/10.1126/science.1242777.

Grotzinger, J.P., Gupta, S., Malin, M.C., Rubin, D.M., Schieber, J., Siebach, K., et al., 2015. Deposition, exhumation, and paleoclimate of an ancient lake deposit, Gale crater, Mars. Science 350. Available from: http://dx.doi.org/10.1126/science.aac7575.

McKay, D.S., Gibson Jr., E.K., Thomas-Keprta, K.L., Vali, H., 1996. Search for past life on Mars: possible relic biogenic activity in Martian meteorite ALH84001. Science 273, 924−930.

McSween, H.Y., Wyatt, M.B., Gellert, R., Bell III, J.F., Morris, R.V., et al., 2006. Characterization and petrologic interpretation of olivine-rich basalts at Gusev Crater, Mars. J. Geophys. Res. 111, E02S10. Available from: http://dx.doi.org/10.1029/2005E02477.

Muyzer, G., Stams, A.J.M., 2008. The ecology and biotechnology of sulphate-reducing bacteria. Nat. Rev. Microbiol. 6, 441−454.

Rothschild, L.J., Mancinelli, R.L., 2001. Life in extreme environments. Nature 409, 1092−1101.

Schwenzer, S.P., Kring, D.A., 2009. Impact-generated hydrothermal systems: capable of forming phyllosilicates on Noachian Mars. Geology 37 (12), 1091−1094. Available from: http://dx.doi.org/10.1130/G30340A.1.

Squyres, S.W., Arvidson, R.E., Bell III, J.F., Brückner, J., Cabrol, N.A., et al., 2004a. The Spirit Rover's Athena science investigation at Gusev Crater, Mars. Science 305, 794−799.

Squires, S.W., Arvidson, R.E., Bell III, J.F., Brückner, J., Cabrol, N.A., Calvin, W., et al., 2004b. The Opportunity Rover's Athena science investigation at Meridiani Planum, Mars. Science 306, 1698−1703.

Squyres, S.W., Knoll, A.H., Arvidson, R.E., Clark, C., Grotzinger, J.P., Jolliff, B.L., et al., 2006. Two years at Meridiani planum: results from the opportunity rover. Science 313, 1403−1407.

Squyres, S.W., Arvidson, R.E., Bollen, D., Bell, J.F., Brückner, J., Cabrol, N.A., et al., 2007. Overview of the Opportunity Mars Exploration Rover Mission to Meridiani Planum: Eagle Crater to Purgatory Ripple. J. Geophys. Res. 111, E12S12. Available from: http://dx.doi.org/10.1029/2006E02771.

Tosca, N.J., McLennan, S.M., Lindsley, D.H., Schoonen, M.A.A., 2004. Acid-sulfate weathering of synthetic Martian basalt: the acid fog model revisited. J. Geophys. Res. 109, E05003. Available from: http://dx.doi.org/10.1029/2003JE002218.

Treiman, A.H., 2003. Submicron magnetite grains and carbon compounds in Martian meteorite ALH84001: inorganic, abiotic formation by shock and thermal metamorphism. Astroniology 3, 369−392.

Treiman, A.H., Essene, E.J., 2011. Chemical composition of magnetite in Martian meteorite ALH 84001: revised appraisal from thermochemistry of phases in Fe−Mg−C−O. Geochim. Cosmochim. Acta 75, 5324−5335.

Vaniman, D.T., Bish, D.L., Ming, D.W., Bristow, T.F., Morris, R.V., Blake, D.F., et al., 2014. Mineralogy of a mudstone on Mars. Science 343, 10.1126/science.1243480.

Williams, R.M.E., Grotzinger, J.P., Dietrich, W.E., Gupta, S., Sumner, D.Y., Mangold, N., et al., 2013. Martian fluvial conglomerates at Gale Crater. Science 340, 1068−1072.

Zolotov, M.J., Mironenko, M.V., 2007. Timing of acid weathering on Mars: a kinetic-thermodynamic assessment. J. Geophys. Res. 112, E07006. Available from: http://dx.doi.org/10.1029/2006JE002882.

Index

Note: Page numbers followed by "*f*" and "*t*" refer to figures and tables, respectively.

Printed in the United States
By Bookmasters